# Heat Transfer Principles and Applications

# Heat Transfer Principles and Applications

Charles H. Forsberg

ACADEMIC PRESS

An imprint of Elsevier

ELSEVIER

Academic Press is an imprint of Elsevier
125 London Wall, London EC2Y 5AS, United Kingdom
525 B Street, Suite 1650, San Diego, CA 92101, United States
50 Hampshire Street, 5th Floor, Cambridge, MA 02139, United States
The Boulevard, Langford Lane, Kidlington, Oxford OX5 1GB, United Kingdom

**Notices**

Knowledge and best practice in this field are constantly changing. As new research and experience broaden our understanding, changes in research methods, professional practices, or medical treatment may become necessary.

Practitioners and researchers must always rely on their own experience and knowledge in evaluating and using any information, methods, compounds, or experiments described herein. In using such information or methods they should be mindful of their own safety and the safety of others, including parties for whom they have a professional responsibility.

To the fullest extent of the law, neither the Publisher nor the authors, contributors, or editors, assume any liability for any injury and/or damage to persons or property as a matter of products liability, negligence or otherwise, or from any use or operation of any methods, products, instructions, or ideas contained in the material herein.

MATLAB® is a trademark of The MathWorks, Inc. and is used with permission.
The MathWorks does not warrant the accuracy of the text or exercises in this book.
This book's use or discussion of MATLAB® software or related products does not constitute endorsement or sponsorship by The MathWorks of a particular pedagogical approach or particular use of the MATLAB® software.

**Library of Congress Cataloging-in-Publication Data**
A catalog record for this book is available from the Library of Congress

**British Library Cataloguing-in-Publication Data**
A catalogue record for this book is available from the British Library

ISBN: 978-0-12-802296-2

For information on all Academic Press publications visit our website at https://www.elsevier.com/books-and-journals

*Publisher:* Katey Birtcher
*Acquisitions Editor:* Stephen Merken
*Editorial Project Manager:* Susan Ikeda
*Production Project Manager:* Sujatha Thirugnana Sambandam
*Cover Designer:* Alan Studholme

Typeset by TNQ Technologies

www.elsevier.com • www.bookaid.org

# Contents

# Unit conversions

| | |
|---|---|
| Length | $1\ m = 39.370\ in = 3.2808\ ft$ |
| | $1\ mile = 5280\ ft = 1.6093\ km$ |
| Area | $1\ m^2 = 10.7639\ ft^2$ |
| Volume | $1\ m^3 = 35.3134\ ft^3$ |
| | $1\ m^3 = 264.17\ gal\ (US)$ |
| Velocity | $1\ m/s = 3.2808\ ft/s$ |
| Density | $1\ kg/m^3 = 0.06243\ lbm/ft^3$ |
| Force | $1\ N = 0.2248\ lbf$ |
| Mass | $1\ kg = 2.20462\ lbm$ |
| | $1\ slug = 1\ lbf\ s^2/ft = 32.174\ lbm$ |
| | $1\ ton = 2000\ lbm$ |
| | $1\ metric\ ton = 1000\ kg$ |
| Pressure | $1\ Pa = 1\ N/m^2 = 1.45038 \times 10^{-4}\ lbf/in^2$ |
| | $1\ MPa = 10^3\ kPa = 10^6\ Pa$ |
| | $1\ bar = 100\ kPa$ |
| | $1\ atm = 101.325\ kPa = 760\ mm\ Hg\ @\ 0\ C$ |
| | $1\ atm = 14.696\ psia = 29.92\ in\ Hg\ @\ 32\ F$ |
| | $1\ torr = 133.322\ Pa$ |
| Energy (heat) | $1\ kJ = 0.94783\ Btu$ |
| | $1\ Btu = 778.169\ ft\ lbf$ |
| | $1\ kWh = 3412.14\ Btu$ |
| Heat flow and power | $1\ W = 1\ J/s = 3.4121\ Btu/h$ |
| | $1\ kW = 3412.1\ Btu/h = 1.341\ hp$ |
| | $1\ hp = 550\ ft\ lbf/s$ |
| Heat flux | $1\ W/m^2 = 0.3171\ Btu/h\ ft^2$ |
| Heat generation rate | $1\ W/m^3 = 0.09662\ Btu/h\ ft^3$ |
| Heat transfer coefficient | $1\ W/m^2\ C = 0.1761\ Btu/h\ ft^2\ F$ |
| Specific heat | $1\ kJ/kg\ C = 0.23885\ Btu/lbm\ F$ |
| Thermal conductivity | $1\ W/m\ C = 0.5778\ Btu/h\ ft\ F$ |
| Absolute viscosity | $1\ kg/m\ s = 2419.1\ lbm/ft\ h$ |
| Kinematic viscosity | $1\ m^2/s = 10.764\ ft^2/s$ |
| Thermal diffusivity | $1\ m^2/s = 10.764\ ft^2/s$ |
| Temperature level | $K = C + 273.15$ |
| | $R = F + 459.67$ |
| | $F = (9/5)\ C + 32$ |
| | $C = (5/9)\ (F - 32)$ |
| Temperature size | $1\ C = 1\ K$ |
| | $1\ F = 1\ R$ |
| | $1\ C = (9/5)\ F$ |
| | $1\ F = (5/9)\ C$ |

# Constants

| | |
|---|---|
| Stefan–Boltzmann | $\sigma = 5.67 \times 10^{-8}$ W/m$^2$ K$^4$ = $0.1714 \times 10^{-8}$ Btu/h ft$^2$ R$^4$ |
| Universal gas constant | $\overline{R} = 8.31446$ kJ/kmol K = $1545.35$ ft lbf/lbm mol R |
| Planck's constant | $h = 6.62607 \times 10^{-34}$ J·s |
| Boltzmann's constant | $k = 1.38065 \times 10^{-23}$ J/K |
| Standard gravity | $g = 9.807$ m/s$^2$ = $32.174$ ft/s$^2$ |
| Speed of light (vacuum) | $c_o = 2.9979 \times 10^8$ m/s |
| Avogadro's number | $N_A = 6.02214 \times 10^{26}$ atoms or molecules/kmol |

# Preface

Over my years of teaching heat transfer, I have watched as the major textbooks in the field have gotten bigger and bigger. They have reached such a massive size that only a small portion of a book can be successfully covered in a semester-length course. Also, the problems in the books have become so complex that many, or most, students have comprehension difficulties. Inordinate amounts of time are needed to just understand the problems, leaving much less time available for their solution.

The overall goal of this book is to assist the student in obtaining the greatest possible benefits from a semester-long undergraduate heat transfer course. The book is of reasonable length. Although it contains more material than can be covered in the course, the extra material is not excessive, and it provides the professor with flexibility in topic selection. The problems in the book are of a practical nature. They require the students to think critically, but are not excessively hard to comprehend.

The book is designed for a one-semester heat transfer course for mechanical engineering students taken in the junior or senior year. It is also beneficial for practicing mechanical engineers and for practicing engineers in other disciplines (e. g., civil and electrical) who encounter heat transfer aspects in their projects. Finally, with a chapter devoted to mass transfer, the book is useful for chemical engineering students and engineers.

The book makes extensive use of Excel and MATLAB® for the solution of numerical problems. Excel *Goal Seek* is used for solution of single nonlinear equations, and *Solver* is used for simultaneous linear and nonlinear equations. The *fzero* function of MATLAB is used for single nonlinear equations, and MATLAB programs incorporating the Gauss–Seidel method are used for simultaneous equations. MATLAB is also used for the solution of transient problems.

There are 12 chapters in the book, plus an appendix. Brief descriptions are as follows:

Chapter 1 introduces the three modes of heat transfer: conduction, convection, and radiation and gives their basic equations. There are also sections regarding the direction of heat flow, temperature continuity and heat balances, unit systems, significant figures, and the recommended approach to problem-solving.

Chapter 2 deals with the differential equations for heat conduction in rectangular, cylindrical, and spherical coordinates. It also discusses boundary conditions for the different geometries.

Chapter 3 discusses the solution of steady-state, one-dimensional heat conduction problems. The resistance concept and the electric–heat analogy are introduced. Heat generation in current-carrying wires is covered, as is the conduction shape factor. Finally, there is a discussion of fins.

Chapter 4 deals with unsteady conduction. It includes the lumped method; the Heisler method for planes, cylinders, spheres, and multidimensional objects; and conduction in semi-infinite solids.

Chapter 5 covers numerical methods applied to both steady and unsteady situations. The appropriate finite-difference equations are derived. Excel and MATLAB are used for the solution of the resulting simultaneous linear and nonlinear equations.

Chapters 6 and 7 deal with convection. Chapter 6 is forced convection for internal and external flows. Chapter 7 is natural (free) convection. Both chapters present the pertinent dimensionless numbers and correlation equations for the heat transfer coefficient. Chapter 7 includes sections on natural convection for enclosed spaces such as rectangular spaces between parallel plates and annular spaces between concentric cylinders and spheres.

Chapter 8 covers heat exchangers. Both the log mean temperature difference and effectiveness–number of transfer unit methods are discussed.

Chapter 9 deals with radiation heat transfer. There are sections on blackbody emission, radiation properties, radiation shape factors, and radiant heat transfer between surfaces of an enclosure. Radiation shields are also discussed.

Chapter 10 is unique. It is designed to enhance the students' problem-solving abilities. Most of the previous chapters were based on a single mode of heat transfer. The problems in this chapter are multi-mode, incorporating more than one mode of heat transfer. The students have to determine what modes are significant, whether the problem is steady state or unsteady, and whether an analytical or numerical solution should be used. They must also search out and provide any necessary information not given in the problem statement.

Chapter 11 covers mass transfer. It discusses concentrations and properties of gas mixtures, Fick's law of diffusion, diffusion coefficients, and Stefan's law. Evaporation, venting from containers, and mass transfer through walls and membranes are also covered. Finally, it discusses the wet-bulb and dry-bulb psychrometer.

Chapter 12 consists of special topics which can be covered in the course if time permits. The Chapter 3 discussion of heat generation in wires is extended to cover generation in plane walls and spheres. Contact resistance and condensation/boiling are also discussed. Finally, there is a discussion of energy usage in buildings, with special emphasis on the degree-day method.

The Appendix includes properties of materials, tables of Bessel and error functions, a discussion of the use of Excel and MATLAB, and listings of MATLAB m-files for the examples of Chapter 5.

A few comments regarding teaching the course: A semester-length course at Hofstra is usually 13 or 14 weeks long with two 80-minute lectures per week. I have found that Chapters 1 through 9 can be readily covered during a semester. It also may be possible to include a couple of topics from Chapters 11 or 12 if the professor so desires. I also recommend devoting a class session to Chapter 10. I typically switch the order of Chapters 4 and 5. I teach Chapter 5 before Chapter 4 because I require the students to submit two computer solutions: one for steady state and the other for unsteady state. Switching the order allows the students to jump-start their computer work. I give three exams: one after Chapter 3, one after Chapter 8, and the final exam. I typically assign three or four homework problems per week. I feel that class attendance is very worthwhile for the students, so my grading is usually 60% exams, 15% homework, 15% attendance, and 10% computer problems.

The following materials are provided to professors adopting the book: A solutions manual for end-of-the-chapter problems; Excel spreadsheets and MATLAB m-files for end-of-the-chapter problems; PowerPoint lecture slides; image bank of figures from the book; To obtain access, qualified instructors should log in, or register and create an account, at www.textbooks.elsevier.com.

# Acknowledgments

There are many people I wish to thank. First, I gratefully thank my parents, Geneva and Harry, who provided me with a wonderful childhood and to whom I owe everything.

The concept for this book started several years ago when I had discussions with publisher's representative Ronde Bradley in my Hofstra University office. I remember discussing the existing heat transfer texts and expressing my interest in possibly writing a text. Ronde passed my ideas to Joe Hayton, currently Publishing Director of Elsevier Science. I had several emails with Joe, but life intervened and the project never got started until a few years ago. Thanks so much, Ronde. And, thanks so much Joe for your insightful comments and your actions to get this project moving.

There are others at Elsevier that I wish to thank. As a new author, my knowledge of the publishing industry was essentially nil. Many thanks to Steve Merken, Senior Acquisitions Editor, who helped me along and gave me much encouragement. Also, many thanks to Susan Ikeda, Senior Editorial Project Manager, for her expert assistance and kindness. Finally, thanks to Sujatha Thirugnana Sambandam, Publishing Services Manager, who has led the production activities.

Regarding my Hofstra University associates, I would like to thank John P. LeGault, former Engineering Labs Supervisor, who provided assistance with MATLAB. Thanks also to the staff of the ILLiad Interlibrary Loan department who responded swiftly and competently to my many requests for journal articles.

Finally, many thanks to my family: Gail, Nancy, Rich, Andrew, Meaghan, and grandchildren Matthew, Abby, Sarah, and Christopher who have given me much joy during this long project and continue to give me much joy. Special thanks to my wonderful wife, Gail, for her encouragement and love. It is to Gail that I dedicate this book.

# Introduction to heat transfer

1

## Chapter outline

## 1.1 Introduction

Let us first discuss the relationship between thermodynamics and heat transfer. Thermodynamics, the first thermal science course in most engineering programs, is a study of equilibrium states of systems. It considers the different types of energy (e.g., mechanical, kinetic, and potential, heat, internal) and the amount of energy transferred during a system's process from one equilibrium state to another. Thermodynamics has two basic laws. The first law deals with conservation of energy and the conversion of energy from one type to another. The second law deals with observed restrictions on system processes. For example, there cannot be a process which solely involves the transfer of heat from a region of lower temperature to a region of higher temperature. Thermodynamics considers the amount of energy being transferred. It does not usually deal with the time needed for this energy transfer.

Heat transfer adds the time dimension to energy processes. It deals with the rate of transfer of heat energy as a system moves from one equilibrium state to another. As an example, let us consider a hot metal object at initial temperature $T_1$ being cooled by a liquid at temperature $T_2$. During the process, the heat flows from the object to the liquid and the temperature of the object decreases. The object finally reaches the temperature $T_2$ of the liquid. Thermodynamics can predict the amount of heat transferred to the liquid during the process. Heat transfer provides further details. It can predict how long it will take for the object's temperature to go from $T_1$ to $T_2$. Heat transfer can also provide the temperature distribution in the object at different times during the process.

In this chapter, we introduce the three modes of heat transfer (conduction, convection, radiation) and present their fundamental equations. We also discuss unit systems and give recommendation regarding approaches to problem solving and the number of significant figures to retain in numerical results.

## 1.2 Modes of heat transfer

Mechanical engineering has two major areas—thermal/fluids and mechanics/machine design. Heat transfer, the subject of this book, is part of the first area.

Transfer of heat occurs as a result of a temperature difference. This is the driving force which causes the heat to flow. If there is no temperature difference, there is no heat flow.

It is also observed that heat flows in the direction from the higher temperature to the lower temperature.

There are three primary modes of heat transfer: *conduction*, *convection*, and *radiation*.

*Heat conduction* occurs if there is a temperature difference between two locations in a solid or two locations in essentially nonmoving liquids or gases. *Heat convection* typically occurs due to a temperature difference between a surface of an object and an adjacent fluid. Change of phase (i.e., evaporation and condensation) is also included in the convection category. *Heat radiation* occurs between two surfaces of different temperatures. Conduction and convection require physical material (i.e., solids or fluids) for their transport. Radiation can occur through a vacuum.

Knowledge of heat transfer fundamentals is essential for design of many devices, systems, and processes. For example, heat transfer knowledge is needed for design of systems for cooling of computers and other electrical devices; heating and cooling equipment for buildings, refrigerators, boilers, and car engines; and heat exchangers for many different industrial applications. Heat transfer is indeed a major subject area for mechanical engineers and also chemical engineers.

We will now introduce the three modes of heat transfer. Later we will discuss unit systems, problem-solving techniques, and significant figures.

## 1.3 Conduction

Transfer of heat by conduction occurs in solids and in essentially nonmoving liquids and gases. It has been observed (through experimentation) that the rate of heat transfer per unit area is proportional to the temperature gradient in the material. That is,

$$\frac{q_n}{A_n} \propto \frac{\partial T}{\partial n} \tag{1.1}$$

where $n$ designates the direction of the heat flow; e.g., $x$, $y$, or $z$ in Cartesian coordinates.

In this equation, $q_n$ is the rate of heat flow in direction $n$, $A_n$ is the cross-sectional area through which the heat flows, and $\partial T/\partial n$ is the temperature gradient in direction $n$. (Note: The area $A$ is the area perpendicular to the heat flow.)

If we introduce a proportionality parameter into Eq. (1.1) and introduce a minus sign so that heat flow in the positive $n$ direction will be numerically positive, we get *Fourier's law* [1,2]:

$$\frac{q_n}{A_n} = -k\frac{\partial T}{\partial n} \tag{1.2}$$

Eq. (1.2), named for French mathematician and physicist Joseph Fourier (1768−1830), gives the heat flow rate per unit area in the $n$ direction at a point in the solid or fluid. Heat flow rate per unit area is called "heat flux." Heat flow is a vector quantity which has direction and magnitude. If we are using Cartesian coordinates, there can be heat flow and heat flux components in all three coordinate directions. The heat flow components are $q_x$, $q_y$, and $q_z$. The heat flux components are

$$\left(\frac{q}{A}\right)_x = -k\frac{\partial T}{\partial x} \quad \left(\frac{q}{A}\right)_y = -k\frac{\partial T}{\partial y} \quad \left(\frac{q}{A}\right)_z = -k\frac{\partial T}{\partial z} \tag{1.3}$$

where $k$ is the *thermal conductivity* of the material. If we are using the SI unit system where $q$ is watts, the area is m$^2$, and the temperature gradient is °C/m, then $k$ has the units of W/m °C. (Note: As the *size* of a Celsius degree is the same as the *size* of a Kelvin degree, the numerical value of $k$ in W/m °C is the same as its value in W/m K. That is, 1 W/m °C = 1 W/m K.)

[*Note*: In this text, we will usually leave out the symbol used for a temperature degree. That is, we will often use W/m C rather than W/m °C for the units of thermal conductivity. And, when we are talking about a temperature of 20 degrees Celsius, we will use 20 C rather than 20°C.]

Thermal conductivity is highest for pure metals and a bit lower for alloys. The conductivity for liquids is generally lower (except for liquid metals) and for gases even lower. Building materials such as wood, plaster, and insulation have low conductivity. Multilayer evacuated insulation used for insulating cryogenic tanks has very low conductivity.

For most materials, thermal conductivity is isotropic; that is, it is the same in all directions. However, for some materials, it varies with direction. For example, although a single average $k$ value is often given for wood, the conductivity of wood is actually different in the across-the-grain and with-the-grain directions.

Table 1.1 gives typical thermal conductivity values for some materials at about 20 C. Appendices A through E give detailed information on the properties of common solids, liquids, and gases.

In general, thermal conductivity decreases with temperature for metals, decreases slightly with temperature for many liquids, and increases with temperature for gases.

## 1.3.1 Conduction through a plane wall

In this section we develop the equation for the rate of heat flow through a plane wall. The equation is very useful for estimating heat flow through large walls of finite thickness, such as the exterior walls of a building. It is also useful in other situations where the heat flow can be considered to be one-dimensional; that is, the heat flow is in a single direction.

Table 1.1 Typical thermal conductivity at 20 C.

| Material | k (W/m C) |
| --- | --- |
| Aluminum | 240 |
| Copper | 400 |
| Gold | 315 |
| Silver | 430 |
| Carbon steel | 40 |
| Stainless steel | 15 |
| Plasterboard | 0.8 |
| Brick | 0.7 |
| Cement (hardened) | 1.0 |
| Hardwoods | 0.16 |
| Softwoods | 0.12 |
| Styrofoam | 0.03 |
| Water | 0.60 |
| Engine oil | 0.14 |
| Air (1 atm pressure) | 0.025 |
| He (1 atm pressure) | 0.15 |

Consider the plane wall shown in Fig. 1.1. The wall has a thickness $L$ and is very large in the $y$ and $z$ directions. The left face of the wall is at temperature $T_1$ and is at location $x = 0$. The right face of the wall is at temperature $T_2$ and is at location $x = L$.

Let us assume that $T_1 > T_2$. The heat flow will be one-dimensional and will be in the positive $x$ direction.

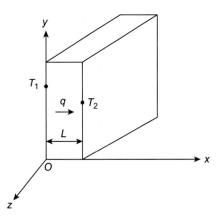

**FIGURE 1.1**

Heat conduction through a plane wall.

As all the heat flow is in the $x$ direction, Eq. (1.2) can be modified by changing "$n$" to "$x$" (or just simply use "$q$") and changing the partial derivative to a total derivative. That is,

$$\frac{q}{A} = -k\frac{dT}{dx} \tag{1.4}$$

Moving $dx$ to the left-hand side of the equation, we can then integrate both sides of the equation:

$$\frac{q}{A}\int_0^L dx = -\int_{T_1}^{T_2} k\, dT \tag{1.5}$$

$$\frac{q}{A}L = \int_{T_2}^{T_1} k\, dT \tag{1.6}$$

In general, the conductivity $k$ is a function of temperature. However, if $k$ is constant or if we use an average $k$ value for the problem, then $k$ may be taken outside of the integral sign in Eq. (1.6), and we have

$$\frac{q}{A}L = k(T_1 - T_2) \tag{1.7}$$

Rearranging this equation, we reach its final form:

$$q = \frac{kA}{L}(T_1 - T_2) \tag{1.8}$$

Details of conduction heat transfer are given in Chapters 2–4.

---

### Example 1.1

#### Heat flow through a plane wall

*Problem*
A large concrete wall is 300 mm thick and has a thermal conductivity of 1.2 W/m C. The heat flux through the wall is 100 W/m². The higher-temperature surface of the wall is at 42 C. What is the temperature of the other surface of the wall?

*Solution*
Heat flux is heat flow per unit area, or q/A. From Eq. (1.8), we have

$$q = \frac{kA}{L}(T_1 - T_2)$$

We want to obtain $T_2$. Rearranging the equation, we get

$$T_2 = T_1 - \left(\frac{q}{A}\right)\left(\frac{L}{k}\right) = 42 - (100)\left(\frac{0.3}{1.2}\right) = 17\ C$$

The temperature of the cooler surface of the wall is 17 C.

---

## 1.4 Convection

Transfer of heat by convection typically occurs between a surface and an adjacent fluid. Convection is a phenomenon involving conduction in the fluid and fluid motion.

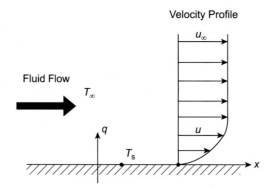

**FIGURE 1.2**

Fluid flow over a surface.

Fig. 1.2 below shows a fluid flowing over a surface. Observations show that the fluid sticks to the surface. As we move away from the surface, the fluid velocity, $u$, increases until it reaches the free stream velocity $u_\infty$. In the free stream, the temperature of the fluid is $T_\infty$. The temperature of the surface is $T_s$.

Let us consider the case where the surface temperature is greater than the fluid temperature; that is, $T_s > T_\infty$. Heat flows by conduction from the surface into the stagnant fluid layer on the surface. The heat then proceeds into the adjacent fluid which is moving. Convection is this combination of conduction at the surface and heat transport into the moving fluid.

There are two main classifications of convection: *forced convection* and *natural (or free) convection*. In forced convection, there is a device causing significant fluid motion. This could be a fan or blower for gases or a pump for liquids. Or, it could be the wind blowing on a building on a windy day. In natural convection, there is no device. The fluid motion is gentler and is caused by buoyancy effects in the fluid caused by thermal expansion.

It has been observed that the rate of heat transfer by convection depends on the difference in temperature of the surface and the fluid. In addition, there are several other factors which influence the heat transfer. These include the motion of the fluid, the thermal and physical properties of the fluid, and the geometry and orientation in space of the surface.

The equation for convection, *Newton's law of cooling*, is

$$q = hA(T_s - T_\infty) \tag{1.9}$$

In this equation, named for English mathematician, physicist, astronomer, and theologian Sir Isaac Newton (1643−1727), $q$ is the rate of heat flow, $A$ is the surface area in contact with the fluid, $T_s$ is the temperature of the surface, $T_\infty$ is the temperature of the fluid, and $h$ is the *convective coefficient*. If we are using the SI unit system, then "$h$" has the units of W/m² C. In the English system, the units are Btu/h ft² F.

Eq. (1.9) looks simple. It really isn't! The convective coefficient $h$ is a complex parameter which depends on the several factors mentioned above. In some cases, it also depends on the temperature difference between the surface and the fluid.

| Table 1.2 Convective coefficient for a heated horizontal cylinder. | |
|---|---|
| **Situation** | **$h$ (W/m$^2$ C)** |
| Natural convection in air | 9 |
| Natural convection in oil | 89 |
| Forced convection with a flow of air at 5 m/s across the cylinder | 40 |
| Forced convection with a flow of oil at 5 m/s across the cylinder | 1500 |

*The cylinder has a diameter of 3 cm. The temperature of the surface of the cylinder is 200 C and the temperature of the adjacent fluid is 20 C.*

In general, $h$ values are greater for liquids than for gases. And, they are greater for forced convection than for natural convection. Table 1.2 illustrates this generality for the situation of convection from a heated horizontal cylinder.

We will learn in Chapters 6 and 7 how to determine the convective coefficient for a variety of situations. Until then, $h$ values will be provided for problems involving convection.

## Example 1.2

### House wall with conduction and convection

*Problem*
A house wall is 8 feet high and 20 feet wide. It is 5 inches thick and has a thermal conductivity of 0.017 Btu/h ft F. It is winter and we want to determine the convective coefficient $h$ on the outside surface of the wall. The temperatures of the inside and outside surfaces of the wall are measured and found to be 72 and 28 F, respectively. The outside air temperature is measured, and it is 25 F. What is the convective coefficient on the outside surface of the wall?

*Solution*
As the wall area is large, we will assume one-dimensional heat transfer, with the heat flow from the inside surface of the wall to the outside air.
The rate of heat conduction through the wall is given by Eq. (1.8):

$$q = \frac{kA}{L}(T_1 - T_2)$$

Dividing the heat flow $q$ by cross-sectional area $A$, we get the heat flux through the wall:

$$\frac{q}{A} = \frac{k}{L}(T_1 - T_2)$$

For this problem,

$$k = 0.017 \text{ Btu / h ft F}$$
$$T_1 = 72 \text{ F}$$
$$T_2 = 28 \text{ F}$$
$$L = 5 \text{ in} \times \frac{1 \text{ ft}}{12 \text{ in}} = 0.41667 \text{ ft}$$

There is convection from the outer surface of the wall to the outside air. The rate of convection is given by Eq. (1.9):

$$q = hA(T_s - T_\infty)$$

For this problem,

$$T_s = 28 \text{ F}$$
$$T_\infty = 25 \text{ F}$$

There is energy conservation at the outer surface of the wall. That is, the heat flux into the surface by conduction equals the heat flux out of the surface by convection. From the above equations,

$$\frac{q}{A} = \frac{k}{L}(T_1 - T_2) = h(T_s - T_\infty)$$

Putting values in this equation, we have

$$\frac{0.017}{0.41667}(72 - 28) = h(28 - 25)$$

Solving for $h$, we get $h = 0.598$ Btu/h ft$^2$ F.
(Note: The h's on the two sides of this result are different. The italic one on the left side is the convective coefficient. The one on the right side is the unit "hours.")
The convective coefficient on the outside surface of the wall is 0.598 Btu/h ft$^2$ F.

## Example 1.3
### Fins on a surface

*Problem*
Fins are often added to a surface to enhance heat transfer. The subject of fins and extended surfaces is discussed in detail in Chapter 3. There are many different geometries of fins. This example deals with a single cylindrical fin, a "pin" fin, as shown in Fig. 1.3.
An aluminum pin fin with a 0.3 cm diameter and a length of 10 cm is attached to a wall whose surface is at 180 C. The conductivity of the fin is 230 W/m C. The convective coefficient between the fin and the room air is 50 W/m$^2$ C and the room air is at 22 C. What is the rate at which heat flows into the fin from the wall, and what is the heat flow rate in the fin at a distance of 3 cm from the wall?
Note: The temperature distribution in the fin can be considered to be one-dimensional in the direction of the fin's axis and, as we will show in Chapter 3, is given by:

$$T(x) = T_\infty + (T_s - T_\infty)e^{-mx} \tag{1.10}$$

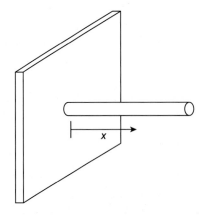

**FIGURE 1.3**
A pin fin.

where $x$ is the distance from the surface to which the fin is attached to the particular location in the fin, $T_s$ is the temperature of the surface to which the fin is attached, and $T_\infty$ is the fluid (i.e., room air) temperature.

$$m = \sqrt{\frac{hP}{kA}}$$

where $h$ is the convective coefficient, and $P$ is the perimeter around the fin. $k$ is the thermal conductivity of the fin, and $A$ is the cross-sectional area of the fin.

*Solution*

This fin problem illustrates a situation including both conduction and convection. At the wall, heat flows into the fin by conduction. The heat is conducted axially through the fin from its point of attachment at the wall to its free end. Heat also goes from the surface of the fin, along its whole length, to the room air by convection.

$$P = \pi D = (\pi)(0.3 \text{ cm}) = 0.9425 \text{ cm} = 0.009425 \text{ m}$$
$$A = \pi R^2 = (\pi)(0.15 \text{ cm})^2 = 0.07069 \text{ cm}^2 = 7.069 \times 10^{-6} \text{ m}^2$$

$$m = \sqrt{\frac{hP}{kA}} = \sqrt{\frac{(50)(0.009425)}{(230)(7.069 \times 10^{-6})}} = 17.025$$

From Eq. (1.4), the heat flux at any $x$ location in the fin is $\frac{q}{A} = -k\frac{dT}{dx}$. Differentiating the $T(x)$ function of Eq. (1.10), we get

$$\frac{dT}{dx} = (T_s - T_\infty)(-m)e^{-mx}$$

At the wall ($x = 0$), $\frac{dT}{dx} = (180 - 22)(-17.025)(1) = -2690 \text{ C/m}$ and the heat flow is

$$q = -kA\frac{dT}{dx} = -(230)(7.069 \times 10^{-6})(-2690) = \textbf{4.37 W}$$

At location $x = 3$ cm, $\frac{dT}{dx} = (180 - 22)(-17.025)e^{-(17.025)(0.03)} = -1614.1 \text{ C/m}$ and the heat flow is $q = -kA\frac{dT}{dx} = -(230)(7.069 \times 10^{-6})(-1614.1) = \textbf{2.62 W}$.

Note: The heat flows are both positive, so they are both in the positive $x$ direction.

The heat flow rate into the fin at the wall is 4.37 W. The heat flow rate in the fin at a distance of 3 cm from the wall is 2.62 W.

Discussion: From the wall to the location 3 cm from the wall, the conduction has decreased from 4.37 to 2.62 W, a difference of 1.75 W. From energy conservation, this 1.75 W is the rate at which heat is flowing by convection from the surface of the fin to the room air over the first 3 cm length of the fin.

# 1.5 Radiation

All surfaces emit thermal radiation. The rate of emission depends on the absolute temperature and characteristics of the surface.

A black surface is an ideal surface which emits the highest possible amount of thermal radiation. For a black surface emitting in accordance with Planck's law, the rate of thermal radiant emission is

$$q_b = \sigma A T^4 \tag{1.11}$$

where $q_b$ is the rate of radiant energy emission, $A$ is the area of the surface, $T$ is the absolute temperature (Kelvin or Rankine) of the surface, and $\sigma$ is the Stefan–Boltzmann constant $= 5.67 \times 10^{-8}$ W/m$^2$ K$^4 = 0.1714 \times 10^{-8}$ Btu/h ft$^2$ R$^4$.

Real surfaces emit less radiation than ideal black surfaces. The surface property related to the emission of real surfaces is the *emissivity*, $\varepsilon$, which is the ratio of the power emitted by a real surface to

| **Table 1.3 Emissivities for various surfaces.** | |
|---|---|
| **Surface** | **Emissivity $\varepsilon$** |
| Aluminum (polished) | 0.05 |
| Aluminum (oxidized) | 0.3 |
| Copper (highly polished) | 0.02 |
| Copper (oxidized) | 0.6 |
| Iron (rusted) | 0.6 |
| Paint (aluminum) | 0.4 |
| Paint (nonaluminum) | 0.9 |
| Brick | 0.94 |
| Wood | 0.9 |

the power emitted by an ideal black surface at the same temperature. For a black surface, $\varepsilon = 1$. It is lower for real surfaces.

Table 1.3 gives typical emissivity values for different surfaces. Additional values are given in Appendix F.

We are usually primarily concerned with the net transfer of radiant energy between surfaces. Like conduction and convection, a temperature difference is needed for heat flow to occur. However, unlike conduction and convection, no physical material is needed to transfer heat via thermal radiation. The radiation is transmitted as an electromagnetic wave. It passes undiminished through a vacuum and is negligibly affected by many gases (e.g., air).

Radiant heat transfer between surfaces depends on the absolute temperatures of the surfaces, the emissivities of the surfaces, the geometries of the surfaces, and the orientation of the surfaces with respect to each other. This can be described by the equation

$$\frac{q}{A} = \sigma F_{G,\varepsilon}\left(T_1^4 - T_2^4\right) \tag{1.12}$$

where $F_{G,\varepsilon}$ is a function of the emissivities, geometries, and orientations of the surfaces.

Radiant heat transfer is a complex subject treated in more detail in Chapter 9. One special case of practical interest is shown in Fig. 1.4. This is the case in which a heated convex object radiates to the surfaces of a large surrounding enclosure.

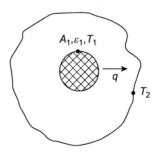

**FIGURE 1.4**

Radiation of a surface to a large enclosure.

If the object has surface area $A_1$, absolute temperature $T_1$, and emissivity $\varepsilon_1$, and the enclosure is at absolute temperature $T_2$, then the radiant heat transfer from the object to the enclosure is

$$q = \varepsilon_1 \sigma A_1 \left( T_1^4 - T_2^4 \right) \tag{1.13}$$

Details of radiative heat transfer are given in Chapter 9.

---

### Example 1.4
**Convection and Radiation from a Cylinder**

*Problem*
A heated aluminum cylinder ($k = 237$ W/m K) is 5 cm diameter and 1 m long. The temperature of its surface is 200 C. The cylinder is in a room where the air temperature is 20 C and the temperature of the room walls is 15 C. The emissivity of the cylinder's surface is 0.5 and the convective coefficient at the surface is 25 W/m$^2$ C. What is the heat flow from the side of the cylinder by convection and radiation?

*Solution*
The temperatures for the convective heat transfer can remain in Celsius (C), but the temperatures must be in absolute temperature Kelvin (K) for the radiant heat transfer as follows:

$T_\infty$ = room air temperature = 20 C.
$T_s$ = surface temperature = 200 C = 200 + 273.15 = 473.15 K.
$T_{surr}$ = temperature of the room walls = 15 C = 15 + 273.15 = 288.15 K.

The area of the side of the cylinder is $A = \pi D L = \pi (0.05 \text{ m}) (1 \text{ m}) = 0.1571 \text{ m}^2$.
The convective heat transfer is

$$q_{conv} = hA(T_s - T_\infty) = (25)(0.1571)(200 - 20) = \mathbf{707.0 \text{ W}}$$

The radiant heat transfer is

$$q_{rad} = \varepsilon \sigma A \left( T_s^4 - T_{surr}^4 \right) = (0.5)\left(5.67 \times 10^{-8}\right)(0.1571)\left(473.15^4 - 288.15^4\right) = \mathbf{192.5 \text{ W}}$$

The heat flow from the side of the cylinder by convection is 707.0 W. The heat flow by radiation is 192.5 W.
Comments: Radiant heat transfer depends on temperature to the fourth power and becomes very significant at high temperatures. As the temperatures in this problem are quite low, the radiation heat transfer is small compared to the convective heat transfer. The radiant heat transfer could be significantly reduced by polishing the cylinder, thereby reducing the emissivity to 0.1 or less. Also, because the convection and radiation occur at the cylinder's surface, the information given about the material of the cylinder and the conductivity are extraneous and not relevant to the problem. (They would have been relevant if the problem had dealt with conduction within the cylinder.) Engineers often encounter situations in which a surplus of information is available. Judgments must be made as to which items of information are relevant to the specific situation. On the other hand, engineers often have to make educated judgments for required information that has not been provided in the problem statement.

---

## 1.6 The direction of heat flow

Heat flow is a vector quantity. It has direction and magnitude. In Eq. (1.2) for conduction, we added a minus sign so that heat flow in the positive $n$ direction would be positive and heat flow in the negative $n$ direction would be negative.

For example, if heat is flowing in the positive $n$ direction, then $\partial T/\partial n$ is negative because heat flows from a higher temperature toward a lower temperature. The minus sign is needed to make $q_n$ positive.

Likewise, if heat is flowing in the negative $n$ direction, then $\partial T/\partial n$ is positive and the minus sign makes $q_n$ negative.

The direction of heat flow must also be considered for convection and radiation. Consider, for example, Eq. (1.9) for convection:

Say that heat is flowing *from* the surface *to* the fluid, as shown in Fig. 1.3.

In this case, $T_s > T_\infty$, and Eq. (1.9) is appropriate because it makes the heat flow positive in the direction it is actually flowing. But, if the heat were flowing *from* the fluid *to* the surface, then the equation should be revised to make the heat flow positive in the direction it is actually flowing. That is, if $T_\infty > T_s$ we should revise Eq. (1.9) to

$$q = hA(T_\infty - T_s) \tag{1.14}$$

Heat flow direction must also be considered when formulating equations for radiative heat transfer.

## 1.7 Temperature continuity and heat balances

Due to the broadness of the subject of heat transfer, it is usual to separate heat transfer into its three modes and study conduction, convection, and radiation separately. While advantageous for the learning process, one has to keep in mind that many practical problems involve not only one but often two or all three of the modes occurring simultaneously.

For example, consider an electric cartridge heater inside a block of metal sitting on a table in a room. This is shown in Fig. 1.5.

When the power to the heater is turned on, heat starts to flow into the block, thereby raising its temperature. The temperature of the block will become greater than that of the air in the room, and there will be heat transfer by convection from the exposed surface of the block to the air. There will also be heat transfer by radiation from the exposed surface of the block to the walls and ceiling of the room. And, because the temperature of the block will become greater than the temperature of the table, there will be conduction from the block into the table.

There will be continuity of temperature at the exposed surface of the block in contact with the air. That is, the temperature of the exposed surface of the block and the temperature of the air at the surface will be the same. Disregarding any contact resistance between the block and the tabletop, the temperature of the block's surface in contact with the tabletop will be the same as that of the tabletop.

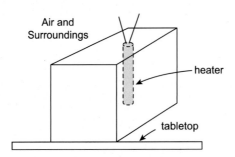

**FIGURE 1.5**

Heater in a metal block.

There will also be continuity of heat flows. The rate of heat flow by conduction reaching the exposed surface of the block will be equal to the rate of heat flow leaving the block by convection and radiation.

Regarding the surface in contact with the tabletop, the rate of heat flow by conduction within the block reaching the tabletop will equal the rate of heat flow by conduction into the table.

If the electric power to the heater remains constant, the temperatures and heat flows will eventually stabilize and reach steady state. When this happens, we can do an energy balance on the block. We can state that the electric power to the heater equals the rate of heat transfer from the block's exposed surface to the air by convection and to the surroundings by radiation plus the rate of heat transfer by conduction from the block to the tabletop.

$$q_{power} = q_{conduction\ to\ table} + q_{convection\ to\ air} + q_{radiation\ to\ surroundings} \qquad (1.15)$$

## Example 1.5
### Heat Balance

*Problem*
Consider again Fig. 1.5 which depicts a metal block with an internal cartridge heater. The block rests on a tabletop. It convects to the room air, radiates to the surroundings, and conducts to the tabletop. The block is 6 inches by 6 inches by 6 inches and has a 300 W internal electric heater. The room air and the surroundings are at 72 F. The convective coefficient is 5 Btu/h ft² F and the emissivity of the block's exposed surfaces is 0.4. It can be assumed that the heat flow by conduction to the tabletop is negligible. What is the average temperature of the exposed surfaces of the block?

*Solution*
Eq. (1.15) gives the heat balance for the block. At steady state, the power into the heater equals the sum of the convection to the room air, the radiation to the surroundings, and the conduction to the tabletop.
The givens of the problem are

$$T_{surr} = 72\ F\ =\ 72\ +\ 459.67\ =\ 531.67\ R\ \text{(Rankine)}$$
$$T_\infty = 72\ F\ =\ 531.67\ R$$
$$h = 5\ \text{Btu}\ /\ \text{h ft}^2\ F$$
$$\varepsilon = 0.4$$
$$q_{power} = 300\ W \times (3.412\ \text{Btu}\ /\ \text{h})\ /\ W\ =\ 1023.6\ \text{Btu}\ /\ \text{h}$$

The exposed area of the block is $A = 5$ sides $\times (6 \times 6) = 180$ in² $= 1.25$ ft².
Let the unknown temperature of the exposed surfaces be $T_s$.
The convective heat transfer is

$$q_{conv} = hA(T_s - T_\infty) = (5)(1.25)(T_s - 531.67) = 6.25(T_s - 531.67)$$

The radiant heat transfer is

$$q_{rad} = \varepsilon\sigma A\left(T_s^4 - T_{surr}^4\right) = (0.4)\left(0.1714 \text{ x } 10^{-8}\right)(1.25)\left(T_s^4 - 531.67^4\right) = 8.57 \times 10^{-10}\left(T_s^4 - 531.67^4\right)$$

Rearranging Eq. (1.15) above, we get

$$q_{conduction\ to\ table} = q_{power} - q_{conv} - q_{rad}$$

From the problem statement, the conduction to the table is zero.
So, we have $q_{power} - q_{conv} - q_{rad} = 0$.

Substituting information from above, we have the following equation:

$$1023.6 - 6.25(T_s - 531.67) - 8.57 \times 10^{-10}(T_s^4 - 531.67^4) = 0$$

Reducing this equation, we get the following equation to solve for $T_s$:

$$8.57 \times 10^{-10} T_s^4 + 6.25 T_s - 4415 = 0$$

We can solve this through several techniques. One way is to use trial and error. A guess is made for $T_s$ and the left-hand side of the equation is evaluated. The result is compared to the right-hand side, which is zero. If the left-hand side is not zero, a new guess is made for $T_s$ and the procedure is repeated. Iterations are made until the left-hand side is zero.
A more efficient solution of the equation is through the use of the *Goal Seek* feature of Microsoft Excel. This useful feature is discussed in Appendix J.
It is found that the average temperature of the exposed surfaces of the block is

$$T_s = 677.51 \text{ R} = 677.51 - 459.67 = 218 \text{ F}.$$

The average temperature of the exposed surfaces of the block is 218 F.

## 1.8 Unit systems

Problems in this book use either the SI unit system or the English Engineering unit system. Most problems use SI. Conversion factors are located on the inside front cover of the book for easy reference.

Unit conversion is really quite simple. One starts with the item to be converted and merely keeps multiplying it by unity factors until the desired units are reached. For example, say we want to convert 10 Watts to BTU/day. We have

$$10 \text{ W} \times \frac{3.412 \dfrac{\text{Btu}}{\text{h}}}{\text{W}} \times 24 \dfrac{\text{h}}{\text{day}} = 819 \dfrac{\text{Btu}}{\text{day}}$$

One common area of confusion pertains to temperature. The confusion is with respect to temperature *levels* versus temperature *units*. Temperature levels are determined by the following equations:

$$\text{Celsius to Kelvin:} \quad K = C + 273.15$$

$$\text{Fahrenheit to Rankine:} \quad R = F + 459.67$$

$$\text{Celsius to Fahrenheit:} \quad F = \frac{9}{5} C + 32$$

$$\text{Fahrenheit to Celsius:} \quad C = \frac{5}{9}(F - 32)$$

Temperature levels in the four temperature scales are usually quite different. One notable exception is that $-40 \text{ C} = -40 \text{ F}$.

Things are quite different when we are considering the *units* of physical and thermal properties. The *size* of a Fahrenheit degree is the same as the *size* of a Rankine degree, and the *size* of a Celsius degree is the same as the *size* of a Kelvin degree. For example, a thermal conductivity of 150 W/m C is same as a conductivity of 150 W/m K. And a specific heat of 0.2 Btu/lbm F is the same as 0.2 Btu/lbm R.

One must remember that there is no need to convert numerical values from Celsius to Kelvin or from Fahrenheit to Rankine when the temperatures are within the *units* of the various material properties. However, it is of course necessary to convert the numerical values when we are considering temperature *levels*.

The size of a Fahrenheit (or Rankine) degree is smaller than the size of a Celsius (or Kelvin) degree. Let us consider the freezing and boiling points of water. Water at atmospheric pressure freezes at 32 F and boils at 212 F. There are 180 Fahrenheit degrees between the two points. For Celsius, water freezes at 0 C and boils at 100 C. There are 100 Celsius degrees between freezing and boiling. Therefore, a difference of 180 Fahrenheit degrees is equivalent to a difference of 100 Celsius degrees. A Fahrenheit degree is (100/180) = (5/9) the size of a Celsius degree. If we have a temperature difference in Fahrenheit degrees, the equivalent temperature difference in Celsius degrees is 5/9 that of the Fahrenheit degree difference.

Finally, we must remember that absolute temperature units (Kelvin or Rankine) must be used for problems involving thermal radiation.

## 1.9 Recommended approach to problem solving

In solving an engineering problem, be it a problem at the end of the chapters in this book or a general problem in engineering, people often have the urge to solve the problem as quickly as possible, and they immediately start writing equations and inserting numbers. This approach should be avoided. Rather, the approach outlined below is recommended. This approach appears to be slower, but it actually is more efficient because it avoids having to go back and redo hastily and ill-performed steps.

### Step 1 - Problem definition

This is perhaps the most important step. We need to clearly understand the objective of the problem. What are we looking for? If we get this step wrong, then we may have a perfect solution, but it is not relevant to the problem we want to solve! Write the objective of the problem in one or two sentences.

### Step 2 - Problem givens

List the information given in the problem statement. In textbook problems, there is usually just enough information given to solve the problem, and there is no extraneous information. This is really unlike problems which you will encounter in the "real world."

In those problems, there is often an excess of information and you will have to determine which parts of this information are relevant to your problem. (Indeed, we included some extraneous information in Example 1.4 above and we will do this occasionally in the end-of-chapter problems.)

### Step 3 - Determine the appropriate equations

Considering the problem definition and the givens, determine the analytical equations appropriate for the problem. For example, what modes of heat transfer are involved? Do we have conduction, convection, or radiation, or a combination of modes? Is the heat transfer one-, two-, or three-dimensional? Is the heat transfer steady state or transient? List the equations.

## Step 4 - Obtain the solution

Put values into the equations and obtain numerical solutions. Make sure that you include *units* with your numerical results! State the result in a sentence or two.

## Step 5 - Review the solution

This final step is of major importance! Consider your solution. Look at the problem definition and assure yourself that you have solved the correct problem. And, is the numerical result reasonable? Are the magnitudes of the results consistent with the magnitudes of the problem givens? For example, if your problem was to determine the temperature at a point in a metal plate, and your answer was 900 C when the surrounding air was at 20 C, and there are no internal heat sources or heaters in the problem, then perhaps something is wrong. Perhaps you made an error in your calculations. Check the equations and the values you put into them. Also, check that the correct units were used.

The above procedure is a methodical way of solving heat transfer problems. Indeed, the procedure is also very applicable to other areas of engineering. It is highly recommended that the procedure be followed as much as possible. Getting into a methodical mindset will serve you well in your engineering career.

# 1.10 Significant figures

There is a great tendency on the part of students (and often engineers) to retain too many significant figures in a final answer. Indeed, calculators provide 10 or more digits and computers have even more. Why not include *all* of the digits in a final answer? Well, inclusion of *all* the digits is usually inappropriate because of the inaccuracies and uncertainties inherent in all engineering calculations. First, simplifications are often made in the problem statement and the analysis technique. For example, surfaces are taken as being perfectly insulated; thermal resistances of thin metal walls are neglected; multidimensional heat flows are assumed to be one-dimensional.

Inaccuracies also occur with the use of tabulated material property data from reference sources. Perhaps such data were obtained under stringent laboratory conditions which are inconsistent with your application. And perhaps you were unable to find the material properties for your particular material and therefore substituted the properties of a similar material. (Of course, these concerns are minimized if tests are conducted on your actual specimens, but such is usually not the case.)

As another example, inaccuracies may be introduced through use of empirical correlations presented in reference sources; e.g., correlations for the convective coefficient "*h*." Such correlations may have been from carefully controlled laboratory experiments. Your conditions may differ considerably from the conditions under which the data were obtained.

In short, one must be careful to not include an excess of significant figures in a final answer. Too many figures imply that you know the result with more precision than you actually do. For example, an answer of 24.373 C implies that the solution is accurate to one-thousandth of a degree. Such is usually not the case. A good rule of thumb for textbook problems and most engineering problems is the following: If an answer begins with "1", give the answer to four significant figures; if it begins with any other digit, give the answer to three significant figures. For example, 1.78425 would become 1.784 and 5.8957 would become 5.90. Of course, this rule of thumb only pertains to the *final* answer of a problem. Intermediate calculations leading up to the final answer should use considerably more significant figures to not introduce inaccuracies through rounding.

Significant figures are discussed in many freshman engineering and data analysis texts. Two such texts are given in Refs. [3,4].

## 1.11 Chapter summary and final remarks

This chapter provided an introduction to the study of heat transfer. The three modes of heat transfer (conduction, convection, and radiation) were discussed, and basic equations for each mode were presented. The concepts of temperature continuity and heat balance were discussed. Finally, there were discussions of unit systems, problem-solving techniques, and significant figures. These last three topics are pertinent not only to heat transfer, but to most other areas of engineering.

We end this chapter with a word of caution. A lot of useful information is available on the Internet. Most is accurate, but some is not. Use material only from reputable sources. Some such sources are government offices (e.g., NIST) and standards/professional organizations (e.g., ASTM, ASME, ASHRAE, IEEE, ANSI). Reputable manufacturing companies are also a useful source of information. It is good to compare information from different Internet sources to gain confidence regarding the accuracy of the information.

## 1.12 Problems

**1-1**  A concrete wall is 20.3 cm thick and has a thermal conductivity of 0.8 W/m C. One surface of the wall is at 20 C and the other surface is at 5 C.
  **(a)** What is the heat flux through the wall (W/m$^2$)?
  **(b)** If the wall is 6 m long and 2.44 m high, what is the heat flow (W) through the wall?

**1-2**  A steel plate is 1.5 m by 1.2 m and 0.5 cm thick. The thermal conductivity of the plate is 60.2 W/m C. If the temperature difference of the plate's surfaces is 0.2 C, what is the heat flow rate (W) through the plate?

**1-3**  A building wall has a layer of fiberglass insulation. The conductivity of the fiberglass is 0.022 Btu/h ft F. The fiberglass is in a channel between wall studs and is 14.5 inches wide, 7.5 feet high, and 3.5 inches thick. If the temperature of the inside surface of the fiberglass is 65 F and that of the outside surface is 40 F, what is the heat flow rate (Btu/h) through the fiberglass layer?

**1-4**  A layer of insulating material is 15.2 cm thick. The temperature difference across the surfaces is 25 C and the heat flux through the layer is 6.6 W/m$^2$. What is the thermal conductivity of the insulating material (W/m C)?

**1-5**  A cup has a diameter of 3 inches and height of 4.5 inches. The wall thickness of the cup is 1/8 inch. The cup is filled with coffee at 140 F. Assume that the inside surface of the cup is at the coffee temperature (140 F) and the outside surface of the cup is at 90 F. The conductivity of the cup's material is 0.8 Btu/h ft F.
  **(a)** What is the heat flow by conduction (Btu/h) through the cylindrical wall of the cup?
  **(b)** If the room air is at 70 F, what is the convective coefficient at the outer surface of the cup (Btu/h ft$^2$ F)?

**1-6** A cup has a diameter of 7.6 cm and height of 11.4 cm. The wall thickness of the cup is 0.32 cm. The cup is filled with coffee at 60 C. Assume that the inside surface of the cup is at the coffee temperature (60 C) and the outside surface of the cup is at 32 C. The conductivity of the cup's material is 1.4 W/m C.
   **(a)** What is the heat flow (W) by conduction through the cylindrical wall of the cup?
   **(b)** If the room air is at 21 C, what is the convective coefficient at the outer surface of the cup (W/m$^2$ C)?

**1-7** A cylindrical electric heater is used to heat water from 15 to 80 C. The heater has a diameter of 1.5 cm and a length of 25 cm. The surface of the heater is at 95 C during the heating process and the convective coefficient between the heater's surface and the water is 800 W/m$^2$ C. What is the convective heat flow (W) from the heater's surface (including the cylindrical side and both ends) when the water temperature is 40 C?

**1-8** A 1000 W cylindrical electric heater has a diameter of 1.5 cm and a length of 25 cm. After using the heater to heat water, the heater has unfortunately been taken out of the water and left "on" in the room air. It is convecting and radiating to the room air and the surroundings. Assume that both the room air and the surroundings are at 20 C. Also, assume the emissivity of the heater's surface is 0.5 and the convective coefficient between the heater and the room air is 10 W/m$^2$ C. What is the temperature of the heater's surface?

**1-9** Cold air at 10 C flows across the outer surface of a 1.5-cm OD tube. The surface of the tube is kept at a constant temperature of 120 C. If the convective coefficient between the outer surface of the tube and the air is 70 W/m$^2$ C, what is the convective heat transfer per meter length of tube (W/m)?

**1-10** A solid aluminum cylinder ($k = 230$ W/m K) has a diameter of 3 cm and a length of 10 cm. It is suspended by a thin wire in the room air. The cylinder has an internal electr-ic heater which keeps the surface of the cylinder at 110 C. The room air is at 25 C and the convective coefficient between the cylinder and the room air is 15 W/m$^2$ C. Assume that the heat flow through the support wire is negligible and that radiation heat transfer is negligible due to the very small emissivity of the surface of the cylinder. What is the heat flow (W) from the cylinder to the room air?

**1-11** A solid copper sphere ($k = 350$ W/m C) has a diameter of 10 cm and is resting on a tabletop. An internal heater keeps the surface of the sphere at 200 C. The surface of the sphere is highly polished and has an emissivity of 0.08. The convective coefficient is 12 W/m$^2$ C. The room air is at 20 C and the surroundings are at 18 C. Assume the heat flow from the sphere to the table by conduction is negligible.
   **(a)** What is the heat transfer rate by convection (W)?
   **(b)** What is the heat transfer rate by radiation (W)?

**1-12** A hot surface at 500 K has an emissivity of 0.7. What is the radiant flux emitted by the surface (W/m$^2$)?

**1-13** A sphere of diameter 10 cm is suspended inside a vacuum chamber by a thin wire from the top of the chamber. The sphere's surface is at 700 C and has an emissivity of 0.3. The vacuum chamber's walls are at 50 C. What is the heat flow (W) from the sphere?

**1-14** Fins are often added to a surface to enhance heat transfer. An aluminum pin fin with a square cross section of 0.5 cm by 0.5 cm and a length of 2 cm is attached to a surface which is at 150 C. The conductivity of the fin is 230 W/m C. The convective coefficient between the fin and the room air is 50 W/m$^2$ C and the room air is at 22 C. What is the heat flow rate (W) in the fin at a distance of 1 cm from the surface to which it is attached?

Note: The temperature distribution in the fin can be considered to be one-dimensional in the direction of the fin's axis and is given by

$$T(x) = T_\infty + \frac{(T_s - T_\infty)\cosh[m(L-x)]}{\cosh(mL)}$$

where $x$ is the distance from the surface to which the fin is attached to the particular location in the fin, $T_s$ is the temperature of the surface to which the fin is attached, and $T_\infty$ is the fluid (i.e., room air) temperature.

$m = \sqrt{\frac{hP}{kA}}$ where $h$ is the convective coefficient, $P$ is the perimeter around the fin, $k$ is the thermal conductivity of the fin, and $A$ is the cross-sectional area of the fin.

For this problem, $P = 4 \times 0.5$ cm $= 2$ cm   and   $A = 0.5$ cm $\times 0.5$ cm $= 0.25$ cm$^2$.

**1-15** A heated steel plate is 0.5 m by 0.5 m by 2 mm thick and has a thermal conductivity of 45 W/m C. The temperature distribution in the plate (C) is given by

$$T(x,y) = 75e^{-(x-0.5)^2} \cdot e^{-2(y-0.5)^2}$$

Note: The $x$ and $y$ axes are in the plane of the plate. Temperature variation in the $z$ direction is negligible. The bottom left corner of the plate is at $x = 0$, $y = 0$ m and the top right corner of the plate is at $x = 0.5$, $y = 0.5$ m.

**(a)** What are the $x$ and $y$ components of the heat flux vector in the plate at the center of the plate ($x = 0.25$, $y = 0.25$ m) and at the location $x = 0.4$, $y = 0.1$ m?

**(b)** For the two points of item (a), combine the $x$ and $y$ components into a vector and determine their magnitudes. Also, determine the direction of the two vectors (i.e., their angle above the positive $x$ axis).

**1-16** A heated steel plate is 1 m by 1 m by 2 mm thick and has a thermal conductivity of 45 W/m C. The surface of the plate is in the $x$-$y$ plane and has the following temperature distribution (C):

$$T(x,y) = 20 + 100\frac{\sin(\pi x) - \sinh(\pi y)}{\sinh(\pi)}$$

Note: The $x$ and $y$ axes are in the plane of the plate. Temperature variation in the $z$ direction is negligible. The bottom left corner of the plate is at $x = 0$, $y = 0$ m and the top right corner of the plate is at $x = 1$, $y = 1$ m.

**(a)** What are the $x$ and $y$ components of the heat flux vector in the plate at location $x = 0.3$, $y = 0.6$ m?

**(b)** Combine the heat flux components of item (a) into a vector and determine the vector's magnitude and direction (i.e., angle above the positive $x$ axis).

**1-17** A kiln produces 1440 W of heat and has interior dimensions of 10 inches by 9 inches by 6.5 inches. The inside surface of the kiln is at 1200 C and the wall of the kiln has a thermal conductivity of 0.24 W/m C. How thick should the wall be so that the temperature of the exterior surface of the kiln is not greater than 50 C?

**1-18** A cylindrical electric heater is 0.5 cm diameter and 10 cm long. It has a power output of 50 W and is immersed in a tank of water. What is the convective coefficient (W/m$^2$ C) when the water temperature is 30 C and the surface of the heater is at 95 C?

**1-19** It is desired to determine the emissivity of a paint. A sphere of diameter 5 cm is coated with the paint and suspended in a large vacuum chamber. The sphere has a 10 W internal electric heater. At steady-state operation, the surface of the sphere is 200 C. If the walls of the chamber are at 25 C, what is the emissivity of the paint?

**1-20** An electronic component is attached to a circuit board. It is 2 cm by 2 cm by 3 mm thick. At steady operation, its power output is 5 W. A fan blows air across the component to keep it cool. The convective coefficient between the component and the air is 80 W/m$^2$ C and the air is at 20 C. If the temperature of the surface of the component cannot be greater than 70 C, what is the rate at which heat must be transferred to the circuit board by conduction (W)? Assume that radiation heat transfer from the component is negligible.

**1-21** The thermal conductivity of a material is 63 W/m C. What is its conductivity in the units of W/m K?

**1-22** The convective coefficient at a surface is 150 W/m$^2$ C. What is the coefficient in the units of W/m$^2$ K?

**1-23** A heater is rated at 1500 W. What is its rating in the units of Btu/h?

**1-24** A metal cylindrical container is filled with ice—water and is on the top of a table. The container is 20 cm diameter and 25 cm high. The container gains heat by convection from the room air and radiation from the room surroundings. Heat flow into the container from the table is negligible. Both the air and the surroundings are at 20 C. The emissivity of the outer surface of the container is 0.1 and the convective coefficient at the outer surface is 20 W/m$^2$ C. The container has a thin wall and it can be assumed that the temperature of the wall is at the ice—water temperature of 0 C.

(a) What is the heat flow rate into the ice—water mixture (W)?

(b) If the heat of fusion of water is 333.5 kJ/kg, how much ice (kg) is melted in 1 hour?

**1-25** Consider the problem of Example 1.5 with the following changes: The average temperature of the exposed surfaces is 175 F. The bottom of the block is not perfectly insulated, and the block transfers heat by conduction to the tabletop. The task is to find the rate of this conduction heat transfer (Btu/h). All other givens in the problem remain the same.

**1-26** A portable electric heater for heating a room has heating elements in the form of thin metal strips. The total heating surface area of the strips is 0.03 m$^2$. The room air and surroundings are at 20 C. It is desired to find the output of the heater given the surface temperature of the heater's elements.

(a) What are the relevant modes of heat transfer for this problem and what are the directions of these modes?

(b) The convective coefficient at the surfaces of the heating elements is 5 W/m$^2$ C and the emissivity of the surfaces is 0.75. If the temperature of the heating surfaces is 950 C, what is the output of the heater (W)?

(c) If we want the heat output to be 4000 W, what is the needed surface temperature of the elements?

**1-27** A pipe transfers cold water through a vacuum chamber. The water is at 4 C and the walls of the chamber are at 20 C. It is desired to find the heat transfer from the pipe per meter length of pipe.

(a) What are the relevant modes of heat transfer for this problem and what are the directions of these modes?

(b) The pipe is 1″ Sch 40 galvanized iron. Its outer surface has an emissivity of 0.8 and a temperature of 4.2 C. What is the heat transfer per meter length of pipe and what is its direction?

**1-28** A can of soda is 12.2 cm high and 6.3 cm diameter. Its temperature is 2.8 C when it is taken out of the refrigerator and placed on a table. The room is at 22 C. It is desired to determine the change in temperature of the soda after 30 min in the room air.

(a) What are the relevant modes of heat transfer for this problem and what are the directions of these modes?

(b) If the convective coefficient at the can's outer surface is 7 W/m$^2$ C and the emissivity of the surface is 0.4, what is the temperature of the soda after 30 min? Assume that soda has the properties of water.

## References

[1] J. Fourier, Theorie analytique de la chaleur, Paris, 1822.

[2] J. Fourier, The Analytical Theory of Heat, Translated by Alexander Freeman, Dover Publications, New York, 1955 (Translation of Reference [1]).

[3] G.C. Beakley, H.W. Leach, Engineering—an Introduction to a Creative Profession, Chapter 6, Macmillan, 1982.

[4] P.F. Dunn, Measurement and Data Analysis for Engineering and Science, Chapter 2, McGraw-Hill, 2005.

# Heat conduction equation and boundary conditions

# 2

## Chapter Outline

## 2.1 Introduction

Chapter 1 briefly described the three modes of heat transfer: conduction, convection, and radiation. In this chapter, we will look closely at the theoretical aspects of conduction. In particular, we will discuss the governing differential equations and boundary condition equations whose solution will give us the temperature distribution in the body and ultimately the heat flow rates at different points in the body.

We will also discuss the initial condition equations that are needed if the problem is time-dependent, i.e., non–steady state.

## 2.2 Heat conduction equation

The heat conduction equation is a partial differential equation for conduction of heat through a body. The solution of the equation gives the temperature distribution in the body at different instances of time. As discussed in Chapter 1, differentiation of the temperature distribution will give the heat flows at the various locations in the body.

The heat conduction equation is derived for the rectangular (i.e., Cartesian) coordinate system. The equation is then presented for cylindrical and spherical coordinate systems.

### 2.2.1 Rectangular coordinates

In this section, we will derive the differential equation describing the temperature distribution in a body. (By "body" we mean a solid or an essentially nonmoving fluid. In this text, "body" is synonymous with "object" or "material".)

Let us consider an infinitesimal element in the object as shown in Fig. 2.1. The element has dimensions $dx$, $dy$, and $dz$ and is located at coordinates $x$, $y$, and $z$, respectively.

Fig. 2.1 shows heat flows into the element of $q_x + q_y + q_z$ and heat flows out of the element of $q_{x+dx} + q_{y+dy} + q_{z+dz}$. For example, for the $x$ direction: At location $x$, the incoming heat flow is $q_x$, and at location $x + dx$, the outgoing heat flow is $q_{x+dx}$.

In addition to heat flowing in and out of the element, there may be heat generated within the element. For example, an electrical current passing through a wire or busbar will cause resistance (or Joule) heating in the material. Water added to cement will result in hydration, an exothermic heat-producing chemical reaction. Heat is also produced through fission in nuclear fuel elements. Such

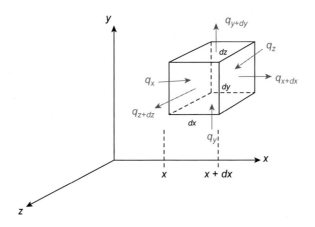

**FIGURE 2.1**

Conductive heat flow through an infinitesimal element.

heat-generating phenomena occur within the material, and the rate of heat generation is usually proportional to the volume of the material.

So, we have heat flowing in and out of the boundaries of the infinitesimal element and possibly also heat input to the material by internal heat generation. If more heat goes into the element than leaves, then the internal energy of the material in the element will increase. If more heat leaves the element than enters, then the internal energy will decrease. Increased internal energy means an increase in temperature, and decreased internal energy means a decrease in temperature. A word equation may be written for the energy balance on the element as follows:

$$\text{Rate of Heat Flow In} - \text{Rate of Heat Flow Out} + \text{Rate of Heat Generation}$$
$$= \text{Rate of Increase in Internal Energy} \tag{2.1}$$

The incoming heat flow rate is $q_x + q_y + q_z$.

The outgoing heat flow rate is $q_{x+dx} + q_{y+dy} + q_{z+dz}$.

The rate of heat generation is $q_{gen}dV$, where $q_{gen}$ is the generation rate per unit volume and $dV$ is the volume of the element (i.e., $dxdy\,dz$).

And the rate of increase in internal energy is mass (i.e., density $\rho$ times volume $dxdydz$) times specific heat $c$ times change of temperature per unit time.

In symbols, Eq. (2.1) is

$$\left(q_x + q_y + q_z\right) - \left(q_{x+dx} + q_{y+dy} + q_{z+dz}\right) + q_{gen}dx\,dy\,dz = \rho(dx\,dy\,dz)c\frac{\partial T}{\partial t} \tag{2.2}$$

For the $x$ direction, from Eq. (1.3), $\left(\frac{q}{A}\right)_x = -k\frac{\partial T}{\partial x}$ and $A_x$ is $dy\,dz$.

In addition, $q_{x+dx}$ is $q_x$ plus the change in $q_x$ over a distance $dx$. Mathematically, this is

$$q_{x+dx} = q_x + \left[\frac{\partial}{\partial x}(q_x)\right]dx$$

This Taylor expansion can be done similarly for the y and z directions.

Substituting these details into Eq. (2.2), canceling $dx\,dy\,dz$, which is in each term, and rearranging the terms, results in the heat conduction equation:

$$\frac{\partial}{\partial x}\left(k\frac{\partial T}{\partial x}\right) + \frac{\partial}{\partial y}\left(k\frac{\partial T}{\partial y}\right) + \frac{\partial}{\partial z}\left(k\frac{\partial T}{\partial z}\right) + q_{gen} = \rho c\frac{\partial T}{\partial t} \tag{2.3}$$

Solution of this equation will give the temperature distribution $T(x,y,z,t)$ in the body at different instances of time. To solve the equation, it has to be combined with appropriate boundary conditions that are discussed in Section 2.3 below. And, if the situation is non–steady state, appropriate initial conditions are needed. These initial conditions are discussed in Section 2.4 below.

### 2.2.1.1 Special cases—rectangular coordinates

The general heat conduction Eq. (2.3) can often be simplified. For example, if the problem is steady state, the $\frac{\partial T}{\partial t}$ term is zero. If there is no heat generation within the material, then $q_{gen}$ is zero.

In addition, in many cases, all three dimensions are not needed to accurately model a problem. For example, the wall of a building can often be modeled as one-dimensional, as shown in Example 2.1:

## Example 2.1
### Heat Flow Through a Wall

*Problem:*
To design air conditioning and heating systems for a building, it is necessary to determine the heating and cooling loads for various components of the building's envelope. Heat flow through walls is often a major contributor to the loads. Let us say we wish to determine the heat flow through the wall shown in Fig. 2.2. We can determine the temperature distribution in the wall by solving a simplified version of Eq. (2.3). The obtained temperature function can then be differentiated to get the heat flow through the wall. The task is to determine the appropriate version of the heat conduction equation for this situation.

*Solution:*
Eq. (2.3) can be simplified as follows: The heat flow is essentially one-dimensional as the wall area is large compared with the wall thickness. In addition, the wall material does not generate heat. Finally, steady-state heat flow is typically used in load calculations.

Therefore, in Eq. (2.3), $\frac{\partial T}{\partial t} = 0$, $q_{gen} = 0$, and the y and z directional terms are zero. Eq. (2.3) thus simplifies to

$$\frac{d}{dx}\left(k\frac{\partial T}{\partial x}\right) = 0 \qquad (2.4)$$

Another usual simplification is that thermal conductivity $k$ is constant. This results in the final equation

$$\frac{d^2 T}{dx^2} = 0 \qquad (2.5)$$

This equation will be solved in Chapter 3.

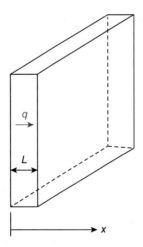

**FIGURE 2.2**

A plane wall.

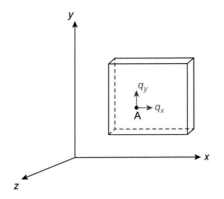

**FIGURE 2.3**

Heat flow in a thin plate.

## Example 2.2

### Heat Flow in a Thin Plate

As another example, let us consider conduction in a thin plate situated in the x-y plane, as shown in Fig. 2.3.

*Problem:*

Determine the heat conduction equation for a thin, flat plate.

*Solution:*

Like Example 2.1, all three dimensions are not needed to properly describe the temperature distribution in a thin plate. There is temperature variation in the x and y directions, but, as the plate is thin, there is negligible temperature variation in the z direction. Only two dimensions are needed to properly describe the temperature distribution. At a point "A" in the plate, there are heat flow components $q_x$ and $q_y$. However, $q_z$ is zero. If the heat transfer is steady, Eq. (2.3) simplifies to

$$\frac{\partial}{\partial x}\left(k\frac{\partial T}{\partial x}\right) + \frac{\partial}{\partial y}\left(k\frac{\partial T}{\partial y}\right) + q_{gen} = 0 \tag{2.6}$$

If there is no internal heat generation in the plate, Eq. (2.6) reduces to

$$\frac{\partial}{\partial x}\left(k\frac{\partial T}{\partial x}\right) + \frac{\partial}{\partial y}\left(k\frac{\partial T}{\partial y}\right) = 0 \tag{2.7}$$

And, if the conductivity is constant, there is a further reduction to

$$\frac{\partial^2 T}{\partial x^2} + \frac{\partial^2 T}{\partial y^2} = 0 \tag{2.8}$$

## 2.2.2 Cylindrical coordinates

Cylindrical coordinates are used for problems involving cylinders. A point in a body in cylindrical coordinates is located by $r$, $\varphi$, and $z$, which are radius, azimuthal angle, and axial distance, respectively. These dimensions for a point "A" are shown in Fig. 2.4.

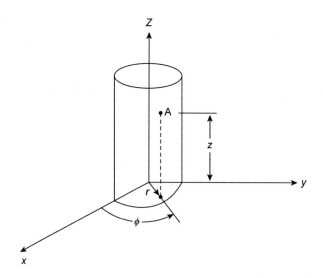

**FIGURE 2.4**

Cylindrical coordinates.

The heat conduction equation in cylindrical coordinates can be obtained from the rectangular coordinate equation, Eq. (2.3), through coordinate transformation using the following relations between the parameters of the two coordinate systems. These relations are

$$x = r \, \cos \varphi$$
$$y = r \, \sin \varphi$$
$$z = z$$

The heat conduction equation in cylindrical coordinates can also be obtained by an energy balance on a cylindrical element, similar to what was done for the rectangular element in Section 2.2.1. A cylindrical element is shown in Fig. 2.5.

The heat conduction equation in cylindrical coordinates is

$$\frac{1}{r}\frac{\partial}{\partial r}\left(kr\frac{\partial T}{\partial r}\right) + \frac{1}{r^2}\frac{\partial}{\partial \varphi}\left(k\frac{\partial T}{\partial \varphi}\right) + \frac{\partial}{\partial z}\left(k\frac{\partial T}{\partial z}\right) + q_{gen} = \rho c \frac{\partial T}{\partial t} \qquad (2.9)$$

With the application of appropriate boundary and initial conditions, this equation can be solved, at least in theory, for the temperature distribution in a cylinder at different instances of time.

## Example 2.3
### Heat Flow in a Wire

*Problem:*
Electrical current is flowing through the wire shown in Fig. 2.6 below. This flow causes uniform heat generation (i.e., Joule heating) due to the wire's resistance. The heat flows and temperatures in the wire are steady, and the conductivity $k$ is constant. As the wire is long, the axial heat flow is negligible.

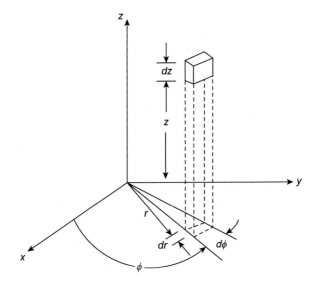

**FIGURE 2.5**

A cylindrical element.

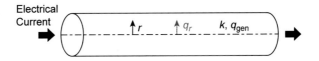

**FIGURE 2.6**

Heat flow in a wire.

(a) Starting with Eq. (2.3), the heat conduction equation in rectangular coordinates, determine the heat conduction equation for this situation in cylindrical coordinates using the coordinate transformation approach.

(b) Determine the heat conduction equation using a heat-balance approach.

*Solution:*

*(a) Transforming the equation from rectangular to cylindrical coordinates.* The heat conduction equation is

$$\frac{\partial}{\partial x}\left(k\frac{\partial T}{\partial x}\right) + \frac{\partial}{\partial y}\left(k\frac{\partial T}{\partial y}\right) + \frac{\partial}{\partial z}\left(k\frac{\partial T}{\partial z}\right) + q_{gen} = \rho c \frac{\partial T}{\partial t} \tag{2.3}$$

As we have steady state, the right side of this equation is zero. As $k$ is a constant, it can be factored out of the first three terms on the left side. Rearrangement of the equation gives

$$\frac{\partial^2 T}{\partial x^2} + \frac{\partial^2 T}{\partial y^2} + \frac{\partial^2 T}{\partial z^2} + \frac{q_{gen}}{k} = 0 \tag{2.10}$$

The $z$ direction is along the center axis of the wire. If the wire is long compared with its diameter, the temperature variation in the wire in the z direction will be negligible. Assuming this usual situation, the third term of Eq. (2.10) will be zero, and the equation becomes

$$\frac{\partial^2 T}{\partial x^2} + \frac{\partial^2 T}{\partial y^2} + \frac{q_{gen}}{k} = 0 \tag{2.11}$$

All we have to do now is to convert the first two terms of Eq. (2.11) to cylindrical coordinates. From Fig. 2.4,

$$r = \sqrt{x^2 + y^2} \tag{2.12}$$

Let us first change the first term of Eq. (2.11) to cylindrical coordinates.

$$\frac{\partial^2 T}{\partial x^2} = \frac{\partial}{\partial x}\left(\frac{\partial T}{\partial x}\right) \tag{2.13}$$

where $r$ is a function of $x$ and $y$, that is, $r = r(x,y)$.
By the chain rule,

$$\frac{\partial T}{\partial x} = \left(\frac{\partial T}{\partial r}\right)\left(\frac{\partial r}{\partial x}\right) \tag{2.14}$$

Through partial differentiation of Eq. (2.12), it can be shown that

$$\frac{\partial r}{\partial x} = \frac{x}{r} \tag{2.15}$$

So, Eq. (2.14) becomes

$$\frac{\partial T}{\partial x} = \left(\frac{x}{r}\right)\left(\frac{\partial T}{\partial r}\right) \tag{2.16}$$

From Eqs. (2.13) and (2.16), we have

$$\frac{\partial^2 T}{\partial x^2} = \frac{\partial}{\partial x}\left[\left(\frac{x}{r}\right)\left(\frac{\partial T}{\partial r}\right)\right] \tag{2.17}$$

Again, from the chain rule, the right side of Eq. (2.17) is

$$\frac{\partial}{\partial x}\left[\left(\frac{x}{r}\right)\left(\frac{\partial T}{\partial r}\right)\right] = \frac{\partial}{\partial r}\left[\left(\frac{x}{r}\right)\left(\frac{\partial T}{\partial r}\right)\right]\left(\frac{\partial r}{\partial x}\right) = \frac{\partial}{\partial r}\left[\left(\frac{x}{r}\right)\left(\frac{\partial T}{\partial r}\right)\right]\left(\frac{x}{r}\right) \tag{2.18}$$

Combining Eqs. (2.17) and (2.18) and performing the differentiation expressed in Eq. (2.18), we get

$$\frac{\partial^2 T}{\partial x^2} = \left(\frac{x}{r}\right)\left[\left(\frac{x}{r}\right)\frac{\partial^2 T}{\partial r^2} + \left(\frac{\partial T}{\partial r}\right)\left(\frac{\partial}{\partial r}\left(\frac{x}{r}\right)\right)\right] \tag{2.19}$$

As $x = \left(r^2 - y^2\right)^{\frac{1}{2}}$, it can be shown that the last term of Eq. (2.19) is

$$\frac{\partial}{\partial r}\left(\frac{x}{r}\right) = \frac{r^2 - x^2}{xr^2} \tag{2.20}$$

and Eq. (2.19) becomes

$$\frac{\partial^2 T}{\partial x^2} = \left(\frac{x}{r}\right)\left[\frac{x}{r}\left(\frac{\partial^2 T}{\partial r^2}\right) + \frac{\partial T}{\partial r}\left(\frac{r^2 - x^2}{xr^2}\right)\right] \tag{2.21}$$

As $r^2 = x^2 + y^2$, Eq. (2.21) becomes

$$\frac{\partial^2 T}{\partial x^2} = \left[\left(\frac{x^2}{r^2}\right)\left(\frac{\partial^2 T}{\partial r^2}\right) + \left(\frac{\partial T}{\partial r}\right)\left(\frac{y^2}{r^3}\right)\right] \tag{2.22}$$

A transformation of the $\frac{\partial^2 T}{\partial y^2}$ term can be done similarly, resulting in

$$\frac{\partial^2 T}{\partial y^2} = \left[\left(\frac{y^2}{r^2}\right)\left(\frac{\partial^2 T}{\partial r^2}\right) + \left(\frac{\partial T}{\partial r}\right)\left(\frac{x^2}{r^3}\right)\right] \tag{2.23}$$

Adding Eqs. (2.22) and (2.23), we get

$$\frac{\partial^2 T}{\partial x^2} + \frac{\partial^2 T}{\partial y^2} = \frac{\partial^2 T}{\partial r^2} + \frac{1}{r}\frac{\partial T}{\partial r} = \frac{1}{r}\frac{\partial}{\partial r}\left(r\frac{\partial T}{\partial r}\right) \tag{2.24}$$

Therefore the cylindrical version of Eq. (2.11) for this problem is

$$\frac{1}{r}\frac{\partial}{\partial r}\left(r\frac{\partial T}{\partial r}\right) + \frac{q_{gen}}{k} = 0 \tag{2.25}$$

**(b) Using the heat-balance approach to obtain the heat conduction equation.** For this problem, the heat flow is solely in the radial $r$ direction. As shown in Fig. 2.7, the infinitesimal volume element is a ring of wall thickness $dr$ and height $dz$.

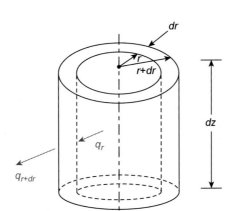

**FIGURE 2.7**

Infinitesimal volume element for Example 2.3.

Heat flows by conduction radially into the volume at $r$ and radially out of the volume at $r + dr$. There is also a heat input to the volume by heat generation (i.e., Joule heating). As the problem is steady state, the heat flow rate into the volume equals the heat flow rate out of the volume, that is, the heat balance on the volume element is

Heat Flow Rate In By Conduction + Heat Flow Rate In By Generation = Heat Flow Rate Out By Conduction

$$(2.26)$$

The incoming heat flow rate by conduction is $q_r$ and the outgoing heat flow rate by conduction is $q_{r+dr}$. The area for heat flow at location $r$ is $dA = 2\pi r\, dz$ and the volume of the element is $dV = 2\pi r\, dr\, dz$. The heat generation rate is $q_{gen}$ per unit volume.

Eq. (2.26) thus becomes

$$q_r + q_{gen}dV = q_{r+dr} \tag{2.27}$$

However, as $q_{r+dr} = q_r + \frac{\partial q_r}{\partial r}\, dr$, Eq. (2.27) becomes

$$-\frac{\partial q_r}{\partial r}\, dr + q_{gen}dV = 0 \tag{2.28}$$

From Eq. (1.2), conduction heat flow in the $r$ direction is $q_r = -kdA\frac{\partial T}{\partial r}$.
Inserting this into Eq. (2.28), we get

$$\frac{\partial}{\partial r}\left(kdA\frac{\partial T}{\partial r}\right)dr + q_{gen}dV = 0 \tag{2.29}$$

Using the relations for $dA$ and $dV$, Eq. (2.29) becomes

$$\frac{\partial}{\partial r}\left(k(2\pi rdz)\frac{\partial T}{\partial r}\right)dr + q_{gen}(2\pi rdrdz) = 0 \tag{2.30}$$

As conductivity $k$ is constant, Eq. (2.30) becomes

$$\frac{\partial}{\partial r}\left(r\frac{\partial T}{\partial r}\right) + \frac{q_{gen}}{k}r = 0 \tag{2.31}$$

Dividing Eq. (2.31) by $r$, we get

$$\frac{1}{r}\frac{\partial}{\partial r}\left(r\frac{\partial T}{\partial r}\right) + \frac{q_{gen}}{k} = 0 \tag{2.32}$$

This is identical to Eq. (2.25), which was obtained by coordinate transformation in part (a) of the solution.

Note that the final heat conduction equation for this problem, Eq. (2.25) or (2.32), could have been obtained directly from the cylindrical heat conduction equation, Eq. (2.9), by eliminating $\varphi$ and $z$ variations and assuming steady state and constant conductivity $k$.

### 2.2.2.1 Special cases—cylindrical coordinates

Depending on the particular problem to be solved, the general Eq. (2.9), repeated here, can often be simplified.

$$\frac{1}{r}\frac{\partial}{\partial r}\left(kr\frac{\partial T}{\partial r}\right) + \frac{1}{r^2}\frac{\partial}{\partial \varphi}\left(k\frac{\partial T}{\partial \varphi}\right) + \frac{\partial}{\partial z}\left(k\frac{\partial T}{\partial z}\right) + q_{gen} = \rho c \frac{\partial T}{\partial t} \tag{2.9}$$

For example, if the problem is steady state, then the $\frac{\partial T}{\partial t}$ term is zero. If there is no heat generation by the material, then $q_{gen}$ is zero. And, if the conductivity $k$ is constant, it can be taken outside the parentheses and then each term of the equation can simply be divided by $k$. Finally, in some problems, the temperature distribution may not depend on all three dimensions. Those terms relating to the nondependent dimensions can then be deleted from Eq. (2.9). For example, in Example 2.3 above, the $\varphi$ and $z$ terms could be deleted, the conductivity $k$ was constant, and the problem was steady state.

## 2.2.3 Spherical coordinates

Spherical coordinates are used for problems involving spheres. A point "A" in a body in spherical coordinates is located by radius $r$, azimuthal angle $\varphi$, and polar angle $\theta$ as shown in Fig. 2.8.

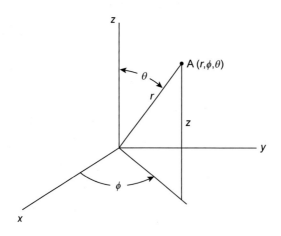

**FIGURE 2.8**

Spherical coordinates.

The heat conduction equation in spherical coordinates can be obtained from the rectangular coordinate equation, Eq. (2.3), through coordinate transformation using the following relations between the parameters of the two coordinate systems:

$$x = r \ \cos \varphi \ \sin \theta$$
$$y = r \ \sin \varphi \ \sin \theta$$
$$z = r \ \cos \theta$$

The heat conduction equation can also be obtained by an energy balance on a spherical element, which is shown in Fig. 2.9.

The heat conduction equation in spherical coordinates is

$$\frac{1}{r^2}\frac{\partial}{\partial r}\left(kr^2\frac{\partial T}{\partial r}\right) + \frac{1}{r^2 \sin^2 \theta}\frac{\partial}{\partial \varphi}\left(k\frac{\partial T}{\partial \varphi}\right) + \frac{1}{r^2 \sin \theta}\frac{\partial}{\partial \theta}\left(k \sin \theta \frac{\partial T}{\partial \theta}\right) + q_{gen} = \rho c \frac{\partial T}{\partial t} \qquad (2.33)$$

With the application of appropriate boundary and initial conditions, this equation can be solved, at least in theory, for the temperature distribution in a sphere at different instances of time.

### 2.2.3.1 Special cases—spherical coordinates

The general heat conduction equation, Eq. (2.33), can often be simplified to fit a specific problem. For example, if the problem is steady state, then the $\frac{\partial T}{\partial t}$ term is zero. If there is no internal heat generation by the material, then $q_{gen}$ is zero. If there is no angular dependence, the terms involving $\varphi$ and $\theta$ can be deleted. Finally, if conductivity $k$ is constant, it can be taken outside the parentheses and then each term of the equation can simply be divided by $k$.

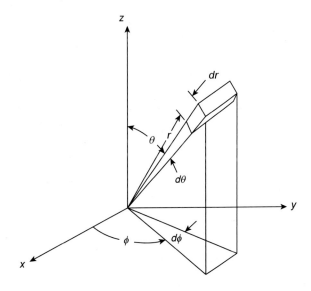

**FIGURE 2.9**

Spherical element.

## Example 2.4

### Radioactive Material in a Spherical Cask

*Problem:*

A spherical cask is filled with radioactive material. The material is initially at temperature $T_1$ and is generating heat uniformly at a constant rate of $q_{gen}$ per unit volume. The conductivity $k$ of the radioactive material is uniform and constant. The cask is thin-walled and has a radius of $r_1$. The outer surface of the cask is convecting heat to the surrounding air with a convective coefficient $h$. The air is at temperature $T_\infty$. Radiative heat transfer from the cask is negligible.

What is the temperature of the material at a distance of $x$ meters from the center of the cask $y$ minutes after the cask is filled?

Note: Only determine the appropriate heat conduction equation for the problem. Do not attempt to solve the problem.

*Solution:*

To solve this problem, the first task is to determine the appropriate heat conduction equation. After this is done, boundary and initial conditions can be applied, and the heat conduction equation can be solved for the temperature distribution in the material at various instances of time.

In this example, we will just determine the heat conduction equation. Solution of the equation is discussed in later sections of this book.

The container is spherical, so the appropriate general heat conduction equation is Eq. (2.33):

$$\frac{1}{r^2}\frac{\partial}{\partial r}\left(kr^2\frac{\partial T}{\partial r}\right) + \frac{1}{r^2\sin^2\theta}\frac{\partial}{\partial\varphi}\left(k\frac{\partial T}{\partial\varphi}\right) + \frac{1}{r^2\sin\theta}\frac{\partial}{\partial\theta}\left(k\sin\theta\frac{\partial T}{\partial\theta}\right) + q_{gen} = \rho c\frac{\partial T}{\partial t} \tag{2.33}$$

Let us tailor this equation for our particular problem, as follows: There is no mention of any angular dependence in the problem description. Hence, the second and third terms of Eq. (2.33) may be deleted, giving

$$\frac{1}{r^2}\frac{\partial}{\partial r}\left(kr^2\frac{\partial T}{\partial r}\right) + q_{gen} = \rho c\frac{\partial T}{\partial t} \tag{2.34}$$

There is indeed heat generation by the radioactive material and the problem deals with time-dependence. However, the conductivity is constant, so it may be taken outside the parentheses in the first term. All of the terms in the equation can then be divided by $k$.

Also, the thermal diffusivity is defined as $\alpha = \frac{k}{\rho c}$.

Eq. (2.34) thus becomes

$$\frac{1}{r^2}\frac{\partial}{\partial r}\left(r^2\frac{\partial T}{\partial r}\right) + \frac{q_{gen}}{k} = \frac{1}{\alpha}\frac{\partial T}{\partial t} \tag{2.35}$$

which is the heat conduction equation for this particular problem.

## 2.3 Boundary conditions

The heat conduction equations of Section 2.2 apply to the interior parts of the body. In this section, we discuss the conditions at the boundaries of the body. The combination of the heat conduction equation and the boundary condition equations for a body may be solved to determine the temperature distribution in the body. (Note: If the problem is unsteady state, initial conditions are also needed for the solution. These conditions are discussed in Section 2.4 below.)

Once the temperature distribution is obtained, it can be differentiated using the equations of Chapter 1 to determine the heat flows at different locations in the body. For example, if the temperature distribution $T(x,y,z)$ is known at a particular instant of time, then, from Eq. (1.3), the heat flow rate per unit area in the $x$ direction at the different points in the body for the particular instant of time is

$$\left(\frac{q}{A}\right)_x = -k\frac{\partial T}{\partial x} \qquad (1.3)$$

In developing the boundary condition equations, two items of information must be provided. These are (a) the location of the boundary using a particular coordinate system and (b) the condition at the boundary (e.g., specified temperature, heat flux rate, convection, radiation, etc.).

### 2.3.1 Rectangular coordinates

Although most of our discussion deals with boundary condition equations for the rectangular coordinate system, similar equations can easily be developed for cylindrical and spherical geometries. These other geometries are discussed in Section 2.3.2.

Let us consider the plane wall of Fig. 2.2 shown earlier and repeated here. The wall has a thickness $L$ in the $x$ direction and large extents in the $y$ and $z$ directions.

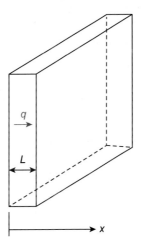

**FIGURE 2.2**

A plane wall.

We will now proceed to obtain boundary equations for several different conditions at the $x = 0$ and $x = L$ boundaries of the wall.

#### 2.3.1.1 Specified temperature

As shown in Fig. 2.10, the left side of the wall is at temperature $T_1$ and the right side is at temperature $T_2$. The boundary condition equations are

$$\text{At } x = 0, \quad T = T_1 \qquad (2.36a)$$

and

$$\text{At } x = L, \quad T = T_2 \qquad (2.36b)$$

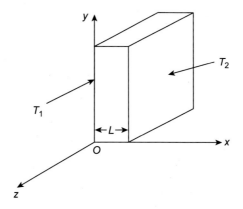

**FIGURE 2.10**

Specified temperature at boundary.

### 2.3.1.2 Specified heat flux

The heat flux $\left(\frac{q}{A}\right)_x$ at the left side of the wall of Fig. 2.11 has a magnitude of $C_1$ and it is in the $+x$ direction. Then the boundary condition equation is

$$\text{At } x = 0, \left(\frac{q}{A}\right)_x = C_1 \tag{2.37}$$

However, the heat conduction equation deals with the temperature, so this boundary equation should be put in terms of temperature. From Eq. (1.3),

$$\left(\frac{q}{A}\right)_x = -k\frac{\partial T}{\partial x} \tag{1.3}$$

So, the boundary condition Eq. (2.37) becomes

$$\text{At } x = 0, -k\frac{\partial T}{\partial x} = C_1 \tag{2.38}$$

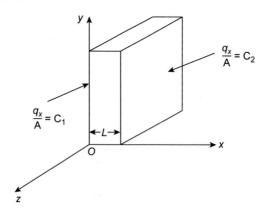

**FIGURE 2.11**

Specified heat flux at boundary.

The heat flux at the right side of the wall has a magnitude of $C_2$ and is in the $-x$ direction. The appropriate boundary equation is

$$\text{At } x = L, \quad k\frac{\partial T}{\partial x} = C_2 \tag{2.39}$$

### 2.3.1.3 Insulated boundary

As shown in Fig. 2.12, the right side of the wall is perfectly insulated and $q_x = 0$.

As $q_x = -kA\frac{\partial T}{\partial x}$ and both $k$ and $A$ are nonzero, the boundary equation is

$$\text{At } x = L, \quad \frac{\partial T}{\partial x} = 0 \tag{2.40}$$

### 2.3.1.4 Convection

As shown in Fig. 2.13, let us say there is a fluid on the left side of the wall. The fluid is at temperature $T_\infty$, and the convective coefficient between the fluid and the wall surface is $h$. The boundary equation can be developed using the concept of heat flow continuity at the boundary. That is, at the boundary, the convective heat flow is equal to the conductive heat flow, or $q_{conv} = q_{cond}$.

By Eq. (1.3), the conductive heat flow is

$$q_{cond} = -kA\frac{\partial T}{\partial x}$$

By Equation (1.9), the convective heat flow is

$$q_{conv} = h \, A \, (T - T_\infty)$$

However, in our situation the heat flow is *from* the fluid *to* the surface. As discussed in Chapter 1, the convective flow at the left surface should be written

$$q_{conv} = h \, A \, (T_\infty - T)$$

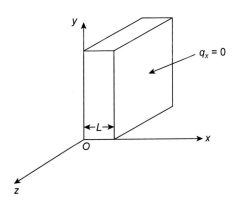

**FIGURE 2.12**

Insulated boundary.

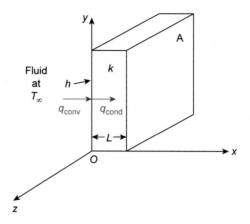

**FIGURE 2.13**

Convection at boundary.

Hence, the boundary equation is, after canceling $A$ from both sides of the equation,

$$\text{At } x = 0, \ h(T_\infty - T) = -k\frac{\partial T}{\partial x} \tag{2.41}$$

### 2.3.1.5 Radiation

Let us say that the wall of Fig. 2.14 is in a vacuum, so there is no convective heat transfer at the wall surfaces. But, let us have radiation heat transfer at both surfaces. The surroundings to the left of the wall are at $T_{surr_1}$ and the left surface of the wall has an emissivity $\varepsilon_1$. The surroundings to the right of the wall are at $T_{surr_2}$, and the right surface of the wall has an emissivity $\varepsilon_2$. Radiation heat transfer is

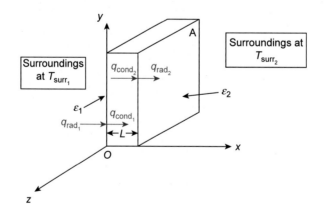

**FIGURE 2.14**

Radiation at boundary.

discussed in detail in Chapter 9. Until then, we will use Eq. (1.13) for problems involving radiation. For this section, that equation is

$$q_{rad} = \varepsilon \sigma A \left( T_s^4 - T_{surr}^4 \right) \tag{2.42}$$

where $T_s$ is the temperature of the surface, and $T_{surr}$ is the temperature of the surroundings. Note that both temperatures must be in absolute temperature units, i.e., Kelvin or Rankine.

Regarding the boundaries at $x = 0$ and $x = L$: At the wall surfaces, there is heat flow continuity. That is, the heat flow by conduction equals the heat flow by radiation. Let us assume that all the conduction and radiation heat flows are in the $+x$ direction as shown in Fig. 2.14. The boundary equations at the two surfaces are then

$$\text{At } x = 0, \quad q_{rad_1} = q_{cond_1}$$

$$\text{At } x = L, \quad q_{cond_2} = q_{rad_2}$$

This gives, after canceling the area $A$ on both sides of the equations,

$$\text{At } x = 0, \quad \varepsilon_1 \sigma \left( T_{surr_1}^4 - T^4 \right) = -k \frac{\partial T}{\partial x} \tag{2.43}$$

$$\text{At } x = L, \quad -k \frac{\partial T}{\partial x} = \varepsilon_2 \sigma \left( T^4 - T_{surr_2}^4 \right) \tag{2.44}$$

Like convection, we must be careful in writing the radiation terms. Looking at Fig. 2.14: At $x = 0$, the assumed radiation flow is *from* the surroundings *to* the wall, so the radiation term is $\varepsilon_1 \sigma \left( T_{surr_1}^4 - T^4 \right)$ and not $\varepsilon_1 \sigma \left( T^4 - T_{surr_1}^4 \right)$. At $x = L$, the assumed radiation flow is *from* the wall *to* the surroundings, so the radiation term is $\varepsilon_2 \sigma \left( T^4 - T_{surr_2}^4 \right)$.

### 2.3.1.6 Convection and radiation

If there is both convection and radiation at a boundary, then the two modes of heat transfer can be added to get the total heat transfer. In Fig. 2.15, there are fluids on both sides of the wall and therefore

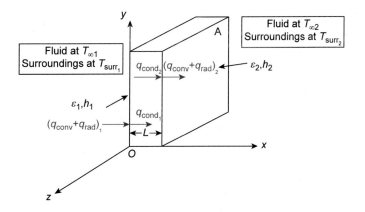

**FIGURE 2.15**

Convection and radiation at boundary.

convective heat transfer on both surfaces. There is also radiative heat transfer between both surfaces and their respective surroundings.

At the left surface, the conduction heat transfer is equal to the sum of the convective and radiative heat transfers. The same continuity of heat flows concept applies to the right surface. The boundary equations are

$$\text{At } x=0, \quad h_1A(T_{\infty_1} - T) + \varepsilon_1\sigma A\left(T_{surr_1}^4 - T^4\right) = -kA\frac{\partial T}{\partial x} \tag{2.45}$$

$$\text{At } x=L, \quad -kA\frac{\partial T}{\partial x} = h_2A(T - T_{\infty_2}) + \varepsilon_2\sigma A\left(T^4 - T_{surr_2}^4\right) \tag{2.46}$$

The area $A$ can be deleted from the two equations as it is in all the terms.

Again, in writing the boundary equations, pay particular attention to the assumed directions of the heat flows so that the terms in the equations will have the correct signs.

Let us continue to discuss this boundary condition where convection occurs together with radiation. $T_s$ is the temperature of the surface, $T_\infty$ is the fluid temperature, and $T_{surr}$ is the temperature of the surroundings. The total heat flow at the boundary is the sum of the convective heat flow $q_{conv}$ and the radiative heat flow $q_{rad}$, where

$$q_{conv} = hA\ (T_s - T_\infty) \tag{2.47}$$

$$\text{and } q_{rad} = \varepsilon\sigma A\left(T_s^4 - T_{surr}^4\right) \tag{2.48}$$

It would be nice to convert Eq. (2.48) into an equation that has a difference in temperature rather than a difference in temperature to the fourth power. This can be done by introducing a new coefficient for radiation, $h_{rad}$, which is defined by the equation

$$q_{rad} = \varepsilon\sigma A\left(T_s^4 - T_{surr}^4\right) = h_{rad}A(T_s - T_{surr}) \tag{2.49}$$

That is,

$$h_{rad} = \frac{\varepsilon\sigma\left(T_s^4 - T_{surr}^4\right)}{T_s - T_{surr}} \tag{2.50}$$

Through algebraic manipulation it can be shown that

$$h_{rad} = \varepsilon\sigma\left(T_s^2 + T_{surr}^2\right)(T_s + T_{surr}) \tag{2.51}$$

Having defined $h_{rad}$, we can now write the total heat flow at the surface as

$$q_{total} = q_{conv} + q_{rad} = hA(T_s - T_\infty) + h_{rad}A(T_s - T_{surr}) \tag{2.52}$$

In many practical situations, the fluid temperature is about the same as the temperature of the surroundings. If this is the case, we can add convective coefficient $h$ and radiative coefficient $h_{rad}$ to get a total coefficient $h_{total} = h + h_{rad}$, and the total heat transfer at the surface becomes

$$q_{total} = h_{total}A(T_s - T_{avg}) \tag{2.53}$$

where $T_{avg}$ is the average of $T_\infty$ and $T_{surr}$.

### 2.3.1.7 Symmetry conditions

There may be situations where the temperature distribution is symmetrical. For example, consider a large plate of thickness $2L$, which has uniform internal heat generation. At the two surfaces, there is convection to the adjacent fluids. If the convective coefficients at both surfaces are the same and the temperatures of the adjacent fluids are the same, then the temperature distribution in the plate will be symmetrical about the center plane. And, the temperature of the center plane will be the maximum temperature in the plate.

If the center plane is at $x = 0$ and the surfaces are at $x = -L$ and $x = +L$, then the symmetry boundary condition is

$$\text{At } x=0, \quad \frac{\partial T}{\partial x} = 0 \tag{2.54}$$

### 2.3.1.8 Interfacial boundary

Figure 2.16 shows two walls in contact. Wall 1 has a conductivity $k_1$ and a temperature distribution $T_1$.

Wall 2 has a conductivity $k_2$ and a temperature distribution $T_2$.

Both the conductivities and the temperatures may vary with $x$.

At the interface between the two walls, i.e., at $x = L_1$, there is continuity of temperature if there is no contact resistance between the walls. We will normally assume this condition of zero contact resistance. (Note: Contact resistance is discussed in Section 12.3.)

From the temperature continuity, we have the boundary equation

$$\text{At } x=L_1, \quad T_1 = T_2 \tag{2.55}$$

There is also continuity of heat flows at the interface, i.e., $q_{cond_1} = q_{cond_2}$. This gives the boundary equation

$$\text{At } x=L_1, \quad -k_1 A \frac{\partial T_1}{\partial x} = -k_2 A \frac{\partial T_2}{\partial x} \tag{2.56}$$

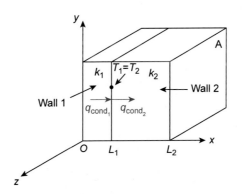

**FIGURE 2.16**

Walls in contact.

Canceling the area $A$ and the minus signs, the boundary equation for the heat flows becomes

$$\text{At } x = L_1, \quad k_1 \frac{\partial T_1}{\partial x} = k_2 \frac{\partial T_2}{\partial x} \tag{2.57}$$

## 2.3.2 Cylindrical and spherical coordinates

Boundary conditions were developed in Section 2.3.1 for the rectangular coordinate system. Similar equations can be developed for cylindrical and spherical coordinates. For cylindrical objects, the common boundaries are at radial and axial locations. There are also boundaries at angular locations if the objects are asymmetrical. If the object is a hollow cylinder, there are boundaries at the inner and outer surfaces and at the two ends. If it is a solid cylinder, there is only one radial boundary and two axial boundaries.

A hollow sphere has boundaries at its inner and outer surfaces. A solid sphere has only one bounding surface.

Possible boundary conditions are similar to those for rectangular coordinates, e.g., specified temperature, specified heat flux, convection, and radiation. If a specific temperature or heat flux is not given for a particular boundary, the boundary equation can be obtained by consideration of heat flow continuity at the boundary. For example, we could have, at a boundary, that conduction equals convection, conduction equals radiation, or conduction equals convection plus radiation.

---

### Example 2.5
#### Boundary Conditions for a Hollow Cylinder

Consider Fig. 2.17 that shows a hollow cylinder of inner radius $r_1$, outer radius $r_2$, and length $L$. The boundaries of this cylinder are the inner surface, the outer surface, and the two ends.

Depending on the specific problem, many different boundary equations can be developed. The two coordinates of interest are the radial distance $r$ and the axial distance $z$. We will set $z = 0$ at the left end of the cylinder. Then the right end of the cylinder is at $z = L$.

Let us say that there is a fluid inside the cylinder at temperature $T_{\infty_1}$ with a convective coefficient at the inner surface of $h_1$ and a fluid outside the cylinder at temperature $T_{\infty_2}$ with a convective coefficient at the outer surface of $h_2$. Let us also say that the two ends of the cylinder are perfectly insulated.

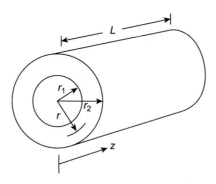

**FIGURE 2.17**

Finite length hollow cylinder.

The conduction heat transfer in the cylinder in the radial direction $r$ is

$$q_{cond} = -kA\frac{\partial T}{\partial r} \qquad (2.58)$$

Area $A$, the area through which heat is flowing, increases linearly with $r$ and is $A = 2\pi rL$.
Thus, from Eq. (2.58), the conduction heat transfer in the $+r$ direction at a distance $r$ from the centerline of the cylinder is

$$q_{cond} = -k(2\pi rL)\frac{\partial T}{\partial r} \qquad (2.59)$$

At the inner surface, i.e., at $r = r_1$, the convective heat transfer equals the conduction heat transfer. Hence, the boundary equation at the inner surface is

$$-k(2\pi r_1 L)\frac{\partial T}{\partial r} = h_1(2\pi r_1 L)(T_{\infty_1} - T) \qquad (2.60)$$

which simplifies to At $r = r_1$, $\quad -k\dfrac{\partial T}{\partial r} = h_1(T_{\infty_1} - T) \qquad (2.61)$

For the outer surface of the cylinder, i.e., at $r = r_2$, the conduction equals the convection. So, the boundary equation, after canceling area $A$ on both sides of the equation, is

$$\text{At } r = r_2, \quad -k\frac{\partial T}{\partial r} = h_2(T - T_{\infty_2}) \qquad (2.62)$$

Note: Again, be careful with the convection terms in Eqs. (2.61) and (2.62). In writing the equations, we have assumed that both the conduction and the convection heat flows are in the $+r$ direction. Hence, for the convection term at the inner surface, we have $(T_\infty - T)$, and for the convection term at the outer surface, we have $(T - T_\infty)$. This was discussed above in this chapter and in Chapter 1.

The two ends of the cylinder are perfectly insulated with no heat flow. Hence, the boundary equations for the ends are

$$\text{At } z = 0, \quad \frac{\partial T}{\partial z} = 0 \qquad (2.63)$$

$$\text{At } z = L, \quad \frac{\partial T}{\partial z} = 0 \qquad (2.64)$$

### 2.3.2.1 Symmetry conditions

If the cylinder or sphere under consideration is solid, then there is only one physical boundary in the radial direction, i.e., at the surface of the cylinder or sphere. However, more than one boundary equation is needed for solution of the heat conduction equation. In many cases, the temperature distribution is symmetrical at the center of the cylinder or sphere. For example, when electricity flows through a wire, there is heating and a symmetrical temperature distribution at the center of the wire. Cooling of a cylinder or sphere can also result in temperature symmetry at the center. For symmetrical heating or cooling, there is a temperature maximum or minimum at the center, and the boundary equation is

$$\text{At } r = 0, \quad \frac{\partial T}{\partial r} = 0 \qquad (2.65)$$

### Example 2.6

#### Cooling of a Heated Sphere

Let us consider the heated sphere of radius $r_1$ shown in Fig. 2.18. The sphere has just come out of an oven. The temperature of the sphere is initially uniform throughout the sphere, and it is higher than the room air and the surroundings.

The thermal conductivity of the sphere is $k$. The surface of the sphere has an emissivity $\varepsilon$ and a convective coefficient $h$. The air is at $T_\infty$ and the surroundings are at $T_{surr}$. The temperature distribution in the sphere is $T$.

There is heat flow continuity at the surface of the sphere. The heat flow by conduction into the surface equals the heat flow by convection and radiation from the surface.

$$\text{That is,} \quad q_{cond} = q_{conv} + q_{rad} \tag{2.66}$$

$$\text{where} \quad q_{cond} = -kA\frac{\partial T}{\partial r} \tag{2.67}$$

$$q_{conv} = hA(T - T_\infty) \tag{2.68}$$

$$q_{rad} = \varepsilon\sigma A\left(T^4 - T_{surr}^4\right) \tag{2.69}$$

Hence, the boundary equation at the surface of the sphere is

$$\text{At} \ r = r_1, \ -k\frac{\partial T}{\partial r} = h(T - T_\infty) + \varepsilon\sigma\left(T^4 - T_{surr}^4\right) \tag{2.70}$$

We will assume that $k$ is uniform throughout the sphere. Then the temperature distribution during cooling will be symmetrical with respect to $r$. Even though $r = 0$ is not a physical boundary, we can write the following equation for the symmetry:

$$\text{At} \ r = 0, \ \frac{\partial T}{\partial r} = 0 \tag{2.71}$$

In closing this discussion of boundary equations, let us consider a situation where we have a boundary condition involving an angular coordinate of a cylinder or sphere. In particular, let us look at the segment of a cylinder shown in Fig. 2.19.

The segment has inner radius $r_1$, outer radius $r_2$, axial length $L$, and includes the angle from $\varphi = 0$ to $\varphi = \varphi_1$. The $\varphi = 0$ surface is perfectly insulated and the $\varphi = \varphi_1$ surface has convection to a fluid at $T_\infty$ with convection coefficient $h$.

For the insulated $\varphi = 0$ surface, the heat flow is zero, so the boundary equation is

$$\text{At} \ \varphi = 0, \ \frac{\partial T}{\partial \varphi} = 0 \tag{2.72}$$

For the $\varphi = \varphi_1$ surface, the conduction in the $+\varphi$ direction equals the convection in that direction. The boundary equation is

$$\text{At} \ \varphi = \varphi_1, \ -kA\frac{\partial T}{\partial \varphi} = hA(T - T_\infty) \tag{2.73}$$

The areas cancel as they are the same. Thus, the final boundary equation is

$$\text{At} \ \varphi = \varphi_1, \ -k\frac{\partial T}{\partial \varphi} = h(T - T_\infty) \tag{2.74}$$

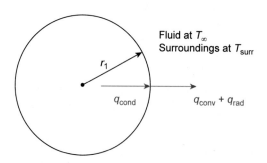

Fluid at $T_\infty$
Surroundings at $T_{surr}$

$r_1$

$q_{cond}$

$q_{conv} + q_{rad}$

**FIGURE 2.18**

Sphere cooled by convection and radiation.

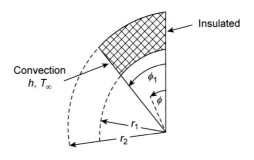

**FIGURE 2.19**

Segment of a cylinder.

Eqs. (2.72) and (2.74) are the boundary equations for the $\varphi$ coordinate. To solve the heat conduction equation for the temperature distribution in the segment, boundary equations will also be needed for the radial and axial bounding surfaces.

## 2.4 Initial conditions

Eqs. (2.3), (2.9), and (2.33) are the general heat conduction equations for rectangular, cylindrical, and spherical coordinates. If the problem is time-dependent, the $\frac{\partial T}{\partial t}$ term in the heat conduction equation is nonzero and initial condition(s) are needed for solution of the equation. Two examples of initial conditions are as follows:

**(1)** If all points in the body are at the same temperature $T_{init}$ at time zero, then the initial condition equation is

$$\text{At } t = 0, \quad T(x, y, z) = T_{init} \tag{2.75}$$

**(2)** If the temperature distribution in the body is a known function $T_{init}(x, y, z)$ at time zero, then the initial condition equation is

$$\text{At } t = 0, \quad T(x, y, z) = T_{init}(x, y, z) \tag{2.76}$$

## 2.5 Chapter summary and final remarks

In this chapter, we discussed the partial differential equation governing heat conduction within a body. The heat conduction equation was presented for rectangular, cylindrical, and spherical coordinate systems. We also discussed boundary and initial condition equations that are used in conjunction with the heat conduction equation to determine the temperature distribution in the body. Differentiation of the temperature distribution will give the heat flows at the various points in the body. Eq. (1.3) showed this for rectangular coordinates. Chapter 3 will give similar equations for cylindrical and spherical coordinates.

In the next chapter we will solve some of the differential equations developed in this chapter for steady-state heat transfer.

## 2.6 Problems

**2-1** In the derivation of the rectangular heat conduction equation, Eq. (2.3), it was assumed that three flows were incoming to the differential element $(q_x, \; q_y, \; q_z)$ and three were outgoing $(q_{x+dx}, \; q_{y+dy}, \; q_{z+dz})$. Redo the derivation assuming that all six flows are incoming to the element. You should obtain the same heat conduction equation.

**2-2** Obtain the cylindrical heat conduction equation, Eq. (2.9), through coordinate transformation of the rectangular heat conduction equation, Eq. (2.3).

**2-3** Obtain the cylindrical heat conduction equation, Eq. (2.9), using the heat-balance approach in conjunction with the cylindrical element of Fig. 2.5.

**For problems 2-4 to 2-8, reduce the rectangular heat conduction equation as much as possible to fit the problem specification. Just give the equation, do not solve it. The general heat conduction equation is**

$$\frac{\partial}{\partial x}\left(k\frac{\partial T}{\partial x}\right) + \frac{\partial}{\partial y}\left(k\frac{\partial T}{\partial y}\right) + \frac{\partial}{\partial z}\left(k\frac{\partial T}{\partial z}\right) + q_{gen} = \rho c \frac{\partial T}{\partial t}$$

**2-4** A rectangular solid is of dimensions 2 cm by 4 cm by 6 cm and is initially at a uniform temperature of 100 C. Its surfaces are suddenly subjected to a fluid at 50 C with a convective coefficient of 2 W/m$^2$ C. What is the temperature of the center of the solid after 30 s? The conductivity of the solid is 150 W/m C. Its density is 2700 kg/m$^3$ and its specific heat is 880 J/kg C. There is no internal heat generation.

**2-5** A long copper bar has a square cross section in the $x$-$y$ plane. There is negligible heat flow in the $z$ direction. The top and bottom of the bar are held at temperature $T_1$, while the other two surfaces are held at $T_2$. Heat transfer in the bar is steady. The material properties, i.e., conductivity, density, and specific heat are uniform throughout the bar and constant. There is no internal heat generation. What is the temperature distribution of the $x$-$z$ plane that is halfway between the top and bottom surfaces of the bar? (Fig. P2.5).

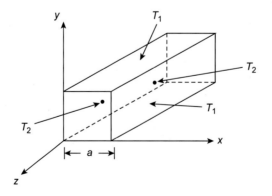

**FIGURE P2.5**

**2-6** A thin rectangular plate is in the $x$-$y$ plane. There is negligible heat transfer in the $z$ direction. The conductivity, density, and specific heat all vary with temperature. There is constant internal heat generation and the heat flow is unsteady.

**2-7** A cube is perfectly insulated from the surroundings. The cube is initially at a uniform temperature and then, at time zero, internal heat generation begins. What is the temperature at a corner of the cube at time $t$ after the heat generation starts? It may be assumed that all material properties are constant.

**2-8** There is steady heat conduction in a rectangular solid. The conductivity of the object is constant and there is no heat generation. What is the temperature at the center of one face of the object?

**2-9** Consider the equation

$$\frac{\partial^2 T}{\partial x^2} + \frac{q_{gen}}{k} = \frac{1}{\alpha}\frac{\partial T}{\partial t}$$

(a) How many dimensions are there (one, two, or three)?
(b) Is the heat conduction steady or unsteady?
(c) Is there internal heat generation?
(d) Is the conductivity constant or variable?

**2-10** Consider the equation

$$\frac{\partial}{\partial y}\left(k\frac{\partial T}{\partial y}\right) + \frac{\partial}{\partial z}\left(k\frac{\partial T}{\partial z}\right) = 0$$

(a) How many dimensions are there (one, two, or three)?
(b) Is the heat conduction steady or unsteady?
(c) Is there internal heat generation?
(d) Is the conductivity constant or variable?

**2-11** Consider the equation

$$\frac{\partial^2 T}{\partial x^2} + \frac{\partial^2 T}{\partial y^2} + \frac{\partial^2 T}{\partial z^2} = 0$$

(a) How many dimensions are there (one, two, or three)?
(b) Is the heat conduction steady or unsteady?
(c) Is there internal heat generation?
(d) Is the conductivity constant or variable?

**2-12** Consider the equation

$$\frac{\partial}{\partial x}\left(k\frac{\partial T}{\partial x}\right) + \frac{\partial}{\partial y}\left(k\frac{\partial T}{\partial y}\right) = \rho c\frac{\partial T}{\partial t}$$

(a) How many dimensions are there (one, two, or three)?
(b) Is the heat conduction steady or unsteady?
(c) Is there internal heat generation?
(d) Is the conductivity constant or variable?

**2-13** Consider the equation

$$\frac{\partial^2 T}{\partial x^2} + \frac{\partial^2 T}{\partial y^2} + \frac{q_{gen}}{k} = 0$$

**(a)** How many dimensions are there (one, two, or three)?
**(b)** Is the heat conduction steady or unsteady?
**(c)** Is there internal heat generation?
**(d)** Is the conductivity constant or variable?
**For problems 2-14 to 2-17, reduce the cylindrical heat conduction equation as much as possible to fit the problem specification. Just give the equation; don't solve it. The general heat conduction equation is**

$$\frac{1}{r}\frac{\partial}{\partial r}\left(kr\frac{\partial T}{\partial r}\right) + \frac{1}{r^2}\frac{\partial}{\partial \varphi}\left(k\frac{\partial T}{\partial \varphi}\right) + \frac{\partial}{\partial z}\left(k\frac{\partial T}{\partial z}\right) + q_{gen} = \rho c\frac{\partial T}{\partial t}$$

**2-14** The problem is steady state with no internal heat generation and no angular dependence. The conductivity is constant.

**2-15** The problem is transient, with constant and uniform internal heat generation. The heat transfer in the axial direction is negligible. The conductivity varies with temperature.

**2-16** The problem is transient, with no internal heat generation. The temperature varies in the radial and axial directions, but there is no angular dependence. The conductivity is constant, but the density and specific heat vary with temperature.

**2-17** The problem is steady state with internal heat generation. The temperature variations in the angular and axial directions are negligible. The thermal conductivity is constant.

**2-18** Consider the equation in cylindrical coordinates

$$\frac{\partial}{\partial r}\left(kr\frac{\partial T}{\partial r}\right) = 0$$

**(a)** How many dimensions are there (one, two, or three)?
**(b)** Is the heat conduction steady or unsteady?
**(c)** Is there internal heat generation?
**(d)** Is the conductivity constant or variable?

**2-19** Consider the equation in cylindrical coordinates

$$\frac{1}{r}\frac{\partial}{\partial r}\left(kr\frac{\partial T}{\partial r}\right) + \frac{\partial}{\partial z}\left(k\frac{\partial T}{\partial z}\right) + q_{gen} = 0$$

**(a)** How many dimensions are there (one, two, or three)?
**(b)** Is the heat conduction steady or unsteady?
**(c)** Is there internal heat generation?
**(d)** Is the conductivity constant or variable?

**2-20** Consider the equation in cylindrical coordinates

$$\frac{\partial^2 T}{\partial r^2} + \frac{1}{r}\frac{\partial T}{\partial r} + \frac{1}{r^2}\frac{\partial^2 T}{\partial \varphi^2} = \frac{1}{\alpha}\frac{\partial T}{\partial t}$$

(a) How many dimensions are there (one, two, or three)?
(b) Is the heat conduction steady or unsteady?
(c) Is there internal heat generation?
(d) Is the conductivity constant or variable?

**2-21** Consider the equation in cylindrical coordinates

$$\frac{\partial^2 T}{\partial r^2} + \frac{1}{r}\frac{\partial T}{\partial r} + \frac{q_{gen}}{k} = \frac{1}{\alpha}\frac{\partial T}{\partial t}$$

(a) How many dimensions are there (one, two, or three)?
(b) Is the heat conduction steady or unsteady?
(c) Is there internal heat generation?
(d) Is the conductivity constant or variable?

**2-22** Consider the equation in cylindrical coordinates

$$\frac{1}{r}\frac{\partial}{\partial r}\left(kr\frac{\partial T}{\partial r}\right) + \frac{1}{r^2}\frac{\partial}{\partial \varphi}\left(k\frac{\partial T}{\partial \varphi}\right) = \rho c\frac{\partial T}{\partial t}$$

(a) How many dimensions are there (one, two, or three)?
(b) Is the heat conduction steady or unsteady?
(c) Is there internal heat generation?
(d) Is the conductivity constant or variable?

**For problems 2-23 to 2-25, reduce the spherical heat conduction equation as much as possible to fit the problem specification. Just give the equation; don't solve it. The general heat conduction equation is**

$$\frac{1}{r^2}\frac{\partial}{\partial r}\left(kr^2\frac{\partial T}{\partial r}\right) + \frac{1}{r^2\sin^2\theta}\frac{\partial}{\partial \varphi}\left(k\frac{\partial T}{\partial \varphi}\right) + \frac{1}{r^2\sin\theta}\frac{\partial}{\partial \theta}\left(k\sin\theta\frac{\partial T}{\partial \theta}\right) + q_{gen} = \rho c\frac{\partial T}{\partial t}$$

**2-23** The problem is steady state with internal heat generation and constant conductivity.

**2-24** The problem is transient, with no internal heat generation and no angular dependence. The conductivity varies with location in the body.

**2-25** The problem is steady state with variable conductivity, no internal heat generation, and no angular dependence.

**2-26** Consider the equation in spherical coordinates

$$\frac{d}{dr}\left(r^2\frac{dT}{dr}\right) = 0$$

(a) How many dimensions are there (one, two, or three)?
(b) Is the heat conduction steady or unsteady?
(c) Is there internal heat generation?
(d) Is the conductivity constant or variable?

**2-27** Consider the equation in spherical coordinates

$$\frac{1}{r^2}\frac{\partial}{\partial r}\left(k\ r^2\frac{\partial T}{\partial r}\right) + \frac{1}{r^2\ \sin^2\theta}\frac{\partial}{\partial\varphi}\left(k\frac{\partial T}{\partial\varphi}\right) + \frac{1}{r^2\ \sin\theta}\frac{\partial}{\partial\theta}\left(k\ \sin\theta\frac{\partial T}{\partial\theta}\right) = 0$$

(a) How many dimensions are there (one, two, or three)?
(b) Is the heat conduction steady or unsteady?
(c) Is there internal heat generation?
(d) Is the conductivity constant or variable?

**2-28** Consider the equation in spherical coordinates

$$\frac{1}{r^2}\frac{\partial}{\partial r}\left(r^2\frac{\partial T}{\partial r}\right) + \frac{1}{r^2\ \sin^2\theta}\frac{\partial^2 T}{\partial\varphi^2} + \frac{1}{r^2\ \sin\theta}\frac{\partial}{\partial\theta}\left(\sin\theta\frac{\partial T}{\partial\theta}\right) = \frac{1}{\alpha}\frac{\partial T}{\partial t}$$

(a) How many dimensions are there (one, two, or three)?
(b) Is the heat conduction steady or unsteady?
(c) Is there internal heat generation?
(d) Is the conductivity constant or variable?

**2-29** Consider the equation in spherical coordinates

$$\frac{\partial^2 T}{\partial r^2} + \frac{2}{r}\frac{\partial T}{\partial r} + \frac{q_{gen}}{k} = \frac{1}{\alpha}\frac{\partial T}{\partial t}$$

(a) How many dimensions are there (one, two, or three)?
(b) Is the heat conduction steady or unsteady?
(c) Is there internal heat generation?
(d) Is the conductivity constant or variable?

**2-30** The thin plate shown in Fig. P2.30 is in the x-y plane. The left and right sides of the plate are perfectly insulated. The top surface of the plate is at 100 C and the bottom is at 20 C. Give the boundary equations for the plate (Fig. P2.30).

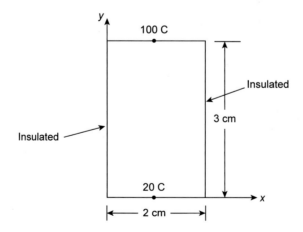

**FIGURE P2.30**

**2-31** For the plate of Problem 2-30, let us say that instead of the left and right sides being perfectly insulated, they both convect to a fluid at $T_\infty$ = 15 C with a convective coefficient of 3 W/m$^2$ C. The conductivity of the plate is 80 W/m C. Give the boundary equations for the left and right sides of the plate.

**2-32** The rectangular solid shown in Fig. P2.32 has dimensions 1 cm by 2 cm by 5 cm. The front and back surfaces are perfectly insulated. The left and right surfaces are at a temperature of 50 C, and the top and bottom surfaces have an emissivity of 0.3 and radiate to the surroundings, which are at 10 C. The object is in a vacuum, so there is no convection to a surrounding fluid. The conductivity of the object is 100 W/m C. Give the boundary equations for the six surfaces of the object (Fig. P2.32).

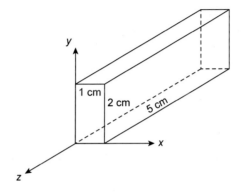

**FIGURE P2.32**

**2-33** The left surface of the copper plate ($k$ = 400 W/m C) shown in Fig. P2.33 absorbs a heat flux from the sun of 200 W/m$^2$. The surface also convects to the adjacent air which is at 25 C. The convective coefficient is 5 W/m$^2$ C. Give the boundary equation for this surface of the plate (Fig. P2.33).

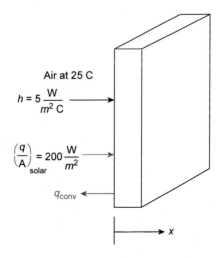

**FIGURE P2.33**

**2-34** The cube shown in Fig. P2.34 has uniform internal heat generation. All six surfaces of the cube convect to the surrounding fluid that is at $T_\infty = 30$ C. The convective coefficient for all the surfaces is $h = 5$ W/m² C and the conductivity of the cube is 40 W/m C. Give the boundary equations for all six surfaces of the cube (Fig. P2.34).

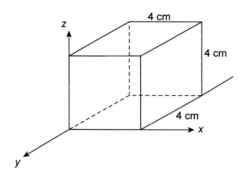

**FIGURE P2.34**

**2-35** For the cube of Problem 2-34: Let us say that the six surfaces of the cube are perfectly insulated rather that being convectors to a surrounding fluid. Give the boundary equations for all six surfaces for this situation.

**2-36** The solid cylinder ($k = 15$ W/m C) shown in Fig. P2.36 has a radius $r_o$ of 6 cm and is 20 cm long. Let us make $z = 0$ at the left end of the cylinder and $z = 20$ cm at the right end. The left end of the cylinder is at 80 C and the right end is at 50 C. The cylindrical surface convects to the surrounding fluid, which is at 30 C with a convective coefficient of 6 W/m² C. Give the boundary equations for this problem (Fig. P2.36).

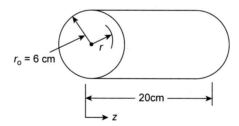

**FIGURE P2.36**

**2-37** A long steel pipe of conductivity 60 W/m C has an inner radius of 0.5 cm and an outer radius of 1.2 cm. Water at 90 C is flowing through the pipe, and there is convection from the hot water to the inner surface of the pipe with a convective coefficient of 50 W/m² C. At the outer surface of the pipe, there is convection to the 20 C room air with a convective coefficient of 2 W/m² C and radiation to the surrounding walls, which are at 15 C. The emissivity of the outer surface of the pipe is 0.2. Give the boundary equations for the inner and outer surfaces of the pipe.

**2-38** An electrical current is flowing through a #12 copper wire. The wire has a diameter of 2.053 mm and a conductivity of 400 W/m C. The heat generated in the wire passes to the surrounding 20 C air with a convective coefficient of 1.5 W/m$^2$ C. Give two boundary equations for this problem.

**2-39** It is desired to determine the temperature distribution in a sphere of radioactive material buried in the ground. The radioactive material is in a thin-walled container that has negligible thermal resistance. The radius of the sphere is $r_o$. The conductivity of the radioactive material is $k_1$ and the conductivity of the soil is $k_2$. The temperature distribution in the radioactive material is $T_1$ and in the soil it is $T_2$. Give the boundary equations at the outer surface of the radioactive material.

**2-40** It is desired to determine the temperature distribution for radioactive material ($k = 0.23$ W/m C) inside a cylindrical container. The thin-walled container is shown in Fig. P2.40. It is on a concrete floor, and it may be assumed that the bottom of the container is perfectly insulated. The top of the container and the curved surface of the container transfer heat to the surrounding air and walls. The air temperature is 20 C and the surrounding walls are also at 20 C. The emissivity of the container's top and side is 0.45 and the convective coefficient is 1.5 W/m$^2$ C. The container is 3 m high and has a diameter of 1 m. Give the heat conduction equation and boundary equations for this problem. The heat transfer is steady (Fig. P2.40).

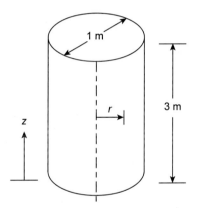

**FIGURE P2.40**

**2-41** A metal cube of dimensions $a$ by $a$ by $a$ and constant conductivity $k$ initially has a uniform temperature $T_i$. The cube is suddenly placed on a hot plate that keeps the bottom surface of the cube at a temperature $T_{box}$, which is greater than $T_i$. The top surface of the cube is perfectly insulated, but the remaining four sides of the cube transfer heat to the surrounding fluid, which is at temperature $T_{fluid}$. The convective coefficient between the vertical sides of the cube and the fluid is $h$. It is desired to find the temperature of the center of the cube after $t_{final}$ seconds. Sketch the cube on an appropriate set of axes. Provide the appropriate heat conduction equation, boundary equations, and the initial condition equation.

**2-42** A large wall of a house has a thickness $L$. The inside surface of the wall convects to the air that is at temperature $T_i$ The convective coefficient is $h_i$. The outside surface convects to the outside air that is at $T_o$ with a convective coefficient $h_o$. The sun is beaming on the wall, giving it a solar heat flux input of $q_{solar}/A$. The wall has conductivity $k$, density $\rho$, and specific heat $c$. It is desired to determine the temperature distribution of the wall. This is a steady heat transfer problem. Sketch the wall on appropriate axes. Give the heat conduction equation and boundary equations.

**2-43** Electric current is flowing through a long wire of radius $r_o$. The flow of current produces internal heat generation of a strength $q_{gen}$ per unit volume. The current is then turned-off and the wire starts to cool due to convection to the surrounding air. The air is at temperature $T_{air}$ and the convective coefficient is $h$. It can be assumed that the wire has a uniform temperature $T_{begin}$ when the current is turned-off. It is desired to find the temperature distribution in the wire at a time $t$ seconds after the current stops. The radial distance from the centerline of the wire is $r$ and the axial coordinate is $z$. Give the heat conduction equation, the boundary equations, and the initial condition equation for this problem.

**2-44** The square plate shown in Fig. P2.44 has thickness $L$ and cross section $a$ by $a$. The $x = 0$ side of the plate has a net heat flux input from a radiant heater of $q_{heater}/A$. The right side of the plate convects to the adjacent fluid at $T_\infty$, with convective coefficient $h$. It also radiates to the surroundings at $T_{surr}$ with a surface emissivity $\varepsilon$. The four edges of the plate are perfectly insulated. It is desired to determine the steady-state temperature distribution in the plate. The plate has conductivity $k$, density $\rho$, and specific heat $c$. Give the heat conduction equation and boundary equations for this problem (Fig. P2.44).

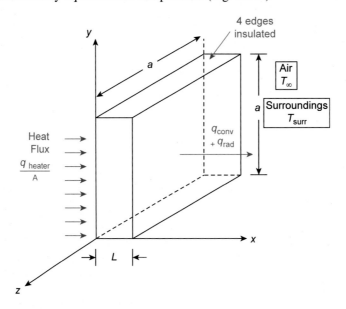

**FIGURE P2.44**

**2-45** A sphere of radius $r_o$ is heated in an oven to a uniform temperature $T_{hot}$. The sphere is then taken out of the oven and placed in a hemispherical metal holder. See Fig. P2.45A and B. The temperature of the holder is $T_{holder}$. The temperature of the holder remains constant, but the sphere cools due to convection from its upper hemispherical area to the room air. The room air is at $T_{air}$ and the convective coefficient is $h$. It is desired to determine the temperature in the sphere as it cools. Give the heat conduction equation, boundary equations, and the initial condition equation for this problem (Fig. P2.45A and B).

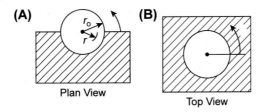

**FIGURE P2.45A,B**

## Uncited references

[1] J.P. Holman, Heat Transfer, ninth ed., McGraw-Hill, 2002.
[2] Y.A. Cengel, A.J. Ghajar, Heat and Mass Transfer Fundamentals & Applications, fourth ed., McGraw-Hill, 2011.
[3] F.P. Incropera, D.P. DeWitt, Fundamentals of Heat and Mass Transfer, fourth ed., Wiley, 1996.
[4] F. Kreith, M.S. Bohn, Principles of Heat Transfer, sixth ed., Thomson Learning, 2001.

# Steady-state conduction

## Chapter outline

## 3.1 Introduction

Chapter 2 discussed the heat conduction equation and the boundary and initial conditions needed to determine the temperature distribution in a body. In this chapter, we reduce the heat conduction equation to its one-dimensional and steady-state form. We discuss practical problems that can be appropriately modeled using only one dimension. Examples of such problems include heat flow through building walls, fluid and heat flow through pipes, and heat generation in electrical wires. The very useful resistance concept from electric theory is discussed, as is the conduction shape factor that can be applied to many unique, practical problems. Finally, there is a discussion of extended surfaces, i.e., fins. Fins are used in many devices to enhance heat flow.

## 3.2 One-dimensional conduction

This section discusses conduction in plane walls, cylindrical shells, and spherical shells.

### 3.2.1 Plane wall

The plane wall shown in Fig. 3.1 has a finite thickness $L$ in the $x$ direction, is very large in the $y$ and $z$ directions, and has a cross-sectional area $A$. The left surface of the wall is at temperature $T_1$ and the right surface is at $T_2$.

Our tasks are to determine the temperature distribution in the wall and the heat flow rate through the wall. This can be done as follows:

The general heat conduction equation in the rectangular coordinate system is

$$\frac{\partial}{\partial x}\left(k\frac{\partial T}{\partial x}\right) + \frac{\partial}{\partial y}\left(k\frac{\partial T}{\partial y}\right) + \frac{\partial}{\partial z}\left(k\frac{\partial T}{\partial z}\right) + q_{gen} = \rho c\frac{\partial T}{\partial t} \tag{3.1}$$

Because the wall is large in the $y$ and $z$ directions, the problem can be reduced to one-dimensional with only the $x$ direction. The problem is also steady state. Eq. (3.1) becomes

$$\frac{\partial}{\partial x}\left(k\frac{\partial T}{\partial x}\right) + q_{gen} = 0 \tag{3.2}$$

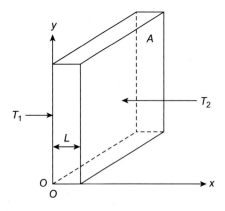

**FIGURE 3.1**

Heat flow through a plane wall.

We will first consider the situation where the conductivity is constant and there is no heat generation. Conductivity $k$ can be removed from the bracket and $q_{gen}$ is zero. Eq. (3.2) then reduces to

$$\frac{d^2T}{dx^2} = 0 \qquad (3.3)$$

Integrating Eq. (3.3) twice with respect to $x$, we obtain

$$T(x) = C_1 x + C_2 \qquad (3.4)$$

where $C_1$ and $C_2$ are constants.

Looking at Eq. (3.4), it is seen that the temperature distribution in the wall is linear, i.e., straight-lined, as shown in Fig. 3.2.

To obtain the values of $C_1$ and $C_2$, we have to apply boundary conditions to Eq. (3.4). For our problem, we have the temperature boundary conditions:

$$\text{At } x = 0, \ T = T_1 \qquad (3.5)$$

$$\text{and} \quad \text{At } x = L, \ T = T_2 \qquad (3.6)$$

Applying these to Eq. (3.4) gives

$$T_1 = C_1(0) + C_2 \qquad (3.7)$$

$$\text{and} \quad T_2 = C_1 L + C_2 \qquad (3.8)$$

Solving Eqs. (3.7) and (3.8) for $C_1$ and $C_2$, we get
$C_1 = (T_2 - T_1)/L$ and $C_2 = T_1$. Substituting this into Eq. (3.4), we get the final temperature distribution in the wall:

$$T(x) = \left(\frac{T_2 - T_1}{L}\right) x + T_1 \qquad (3.9)$$

To find the heat flow rate through the wall, we use Eq. (1.3), which is

$$\left(\frac{q}{A}\right)_x = -k\frac{dT}{dx} \qquad (1.3)$$

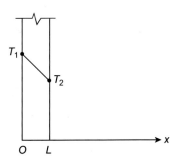

**FIGURE 3.2**

Temperature distribution in a plane wall.

From Eq. (3.9), $\frac{dT}{dx}$ is $\frac{T_2-T_1}{L}$ so the heat flow rate is

$$q_x = \frac{kA}{L}(T_1 - T_2) \tag{3.10}$$

From Fig. 3.2, temperature $T_1$ on the left surface of the wall is greater than temperature $T_2$ on the right surface. Hence, $q_x$ in Eq. (3.10) is positive. In Chapter 1, we discussed the sign convention for heat flow. Positive $q_x$ means that flow is in the positive $x$ direction. This makes sense for our problem. Heat flows from high temperature to low temperature. The left surface of the wall is at a higher temperature than the right surface, and the heat is indeed flowing in the $+x$ direction.

### 3.2.1.1 Multilayered Walls

Sometimes walls consist of multiple layers. Consider, for example, the wall of Fig. 3.3, which has four layers. The thicknesses of the layers are $L_1$, $L_2$, $L_3$, and $L_4$. The respective conductivities are $k_1$, $k_2$, $k_3$, and $k_4$. As shown in the figure, the temperatures at the interfaces and boundaries are labeled $T_1$ through $T_5$. Finally, the cross-sectional area of the wall is $A$.

Let us say that there are temperature boundary conditions at the outer surfaces of the wall. The left surface is held at $T_1$ and the right surface is held at $T_5$. If $T_1$ equals $T_5$, there is no heat flow. If $T_1$ is greater than $T_5$, heat flows from left to right. And, if $T_5$ is greater than $T_1$, heat flows from right to left. Let us make $T_1$ greater than $T_5$. Then the heat flow will be from left to right in each layer.

Because of continuity of heat flow, the heat flow $q$ through each of the four layers will be the same, and Eq. (3.10) can be written for each layer:

$$q = \frac{k_1 A}{L_1}(T_1 - T_2) \quad q = \frac{k_2 A}{L_2}(T_2 - T_3) \quad q = \frac{k_3 A}{L_3}(T_3 - T_4) \quad q = \frac{k_4 A}{L_4}(T_4 - T_5) \tag{3.11}$$

These four equations can be rearranged so that the temperature differences are on the right-hand sides of each equation, as follows:

$$q\left(\frac{L_1}{k_1 A}\right) = (T_1 - T_2) \quad q\left(\frac{L_2}{k_2 A}\right) = (T_2 - T_3) \quad q\left(\frac{L_3}{k_3 A}\right) = (T_3 - T_4) \quad q\left(\frac{L_4}{k_4 A}\right) = (T_4 - T_5) \tag{3.12}$$

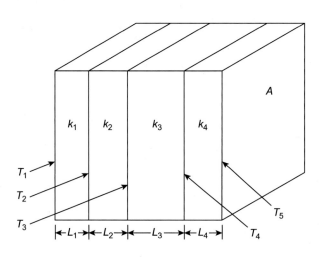

**FIGURE 3.3**

Multilayered wall.

We can now sum all the left-hand sides of the equations and the right-hand sides and make the two resulting sums equal. Doing this, and factoring out $q$, we get

$$q\left(\frac{L_1}{k_1A} + \frac{L_2}{k_2A} + \frac{L_3}{k_3A} + \frac{L_4}{k_4A}\right) = T_1 - T_5 \tag{3.13}$$

Rearranging this, we finally arrive at the equation for the rate of heat flow through the wall:

$$q = \frac{T_1 - T_5}{\left(\dfrac{L_1}{k_1A} + \dfrac{L_2}{k_2A} + \dfrac{L_3}{k_3A} + \dfrac{L_4}{k_4A}\right)} \tag{3.14}$$

This result leads directly to a discussion of the electric-heat analogy, as follows:

### 3.2.1.2 Electric-heat analogy and the resistance concept

There is an analogy between the flow of electric current and the flow of heat. The flow of electric current is caused by a potential difference, and the flow of heat is caused by a temperature difference. Ohm's law states that the flow of electric current $I$ in a conductor is proportional to the potential difference $E$ across the conductor. That is,

$$I = \frac{E}{R} \quad \text{Ohm's Law} \tag{3.15}$$

where $R$ is the resistance of the conductor.

Let us look at Eq. (3.10) that is for the flow of heat through a plane wall. The equation can be rearranged to

$$q = \frac{T_1 - T_2}{\left(\dfrac{L}{kA}\right)} \tag{3.16}$$

Let us compare Eq. (3.15) that is for electric flow and Eq. (3.16) that is for heat flow. It is seen that the left-hand sides of the equations are the flow rates and the numerators on the right-hand sides are the potentials causing the flows. Applying the electric-heat analogy, the denominator of the right-hand side of Eq. (3.15) is the electrical resistance, so the denominator of the right-hand side of Eq. (3.16) must be the heat flow resistance. That is, the thermal resistance of a plane wall is the thickness of the wall divided by the product of the conductivity and the area of the wall, or

$$R_{wall} = \frac{L}{kA} \tag{3.17}$$

Let us look again at Eq. (3.14) for the heat flow through our four-layer wall.

$$q = \frac{T_1 - T_5}{\left(\dfrac{L_1}{k_1A} + \dfrac{L_2}{k_2A} + \dfrac{L_3}{k_3A} + \dfrac{L_4}{k_4A}\right)} \tag{3.14}$$

The left-hand side of the equation is the heat flow, and the numerator on the right-hand side is the temperature difference or potential that is causing the flow. The four terms in the denominator of the right-hand side are the resistances of the four layers of the wall. The sum of these four resistances is the total thermal resistance of the wall, i.e., the sum of the resistances between the two temperatures causing the flow. This can be generalized to the following equation, which will be very useful in solving later problems:

$$q = \frac{\Delta T_{overall}}{\sum R}$$

(3.18)

That is, the heat flow rate is the overall temperature difference divided by the sum of the resistances between the two temperatures.

---

## Example 3.1
### Heat Flow Through a Wall

*Problem:*
The house wall shown in Fig. 3.4 below consists of three layers. The layers, from outside to inside, are T-111 siding (5/8 inch thick, $k = 0.8$ Btu in/h ft$^2$ F), fiberglass insulation (3-1/2 inch thick, $k = 0.27$ Btu in/h ft$^2$ F), and plasterboard (1/2 inch thick, $k = 1.2$ Btu in/h ft$^2$ F). The wall is 8 ft high by 15 ft wide. The temperature of the outside surface of the wall is 35 F, and the temperature of the inside surface of the wall is 72 F. Determine the rate of heat flow through the wall (Btu/h).

*Solution:*
Let us call the T-111 layer "1", the fiberglass layer "2", and the plasterboard layer "3." This is a problem in English units. The first thing to do is to make sure the units are consistent. Usually the best approach is to change all lengths to feet if the problem is in the English system and all lengths to meters if the problem is SI. For this problem, however, the conductivities are in units of Btu in/h ft$^2$ F and it is not necessary to change the thicknesses of the layers to feet. They can remain in inches. However, if the conductivities were given in the usual units of Btu/h ft F, we would have to change the thicknesses to feet in order for the heat flow to have the units of Btu/h.
To get the heat flow through the wall, we use Eq. (3.18)

$$q = \frac{\Delta T_{overall}}{\sum R}$$

(3.18)

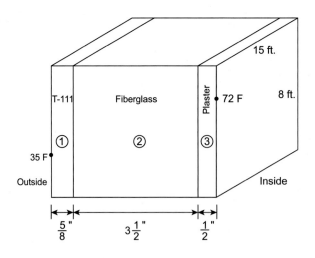

**FIGURE 3.4**

A house wall.

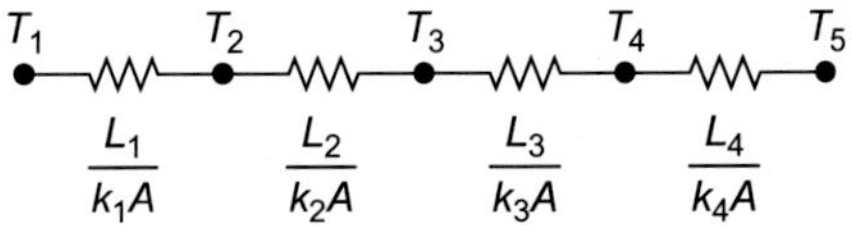

**FIGURE 3.5**

Circuit for four-layer wall.

For a wall, the resistance is $L/k\,A$. So, for our problem Eq. (3.18) is

$$q = \frac{\Delta T_{overall}}{R_1 + R_2 + R_3} \tag{3.19}$$

where

$$R_1 = \frac{L_1}{k_1\,A} = \frac{5/8}{0.8\,A} = \frac{0.781}{A}$$

$$R_2 = \frac{L_2}{k_2\,A} = \frac{3.5}{0.27\,A} = \frac{12.963}{A}$$

$$R_3 = \frac{L_3}{k_3\,A} = \frac{0.5}{1.2\,A} = \frac{0.417}{A}$$

Applying Eq. (3.19), we have

$$q = \frac{72 - 35}{\left(\dfrac{0.781}{A} + \dfrac{12.963}{A} + \dfrac{0.417}{A}\right)} = 2.613\,A \tag{3.20}$$

The wall is 8 ft high and 15 ft wide, so $A = 8 \times 15 = 120$ ft$^2$
We therefore arrive at the result $q = 2.613 \times 120 = 314$ Btu/h.
The heat flow through the wall is **314 Btu/h**.

In Example 3.1 and also the four-layered wall discussed earlier, the same amount of heat flows through each layer of the wall. The situation is analogous to an electric current flowing through several resistances in series. Let us return to the 4-layer wall of Fig. 3.3. For that wall, the heat flow was

$$q = \frac{T_1 - T_5}{\left(\dfrac{L_1}{k_1 A} + \dfrac{L_2}{k_2 A} + \dfrac{L_3}{k_3 A} + \dfrac{L_4}{k_4 A}\right)} \tag{3.14}$$

As discussed above, heat flow is analogous to electric flow. Hence, the 4-layer wall can be modeled as a circuit with four resistances in series between the temperature potentials $T_1$ and $T_5$ (Fig. 3.5).
As noted above in Eq. (3.17), the resistance of a wall is $L/k\,A$. Heat flow by convection is $q_{conv} = h\,A\,(T_s - T_\infty)$, or $q_{conv} = \frac{T_s - T_\infty}{\left(\frac{1}{h\,A}\right)}$, so the thermal resistance of convection is

$$R_{conv} = \left(\frac{1}{hA}\right) \tag{3.21}$$

In Example 3.1, we had a house wall with temperature boundary conditions at both the inside and outside surfaces. Let us redo Example 3.1 but with convection boundary conditions at the surfaces.

---

## Example 3.2
### Heat Flow Through a Wall with Convection

*Problem:*
The house wall of Example 3.1 is now modified to have convection boundary conditions rather than temperature boundary conditions, as shown in Fig. 3.6 below. The outside air is at 31 F with $h_o = 2$ Btu / h ft$^2$ F. The inside air is at 74 F with $h_i = 0.7$ Btu/h ft$^2$ F.

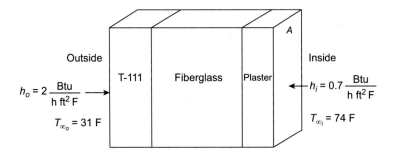

**FIGURE 3.6**

House wall with convection.

**(a)** What is the rate of heat flow through the wall?

**(b)** What is the temperature at the interface between the T-111 and the fiberglass?

*Solution:*

Rate of Heat Flow

We can again use Eq. (3.18), but now we have five resistances in the denominator, three for the wall layers and two for the convections at the wall surfaces.

$$q = \frac{\Delta T_{overall}}{\sum R} \tag{3.18}$$

where $\Delta T_{overall}$ is the temperature difference across the whole system, which now includes the convection. Thus, the overall temperature difference is the difference between the temperature of the room air and that of the outside air. Eq. (3.20) is revised to

$$q = \frac{74 - 31}{\left(\dfrac{1}{2A} + \dfrac{0.781}{A} + \dfrac{12.963}{A} + \dfrac{0.417}{A} + \dfrac{1}{0.7A}\right)} = 2.673A \tag{3.22}$$

From Example 3.1, $A = 120$ ft$^2$, so $q = 2.673 \times 120 =$ **321 Btu/h.**

Interface Temperature

The series electrical circuit for the problem is shown in Fig. 3.7. The heat flow through each layer is the same. Once we have determined $q$, we can determine the temperature at any location in the system by using Eq. (3.18).

We wish to find temperature $T_2$. We can apply Eq. (3.18) between $T_2$ and the outside air temperature $T_{\infty_o}$. Between these two temperatures there is one conductive resistance (the T-111 layer) and the outside convective resistance. Hence, Eq. (3.18) becomes

$$q = \frac{T_2 - T_{\infty_o}}{\left(\dfrac{L}{kA}\right)_{T-111} + \dfrac{1}{h_o A}} \tag{3.23}$$

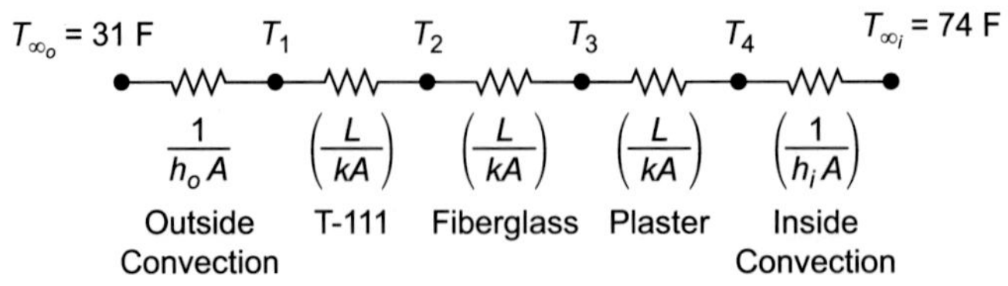

**FIGURE 3.7**

Electrical circuit for Example 3.2.

Putting in values in Eq. (3.23), we have

$$321 = \frac{T_2 - 31}{\left(\dfrac{0.781}{120}\right)_{T-111} + \dfrac{1}{2 \cdot 120}} \tag{3.24}$$

Solving for $T_2$ we have $T_2 = 34.4$ F.

We could also have found $T_2$ by applying Eq. (3.18) between $T_2$ and the inside air temperature $T_{\infty_i}$. Between these two temperatures there are two conductive resistances (the fiberglass and plasterboard layers) and the inside convective resistance. The equations for this approach are

$$q = \frac{T_{\infty_i} - T_2}{\left(\dfrac{L}{kA}\right)_{fiberglass} + \left(\dfrac{L}{kA}\right)_{plaster} + \dfrac{1}{h_i A}} \tag{3.25}$$

$$321 = \frac{74 - T_2}{\left(\dfrac{12.963}{120}\right)_{fiberglass} + \left(\dfrac{0.417}{120}\right)_{plaster} + \dfrac{1}{0.7 \cdot 120}} \tag{3.26}$$

As expected, solving Eq. (3.26) for $T_2$, we get the same $T_2 = 34.4$ F.

We can also get the interface temperature $T_2$ without first obtaining heat flow $q$, as follows:

As the same amount of heat flows through each of the five components of this problem (i.e., the two convective components and the three physical layers), we can apply the resistance technique between any two locations in the system. In particular, we can apply the technique between $T_2$ and the outside air and between the inside air and $T_2$. That is,

$$\left(\frac{\Delta T_{overall}}{\Sigma R}\right)_{T_2 \text{ to outside air}} = \left(\frac{\Delta T_{overall}}{\Sigma R}\right)_{\text{inside air to } T_2} \tag{3.27}$$

Looking at the resistances between the particular locations, we get

$$\frac{T_2 - T_{\infty_o}}{\dfrac{1}{h_o A} + \left(\dfrac{L}{kA}\right)_{T-111}} = \frac{T_{\infty_i} - T_2}{\left(\dfrac{L}{kA}\right)_{fiberglass} + \left(\dfrac{L}{kA}\right)_{plaster} + \dfrac{1}{h_i A}} \tag{3.28}$$

The only unknown in Eq. (3.28) is $T_2$. When we put the values for the other variables in the equation and solve for $T_2$, we get $T_2 = 34.4$ F, which is the same answer as from the other two approaches.

In the above examples we had series heat flow. That is, the same amount of heat flows through all components of the system. In addition to series heat flow, we can also have parallel heat flow. Consider Fig. 3.8 that shows a situation where heat flow is both in series and in parallel.

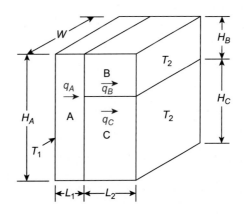

**FIGURE 3.8**

Series and parallel heat flow.

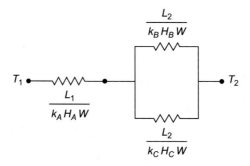

**FIGURE 3.9**

Circuit diagram for series and parallel heat flow.

There are two layers in the wall shown in Fig. 3.8. The first layer is of material A with conductivity $k_A$, thickness $L_1$, height $H_A$, and width $W$. The second layer has two materials: B and C. Material B has conductivity $k_B$, thickness $L_2$, height $H_B$, and width $W$. Material C has conductivity $k_C$, thickness $L_2$, height $H_C$, and width $W$. The left surface of the wall is at temperature $T_1$ and the right surface is at $T_2$.

The heat flow through the first layer is $q_A$. In the second layer, there are two heat flows, $q_B$ and $q_C$, which are parallel to each other. From continuity of heat flow, $q_A = q_B + q_C$.

The circuit diagram for this situation is shown in Fig. 3.9.

In this circuit, there are two resistances in parallel. Let us review circuit theory to get the equivalent resistance for two resistances in parallel. If we have two resistances $R_1$ and $R_2$ in parallel, then, by circuit theory, the equivalent resistance is

$$R_{equiv} = \frac{1}{\dfrac{1}{R_1} + \dfrac{1}{R_2}} \tag{3.29}$$

Using this information and also the resistance technique of Eq. (3.18), we can easily write the equation for the heat flow through the wall, as follows:

$$q = q_A = \frac{T_1 - T_2}{\dfrac{L_1}{k_A H_A W} + \dfrac{1}{\frac{k_B H_B W}{L_2} + \frac{k_C H_C W}{L_2}}} \tag{3.30}$$

## Example 3.3

### Heat Flow Through a Wall with Studs

Unlike the house wall of Example 3.2, a real house wall has studs. This example will determine the effect of the studs on the heat flow through the wall. We will consider the same wall as Example 3.2 except the middle fiberglass layer will now have fiberglass batts with studs. The inclusion of studs causes the heat flow in the middle layer to be one of parallel flow rather than series flow. Some of the heat goes through the fiberglass, and the remainder goes through the studs. A top view of the wall is shown in Fig. 3.10.

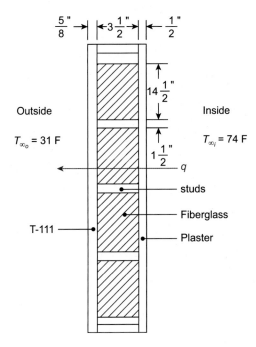

**FIGURE 3.10**

Top view of house wall with studs.

*Problem:*

The house wall shown in Fig. 3.10 consists of three layers. The first layer is T-111 siding (5/8 inch thick, $k = 0.8$ Btu in/h ft$^2$ F). The second layer consists of fiberglass batts (3-1/2 inch thick, $k = 0.27$ Btu in/h ft$^2$ F) separated by studs on 16 inch centers. The studs are 2 × 4s. (Their actual dimensions are 1-1/2 inch by 3-1/2 inch.) Like the T-111, the studs are soft-wood, so we will use the same conductivity for them as the T-111. The third layer is plasterboard (1/2 inch thick, $k = 1.2$ Btu in/h ft$^2$ F). The wall is 8 ft high by 15 ft wide. The outside air is at 31 F and $h_o = 2$ Btu/h ft$^2$ F. The inside air is at 74 F and $h_i = 0.7$ Btu/h ft$^2$ F.

Determine the rate of heat flow through the wall.

*Solution:*

Fig. 3.10 shows four fiberglass batts in the middle layer of the wall. For a wall that is 15 feet long, there will actually be 11 such batts, each 14-1/2 inches wide. Also, as shown in the figure, we will assume that there are double studs at both ends of the wall. For a 15 foot wall there will be 14 studs. Let us check that 11 batts and 14 studs will equal 15 feet. We have (11) (14.5) + (14) (1.5) = 180.5 inches = 15.04 ft. So, we are OK. Let us combine all the studs together and all the fiberglass batts together to result in two parallel heat flows through the center layer of the wall. The wall is 8 feet high and 15 feet wide. For the T-111 and plaster layers, the heat flow area is 8 × 15 = 120 ft$^2$. For the middle layer, the flow area for the studs is 14 studs x 1.5/12 ft per stud x 8 ft = 14 ft$^2$. The flow area for the fiberglass is then 120− 14 = 106 ft$^2$.

The circuit diagram for the wall is shown in Fig. 3.11.

Using the circuit diagram, we can write the equation for the heat flow through the wall as follows:

$$q = \frac{T_{\infty_i} - T_{\infty_o}}{\dfrac{1}{h_o A} + \left(\dfrac{L}{kA}\right)_{T-111} + \dfrac{1}{\left(\dfrac{kA}{L}\right)_{FG} + \left(\dfrac{kA}{L}\right)_{studs}} + \left(\dfrac{L}{kA}\right)_{plaster} + \dfrac{1}{h_i A}} \qquad (3.31)$$

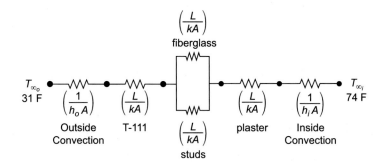

**FIGURE 3.11**

Circuit diagram for Example 3.3.

Putting values into this equation, we get

$$q = \frac{74 - 31}{\dfrac{1}{2 \cdot 120} + \left(\dfrac{0.781}{120}\right)_{T-111} + \dfrac{1}{\left(\dfrac{106}{12.963}\right)_{FG} + \left(\dfrac{14}{4.375}\right)_{studs}} + \left(\dfrac{0.417}{120}\right)_{plaster} + \dfrac{1}{0.7 \cdot 120}} \tag{3.32}$$

$q = $ **377 Btu/h.**

We got a result of 321 Btu/h for Example 3.2. Hence, the inclusion of studs in the wall increased the heat transfer by 17%.

We have discussed the thermal resistance of a plane wall, $R_{wall}$, and the thermal resistance of convection at a surface, $R_{conv}$. They are, from Eqs. (3.17) and (3.21).

$$R_{wall} = \frac{L}{kA} \tag{3.17}$$

$$R_{conv} = \frac{1}{hA} \tag{3.21}$$

Let us now discuss the thermal resistance of a surface having radiative heat transfer. Similar to the convective coefficient $h$, we will define the radiative coefficient $h_{rad}$ by

$$q = h_{rad}(T_s - T_{surr}) \tag{3.33}$$

where $T_s$ is the surface temperature and $T_{surr}$ is the temperature of the surroundings.

From Eq. (1.13), the radiant heat transfer from a surface to a large enclosure is

$$q = \varepsilon \sigma A \left(T_s^4 - T_{surr}^4\right) \tag{3.34}$$

Equating the right sides of Eqs. (3.33) and (3.34), we get the following for $h_{rad}$:

$$h_{rad} = \frac{\varepsilon \sigma \left(T_s^4 - T_{surr}^4\right)}{T_s - T_{surr}} = \varepsilon \sigma \left(T_s^2 + T_{surr}^2\right)(T_s + T_{surr}) \tag{3.35}$$

From Eq. (3.33), the thermal resistance of a surface for radiation is

$$R_{rad} = \frac{1}{h_{rad}A} \tag{3.36}$$

Defining the radiation coefficient $h_{rad}$ significantly simplifies electrical circuit diagrams. The nonlinearities are in the coefficient. There are no $T^4$ nodes.

In many situations, there is concurrent convection from a surface to an adjacent fluid and radiation from the surface to the surroundings. The total heat transfer from the surface is then

$$q_{total} = q_{conv} + q_{rad} = hA(T_s - T_\infty) + h_{rad}A(T_s - T_{surr}) \qquad (3.37)$$

If fluid temperature $T_\infty$ and surroundings temperature $T_{surr}$ are equal, a combined coefficient $h_{combined}$ may be defined, which includes both the convective and radiative heat transfers.

$$h_{combined} = h + h_{rad} \qquad (3.38)$$

If this is done, the heat transfer at the surface is

$$q_{total} = h_{combined}A(T_s - T_\infty) \qquad (3.39)$$

and the resistance for the combined convection and radiation is

$$R_{conv+rad} = \frac{1}{h_{combined}A} \qquad (3.40)$$

### 3.2.1.3 Overall heat transfer coefficient and R-Value

Two terms often used in heat transfer are the Overall Heat Transfer Coefficient, $U$, and the *R-value*. The overall heat transfer coefficient is defined by the equation

$$q = UA\Delta T \qquad (3.41)$$

From this equation, $U$ is the heat flow rate per unit area per unit temperature difference. $U$ is also sometimes called the Conductance, $C$.

Let us return to the four-layer wall that we considered earlier. For that wall, the heat flow was given by Eq. (3.14): $q = \dfrac{T_1 - T_5}{\left( \dfrac{L_1}{k_1 A} + \dfrac{L_2}{k_2 A} + \dfrac{L_3}{k_3 A} + \dfrac{L_4}{k_4 A} \right)}$.

Comparing Eqs. (3.14) and (3.41), we can see that, for the four-layer wall, the overall heat transfer coefficient $U$ is

$$U = \frac{1}{\dfrac{L_1}{k_1} + \dfrac{L_2}{k_2} + \dfrac{L_3}{k_3} + \dfrac{L_4}{k_4}} \qquad (3.42)$$

The *R-value* is another term that is frequently used in heat transfer. It is the reciprocal of the Overall Heat Transfer Coefficient $U$.

*R-value* is commonly used in the specification of thermal insulation. The higher the *R-value*, the more resistance the insulation has to heat flow. As an example, 3-1/2 inch thick fiberglass batts used between studs in a 2 × 4 wall typically have an *R-value* of R-11 or R-13. For a 2 × 6 wall, the fiberglass insulation would be 5-1/2 inch thick and the *R-value* would be R-19. *R-values* are also used by states in their energy codes. For example, in New York, the energy code specifies the minimum *R-value* requirements for walls, floors, and ceilings of buildings.

The heat flow through a wall of conductivity $k$, thickness $L$, and area $A$ is $q = \frac{kA}{L}\Delta T$. From Eq. (3.41), $q = UA\Delta T$. Hence, for a wall, the overall heat transfer coefficient is $U = k/L$. The *R-value* is the reciprocal of $U$, so the *R-value* for the wall is $L/k$.

Let us determine the *R-value* for the fiberglass insulation used in the above examples. The fiberglass is 3-1/2 inches thick and its conductivity is $k = 0.27$ Btu in/h ft$^2$ F. The *R-value* is $L/k = 3.5$ in/(0.27 Btu in/h ft$^2$ F) $= 13$ h ft$^2$ F/Btu. When purchasing this insulation in a store, you will only see R-13 on the insulation. The units will not be included in the labeling.

## 3.2.2 Cylindrical shell

The cylindrical shell (or hollow cylinder) shown in Fig. 3.12 has an inner radius $r_i$, an outer radius $r_o$, and a length $L$. The thermal conductivity is $k$. The cylinder has temperature boundary conditions with the inner surface at $T_i$ and the outer surface at $T_o$. Heat conduction through the wall is steady and is only in the radial $r$ direction. We want to determine the temperature distribution in the cylindrical wall and the heat flow rate through the wall.

From Chapter 2, the heat conduction equation is

$$\frac{1}{r}\frac{\partial}{\partial r}\left(kr\frac{\partial T}{\partial r}\right) + \frac{1}{r^2}\frac{\partial}{\partial \varphi}\left(k\frac{\partial T}{\partial \varphi}\right) + \frac{\partial}{\partial z}\left(k\frac{\partial T}{\partial z}\right) + q_{gen} = \rho c \frac{\partial T}{\partial t} \tag{3.43}$$

We have steady state and flow only in the $r$ direction, so Eq. (3.43) reduces to

$$\frac{1}{r}\frac{d}{dr}\left(kr\frac{dT}{dr}\right) + q_{gen} = 0 \tag{3.44}$$

In the future, we will consider heat generation, but for now let us make $q_{gen}$ zero. Also, let us assume that the conductivity is constant. Then Eq. (3.44) becomes

$$\frac{d}{dr}\left(r\frac{dT}{dr}\right) = 0 \tag{3.45}$$

Integrating Eq. (3.45) with respect to $r$, we get

$$r\frac{dT}{dr} = C_1 \tag{3.46}$$

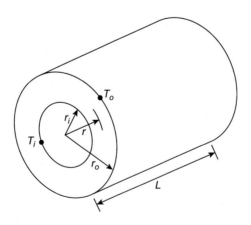

**FIGURE 3.12**

Cylindrical shell.

Integrating again, we get the solution

$$T(r) = C_1 \ln r + C_2 \tag{3.47}$$

To get constants $C_1$ and $C_2$, we apply, to Eq. (3.47), the boundary conditions:

$$\text{At} \quad r = r_i, \quad T = T_i \tag{3.48}$$

$$\text{and} \quad \text{At} \quad r = r_o, T = T_o \tag{3.49}$$

$$\text{This gives} \quad T_i = C_1 \ln r_i + C_2 \tag{3.50}$$

$$\text{and} \quad T_o = C_1 \ln r_o + C_2 \tag{3.51}$$

Eqs. (3.50) and (3.51) are solved simultaneously to give

$$C_1 = \frac{T_o - T_i}{\ln\left(\dfrac{r_o}{r_i}\right)} \quad \text{and} \quad C_2 = T_i - \frac{T_o - T_i}{\ln\left(\dfrac{r_o}{r_i}\right)} \ln r_i \tag{3.52}$$

Putting these constants into Eq. (3.47) and doing algebraic manipulations, we arrive at the final equation for the temperature distribution in the cylinder:

$$T(r) = T_i - \frac{\ln\left(\dfrac{r}{r_i}\right)}{\ln\left(\dfrac{r_o}{r_i}\right)} (T_i - T_o) \tag{3.53}$$

Heat flow through the cylinder's wall is in the $r$ direction. Hence, to get the heat flow we can use the relation

$$q = -kA\frac{dT}{dr} \tag{3.54}$$

$A$ is the area through which the heat flows. This area varies with the distance from the centerline of the cylinder. At radial distance $r$ from the centerline, the area is the circumference times the length of the cylinder or $A = 2\pi r L$.

$dT/dr$ may be obtained by differentiating Eq. (3.53) or by using Eqs. (3.46) and (3.52).

Putting the expressions for $A$ and $dT/dr$ into Eq. (3.54), we arrive at the expression for the heat flow through the cylinder's wall:

$$q = \frac{2\pi kL}{\ln\left(\dfrac{r_o}{r_i}\right)} (T_i - T_o) \tag{3.55}$$

Inspection of Eq. (3.55) shows that we can define the thermal resistance for a cylindrical shell as

$$R_{cyl} = \frac{\ln\left(\dfrac{r_o}{r_i}\right)}{2\pi kL} \tag{3.56}$$

This thermal resistance is very useful in problems involving concentric cylinders. One such problem is an insulated pipe, as illustrated in Example 3.4 below.

Before leaving this discussion of cylindrical shells, let us discuss the overall heat transfer coefficient $U$: Unlike a plane wall, a cylinder has no single area through which heat is flowing. The area changes with $r$. At the inner surface of the cylinder, the area is $A_i = 2\pi r_i L$ and at the outer surface it is $A_o = 2\pi r_o L$. Because there is no unique area, the value of $U$ must be stated with respect to a specific area. For example, we can give the value of $U$ with respect to the inner area or the value of $U$ with respect to the outer area. From Eq. (3.41) above, $U$ is defined by the equation $q = UA \ \Delta T$. For a cylinder, the two physical areas are $A_i$ and $A_o$, and the corresponding $U$ values are $U_i$ and $U_o$. Thus, for a cylinder,

$$q = U_i A_i \Delta T = U_o A_o \Delta T \tag{3.57}$$

From Eq. (3.57), $U_i A_i = U_o A_o$ and $U_i = (A_o / A_i) \ U_o$. Putting in the relations for $A_i$ and $A_o$, we arrive at

$$U_i = (r_o / r_i) \ U_o \tag{3.58}$$

## Example 3.4

### Heat Transfer from an Insulated Pipe

*Problem:*
Steam at 150 C is flowing through a 100 mm nominal size standard metric pipe. The pipe is carbon steel ($k = 60$ W/m C) and is located in a factory where the air temperature is 20 C. The convective coefficient between the pipe and the room air is 4 W/m$^2$ C. The pipe is 50 m long and is uninsulated.

**(a)** What is the rate of heat transfer from the bare pipe to the room air?

**(b)** Insulation is added to the pipe to reduce the heat transfer. The insulation is 51 mm thick and has a conductivity of 0.05 W/m C. What is the new rate of heat transfer?

You may make the following assumptions: Radiation heat transfer from the pipe is negligible. The insulated pipe has the same convective coefficient at its outer surface as the uninsulated pipe. The temperature of the inside surface of the pipe is essentially the same as the steam temperature.

*Solution:*
The uninsulated pipe is shown in Fig. 3.13.
From an Internet search, the outside diameter of the pipe is 114.3 mm and the inside diameter is 102.26 mm. Therefore, $r_1 = 0.05113$ m and $r_2 = 0.05715$ m. There are two resistances between the two known temperatures $T_1$ and $T_{\infty_o}$: the resistance of the pipe wall and the convective resistance at the outside surface of the pipe. The circuit for the problem is shown in Fig. 3.14.
From the circuit drawing and the resistance technique, we have

$$q = \frac{\Delta T_{overall}}{\Sigma R} = \frac{T_1 - T_{\infty_o}}{\dfrac{\ln(r_2/r_1)}{2\pi k L} + \dfrac{1}{h_o(2\pi r_2 L)}} \tag{3.59}$$

Putting values into Eq. (3.59), we have

$$q = \frac{150 - 20}{\dfrac{\ln(0.05715/0.05113)}{2 \cdot \pi \cdot 60 \cdot 50} + \dfrac{1}{4 \cdot (2 \cdot \pi \cdot 0.05715 \cdot 50)}} = 9332 \text{ W}$$

The insulated pipe is shown in Fig. 3.15. There is an added 51 mm layer of insulation, making $r_3 = r_2 + 0.051 = 0.10815$ m. The area at the outer surface of the insulation is now $2\pi r_3 L$ rather than $2\pi r_2 L$. The added layer of insulation adds a term to the denominator of Eq. (3.59). The new equation is

$$q = \frac{T_1 - T_{\infty_o}}{\dfrac{\ln(r_2/r_1)}{2\pi k_{pipe} L} + \dfrac{\ln(r_3/r_2)}{2\pi k_{ins} L} + \dfrac{1}{h_o(2\pi r_3 L)}} \tag{3.60}$$

When values are entered into Eq. (3.60), it is seen that the new heat transfer rate is 2710 W. Addition of insulation has decreased the heat transfer from 9332 to 2710 W, a decrease of 71%.

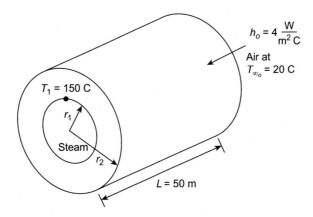

**FIGURE 3.13**

Original pipe of Example 3.4.

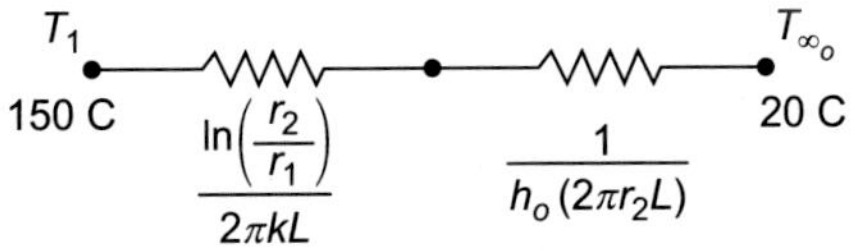

$$T_1 \quad \underset{150\ C}{\bullet} \quad \underset{\dfrac{\ln\left(\dfrac{r_2}{r_1}\right)}{2\pi kL}}{\text{WW}} \quad \bullet \quad \underset{\dfrac{1}{h_o\,(2\pi r_2 L)}}{\text{WW}} \quad \underset{20\ C}{\overset{T_{\infty_o}}{\bullet}}$$

**FIGURE 3.14**

Circuit for bare pipe of Example 3.4.

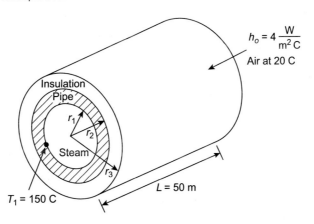

**FIGURE 3.15**

Insulated pipe of Example 3.4.

As a final observation, we assumed that the temperature of the inside surface of the pipe was essentially the same as the steam temperature. This is often a very good assumption. The convective coefficient, $h$, for steam on a surface is very large. The convective heat transfer is $q = h\,A\,(T_s - T_\infty)$. Rearranging this, we have $T_s - T_\infty = \frac{q}{h\,A}$. So, for very large $h$, the difference between the surface temperature and the fluid temperature is often very small.

### 3.2.3 Spherical shell

The spherical shell (or hollow sphere) shown in Fig. 3.16 has an inner radius $r_i$, and an outer radius $r_o$. The thermal conductivity is $k$. The sphere has temperature boundary conditions with the inner surface at $T_i$ and the outer surface at $T_o$. Heat conduction through the wall is steady and is only in the radial $r$ direction. We want to determine the temperature distribution in the spherical wall and the heat flow rate through the wall.

From Chapter 2, the heat conduction equation is

$$\frac{1}{r^2} \frac{\partial}{\partial r}\left(k\,r^2\frac{\partial T}{\partial r}\right) + \frac{1}{r^2 \sin^2\theta}\frac{\partial}{\partial\varphi}\left(k\,\frac{\partial T}{\partial\varphi}\right) + \frac{1}{r^2\,\sin\theta}\frac{\partial}{\partial\theta}\left(k\,\sin\theta\,\frac{\partial T}{\partial\theta}\right) + q_{gen}$$
$$= \rho\,c\,\frac{\partial T}{\partial t}$$

(3.61)

We have steady state and flow only in the $r$ direction, so Eq. (3.61) reduces to

$$\frac{1}{r^2}\frac{d}{dr}\left(k\,r^2\frac{dT}{dr}\right) + q_{gen} = 0$$

(3.62)

Let us assume that there is no internal heat generation, i.e., $q_{gen} = 0$ and that the thermal conductivity is constant. Then Eq. (3.62) becomes

$$\frac{d}{dr}\left(r^2\frac{dT}{dr}\right) = 0$$

(3.63)

Integrating Eq. (3.63) with respect to $r$, we get

$$r^2\frac{dT}{dr} = C_1$$

(3.64)

Integrating again, we get the solution

$$T(r) = C_2 - \frac{C_1}{r}$$

(3.65)

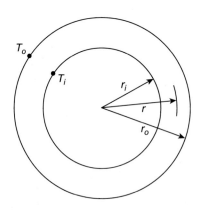

**FIGURE 3.16**

Spherical shell.

To get constants $C_1$ and $C_2$ we apply, to Eq. (3.65), the boundary conditions:

$$\text{At} \quad r = r_i, \quad T = T_i \tag{3.66}$$

$$\text{and} \quad \text{At} \quad r = r_o, T = T_o \tag{3.67}$$

$$\text{This gives} \quad T_i = C_2 - C_1/r_i \tag{3.68}$$

$$\text{and} \quad T_o = C_2 - C_1/r_o \tag{3.69}$$

Eqs. (3.68) and (3.69) are solved simultaneously to give

$$C_1 = \frac{T_i - T_o}{\left(\dfrac{1}{r_o} - \dfrac{1}{r_i}\right)} \quad \text{and} \quad C_2 = T_i - \frac{T_i - T_o}{\left(1 - \dfrac{r_i}{r_o}\right)} \tag{3.70}$$

Putting these constants into Eq. (3.65) and doing algebraic manipulations we arrive at the final equation for the temperature distribution in the sphere:

$$T(r) = T_i - (T_i - T_o)\left[\frac{1 - \left(\dfrac{r_i}{r}\right)}{1 - \left(\dfrac{r_i}{r_o}\right)}\right] \tag{3.71}$$

Heat flow through the sphere's wall is in the $r$ direction. Hence, to get the heat flow we can use the relation

$$q = -kA\frac{dT}{dr} \tag{3.72}$$

$A$ is the area through which the heat flows. This area varies with the distance from the center of the sphere. At radial distance $r$ from the center, the area is the surface area of a sphere or $A = 4\pi r^2$.
$dT/dr$ may be obtained by differentiating Eq. (3.71) or by using Eqs. (3.64) and (3.70).

Putting the expressions for $A$ and $dT/dr$ into Eq. (3.72), we arrive at the expression for the heat flow through the sphere's wall:

$$q = \frac{4\pi k}{\left(\dfrac{1}{r_i} - \dfrac{1}{r_o}\right)}(T_i - T_o) \tag{3.73}$$

From Eq. (3.73), we can see that the thermal resistance for a spherical shell is

$$R_{sphere} = \frac{\left(\dfrac{1}{r_i} - \dfrac{1}{r_o}\right)}{4\pi k} \tag{3.74}$$

This thermal resistance is very useful in problems involving concentric spheres as was done earlier for concentric cylinders.

As with cylinders, the overall heat transfer coefficient $U$ for spheres must be related to a specific area.

## 3.3 Critical insulation thickness

Contrary to intuition, addition of insulation to an object can sometimes *increase* the heat transfer from the object rather than decrease it. However, this phenomenon occurs only in limited situations.

Fig. 3.17 shows a pipe or wire covered with a layer of insulation.

The bare pipe or wire has an outer radius of $r_i$, and the radius to the outer surface of the insulation is $r_o$. As insulation is added to the pipe or wire, $r_o$ goes from $r_i$ to its final value when addition of insulation is stopped. The length of the pipe or wire is $L$.

Let us look at the heat transfer from the pipe or wire to the surrounding fluid. The temperature at the interface between the pipe or wire and the insulation is $T_1$. The insulation has conductivity $k$ and the surrounding fluid is at temperature $T_\infty$. The convective coefficient at the outer surface of the insulation is $h$.

Between $T_1$ and $T_\infty$ are two resistances: the conductive resistance of the insulation and the convective resistance between the outer surface of the insulation and the fluid. The expression for the heat flow through the insulation is

$$q = \frac{T_1 - T_\infty}{\left( \dfrac{1}{h(2\pi r_o L)} + \dfrac{\ln\left(\frac{r_o}{r_i}\right)}{2\pi k L} \right)} \tag{3.75}$$

Looking at Eq. (3.75), we see that, as insulation is added to the pipe or wire, i.e., as $r_o$ increases, the first term in the denominator decreases and the second term increases. Because of this, there is a certain value of $r_o$ for which the heat flow is a maximum. To find this value, called the "critical radius $r_{oc}$", we can differentiate the expression for $q$ in Eq. (3.75) with respect to $r_o$ and set the result to zero. When this is done, it is found that the critical radius is $k/h$.

$$r_{oc} = \frac{k}{h} \tag{3.76}$$

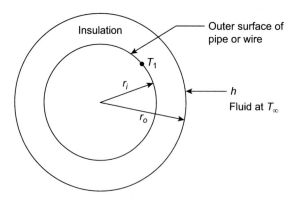

**FIGURE 3.17**

Pipe or wire with insulation.

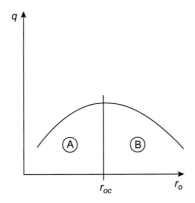

**FIGURE 3.18**

Heat flow versus outer radius.

So, the maximum heat flow occurs when the outer radius of the insulation equals $k/h$.

Fig. 3.18 shows a typical plot of $q$ versus $r_o$. We have indicated two regions on the plot: Region A where $r_o$ is less than $r_{oc}$ and Region B where $r_o$ is greater than $r_{oc}$.

Let us say that a bare pipe or wire has an outer radius $r_i$ that is less than $r_{oc}$. This is in Region A. As insulation is added, $r_o$ increases and the heat transfer increases until $r_o$ reaches $r_{oc}$. Further addition of insulation will cause the heat transfer to decrease. On the other hand, if the bare pipe or wire has a radius $r_i$ that is equal to or greater than $r_{oc}$, i.e., $r_i$ is in Region B; then any amount of added insulation will decrease the heat transfer.

## Example 3.5
### Heat Flow from an Insulated Wire

*Problem:*
Please refer to the previous Fig. 3.17 that shows an insulated wire. We have a copper #12 AWG wire with PTFE insulation. The bare wire has a diameter of 2.052 mm and the insulation is 15 mils thick. The thermal conductivity of the insulation is $k = 0.25$ W/m C. The temperature $T_1$ at the interface between the wire and the insulation is kept at 100 C, and the surrounding air is at 20 C. The convective coefficient $h$ at the outer surface of the insulation is 10 W/m² C.
(a) What is the heat transfer rate from the wire per meter length before additional insulation is added?
(b) If more insulation is added to the wire, will the heat transfer rate increase or decrease?
(c) If the answer to Part (b) is that the heat transfer will increase, then how much insulation has to be added for the heat transfer to reach a maximum? Also, what is the value of this maximum heat transfer rate?

*Solution:*
(a) For the original wire with the original insulation, $r_i = 0.001026$ m and $r_o = 0.001407$ m. (Note: 1 mil is 1/1000 inch)
Putting the values of this problem into Eq. (3.75), we get that the original heat transfer is $q/L = 6.95$ W/m.
(b) Maximum heat transfer occurs when $r_o = r_{oc} = k/h$.

$$r_{oc} = \frac{0.25 \text{ W} / \text{m C}}{10 \text{ W} / \text{m}^2 \text{ C}} = 0.025 \text{ m}$$

Because our initial $r_o$ is 0.001407 m, which is less than $r_{oc}$, addition of insulation will increase the heat transfer until $r_o$ reaches $r_{oc}$.

**(c)** The thickness of insulation that has to be added to reach $r_{oc}$ is $0.025-0.001407 = 0.0236$ m $= 23.6$ mm. Using $r_{oc} = 0.025$ m in Eq. (3.75), it is found that the maximum heat transfer rate is 29.97 W/m.
The below graph shows $q/L$ versus $r_o$ for this example. It is seen that the maximum heat flow indeed occurs at $r_o = 0.025$ m and the maximum heat flow is 29.97 W/m.

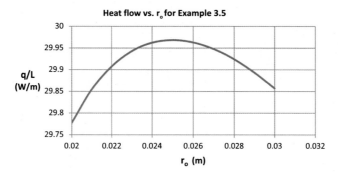

One final note on this example: The inner surface of the insulation had a temperature boundary condition. The temperature was held constant at 100 C. As insulation was added to the wire, up to $r_{oc}$, the heat transfer from the wire increased. To keep the temperature constant as the heat transfer increased, the electrical current would have to be increased to increase the heat generation of the wire. If, instead of the constant temperature boundary condition, we had a constant power boundary condition, then, as the insulation was added, the increased heat transfer would have caused a drop in temperature at the inner surface of the insulation.

At the beginning of this section, we mentioned that the critical insulation phenomenon (i.e., an increase in heat flow with increased insulation) is only important in limited situations. Indeed, for common values of $k$ and $h$, the phenomenon is only relevant when the diameter of the pipe or tube being insulated is very small or, as in Example 3.5, we are insulating a thin wire.

To illustrate this, let us consider a practical situation where we are putting fiberglass insulation on a 1/2 inch Schedule 40 pipe. Fiberglass has a $k$ value of about 0.25 Btu in/h ft$^2$ F. And, we will assume that the pipe is in still air where $h$ is about 2 Btu/h ft$^2$ F. For a 1/2 inch Schedule 40 pipe, the OD is 0.84 inch. This makes $r_i$ equal to 0.42 inch.

The critical radius $k/h$ is 0.25/2 or 0.125 inch. However, $r_o = r_i +$ the thickness of the insulation. So, $r_o$ is always greater than 0.42 inch, which is greater than $r_{oc}$. Addition of any amount of insulation to the pipe will decrease the heat transfer.

## 3.4 Heat generation in a cylinder

Fig. 3.19 shows a cylinder of radius $r_o$ and length $L$. The cylinder has a uniformly distributed heat source at strength $q_{gen}$ per unit volume. The heat flow is steady and one-dimensional, i.e., $T = T(r)$, and there is a temperature boundary condition at the surface of the cylinder.

The heat conduction equation for this situation is Eq. (3.44), repeated here:

$$\frac{1}{r}\frac{d}{dr}\left(kr\frac{dT}{dr}\right) + q_{gen} = 0 \tag{3.44}$$

Let us assume that the conductivity is constant. Then, Eq. (3.44) becomes

$$\frac{d}{dr}\left(r\frac{dT}{dr}\right) + \frac{q_{gen}r}{k} = 0 \tag{3.77}$$

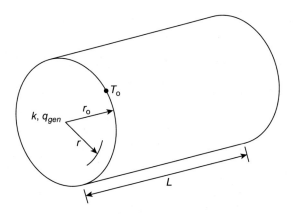

**FIGURE 3.19**

Cylinder with uniform heat generation.

Integrating Eq. (3.77) with respect to $r$, we get

$$r\frac{dT}{dr} = -\frac{q_{gen}r^2}{2k} + C_1 \qquad (3.78)$$

Dividing by $r$ and integrating again, we get the following general solution for the temperature distribution:

$$T(r) = \frac{-q_{gen}r^2}{4k} + C_1 \ln r + C_2 \qquad (3.79)$$

We apply boundary conditions to get constants $C_1$ and $C_2$. These conditions are:

**(1)** At $r = 0$, the temperature must be finite. Therefore, $C_1$ must be zero and Eq. (3.79) becomes

$$T(r) = \frac{-q_{gen}r^2}{4k} + C_2 \qquad (3.80)$$

**(2)** At $r = r_o$, $T = T_o$. Applying this to Eq. (3.80) and solving for $C_2$, we have

$$C_2 = T_o + \frac{q_{gen}r_o^2}{4k} \qquad (3.81)$$

Applying $C_1$ and $C_2$ to Eq. (3.79), we get the final expression for the temperature distribution in the cylinder:

$$T(r) = T_o + \frac{q_{gen}r_o^2}{4k}\left[1 - \left(\frac{r}{r_o}\right)^2\right] \qquad (3.82)$$

Inspection of Eq. (3.82) shows that the maximum temperature in the cylinder is on the centerline ($r = 0$) and is

$$T_{\max} = T_o + \frac{q_{gen}r_o^2}{4k} \tag{3.83}$$

The heat flow in the cylinder is in the radial direction and can be determined using the equation

$$q = -kA\frac{dT}{dr} \tag{3.84}$$

At distance $r$ from the centerline, the area through which heat flows is $A = 2\pi rL$. The derivative $dT/dr$ can easily be obtained from Eq. (3.78) as $C_1 = 0$. Putting this in Eq. (3.84), we get the expression for the heat flow:

$$q(r) = \pi r^2 L q_{gen} \tag{3.85}$$

Looking at the right side of Eq. (3.85): $\pi r^2 L$ is the volume of the cylinder from $r = 0$ to $r = r$. Let us call this volume the "control volume" When the volume is multiplied by heat generation rate $q_{gen}$, we have the rate of heat generation in the control volume. For steady state, the heat inflow to this volume must equal the heat outflow from this volume. The heat generated in the control volume is removed by heat flow through the $r = r$ boundary of the volume.

The heat generation for the whole cylinder is $\pi r_o{}^2 L q_{gen}$. Again, as we have steady state, this heat generation is removed by heat flow out of the cylinder at the $r = r_o$ surface.

Let us say that we have convection at the surface of the cylinder with an $h$ and a $T_\infty$. This is shown in Fig. 3.20 below. The area of the surface of the cylinder is $2\pi r_o L$. Because of continuity of heat flow, the heat generated in the cylinder equals the heat into the surrounding fluid by convection. If the temperature of the surface is $T_o$, then the boundary equation for the surface of the cylinder is

$$q(r_o) = \pi r_o^2 L q_{gen} = h(2\pi r_o L)(T_o - T_\infty) \tag{3.86}$$

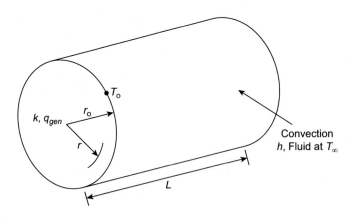

**FIGURE 3.20**

Cylinder with heat generation and convection.

## Example 3.6
### Heat Generation in a Wire

*Problem:*

A #22 AWG copper wire shorts out across a 0.15 V source. This causes the wire to heat up significantly. The wire has a length of 0.2 m and is located in 20 C room air. The wire's thermal conductivity is $k = 385$ W/m C. Finally, the convective coefficient at the surface of the wire is 15 W/m$^2$ C. For steady-state conditions,

**(a)** What is the power generated in the wire?
**(b)** What is the heat generation rate $q_{gen}$ (W/m$^3$)?
**(c)** What are the temperatures at the center and surface of the wire?

*Solution:*

**(a)** The power generated in the wire is $P = i^2 R$, where $i$ is the current through the wire and $R$ is the resistance of the wire. The resistance of the wire can be obtained from the relation

$$R = \frac{\rho L}{A} \tag{3.87}$$

where $\rho$ is the resistivity of the wire, $L$ is the length of the wire, and $A$ is the cross-sectional area of the wire. From the Internet, the resistivity of copper is $1.68 \times 10^{-8}$ $\Omega$ m and the diameter of the wire is 0.644 mm. The cross-sectional area of the wire is $\pi r_o^2 = 3.257 \times 10^{-7}$ m$^2$. The length of the wire was given as 0.2 m. Putting these values into Eq. (3.87) gives a resistance $R = 0.010315$ $\Omega$.

The current in the wire by Ohm's law is $i = \frac{E}{R}$. We were given that $E = 0.15$ V and we determined that $R = 0.010315$ $\Omega$. Hence, the current is $i = 0.15/0.010315 = 14.54$ A.

The power generated in the wire is $P = i^2 R = (14.54)^2 (0.010,315) = 2.181$ W.

**(b)** $q_{gen}$ is the heat generation rate per unit volume. The volume of the wire is cross-sectional area times length. Using values above, we have $V = (3.257 \times 10^{-7}) (0.2) = 6.514 \times 10^{-8}$ m$^3$.

Therefore,

$$q_{gen} = \frac{P}{V} = \frac{2.181}{6.514 \times 10^{-8}} = 3.348 \times 10^7 \text{ W / m}^3$$

**(c)** Rearranging Eq. (3.86), we have

$$T_o = T_\infty + \left(\frac{r_o}{2h}\right) q_{gen} \tag{3.88}$$

Putting in the values, we get, for the surface temperature

$$T_o = 20 + \left(\frac{0.322 \times 10^{-3}}{2 \cdot 15}\right) (3.348 \times 10^7) = 379.4 \text{ C}$$

We can get the center temperature from Eq. (3.83)

$$T_{max} = T_o + \frac{q_{gen} r_o^2}{4k} = 379.4 + \frac{(3.348 \times 10^7)(0.322 \times 10^{-3})^2}{4 \cdot 385} = 379.402 \text{ C}$$

It is seen that the difference between the center and surface temperatures is only 0.002 C. This very small difference is due to the high conductivity of copper and the small diameter of the wire.

As a final item for this section, let us say that, instead of a solid cylinder, we have a hollow cylinder that has uniform heat generation, as shown in Fig. 3.21. We will also have temperature boundary conditions at the inner and outer surfaces.

For this situation, the heat conduction equation, Eq. (3.77), is the same as that for the solid cylinder, and the general solution, Eq. (3.79), is also the same as for the solid cylinder. When we applied the boundary conditions for the solid cylinder, we made $C_1$ equal to zero due to the existence of material at

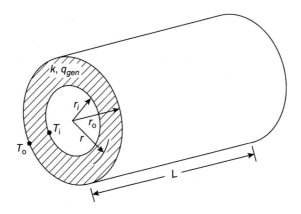

**FIGURE 3.21**

Hollow cylinder with heat generation.

the center of the cylinder and the logarithmic term. For the hollow cylinder, there is no material at $r = 0$ and we cannot eliminate the $C_1 \ln r$ term. The boundary conditions for the hollow cylinder situation are applied to Eq. (3.79) in its entirety.

Heat generation for other geometries is given in Chapter 12.

## 3.5 Temperature-dependent thermal conductivity

The thermal conductivity $k$ of materials varies with temperature. However, in many cases, this variation is small and the conductivity can be taken as being constant at the average temperature of the problem.

Let us consider conduction through the plane wall of Fig. 3.22, which has conductivity varying with temperature. The left face of the wall is at temperature $T_1$ and the right face is at $T_2$.

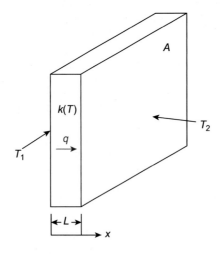

**FIGURE 3.22**

Plane wall with variable conductivity.

The conductive heat transfer equation is

$$q = -kA\frac{dT}{dx} \tag{3.89}$$

Rearranging and integrating across the wall, we have

$$\frac{q}{A}\int_0^L dx = -\int_{T_1}^{T_2} k\,dT \tag{3.90}$$

$$\text{or,}\quad \frac{q}{A} = \frac{1}{L}\int_{T_2}^{T_1} k\,dT \tag{3.91}$$

The functional relation $k\ (T)$ can be put into the integral, the integral evaluated, and the final result for the heat flux $(q/A)$ obtained.

As a special case, let us assume that the $k-T$ relation is linear. That is,

$$k = a + bT \tag{3.92}$$

where $a$ and $b$ are constants.

Putting this relation in Eq. (3.91), we have

$$\frac{q}{A} = \frac{1}{L}\int_{T_2}^{T_1} (a + bT)\,dT \tag{3.93}$$

Performing the integration,

$$\frac{q}{A} = \frac{1}{L}\left[aT + \frac{bT^2}{2}\right]_{T_2}^{T_1} \tag{3.94}$$

$$\frac{q}{A} = \frac{1}{L}\left[a(T_1 - T_2) + \frac{b}{2}(T_1^2 - T_2^2)\right] \tag{3.95}$$

$$\frac{q}{A} = \frac{1}{L}\left[a(T_1 - T_2) + \frac{b}{2}(T_1 - T_2)(T_1 + T_2)\right] \tag{3.96}$$

Factoring out $(T_1 - T_2)$, we have

$$\frac{q}{A} = \frac{1}{L}(T_1 - T_2)\left[a + b\left(\frac{T_1 + T_2}{2}\right)\right] \tag{3.97}$$

But $\frac{T_1 + T_2}{2}$ is the average, or mean, of the two surface temperatures, so the bracketed term is the conductivity at this mean temperature $k_m$. Therefore, the equation for the heat flow $q$ through the wall is

$$q = \frac{k_m A}{L}(T_1 - T_2) \tag{3.98}$$

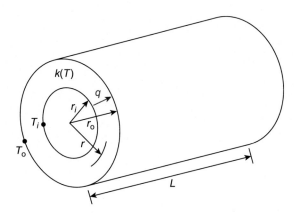

**FIGURE 3.23**

Hollow cylinder with variable conductivity.

Comparing this with Eq. (3.10), it is seen that we can use the constant conductivity equation if we use the conductivity value $k_m$ in the equation.

Similar results can be obtained for heat flow through cylinders and spheres when the conductivity is linear with temperature. For example, consider heat flow through the hollow cylinder shown in Fig. 3.23.

If the conductivity of the cylinder is linear with temperature, it can be shown that the heat flow $q$ through the wall is

$$q = \frac{2\pi k_m L}{\ln\left(\dfrac{r_o}{r_i}\right)} (T_i - T_o) \tag{3.99}$$

where $k_m$ is the conductivity at the mean temperature $\left(\frac{T_i + T_o}{2}\right)$.

This is the same equation as Eq. (3.55) derived above for the case of constant conductivity. The constant conductivity $k$ has just been replaced with $k_m$.

## Example 3.7
### Plate With Temperature-Dependent Conductivity

*Problem:*
A plate of pure copper is 15 cm by 20 and 1 cm thick. One side of the plate is at 200 C and the other side is at 600 C. Determine the heat flow rate through the plate (W). Consider the temperature dependence of the thermal conductivity.

*Solution:*
The plate is like the wall of Fig. 3.22. The cross-sectional area $A$ is $0.15 \times 0.20$ m $= 0.03$ m$^2$. The thickness L is 0.01 m. From an Internet search, the conductivity of the copper varies with temperature as shown in Table 3.1. Fig. 3.24 is an Excel chart of the data of Table 3.1.

| Table 3.1 Conductivity of pure copper. | |
| --- | --- |
| **T (K)** | **k (W/m K)** |
| 400 | 392 |
| 600 | 383 |
| 800 | 371 |
| 1000 | 357 |
| 1200 | 342 |

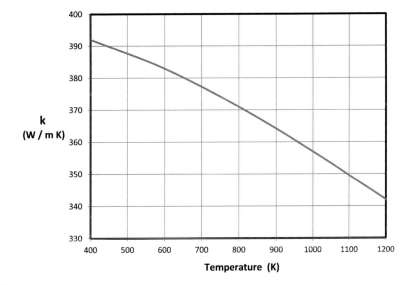

**FIGURE 3.24**

Conductivity of pure copper.

At first glance, the figure appears to show that the conductivity varies significantly with temperature. However, the figure is misleading due to the scaling on the vertical axis. If the vertical axis had started at zero, then the figure would have shown a much more gradual variation of conductivity with temperature. This is shown in Fig. 3.25.

Indeed, the conductivity varies only 13.6% over the entire 800 K temperature range. And, the temperature relationship is quite linear. Using Excel, the linear trendline fit of Table 3.1 data is $k = 419.4 - 0.063\ T$ with an $R^2$ of 0.991. (Note: A second-degree polynomial fit is even better. It is $k = 0.000025\ T^2 - 0.023\ T + 405.4$ with an $R^2$ of 0.998.) The units in these equations are $k$ (W/m K) and $T$ (K).

If we assume that $k$ varies linearly with $T$, then we can use Eq. (3.98):

$$q = \frac{k_m A}{L}(T_1 - T_2) \tag{3.98}$$

For this problem, $k_m$ is $k$ at the mean temperature $(200 + 600)/2 = 400\ C = 673\ K$.

Using the Excel linear trendline equation $k = 419.4 - 0.063T$, we have

$k_m = 419.4 - 0.063\ (673) = 377\ W\ /\ m\ K$.　From　Eq.　(3.98),　the　heat　flow　through　the　wall　is

$q = \frac{377\ (0.03)}{0.01}\ (600 - 200)\quad = \quad 4.524\ x\ 10^5\ W.$

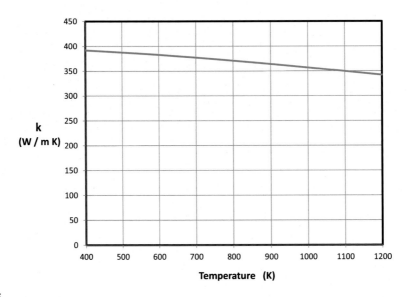

**FIGURE 3.25**

Conductivity of pure copper (revised scale).

Alternatively, we could have used the equation from the second-degree polynomial fit of the conductivity and perform the integration indicated in Eq. (3.91), as follows:

$$q = \frac{A}{L} \int_{T_2}^{T_1} k \, dT$$

$$q = \frac{0.03}{0.01} \int_{200+273}^{600+273} (0.000025T^2 - 0.023T + 405.4) \, dT$$

$$q = \frac{0.03}{0.01} \left[ \frac{0.000025}{3} T^3 - \frac{0.023}{2} T^2 + 405.4T \right]_{473}^{873}$$

$$q = \frac{0.03}{0.01} \left[ \frac{0.000025}{3} (873^3 - 473^3) - \frac{0.023}{2} (873^2 - 473^2) + 405.4(873 - 473) \right]$$

$$q = 4.819 \times 10^5 \text{ W}.$$

The difference between heat flows using these two data fits is only 6%.

An even quicker solution would be to merely look at Fig. 3.24 and use the $k$ value at the mean temperature of $400 \, C = 673 \, K$. Eyeballing the figure, we get that $k$ is about 377 W/m K. Hence,

$$q = \frac{kA}{L}(T_1 - T_2) = \frac{377(0.03)}{0.01}(400) = 4.52 \times 10^5 \text{ W}.$$

The closeness of these three results for $q$ is due to pure copper's small variation of conductivity with temperature and the essentially linear characteristic of the $k$–$T$ function.

Let us next consider a problem involving a material having significant temperature-dependent conductivity.

## Example 3.8
### Cylinder With Temperature-Dependent Conductivity

*Problem:*
A long aluminum oxide cylinder has an inner diameter of 1 cm and an outer diameter of 2 cm. The temperature of the inner surface is 500 C and that of the outer surface is 100 C. Determine the heat flow rate through the cylinder's wall per meter length of the cylinder. Consider the temperature dependence of the thermal conductivity.

*Solution:*
Consider Fig. 3.23 above. For this problem, $r_i = 0.005$ m, $r_o = 0.01$ m, $T_i = 500$ C $= 773$ K, and $T_o = 100$ C $= 373$ K. From an Internet search, the conductivity of the aluminum oxide varies with temperature as shown in Table 3.2. Fig. 3.26 is an Excel chart of the data of Table 3.2.

It is seen that the conductivity of aluminum oxide varies significantly with temperature and is quite nonlinear.

Let us look again at Fig. 3.23 and consider conduction through a cylindrical shell having variable thermal conductivity. The heat conduction for our problem is radially outward. The heat flow at location $r = r$ is $q = -k\,A\,\frac{dT}{dr}$ and the area through which heat is flowing at this location is $A = 2\pi r L$. Hence, at $r = r$,

$$q = -k(2\pi r L)\frac{dT}{dr} \tag{3.100}$$

Rearranging Eq. (3.100) and integrating between the inner and outer surfaces of the cylinder, we have

$$q \int_{r_i}^{r_o} \frac{dr}{r} = -2\pi L \int_{T_i}^{T_o} k\,dT \tag{3.101}$$

Note: $q$ does not vary with $r$. From heat flow continuity, it is the same at all radial locations. We therefore were able to move it outside the left integral in Eq. (3.101).
Continuing the analysis:

$$q \ln\left(\frac{r_o}{r_i}\right) = 2\pi L \int_{T_o}^{T_i} k\,dT \tag{3.102}$$

And finally,

$$q = \frac{2\pi L}{\ln\left(\dfrac{r_o}{r_i}\right)} \int_{T_o}^{T_i} k\,dT \tag{3.103}$$

To perform the integration, we need to know function $k(T)$. It is often possible to accurately fit thermal conductivity data to a polynomial of second, third, or, at most, fourth degree. Using Excel, it is found that the following third degree polynomial is an excellent fit to the data of Table 3.2:

$$k = -4 \times 10^{-8} T^3 + 0.0001\ T^2 - 0.1537\ T + 69.052 \tag{3.104}$$

### Table 3.2 Conductivity of aluminum oxide.

| T (K) | k (W/m K) |
|-------|-----------|
| 400 | 26 |
| 600 | 15.2 |
| 800 | 10 |
| 1000 | 8 |
| 1200 | 6.5 |
| 1400 | 5.8 |

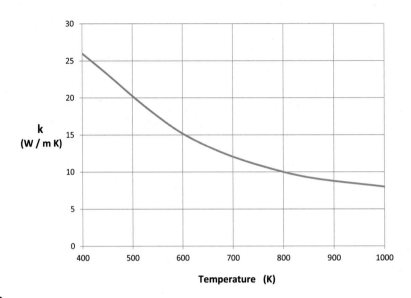

**FIGURE 3.26**

Conductivity of aluminum oxide.

Inserting this function and other information into Eq. (3.103), we have

$$\frac{q}{L} = \frac{2\pi}{\ln\left(\dfrac{0.01}{0.005}\right)} \int_{373}^{773} \left(-4 \times 10^{-8}\ T^3 + 0.0001\ T^2 - 0.1537\ T + 69.052\right) dT$$

Performing the integration and calculation, we have the result

$$\frac{q}{L} = 5.32 \times 10^4\ \text{W/m}$$

It is always good to check a final answer, especially if the calculation involves several steps as in the above integration. So, let us look at Fig. 3.26 and get an average value of $k$ for this problem. The average temperature of the two surfaces is $(373 + 773)/2 = 573$ K. Looking at Fig. 3.26, $k$ is about 16 W/m K for this temperature. If we use an average $k$ value, Eq. (3.103) becomes

$$\frac{q}{L} = \frac{2\pi}{\ln\left(\dfrac{r_o}{r_i}\right)} k_{avg}(T_i - T_o) \tag{3.105}$$

Putting in values for our problem, we have

$$\frac{q}{L} = \frac{2\pi}{\ln(2)}(16)(773 - 373) = 5.80 \times 10^4\quad \text{W/m.}$$

This approximate solution is only 10% different from our earlier value of $5.32 \times 10^4$ so we are quite confident that no calculation errors were made in our earlier solution.

## 3.6 Multi-dimensional conduction

The above sections discussed one-dimensional conduction. Multi-dimensional problems (i.e., two and three dimensions) can be solved using a variety of methods: classical analytical techniques, conduction shape factor, numerical methods, or packaged software developed specifically for such problems.

Classical analytical techniques for solving the partial differential equations of heat transfer have been in existence for many years. Several books have been written on such solutions, including those by Carslaw and Jaeger [1], Schneider [2], Myers [3], and Arpaci [4]. Analytical solutions were very important and necessary before the advent of calculators and computers. They are also important for research and mathematical investigations today. However, analytical solutions are limited in scope and applicability and are often very complex. It is felt that the conduction shape factor and numerical methods are much more useful for students and practicing engineers. Therefore, analytical techniques such as separation of variables are not covered in this book. Shape factors are discussed in the next section, and numerical methods in Chapter 5.

Also, several companies have developed software packages for engineering design and heat transfer applications. These include COMSOL [5], Solidworks [6], Ansys [7], and CD-Adapco (acquired by Siemens) [8].

## 3.7 Conduction shape factors

The conduction shape factor is very useful in solving specialized two- and three-dimensional problems involving two temperatures. Examples of applicability include heat transfer from buried pipes and heat transfer through rectangular enclosures such as kilns.

The heat transfer equation for this method is

$$q = kS(T_1 - T_2) \tag{3.106}$$

where $q$ is the rate of heat transfer between the two surfaces at $T_1$ and $T_2$

$k$ is the thermal conductivity of the medium between $T_1$ and $T_2$

$S$ is the conduction shape factor.

From Eq. (3.106), the resistance for shape factors is $\frac{1}{kS}$. This resistance can be used in solving problems by means of the resistance concept. It is particularly useful for problems involving multi-layered objects or surfaces with convection.

Some of the more useful conduction shape factors are given in Table 3.3. More shape factors can be obtained from Refs. [9,10] and [11].

**Table 3.3 Conduction shape factors "S."**

**Shape Factor (1) Cylinder buried in a semi-infinite medium**

$$S = \frac{2\pi L}{\cosh^{-1}\left(\dfrac{2d}{D}\right)} \quad \text{if } L \gg D$$

$$S = \frac{2\pi L}{\ln\left(\dfrac{4d}{D}\right)} \quad \text{if } L \gg D \text{ and } d > 1.5D$$

**Shape Factor (2) Two Parallel Cylinders in an infinite medium**

$$S = \frac{2\pi L}{\cosh^{-1}\left(\dfrac{4w^2 - D_1^2 - D_2^2}{2D_1 D_2}\right)} \quad \text{if } L \gg w,\ L \gg D_1 \text{ and } L \gg D_2$$

**Shape Factor (3) Equally-Spaced Parallel Cylinders buried in a semi-infinite medium**

$$S(\text{per cylinder}) = \frac{2\pi L}{\ln\left(\dfrac{2w}{\pi D}\sinh\left(\dfrac{2\pi d}{w}\right)\right)} \quad \text{if } L \gg d,\ L \gg D, \text{ and } w > 1.5D$$

**Shape Factor (4) Vertical Cylinder buried in a semi-infinite medium**

$$S = \frac{2\pi L}{\ln\left(\dfrac{4L}{D}\right)} \quad \text{if } L \gg D$$

**Shape Factor (5) Cylinder at center of a square solid bar**

$$S = \frac{2\pi L}{\ln\left(\frac{1.08w}{D}\right)}$$

**Shape Factor (6) Sphere buried in a semi-infinite medium**

$$S = \frac{2\pi D}{1 - \frac{D}{4d}}$$

**Shape Factor (7) Square Flow Channel**

$$S = \frac{2\pi L}{0.785 \ln\left(\frac{a}{b}\right)} \quad \text{if } \frac{a}{b} < 1.4$$

$$S = \frac{2\pi L}{0.93 \ln\left(0.948\frac{a}{b}\right)} \quad \text{if } \frac{a}{b} > 1.4$$

**Shape Factor (8) Large Plane Wall**

$$S = \frac{A}{L}$$

*Continued*

## Table 3.3 Conduction shape factors "S."—cont'd

**Shape Factor (9) Edge of Two Adjoining Walls**

$S = 0.54w$

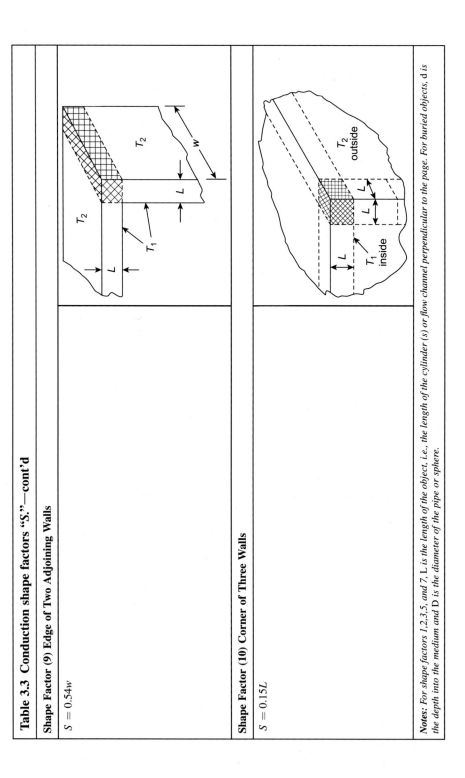

**Shape Factor (10) Corner of Three Walls**

$S = 0.15L$

*Notes: For shape factors 1,2,3,5, and 7, L is the length of the object, i.e., the length of the cylinder (s) or flow channel perpendicular to the page. For buried objects, d is the depth into the medium and D is the diameter of the pipe or sphere.*

## Example 3.9
### Heat Flow Through a Kiln Wall

*Problem:*
A rectangular electric kiln for pottery has a maximum operating temperature of 1000 C. Its inside dimensions are 60 cm wide by 50 cm deep by 30 cm high. The walls of the kiln are 8 cm thick and of refractory firebrick having a conductivity of 1.2 W/m C. During maximum operation, the inner surface of the kiln is at 700 C and the outer surface is at 375 C.
**(a)** Determine the heat flow through the wall of the kiln (W) for the kiln at its maximum operating level
**(b)** The factory has 240V single-phase electric service. What is the maximum current draw of the kiln (A)? What size wiring would you use for the kiln? Finally, what size circuit breaker would you use in the electric panel?

*Solution:*
**(a)** If the kiln were spherical, we could use the earlier derived equation for heat flow through a spherical shell. However, the kiln is rectangular so we have to take into account heat flow at edges and corners of the kiln. The conduction shape factor technique is excellent for this problem. Indeed, there are shape factors in Table 3.3 which cover our situation; namely shape factors no. 8 (wall), 9 (edge), and 10 (corner). The kiln has 6 plane walls, 12 edges, and 8 corners. From Table 3.3, the shape factor calculations are

$$S_{walls} = 2(60 \times 50)/8 \ + \ 2(60 \times 30)/8 \ + \ 2(50 \times 30)/8 = 1575 \text{ cm}$$

$$S_{edges} = \ 4 \ (0.54 \times 60) + 4(0.54 \times 50) + 4(0.54 \times 30) = 302.4 \ \text{ cm}$$

$$S_{corners} = \ 8 \ (0.15) \ (8) \ = \ 9.6 \text{ cm}$$

$$S_{total} = \ 1575 \ + \ 302.4 \ + \ 9.6 \ = \ 1887 \text{ cm} \ = \ 18.87 \text{ m}$$

The heat flow rate through the kiln's wall is

$$q = kS_{total}\Delta T = 1.2(18.87)(700 - 375) = 7360 \quad \text{W}$$

Although the shape factor technique is one way to model this problem, there are other possible ways. For example, let us replace the rectangular wall of the kiln with a spherical wall and use the equation for conduction through a spherical wall. The inner surface area of the kiln is 2 (60 × 50) + 2 (60 × 30) = 2 (50 × 30) = 12,600 cm$^2$.
The surface area of a sphere is $4\pi r^2$, so this inner surface area of 1.26 m$^2$ corresponds to a spherical surface with a radius of 0.3167 m.
The thickness of the kiln wall is 8 cm. Therefore, let us consider conduction through a spherical wall of inner radius 0.3167 m and outer radius 0.3167 + 0.08 = 0.3967 m.
Conduction through a spherical wall is given by Eq. (3.73).

$$q = \frac{4\pi k}{\left(\dfrac{1}{r_i} - \dfrac{1}{r_o}\right)} (T_i - T_o) \tag{3.73}$$

Putting in the values for our problem, we have

$$q = \frac{4\pi(1.2)}{\left(\dfrac{1}{0.3167} - \dfrac{1}{0.3967}\right)} (700 - 375) = 7700 \quad \text{W}$$

There is only a 4.6% difference from the shape factor result.
**(b)** Power = Voltage × Current for single-phase service. For 240 V single-phase service, the 7360 W result from the shape factor technique corresponds to a current of 7360/240 = 30.67 A. From the Internet, AWG #10 copper wire has an ampacity of 30 A, AWG #8 wire has 40 A, and #6 has 55 A. Therefore, #8 AWG wire should be used. Finally, a 40 A circuit breaker would be appropriate for the circuit.

   The conduction shape factor is very useful for calculating heat transfer from buried pipes. The following example illustrates this situation.

## Example 3.10
### Heat Loss from a Buried Pipe

*Problem:*
This problem is based on a study made by the author of heat loss from a buried hot water pipe at a dog kennel in New York. Water at 190 F flows at a rate of 5 gpm through a buried one-inch Schedule 40 steel pipe from one building to another. The pipe is horizontal and uninsulated. It is 60 feet long and is buried at a depth of 2 feet. The soil has a conductivity of 0.833 BTU/h ft F, and the ground surface temperature is 35 F.
Determine the rate of heat loss from the pipe (BTU/h).

*Solution:*
The outside diameter of 1 inch Sch 40 pipe is 1.315 inch. We will assume that the outer surface of the pipe is at the water temperature of 190 F. This is a reasonable assumption as the pipe is metal, relatively thin-walled, and the convective coefficient at the water/pipe interface is quite large.
The appropriate shape factor is #1 from Table 3.3.

$$S = \frac{2\pi L}{\ln\left(\frac{4d}{D}\right)} \tag{3.107}$$

Putting in the values for this problem, we have

$$S = \frac{2\pi(60)}{\ln\left(\frac{4 \times 2}{1.315/\ 12}\right)} = 87.87 \ \text{ft}$$

(Note: We could have used the other equation in Table 3.3 to get the shape factor. The value for $S$ would have been the same.)
The rate of heat loss from the pipe is

$$q = kS\Delta T = (0.833)\ (87.87)\ (190 - 35) = 11340 \quad \text{Btu/h}$$

(Note: The flow rate of 5 gpm was extraneous information not used in the solution.)

## 3.8 Extended surfaces (fins)

This section gives an introduction to the topic of extended surfaces or *fins*. Much research has been done on this topic, and references [12−14] may be consulted for more detailed information.

As shown in Chapter 1, the heat transfer rate is proportional to heat flow area for all three modes of heat transfer. One way of increasing heat transfer from a surface is to add extended surfaces (or "fins") to the surface. This increases the heat transfer area and, in turn, the rate of heat transfer.

Fins are used on a variety of devices. Some applications include coils for air conditioning and refrigeration; baseboard heaters for residential and commercial buildings; heat sinks for electronic devices, car radiators, and engine and air compressor cylinders. Some examples of these are shown in Fig. 3.27:

Let us first consider fins of constant cross-sectional area. We will derive the governing differential equation for the temperature distribution in the fin and discuss the appropriate boundary condition equations. We will see how the addition of fins enhances the heat transfer from a surface. Common fins of varying cross-sectional area will then be discussed. Examples will be given for a single fin on a surface and also for multiple fins on a surface.

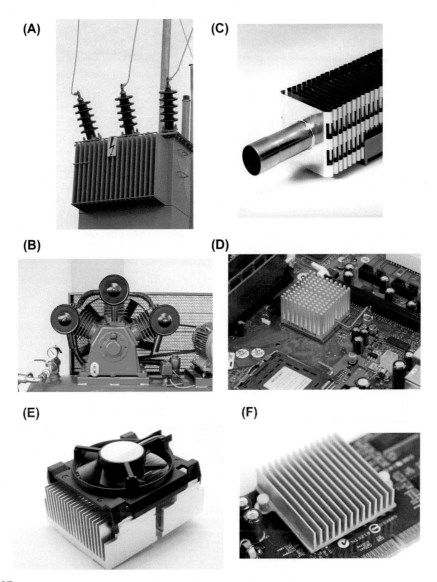

**FIGURE 3.27**

Some devices with fins. (A) Power transformer with straight fins. (B) Reciprocating air compressor with circumferential fins. (C) Baseboard heating tube with circumferential fins. (D) Heat sink with pin fins on a circuit board. (E) Heat sink with fan and straight fins. (F) Heat sink for computer video card. Many thanks to Slant Fin Corporation of Greenvale, NY for providing Figure 3.27 (c).

### 3.8.1 Fins of constant cross section

#### 3.8.1.1 The governing differential equation and boundary conditions

Let us consider the fin of square cross section shown in Fig. 3.28. The cross section is of dimension $a$ x $a$ and the fin has a length $L$. It is attached to a surface that is at temperature $T_s$. The surrounding fluid is at temperature $T_\infty$ and the convective coefficient at the side surfaces and end of the fin is $h$. Finally, the heat transfer is steady state.

Our analytical model is one-dimensional. This assumes that the temperature in the fin only depends on $x$, the distance from the surface on which the fin is attached. This 1D model is usually very acceptable as fins are normally thin and of high thermal conductivity.

To derive the equation for the temperature variation $T(x)$ of the fin, let us consider the heat flow in and out of a differential vertical slice of the fin. The slice (or "celement") is from $x=x$ to $x=x+dx$ and has a thickness $dx$. There is conduction $q_x$ entering the left side of the element and conduction $q_{x+dx}$ leaving the right side. There is also convection $q_{conv}$ from the element to the adjacent fluid. For steady state, the rate of heat flow into the element equals the rate of heat flow out of the element or

$$q_x = q_{x+dx} + q_{conv} \qquad (3.108)$$

$$\text{However,} \quad q_x = -kA\frac{dT}{dx} \qquad (3.109)$$

$$q_{x+dx} = q_x + \frac{dq_x}{dx}\ dx \qquad (3.110)$$

$$\text{and} \quad q_{conv} = hA(T - T_\infty) \qquad (3.111)$$

Note that the $A$'s in Eqs. (3.109) and (3.111) are different. In Eqs. (3.109), $A$ is the area for conduction heat transfer or $a^2$ for our square fin. In Eqs. (3.111), $A$ is the surface area for convection or $P\ dx$, where $P$ is the perimeter around the fin. For our square fin of side "$a$," the perimeter is $4\ a$.

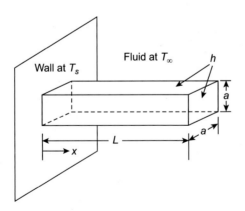

**FIGURE 3.28**

Single fin of square cross section.

Putting Eqs. (3.109–3.111) into the main Eq. (3.108) and rearranging, we have

$$\frac{d}{dx}\left(kA\frac{dT}{dx}\right) - hP(T - T_\infty) = 0 \tag{3.112}$$

For our fin, the cross-sectional area is constant. We will also assume that conductivity $k$ is constant. Eq. (3.112) then becomes

$$\frac{d^2T}{dx^2} - \frac{hP}{kA}(T - T_\infty) = 0 \tag{3.113}$$

We will assume that $T_\infty$ is constant. Then Eq. (3.113) can be changed to

$$\frac{d^2(T - T_\infty)}{dx^2} - \frac{hP}{kA}(T - T_\infty) = 0 \tag{3.114}$$

This is the differential equation to be solved for the temperature distribution $T(x)$ in the fin. The general solution for Eq. (3.114) is

$$T(x) - T_\infty = C_1 e^{mx} + C_2 e^{-mx} \tag{3.115}$$

where $m = \sqrt{\frac{hP}{kA}}$.

We need two boundary conditions to determine constants $C_1$ and $C_2$ in Eq. (3.115). The boundaries are at the wall ($x = 0$) and at the end of the fin ($x = L$). The temperature at the wall is $T_s$ so the boundary condition there is

$$\text{At } x=0, \quad T - T_\infty = T_s - T_\infty \tag{3.116}$$

There is convection from the fin to the fluid at the $x = L$ end of the fin. At this location, the heat flow by conduction equals the heat flow by convection. The boundary condition is therefore

$$\text{At } x=L, \quad -kA\frac{dT}{dx} = hA(T - T_\infty)$$

The $A$'s are the same for conduction and convection at the end of the fin, and the fluid temperature $T_\infty$ is constant. Hence, the boundary condition at the end of the fin is

$$\text{At } x=L, \quad \frac{d(T - T_\infty)}{dx} = -\frac{h}{k}(T - T_\infty) \tag{3.117}$$

### 3.8.1.2 The solution for temperature distribution and heat flow

It can be shown that the solution of differential Eq. (3.114) with boundary condition Eqs. (3.116) and (3.117) is

$$T(x) - T_\infty = (T_s - T_\infty)\frac{\cosh[m(L - x)] + \left(\frac{h}{mk}\right)\sinh[m(L - x)]}{\cosh(mL) + \left(\frac{h}{mk}\right)\sinh(mL)} \tag{3.118}$$

where $m = \sqrt{\frac{hP}{kA}}$.

Eq. (3.118) gives the temperature distribution along the total length of the fin, i.e., from $x = 0$ to $x = L$. However, the objective in most situations is to determine the heat transfer from the fin to the fluid. This can be done in two different ways. One way involves differentiation of the $T(x)$ function of Eq. (3.118). The other way involves integration of the $T(x)$ function.

For the first way, we consider the heat transfer into the fin at location $x = 0$. This is conduction heat transfer and is given by

$$q = -kA \left(\frac{dT}{dx}\right)_{x=0} \tag{3.119}$$

The temperature function of Eq. (3.118) is differentiated with respect to $x$ and the result is evaluated at $x = 0$. The result is then used in Eq. (3.119) to obtain the heat flow into the fin.

In the second way, we look at the convection heat transfer from the fin's surfaces into the fluid. In so doing, we integrate the function of Eq. (3.118) to get the heat transfer from the sides of the fin into the fluid and use the temperature at the $x = L$ end of the fin to get the heat transfer from the end of the fin into the fluid. The total convection heat transfer to the fluid is given by

$$q = \int_{0}^{L} hP(T - T_\infty)dx + hA(T_{x=L} - T_\infty) \tag{3.120}$$

As we have steady-state conditions, the heat flow into the fin at $x = 0$ equals the heat flow into the fluid by convection from the fin's surfaces. That is, the $q$'s in Eqs. (3.119) and (3.120) are equal. Regardless of whether the first or second approach is used to get the heat flow, the result is

$$q = \sqrt{h \, P \, k \, A} \ (T_s - T_\infty) \ \frac{\sinh(mL) + \left(\dfrac{h}{mk}\right)\cosh(mL)}{\cosh(mL) + \left(\dfrac{h}{mk}\right)\sinh(mL)} \tag{3.121}$$

where $m = \sqrt{\frac{hP}{kA}}$.

The equations for the temperature distribution in the fin (Eq. 3.118) and for the heat flow from the fin (Eq. 3.121) involve hyperbolic trigonometric functions. These functions are readily available on scientific calculators and in software packages. However, such was not always the case. In years past, evaluation of the equations was very tedious and time-consuming as slide rules and tables had to be used to evaluate the exponential and trigonometric functions. To speed up calculations, two frequently used approximations were developed. These are discussed in the next two sections. Indeed, they will speed up hand calculations. However, one must keep in mind that they are approximations and should be used only in situations consistent with the assumptions made in their derivation.

### 3.8.1.3 Very-long-fin approximation
The first approximation assumes that the fin is very long. This means that the temperature at the end of the fin approaches the fluid temperature. Mathematically, the boundary condition at the end of the fin is

$$\text{As} \quad x \rightarrow \infty, T \rightarrow T_\infty \tag{3.122}$$

The previous differential equation (Eq. 3.114) and boundary condition equation at the base (Eq. 3.116) remain applicable. The solution for the temperature distribution in the very long fin is

$$T(x) - T_\infty = (T_s - T_\infty)e^{-mx} \quad \text{where } m = \sqrt{\frac{hP}{kA}} \tag{3.123}$$

and the solution for the heat transfer to the fluid is

$$q = \sqrt{hPkA}(T_s - T_\infty) \tag{3.124}$$

### 3.8.1.4 Insulated-at-end fin approximation

The other approximation assumes that there is negligible heat transfer from the tip of the fin. This is often an excellent approximation as most fins have a lateral area much larger than the tip area. With no heat flow from the tip, the boundary condition equation at the end of the fin is

$$q = 0 = -kA\frac{dT}{dx}$$

which simplifies to

$$\text{At } x = L, \quad \frac{dT}{dx} = 0 \tag{3.125}$$

The differential equation for the temperature distribution in the fin (Eq. 3.114) and boundary condition equation at the base (Eq. 3.116) remain applicable. The solution for the temperature distribution in the insulated-at-end fin is

$$T(x) - T_\infty = (T_s - T_\infty)\frac{\cosh[m(L - x)]}{\cosh(mL)} \tag{3.126}$$

and the solution for the heat transfer to the fluid is

$$q = \sqrt{hPkA}(T_s - T_\infty)\tanh(mL) \tag{3.127}$$

$$\text{As before, } m = \sqrt{\frac{hP}{kA}}$$

As mentioned above, this approximation often gives excellent results. To make the results even better, the lateral area can be increased to compensate for the neglected heat transfer area at the tip. The lateral area is increased by adding an amount equal to $A/P$ to the length of the fin. This "corrected" length $L_c$ is used instead of $L$ in Eqs. (3.126) and (3.127). For example, consider a pin fin (i.e., a cylindrical fin of circular cross section) of length $L$ and diameter $D$. The corrected length is

$$L_c = L + A/P = L + \frac{\pi D^2/4}{\pi D} = L + \frac{D}{4}$$

Similarly, a fin of square cross section of side $a$ will have a corrected length

$$L_c = L + A/P = L + \frac{a^2}{4a} = L + \frac{a}{4}$$

Finally, a thin fin of rectangular cross section with width $w$ and thickness $t$ with $t << w$ will have a corrected length $L_c = L + \frac{t}{2}$.

## 3.8.2 Fin efficiency

Fin Efficiency is the ratio of the heat transferred by a fin to the heat that would have been transferred if the total fin surface were at the base (or wall) temperature. That is,

$$\eta_f = \frac{q_{fin}}{\text{heat transferred if entire fin surface were at } T_s} \tag{3.128}$$

This definition is very useful. It is easy to calculate the denominator in the definition. Then, if the efficiency of a fin is known, the heat transferred by the fin can readily be determined. As shown below in Section 3.8.4, fin efficiency graphs are available for several common types of fins.

## 3.8.3 Fin effectiveness

Fin Effectiveness indicates how successful the addition of a fin to a surface has been in enhancing the heat transfer from the surface. It is the ratio of the heat transferred by the fin to the heat transfer from the base area covered by the fin if the fin had not been attached. If $A_b$ is the area of fin attachment, then the effectiveness is

$$\varepsilon_{fin} = \frac{q_{fin}}{hA_b(T_s - T_\infty)} \tag{3.129}$$

Determination of fin efficiency and fin effectiveness is illustrated in Example 3.11.

## Example 3.11
### Heat Transfer from a Pin Fin

*Problem:*
A pin fin ($k = 250$ W/m C) is attached to a wall. The fin's diameter is 1.5 cm and it is 10 cm long. The wall is at 200 C. The surrounding air is at 15 C and the convective coefficient is 9 W/m² C. (Note: The fin is like that shown in Fig. 3.28 except the cross section is circular rather than square.)
**(a)** Determine the rate of heat transfer from the fin to the surrounding air.
**(b)** Compare the result of part (a) with the result from the very-long-fin approximation.
**(c)** Compare the result of part (a) with the insulated-at-end approximation.
**(d)** Determine the efficiency of the fin.
**(e)** Determine the effectiveness of the fin.

*Solution:*
**(a)** The heat transfer from the fin is given by Eq. (3.121).

$$q = \sqrt{hPkA}(T_s - T_\infty) \frac{\sinh(mL) + \left(\dfrac{h}{mk}\right)\cosh(mL)}{\cosh(mL) + \left(\dfrac{h}{mk}\right)\sinh(mL)} \tag{3.121}$$

where $m = \sqrt{\frac{hP}{kA}}$.
For our problem, $P = \pi D = \pi(0.015) = 0.0471$ m.

$$A = \pi D^2/4 = \pi(0.015)^2/4 = 1.767 \times 10^{-4}\,\text{m}^2$$

$$m = \sqrt{\frac{hP}{kA}} = \sqrt{\frac{(9)(0.0471)}{(250)(1.767 \times 10^{-4})}} = 3.098$$

$$mL = (3.098)(0.1) = 0.3098$$

$$\sqrt{hPkA} = \sqrt{(9)(0.0471)(250)(1.767 \times 10^{-4})} = 0.1368$$

$$h/mk = \frac{9}{(3.098)(250)} = 0.01162$$

Putting this information into Eq. (3.121), we have

$$q = (0.1368)(200 - 15)\frac{\sinh(0.3098) + (0.01162)[\cosh(0.3098)]}{\cosh(0.3098) + (0.01162)[\sinh(0.3098)]} = 7.87 \text{ W}$$

**(b)** The heat transfer using the very-long-fin approximation is given by Eq. (3.124).

$$q = \sqrt{hPkA}(T_s - T_\infty) \tag{3.124}$$

Therefore, $q = (0.1368)(200 - 15) = 25.3$ W.

Comparing this result with that of Part (a), it is seen that the very-long-fin approximation is totally inappropriate for this fin. The fin is much too short to apply the very-long-fin approximation. Indeed, the fin would have to be 49 cm long before the very-long-fin approximation gives a heat transfer within 10% of the actual heat transfer rate of the fin.

**(c)** The heat transfer using the insulated-at-end approximation is given by Eq. (3.127).

$$q = \sqrt{hPkA}(T_s - T_\infty)\tanh(mL) \tag{3.127}$$

Therefore $q = (0.1368)(200 - 15)\tanh(0.3098) = 7.60$ W.

It is seen that the insulated-at-end fin approximation gives very good results. The heat transfer is 7.60 W, which is only 3.4% different from the 7.87 W result of Part (a). Let's apply the corrected-length concept and see what the result is. For our fin, the corrected length is

$$L_c = L + \frac{D}{4} = 0.1 + 0.015/4 = 0.10375 \text{ m}$$

Using the corrected length in Eq. (3.127) gives

$$q = (0.1368)(200 - 15)\tanh[(3.098)(0.10375)] = 7.87 \text{ W}$$

This is the same result that we got for Part (a). We conclude that using the insulated-at-end approximation with the corrected length technique gives excellent results.

**(d)** The fin efficiency is given by Eq. (3.128).

$$\eta_f = \frac{q_{fin}}{\text{heat transferred if entire fin surface were at } T_s} \tag{3.128}$$

$$\eta_f = \frac{q_{fin}}{h[(\pi DL) + A](T_s - T_\infty)} = \frac{7.87}{(9)[\pi(0.015)(0.1) + 1.767 \times 10^{-4}](200 - 15)} = 0.967$$

**(e)** The fin effectiveness is given by Eq. (3.129).

$$\varepsilon_{fin} = \frac{q_{fin}}{hA_b(T_s - T_\infty)} \tag{3.129}$$

$$\varepsilon_{fin} = \frac{7.87}{(9)(1.767 \times 10^{-4})(200 - 15)} = 26.8$$

It is seen that the fin is very effective. Heat transfer from the bare, unfinned area has increased by a factor of 26.8.

### 3.8.4 Fins of varying cross section

Section 3.8.1 discussed fins of constant cross-sectional area. Example 3.11 also dealt with a fin of constant cross-sectional area. However, some types of fins have cross-sectional areas that vary along their length. We discuss three types of these fins: circumferential fins, triangular straight fins, and conical pin fins.

#### 3.8.4.1 Circumferential fins

Circumferential (or "annular") fins are often used for cooling of engine cylinders and for enhancement of heat transfer from heating pipes in buildings. Examples of circumferential fins are in Fig. 3.27 above.

A circumferential fin is shown in Fig. 3.29. Its efficiency is given by Eq. (3.130).

$$\eta_f = \frac{2r_1}{m(r_{2c}^2 - r_1^2)} \frac{K_1(mr_1)I_1(mr_{2c}) - I_1(mr_1)K_1(mr_{2c})}{I_0(mr_1)K_1(mr_{2c}) + K_0(mr_1)I_1(mr_{2c})} \tag{3.130}$$

where : $m = \sqrt{\frac{2h}{kt}}$ and $r_{2c} = r_2 + \frac{t}{2}$

Functions $I$ and $K$ in Eq. (3.130) are the modified Bessel functions. The functions are available in software such as Excel and Matlab, and a table of the functions is in Appendix H.

The fin efficiency is also shown in Figs. 3.30 and 3.31. Fig. 3.30 gives the overall graph. It is seen that the upper left corner of the graph is hard to read. Thus, we have included a second graph, Fig. 3.31, which expands that portion of the graph. In these graphs, $L_c = L + \frac{t}{2}$ and

$$r_{2c} = r_2 + \frac{t}{2}.$$

After the fin efficiency is determined, the heat transfer from the fin can be determined using Eq. (3.128) above. This is illustrated in Example 3.12.

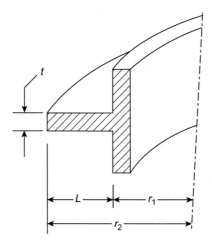

**FIGURE 3.29**

Circumferential fin.

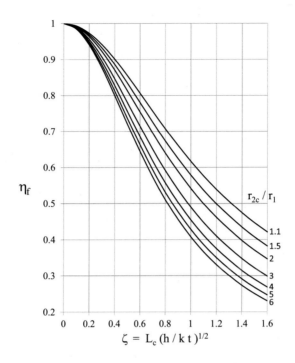

**FIGURE 3.30**

Efficiency of circumferential fins.

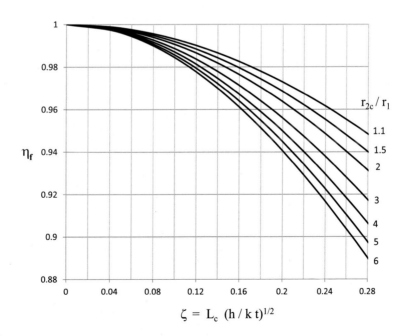

**FIGURE 3.31**

Efficiency of circumferential fins (expanded portion).

## Example 3.12
### Heat Transfer For a Baseboard Heater

*Problem:*
The finned tube for a baseboard heater is shown in Fig. 3.32. It consists of one-inch Type L copper tubing with square circumferential fins. The fins are 4.25 inch by 4.25 inch and 0.02 inch thick. There are 40 fins per foot length of tubing. The conductivity of the fins is 250 W/m C and the room air is at 65 F. The water flowing through the heater is at 180 F. It may be assumed that the convective heat coefficient is 7.5 W/m$^2$ C.
Determine the heat transfer rate from the heater (Btu/h) per foot length of the heater.

*Solution:*
Figs. 3.30 and 3.31 give the fin efficiency for a circular circumferential fin. However, we have square fins rather than cir-cular fins. As an approximation, let us determine the dimensions of a circular fin which has the same heat transfer area as our square fin.
Our fin is 4.25 inch by 4.25 inch. One side of the fin has an area 4.25 × 4.25 = 18.063 in$^2$ minus the area cut by the copper tubing. One-inch copper tubing has an outside diameter of 1.125 in. One side of the fin thus has a heat transfer area of
$A = 18.063$ in$^2$ - $\pi/4$ $(1.125)^2 = 17.069$ in$^2$. Converting to square meters, we have
$A = 17.069$ in$^2$ × (2.54 cm/in × 1 m/100 cm)$^2 = 0.011012$ m$^2$
We need to find the outside diameter for an equivalent circular fin. Looking at Fig. 3.33,

$$D_1 = 1.125 \text{ in} = 1.125 \text{ in} \times \frac{0.0254 \text{ m}}{\text{in}} = 0.028575 \text{ m}$$

The area for one side of the fin is $A = \frac{\pi}{4}\left(D_2^2 - D_1^2\right)$. From above, $A = 0.011012$ m$^2$ and $D_1 = 0.028575$ m.
Therefore, $A = 0.011012 = \frac{\pi}{4}\left(D_2^2 - 0.028575^2\right)$ and $D_2 = 0.1218$ m.
The efficiency of the fin is obtained from Fig. 3.30. The values for this problem are

$$\text{thickness t} = 0.02 \text{ in} = 0.000508 \text{ m}$$

$$\text{inner radius} = r_1 = D_1/2 = 0.014288 \text{ m}$$

$$\text{outer radius} = r_2 = D_2/2 = 0.0609 \text{ m}$$

$$\text{corrected outer radius} = r_{2c} = r_2 + t/2 = 0.061154 \text{ m}$$

$$\text{length} = L = r_2 - r_1 = 0.04661 \text{ m}$$

$$\text{corrected length} = L_c = L + t/2 = 0.04687 \text{ m}$$

$$A_p = L_c t = 0.000023808 \text{ m}^2$$

The different curves on the figure are for different values of $r_{2c}/r_1$. For our problem, $r_{2c}/r_1 = 4.28$. The value for the bottom axis is 0.360.
Going into Fig. 3.30 with these values of $r_{2c}/r_1$ and $\xi$ we see that the fin efficiency $\eta_f$ is 0.82.

**FIGURE 3.32**

Finned tube for baseboard heater (Example 3.12). Many thanks to Slant Fin Corporation of Greenvale, NY for providing this figure.

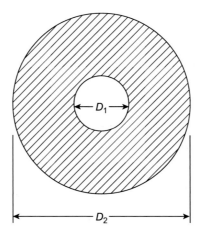

**FIGURE 3.33**

Equivalent circular fin for Example 3.12.

From Eq. (3.128), the heat transferred by a fin is the fin efficiency multiplied by the heat that would have been transferred if the entire fin surface were at the base temperature or

$$q_{fin} = \eta_f [hA_f (T_s - T_\infty)] \tag{3.131}$$

$A_f$ is the total surface area of the fin, i. e., the top and bottom surfaces plus the edge surface.

$$A_f = 2\left[\frac{\pi}{4}(D_2^2 - D_1^2)\right] + t\pi D_2 = 0.022218 \text{ m}^2$$

We will assume that the temperature of the outer surface of the copper tube is essentially the same as the temperature of the water flowing in the tube. Then,

$$T_s - T_\infty = 180 - 65 = 115 \text{ F}^\circ \times \frac{5 \text{ C}^\circ}{9 \text{ F}^\circ} = 63.89 \text{ C}^\circ$$

Returning to Eq. (3.131), we have

$$q_{fin} = (0.82)(7.5)(0.022218)(63.89) = 8.73 \text{ W / fin}$$

From the problem statement, there are 40 fins per foot length of tube. This is equivalent to 131 fins per meter length:

$$40 \text{ fins / ft} \times 3.2808 \text{ ft / m} = 131 \text{ fins/m}$$

The total heat transfer for a one-meter length of finned tube is the heat transfer from the 131 fins plus the heat transfer from the bare tube area between the fins. The bare tube area per meter length is

$$A_{bare} = \pi D_1 L_{bare} = \pi D_1 (1 - 130t) = \pi(0.028575)(1 - 130(0.000508)) = 0.08384 \text{ m}^2$$

So,

$$q_{total} = hA_{bare}(T_s - T_\infty) + (\text{no. of fins})(q_f)$$

$$q_{total} = (7.5)(0.08384)(63.89) + (131)(8.73) = 1184 \text{ W per meter of finned tube}$$

Converting this to Btu/h per foot of finned tube, we have

$$q_{total}/L = 1184 \text{ W/m} \times \frac{1 \text{ m}}{3.2808 \text{ ft}} \times \frac{3.412 \text{ Btu / h}}{1 \text{ W}} = 1230 \text{ Btu / h per foot of finned tube}$$

This example was based on a bare elements heat exchanger from a well-known manufacturer. The company's website gives the heat transfer for this exchanger as 1070 Btu/h per foot length. Our calculated result of 1230 Btu/h per foot is only 15% different from this.

### 3.8.4.2 Straight triangular fins

This section compares the fin efficiency of the variable-area straight triangular fin with the efficiency of the constant-area straight rectangular fin.

A straight rectangular fin is shown in Fig. 3.34. Its efficiency is given by Eq. (3.132).

$$\eta_f = \frac{\tanh(mL_c)}{mL_c} \tag{3.132}$$

$$\text{where: } m = \sqrt{\frac{2h}{kt}} \quad \text{and} \quad L_c = L + \frac{t}{2}$$

A straight triangular fin is shown in Fig. 3.35. Its efficiency is given by Eq. (3.133).

$$\eta_f = \frac{1}{mL} \frac{I_1(2mL)}{I_0(2mL)} \tag{3.133}$$

where $m = \sqrt{\frac{2h}{kt}}$ and $I_0$ and $I_1$ are modified Bessel functions.

The efficiencies of the rectangular and triangular fins are shown in Fig. 3.36.

After the fin efficiency is determined, the heat transfer from the fin can be determined using Eq. (3.128) above.

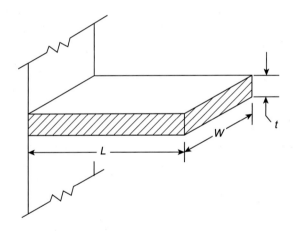

**FIGURE 3.34**

Straight rectangular fin.

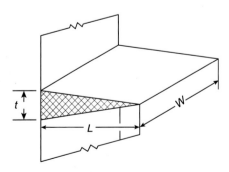

**FIGURE 3.35**

Straight triangular fin.

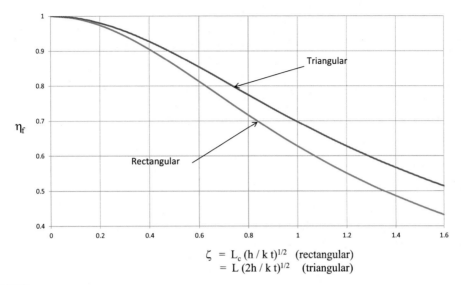

$$\zeta = L_c (h/k t)^{1/2} \quad \text{(rectangular)}$$
$$= L (2h/k t)^{1/2} \quad \text{(triangular)}$$

**FIGURE 3.36**

Efficiency of straight fins.

### 3.8.4.3 Conical pin fins

This section compares the fin efficiency of the variable-area conical pin fin with the efficiency of the constant-area circular pin fin.

A circular pin fin is shown in Fig. 3.37. Its efficiency is given by Eq. (3.134).

$$\eta_f = \frac{\tanh(mL_c)}{mL_c} \tag{3.134}$$

where $m = \sqrt{\dfrac{4h}{kD}}$ and $L_c = L + \dfrac{D}{4}$.

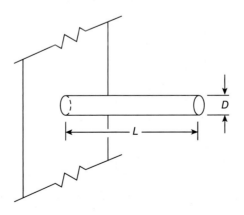

**FIGURE 3.37**

Circular pin fin.

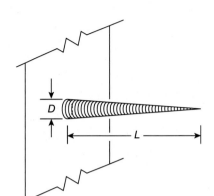

**FIGURE 3.38**

Conical pin fin.

A conical pin fin is shown in Fig. 3.38. Its efficiency is given by Eq. (3.135).

$$\eta_f = \frac{2}{mL} \frac{I_2(2mL)}{I_1(2mL)}$$ (3.135)

where $m = \sqrt{\frac{4h}{kD}}$ and $I_1$ and $I_2$ are modified Bessel functions.

The efficiencies of the circular and conical fins are shown in Fig. 3.39. For the circular fin, $L$ is the corrected length $L_c$.

After the fin efficiency is determined, the heat transfer from the fin can be determined using Eq. (3.128) above.

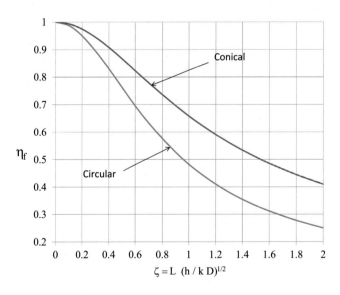

**FIGURE 3.39**

Efficiency of pin fins.

### 3.8.5 Closing comments on fins

Selecting fins to enhance heat transfer is a complex matter. In general, the fins should have high conductivity. Fins are also more effective for smaller values of the convective coefficient. That is, they are more effective in gases than in liquids and for situations of natural convection rather than forced convection. There are many different types of fins and many different materials to choose from. And, in addition to heat transfer capability, there may be size and weight constraints. Selection of fins for a particular application is often an economic decision based on the costs of manufacture and material.

In this Section 3.8, we have considered the enhancement of heat transfer from a surface through the addition of fins. We have derived the governing differential equation for the temperature distribution in a fin and have solved the equation for a fin of constant cross-sectional area. We have also obtained or set forth expressions for the heat transfer from several common types of fins. It should be emphasized that the material in this section is theoretical and has been based on several assumptions. These include

**(1)** The temperatures in the fin and the heat flows in and from the fin are steady state, i.e., they do not change with time.

**(2)** The temperature of the medium surrounding the fin is constant and uniform over all surfaces of the fin.

**(3)** The fin is thin and the heat flow and temperature distribution in the fin is one-dimensional.

**(4)** There is no contact resistance where the fin is attached to the surface.

**(5)** Heat transfer from the fin to the surrounding medium is by convection. There is no radiative heat transfer.

**(6)** The thermal conductivity is homogeneous and is the same at all locations in the fin. The conductivity is also independent of direction.

**(7)** The convective coefficient is the same over all surfaces of the fin and is accurately known.

Some of these assumptions are more appropriate and realistic than others. The worst assumption is the last one. We will see in later chapters that the convective coefficient often depends significantly on surface orientation and surface dimensions. For example, the coefficient is usually quite different for horizontal and vertical surfaces. The only really accurate determination of $h$ is through experimentation on the particular devices and systems involved, during their actual operation. Empirical relations and $h$-values developed through carefully controlled laboratory experiments are often significantly different from results obtained during actual component and system use in practice. In short, the prediction of fin performance depends greatly on the choice of $h$-value. The best approach would be to measure $h$ during actual device operation. If this is not possible, which is often the case, then one should use their best judgment in picking an appropriate $h$ to use in the equations of this chapter.

## 3.9 Chapter summary and final remarks

In this chapter, we covered several steady-state situations in which the temperature distributions and heat flows did not vary with time. These included the geometries of flat plates, cylindrical shells, and spherical shells. We discussed practical multilayered situations such as heat flow through a building's walls and from pipes covered with insulation. We also discussed special cases where thermal conductivity was not constant, but rather a function of temperature, and where a material had internal heat generation, such as a wire carrying electrical current. The conduction shape factor technique was discussed. This technique is very useful in solving problems involving unique geometries such as buried pipes and spheres. And, finally, we provided an introduction to extended surfaces and fins, which are often used in devices to enhance heat transfer from surfaces.

In our discussion of fins, we briefly discussed the difficulty in accurately determining the convective coefficient $h$. For now, values of $h$ are provided as givens for all problems involving convection. In later chapters, we will discuss how to determine $h$ for ourselves.

We now move on to the next chapter and consider situations where the temperature distributions and heat flows vary with time.

## 3.10 Problems

Notes:

- If needed information is not given in the problem statement, then use the Appendix for material properties and other information. If the Appendix does not have sufficient information, then use the Internet or other reference sources. Some good reference sources are mentioned at the end of Chapter 1. If you use the Internet, double-check the validity of the information by using more than one source.
- Your solutions should include a sketch of the problem.

**3-1** The exposed part of a basement wall is 25.4-cm-thick poured concrete ($k = 1.4$ W/m C). The outside air temperature is 5 C and the inside air temperature is 20 C. The outside convective coefficient is 15 W/m$^2$ C and the inside convective coefficient is 5 W/m$^2$ C.
What is the heat flow through the basement wall per square meter of wall surface (W/m$^2$)?

**3-2** Air at 100 C is blowing over one side of a steel plate ($k = 40$ W/m K). The plate is 0.2 m by 0.4 m by 4 mm thick. The convective coefficient at the surface is 70 W/m$^2$ K. There is stagnant air at 30 C on the other side of the plate. For this side of the plate, the convective coefficient is 7 W/m$^2$ C.
**(a)** What is the rate of heat flow (W) through the plate?
**(b)** What is the temperature of the surface of the plate that is in contact with the stagnant air?

**3-3** We wish to determine the heat flow through the ceiling of a 12 ft by 14 ft bedroom. The ceiling has R-38 insulation and 3/4 inch plasterboard ($k = 1.2$ Btu in/h ft$^2$ F). It is summer and the air temperature in the attic is 140 F. The air temperature in the bedroom is maintained at 70 F by the air conditioning system. You may assume that the exposed surface of the insulation is at 140 F and the exposed surface of the plasterboard is at 70 F.
**(a)** What is the heat flow rate (Btu/h) through the bedroom ceiling?
**(b)** In this problem we neglected the convective resistances on the top and bottom surfaces of the ceiling. That is, we assumed that the surface temperatures were the same as the respective air temperatures. If we had included the convective resistances, would this have (i) increased or (ii) decreased the result obtained in Part (a)?

**3-4** A sandwich of four square plates is on a hot plate. The plates are all 15 cm by 15 and 0.4 cm thick. Starting at the hot plate, we have carbon steel ($k = 50$ W/m C), insulating board ($k = 0.02$ W/m C), aluminum alloy ($k = 180$ W/m C), and AISI 304 stainless steel ($k = 18$ W/m C). The hot plate side of the carbon steel is at 800 C. The exposed surface of the stainless steel convects to the 20 C room air with a convective coefficient of 8 W/m$^2$ C.

What is the temperature at the interface between the insulating board and the aluminum? (Note: There is no need to determine the heat flow through the sandwich. We only need the interface temperature.)

**3-5** A 2-inch Schedule 40 steel pipe ($k = 43$ W/m C) has its inner surface kept at 200 C and its outer surface kept at 30 C.

What is the rate of heat conduction through the pipe per meter length of pipe (W/m)?

**3-6** Water at a temperature of 60 C and a velocity of 1 m/s is flowing through a copper pipe. The pipe has an outside diameter of 35 mm and a wall thickness of 1.5 mm. The copper has a conductivity of 400 W/m C, and the convective coefficient at the inner surface of the pipe is 150 W/m² C. Quiet room air at 20 C is on the outside of the pipe. The convective coefficient at this outer surface is 4 W/m² C.

(a) What is the rate of heat flow from the hot water to the room air per meter length of pipe (W/m)?

(b) What are the temperatures of the inner and outer surfaces of the pipe?

**3-7** Foam insulation ($k = 0.039$ W/m C) is put on the water pipe of Problem 3–6. The insulation is 1.27 cm thick.

Assuming that the convective coefficient at the outer surface of the insulation remains at 4 W/m² C, what is the new rate of heat flow?

**3-8** A hollow cylinder has an inner radius $r_1$, an outer radius $r_2$, and conductivity $k$. A heat source inside the cylinder gives a heat flux input of $HF$ (W/m²) to the inner surface. The outer surface of the cylinder convects to the surrounding fluid. The convective coefficient at the outer surface is $h$ and the surrounding fluid is at $T_\infty$.

Determine the expression for the temperature distribution $T(r)$ in the cylinder.

**3-9** A thin-walled cylindrical container is 61 cm diameter and 150 cm high. It is filled with liquid helium (LHe). The temperature of the LHe is 4.2 K. The container is insulated on top, bottom, and sides with 30 layers of multilayer insulation having a total thickness of 1/2 inch. The conductivity of the insulation is $1 \times 10^{-4}$ W/m K. The room air is at 20 C and the convective coefficient at the exposed surface of the insulation is 12 W/m² C. As heat flows into the LHe, it evaporates. The heat of vaporization of LHe is 20.28 kJ/kg and the density of LHe is 125 kg/m³. You may assume that the surfaces of the container are at the LHe temperature of 4.2 K during evaporation of the LHe.

(a) What is the rate of loss of LHe from the container (kg/h)?

(b) How long will it take for all the LHe in the container to be lost?

**3-10** A hollow sphere ($k = 20$ W/m C) has an inside diameter of 6 cm and an outside diameter of 12 cm. The inner surface of the sphere is at 300 C and the outer surface convects to the surrounding 25 C air with a convective coefficient of 15 W/m² C.

(a) What is the rate of heat flow through the wall of the sphere?

(b) If insulation is added to the outer surface of the sphere, what thickness of insulation (k = 0.035 W/m C) is needed to reduce the heat flow of Part (a) by 50%?

**3-11** A thin-walled spherical container of 15 cm diameter is filled with a liquid at 75 C. The convective coefficient between the liquid and the container is 80 W/m² C. The container is covered with two layers of insulation. The first layer is 0.5 cm thick and has a conductivity of 0.01 W/m C. The second layer is 1 cm thick and has a conductivity of 0.04 W/m C. The outer surface of the insulation is exposed to the room air and is essentially at the room air temperature of 22 C.

**(a)** What is the rate of heat flow to the room air (W)?

**(b)** What is the temperature at the interface between the two layers of insulation?

**3-12** Electric current is flowing through an uninsulated #8 AWG copper wire that has a diameter of 3.2636 mm. The temperature of the surface of the wire is 80 C and the convective coefficient at the surface is 5 W/m$^2$ C.

**(a)** If the room air is at 15 C, what is the rate of convective heat transfer to the air per meter length of the wire (W/m)?

**(b)** If the surface of the wire has an emissivity of 0.05, what is the radiative heat transfer to the room walls per meter length of the wire (W/m)? Compare the result of this Part (b) with the result of Part (a) and make a conclusion about the relative magnitudes of the convective and radiative heat flows.

**3-13** A refrigerator has 1/4″ OD copper refrigerant tubing. Some of the tubing is bare and some has a 1/8″ layer of insulation ($k = 0.15$ Btu/h ft F) added to it. It may be assumed that the convective coefficient between the room air and the tubing is 0.9 Btu/h ft$^2$ F for both the bare and the insulated portions of the tubing.

**(a)** Will the addition of insulation to the tubing (a) increase or (b) decrease the heat transfer rate between the tubing and the room air?

**(b)** If the temperature of the outer surface of the bare tubing is 15 F and the room air is 70 F, what is the rate of heat flow to the tubing (W/m length) for both the bare and insulated portions of the tubing?

**3-14** The *critical insulation thickness* concept discussed in Section 3.3 also applies to spheres as well as pipes and wires. Modify the discussion of Section 3.3 to treat a sphere with the addition of insulation. That is, obtain an equation like Eq. (3.75) for the heat transfer from the surface of the sphere to the surrounding fluid. There will be two resistances in the equation: the conductive resistance of the insulation and the convective resistance between the outer surface of the insulation and the surrounding fluid. After the equation for $q$ for the sphere is determined, then differentiate it, set the result to zero, and determine the critical radius for the sphere.

**3-15** A long cylinder of 1.5 cm radius has a conductivity of 5 W/m C and uniform internal heat generation of $8 \times 10^5$ W/m$^3$. The surface of the cylinder is at 50 C.

What is the maximum temperature in the cylinder and where does it occur?

**3-16** An uninsulated #12 AWG copper wire ($k = 385$ W/m C) carries a current of 20 A. The diameter of the wire is 2.052 mm and its resistivity of $1.68 \times 10^{-8}$ Ω · m. The surrounding air is at 22 C, and the convective coefficient at the surface of the wire is 10 W/m$^2$ C.

What is the temperature of the surface of the wire?

**3-17** The wire of Problem 3–16 is now covered with 0.635 mm thick PTFE insulation ($k = 0.25$ W/m C). Like Problem 3–16, a current of 20 A flows through the wire and the surrounding air is at 22 C. We will assume that the addition of insulation has not changed the convective coefficient.

**(a)** What is the temperature at the interface between the wire and the insulation?

**(b)** What is the temperature at the exposed surface of the insulation?

**3-18** A long, hollow cylinder has a uniform internal heat source that generates heat at a rate of $q_{gen}$ per unit volume. The inner radius of the cylinder is $r_i$, the outer radius is $r_o$, and the conductivity is $k$. The inner surface of the cylinder is perfectly insulated, i.e., $q = 0$ at the surface. The outer surface convects to the surrounding fluid. The fluid is at $T_\infty$ and the convective coefficient is $h$.

Determine an expression for the temperature at the outer surface of the cylinder based on the parameters of the problem.

**3-19** The general solution for the temperature distribution in a long cylinder with uniform heat generation is $T(r) = \frac{-q_{gen}r^2}{4k} + C_1 \ln r + C_2$. Consider the radial heat conduction in a hollow cylinder of inner radius $r_1$ and outer radius $r_2$. The cylinder has temperature boundary conditions: At $r = r_1$, $T = T_1$, and at $r = r_2$, $T = T_2$.

**(a)** Determine the expression for the temperature distribution $T(r)$ in the hollow cylinder.

**(b)** Using the result of Part (a), determine the expression for the heat flow $q(r)$ in the cylinder. (Remember that heat flow is $q = -kA\frac{dT}{dr}$ and that $A = 2\pi rL$ for a cylinder.)

**3-20** If the heat generation rate in a hollow cylinder is big enough, there will be a location in the cylinder where the temperature is a maximum.

**(a)** Find the expression for the location of the maximum temperature by differentiating the expression for $T(r)$ obtained in Problem 3.19 (a) with respect to $r$ and setting the result to zero.

**(b)** A hollow cylinder has an inner radius of 2 cm and an outer radius of 8 cm. The cylinder has a conductivity of 8 W/m C and generates heat at a rate of $3 \times 10^6$ W/m$^3$. The inner surface is at 30 C and the outer surface is at 200 C.

Determine the highest temperature in the cylinder and where it occurs.

**(c)** Do the same problem as Part (b) but make the generation rate $3 \times 10^4$ W/m$^3$.

Determine the highest temperature in the cylinder and where it occurs.

**(d)** What do you conclude about the impact of the heat generation rate on the temperature distribution?

**3-21** A hollow cylinder with uniform heat generation has an inner radius of 5 cm and an outer radius of 11 cm. The heat generation rate is $5 \times 10^4$ W/m$^3$. The inner surface is perfectly insulated. That is, $q = 0$ at $r = r_1$. The outer surface has convection to the surrounding fluid. The fluid is at 80 C and the convective coefficient is 15 W/m$^2$ C.

What is the temperature of the outer surface of the cylinder?

**3-22** A hollow cylinder with uniform heat generation has an inner radius of 2 inches and an outer radius of 5 inches. The heat generation rate is 5000 Btu/h ft$^3$. The inner surface is perfectly insulated. That is, $q = 0$ at $r = r_1$. The outer surface has convection to the surrounding fluid. The fluid is at 175 F and the convective coefficient is 2.8 Btu/h ft$^2$ F.

What is the temperature of the outer surface of the cylinder?

**3-23** The general solution for the temperature distribution in a long cylinder with uniform heat generation is $T(r) = \frac{-q_{gen}r^2}{4k} + C_1 \ln r + C_2$. Consider the radial heat conduction in a hollow cylinder of inner radius $r_1$ and outer radius $r_2$. The cylinder has convection boundary conditions at both surfaces: At $r = r_1$, the convective coefficient is $h_1$ and the fluid temperature is $T_{\infty_1}$. At $r = r_2$, the convective coefficient is $h_2$ and the fluid temperature is $T_{\infty_2}$.

**(a)** Determine the expression for the temperature distribution $T(r)$ in the hollow cylinder.

**(b)** Using the result of Part (a), determine the expression for the heat flow $q(r)$ in the cylinder. (Remember that heat flow is $q = -kA\frac{dT}{dr}$ and that $A = 2\pi r L$ for a cylinder.)

**3-24** The heat conduction equation for steady radial heat conduction in a sphere with uniform heat generation at a rate of $q_{gen}$ per unit volume is

$$\frac{1}{r^2}\frac{d}{dr}\left(r^2\frac{dT}{dr}\right) + \frac{q_{gen}}{k} = 0$$

The sphere has an outer radius of $r_0$. The boundary condition at the surface of the sphere is of the temperature type. That is, $T = T_0$ at $r = r_0$.

**(a)** Determine the expression for the temperature distribution $T(r)$ in the sphere.

**(b)** Using the result of Part (a), determine the expression for the heat flow $q(r)$ in the sphere. (Remember that heat flow is $q = -kA\frac{dT}{dr}$ and that $A = 4\pi r^2$ for a cylinder.)

**3-25** A sphere of conductivity $k = 0.8$ W/m C has a diameter of 10 cm. It has uniform internal heat generation at a rate of $q_{gen} = 4 \times 10^4$ W/m³. The temperature of the surface of the sphere is 60 C.

What is the maximum temperature in the sphere and where does it occur?

**3-26** A sphere of conductivity $k = 0.5$ Btu/h ft F has a diameter of 6.5 inches. It has uniform internal heat generation at a rate of $q_{gen} = 4 \times 10^3$ Btu/h ft³. The temperature of the surface of the sphere is 200 F.

What is the maximum temperature in the sphere and where does it occur?

**3-27** A sphere of conductivity $k = 1.5$ W/m C has a radius of 4 cm and a heat generation rate of $6 \times 10^4$ W/m³. The surface convects to the surrounding fluid that is at 30 C. The surface of the sphere has a temperature of 55 C.

**(a)** What is the convective coefficient (W/m² C) at the surface of the sphere?

**(b)** What is the rate of heat flow (W) to the surrounding fluid?

**3-28 to 3-31 These four problems deal with variable thermal conductivity. For all of them, the material has the following temperature/conductivity characteristic:**

| T (C) | k (W/m C) |
|-------|-----------|
| 0     | 17        |
| 50    | 20        |
| 100   | 24        |
| 150   | 30        |
| 200   | 39        |
| 250   | 55        |
| 300   | 70        |

**3-28** Sketch, by hand, a graph of $k$ versus $T$ with $k$ on the vertical axis and $T$ on the horizontal axis. Then draw, by hand, the best straight line through the data points. Choose two points on the straight line and, by hand, determine the equation of the linear function $k = a + bT$ for the line. Using this information, determine the rate of heat flow (W) through a vertical flat plate of size 5 cm high by 8 cm wide by 0.5 cm thick having the left side of the plate kept at 55 C and the right side at 275 C.

**3-29** Use a software package such as Excel or Matlab to determine the best linear fit $k = a + bT$ for the data. Using this information, find the heat flow through the plate of Problem 3–28.

**3-30** Use a software package such as Excel or Matlab to determine the best third degree polynomial fit ($k = a + bT + cT^2 + dT^3$) for the data. Using this information, find the heat flow through the plate of Problem 3–28.

**3-31** A hollow cylinder having conductivity as shown above has an inner radius of 2 cm and an outer radius of 12 cm. The inner surface of the cylinder is at 250 C. The cylinder is covered by a 1 cm layer of insulation ($k = 0.04$ W/m C). The outer surface of the insulation transfers heat by convection to the surrounding fluid. The convective coefficient is 30 W/m$^2$ C and the fluid is at 50 C.

What is the rate of heat flow to the fluid per meter length of the cylinder (W/m)?

**3-32** A concrete bar ($k = 0.6$ Btu/h ft F) of square cross section 6 inch by 6 inch and 8 foot length has a copper heating pipe of 2-inch outside diameter running through its center. The temperature of the outer surface of the pipe is 160 F. The outer surfaces of the concrete bar are at 70 F.

What is the rate of heat transfer from the pipe (Btu/h)?

**3-33** This is the same Problem 3–32 except the concrete bar has a rectangular cross section of 4 inch by 8 inch.

What is the rate of heat transfer from the pipe (Btu/h)?

**3-34** Two long copper pipes are buried in the ground. The soil has a conductivity of 0.5 W/m C. The pipes are buried parallel to each other at a depth of 0.8 m from the surface of the ground. One pipe has an OD of 5 cm and an outer temperature of 90 C. The other pipe has an OD of 8 cm and an outer temperature of 15 C. The pipes are 5 cm apart.

What is the rate of heat flow from the hot pipe to the cold pipe per meter length (W/m)?

**3-35** A thin-walled sphere of 7 cm diameter is buried in the ground. The top of the sphere is 20 cm from the surface of the ground, and the soil conductivity is 0.45 W/m C.

If the temperature of the wall of the sphere is 25 C and the temperature of the ground is 5 C, what is the rate of heat flow from the sphere (W)?

**3-36** A long steel horizontal pipe ($k = 45$ W/m C) is buried in the ground ($k = 0.5$ W/m C) at a centerline depth of 20 cm. The pipe has an inner diameter of 2 cm and an outer diameter of 4 cm. The temperature of the inner surface of the pipe is 40 C and the temperature of the outer surface of the pipe is 39.6 C. The surface of the ground has a temperature of 25 C.

What is the rate of heat loss from the pipe per meter length of pipe (W/m)?

**3-37** A long horizontal PVC pipe ($k = 0.19$ W/m C) is buried in the ground ($k = 0.8$ W/m C) at a centerline depth of 20 cm. The pipe has a nominal size of one inch and is Schedule 80. The pipe carries hot water at 85 C and the convective coefficient between the water and the inner surface of the pipe is 150 W/m$^2$ C. The surface of the ground has a temperature of 15 C.

What is the rate of heat loss from the water per meter length of pipe (W/m)?

**3-38** Five cylinders of 2 cm diameter and 6 m length are buried in soil ($k = 0.3$ W/m C) at a depth of 50 cm from the surface of the soil. The cylinders are horizontal and parallel to each other with a centerline spacing between them of 5 cm. The surface of the soil is at 20 C and the surfaces of all five cylinders are 175 C.

What is the total heat flow from the five cylinders (W)?

**3-39** A heating duct is of sheet metal and has a square cross section of 25 cm by 25 cm. The duct is covered by 3 cm of insulation ($k = 0.05$ W/m C). The duct goes through a room where the room air is at 18 C. The inside surface of the duct is 60 C and you can assume that the outer surface of the insulation is the same as the room air temperature.

What is the rate of heat transfer through the insulation per meter length of duct (W/m)?

**3-40** A long sheet-metal duct carries 60 C air. The flow area is square, 30 cm by 30 cm, and the duct is covered by 3 cm of insulation ($k = 0.05$ W/m C). The duct passes through a room where the air is at 20 C. The convective coefficient at the inner surface of the duct is 25 W/m$^2$ C and the convective coefficient between the room air and the outer surface of the insulation is 5 W/m$^2$ C.

**(a)** What is the rate of heat loss per meter length from the hot air in the duct (W/m)?

**(b)** What is the temperature of the outer surface of the insulation?

**(c)** What percent reduction in heat loss from the air in the duct was achieved by the addition of insulation to the duct?

**3-41** A cylinder of 15 cm diameter and 200 cm length is buried vertically in the ground. The soil has a conductivity of 0.9 W/m C. The top of the cylinder is at the surface of the ground. The cylinder's surface temperature is 100 C and the ground's surface temperature is 25 C.

What is the rate of heat flow from the cylinder (W)?

**3-42** A rod of square cross section (4 inch by 4 inch) and 10 feet long is buried in the ground. The rod is vertical, with one end located at the surface of the ground. The conductivity of the soil is 0.5 Btu/h ft F. The surface of the rod is at 150 F, and ground's surface is at 70 F.

What is the rate of heat flow from the rod to the ground (Btu/h)?

**3-43** A circular pin fin of constant diameter $D$ and length $L$ is attached to a surface whose temperature is $T_s$. The temperature in the fin is one-dimensional. It is only a function of the distance $x$ from the surface of attachment. That is, $T = T(x)$. The fin is in a vacuum so the only mode of heat transfer from its surface is by radiation. The emissivity of the fin surface is $\varepsilon$ and the fin radiates to the surroundings which are at temperature $T_{surr}$.

**(a)** Derive the differential equation for the temperature distribution $T(x)$ in the fin. Also give the boundary condition equations. Do not try to solve the differential equation. (Remember: Heat transfer by radiation is $q = \varepsilon\sigma A\left(T^4 - T_{surr}^4\right)$, where $A$ and $T$ are the surface area and temperature of the surface which radiates and $\sigma$ is the Stefan–Boltzmann constant.)

**(b)** The heat flow from the fin to the surroundings can be determined once the temperature distribution is known. Give two expressions for the heat flow from the fin to the surroundings. (Hint: One expression has a derivative; the other has an integral.)

**3-44** The same as Problem 3–43 except there is both convection and radiation from the surface of the fin. The convective coefficient at the fin surface is $h$ and the surrounding fluid is at temperature $T_\infty$.

**3-45** Consider a fin of constant cross-sectional area $A$ and length $L$ attached to a surface that is at temperature $T_s$. There is convection to the surrounding fluid from the lateral surface of the fin, but the tip is perfectly insulated. That is, the heat transfer from the tip is zero. The surrounding fluid is at $T_\infty$ and the convective coefficient is $h$ at the lateral surface.

Using the definition of fin efficiency, show that the fin efficiency $\eta_f$ is $\frac{1}{mL}\tanh(mL)$, where $m = \sqrt{\frac{hP}{kA}}$ and $P$ is the perimeter of the fin.

**3-46** Consider a fin attached to a surface which is at temperature $T_s$. The fin has an area of attachment of $A$. The surrounding fluid is at $T_\infty$ and the convective coefficient at the fin surface is $h$. The total fin surface area in contact with the fluid is $A_{surf}$.

From the definitions of fin efficiency $\eta_f$ and fin effectiveness $\varepsilon_{fin}$, show that

$$\varepsilon_{fin} = \frac{A_{surf}}{A}\eta_f$$

**3-47** A pure aluminum circular pin fin is 0.5 cm diameter and 20 cm long. The fin is attached to a wall that is at 250 C. The surrounding fluid is at 20 C, and the convective coefficient is 15 W/m² C.
   **(a)** What is the fin efficiency?
   **(b)** What is the rate of heat transfer from the fin to the surrounding fluid (W)?
   **(c)** What is the fin effectiveness?

**3-48** A pure aluminum conical pin is attached to a wall that is at 250 C. The diameter of the fin at the location of attachment is 0.5 cm and the fin is 20 cm long. The surrounding fluid is at 20 C, and the convective coefficient is 15 W/m² C.
   **(a)** What is the fin efficiency?
   **(b)** What is the rate of heat transfer from the fin to the surrounding fluid (W)?
   **(c)** What is the fin effectiveness?

**3-49** Many circular pin fins are added to a 0.4 m by 0.4 m plate to enhance heat transfer. The fins are 0.5 cm diameter and 12 cm long and have a center-to-center spacing of 1.5 cm. The fin material has a conductivity of 400 W/m C. The plate is at 250 C. There is convection to the surrounding fluid from the surfaces of the fins and from the unfinned portions of the plate. The fluid is at 40 C and the convective coefficient is 50 W/m² C.
   **(a)** How many fins can fit on the plate?
   **(b)** What is the total heat transfer rate from the finned plate to the fluid (W)?
   **(c)** Compare the heat transfer of the finned plate to that of the bare plate without fins. What percent increase in heat flow rate was achieved by the addition of the fins?

**3-50** As shown in the below figure, circular circumferential fins of rectangular profile have been added to an engine cylinder. The fins have a conductivity of 180 W/m C. The engine cylinder has an outer diameter of 10 cm. The fins are attached to this outer surface that is at 550 C. The fins have a thickness of 2 mm and an outer diameter of 12 cm. The surrounding air is at 40 C and the convective coefficient is 45 W/m² C. There are 30 fins on the engine cylinder, with a center-to-center spacing of 5 mm.

What is the heat transfer from the cylinder to the surrounding air (W)? Include heat transfer from both the fins and the unfinned areas between the fins.

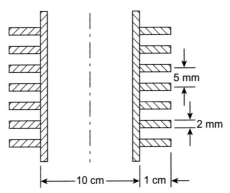

**3-51** A pin fin has the constant cross section of an equilateral triangle with a 0.4 cm side. The fin is of high conductivity ($k = 350$ W/m C) and has a length of 25 cm. The fin is attached to a wall at 150 C and the surrounding fluid is at 10 C. The convective coefficient is 12 W/m$^2$ C.
**(a)** What is the rate of heat transfer from the fin to the surrounding fluid (W)?
**(b)** What is the fin efficiency?
**(c)** What is the fin effectiveness?
**(d)** Is the fin long enough to use the very-long-fin approximation to determine the heat transfer rate?

**3-52** A metal rod ($k = 50$ W/m C) has a diameter of 1 cm. It is attached to two surfaces as shown below. The surfaces are 60 cm apart. One surface is at 200 C; the other is at 50 C. The fluid surrounding the rod is at 20 C and the convective coefficient is 5 W/m$^2$ C.
**(a)** What is the temperature at the midpoint of the rod?
**(b)** What is the rate of heat transfer from the rod to the fluid (W)?

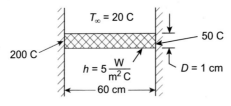

**3-53** A straight rectangular fin has a length of 3 cm, a thickness of 2.5 mm, and a width of 15 cm. It is attached to a wall that is at 300 C. The fin has a conductivity of 250 W/m C. The fin convects to a surrounding fluid that is at 18 C. The convective coefficient at the fin's exposed surface is 7 W/m$^2$ C.
What is the rate of heat transfer from the fin to the fluid (W)?

**3-54** Same as Problem 3–53 except the fin is a straight triangular fin rather than a straight rectangular fin.

**3-55** Hot air at 1150 C is inside a kiln. The wall of the kiln is 7.6 cm thick and has a thermal conductivity of 0.35 W/m C. The convective coefficient on the inner surface of the kiln is 15 W/m$^2$ C. The outer surface has a convective coefficient of 3 W/m$^2$ C and an emissivity of 0.75. The room air is at 25 C and the walls, ceiling, and floor of the room are at 30 C.

  **(a)** Give the thermal circuit for the heat transfer from the air in the kiln to the room air and surroundings. Use thermal resistances $R_{wall}$, $R_{conv}$, and $R_{rad}$ in the circuit.

  **(b)** Using the thermal circuit of Part (a), determine the heat flux through the kiln wall.

  **(c)** Revise the circuit of Part (a) by using a combined resistance $R_{combined} = (1/h_{combined})$ between the outer surface of the wall and fluid temperature $T_\infty$.

  **(d)** Determine the heat flux through the wall for the revised circuit and compare it with the result of Part (b).

**3-56** A homeowner is annoyed by water dripping due to condensation from cold water pipes in his basement during hot, humid summers. The water drips on his head as he walks through the basement and, even worse, it drips on cardboard boxes he has stored in the basement. The son of his neighbor is an engineering student who tells him that he should insulate the pipes. Condensation occurs if the temperature of the outer surface of the pipes is below the dew point temperature of the air in the basement.

  **(a)** If the basement air temperature is 75 F and the relative humidity is 60%, what is the dew point temperature? (Note: This can be obtained from a psychrometric chart from an HVAC text or from various internet sites.)

  **(b)** The cold water flowing through the water lines is at about 52 F and the water lines are 3/4″ Type L copper pipe. If the homeowner adds fiberglass insulation to the lines, what is the minimum thickness of insulation needed to prevent the condensation? Assume the convective coefficient at the outer surface of the insulation is 0.7 Btu/h ft$^2$ F.

**3-57** A house has a 40 gallon hot water heater. The water tank of the heater is 115 cm high and has a diameter of 42 cm. It is covered with 5 cm thick insulation ($k = 0.035$ W/m C). The water in the tank is maintained at 62 C. The steel tank ($k = 45$ W/m C) has a wall thickness of 3 mm. The room air is at 10 C. The convective coefficient at the inner surface of the tank is 55 W/m$^2$ C and is 6 W/m$^2$ C at the outer surface of the insulation.

  **(a)** What is the needed power (W) to the heater to keep the water at 62 C? The heat loss from the bottom of the tank should be small. Consider only the heat loss from the top and the side of the tank.

  **(b)** If the heater has a recovery rate of 41 GPH for a 90 F recovery, what size heat source (gas, electric) to the water is needed?

**3-58** Two large plates have a thin electric resistance heater sandwiched between them. Plate A is 1.5 cm thick and has a conductivity of 0.3 W/m C. Plate B is 3 cm thick and has a conductivity of 6 W/m C. The two exposed surfaces of the arrangement convect to the adjacent air and radiate to the surroundings. For Plate A, the exposed surface has convective coefficient 12 W/m$^2$ C and emissivity 0.8. For Plate B, the exposed surface has convective coefficient 2 W/m$^2$ C and emissivity 0.4. The adjacent fluids and surroundings are at 35 C. The total heat flux provided by the heater is 1800 W/m$^2$.

  **(a)** What is the temperature of the heater?

  **(b)** What are the heat fluxes through the individual plates?

**3-59** Same as Problem 3–58, but there is no radiation. There is only convection at the exposed surfaces.

# References

[1] H.S. Carslaw, J.C. Jaeger, Conduction of Heat in Solids, Oxford University Press, 1959.

[2] P.J. Schneider, Conduction Heat Transfer, Addison-Wesley, 1955.

[3] G.E. Myers, Analytical Methods in Conduction Heat Transfer, McGraw-Hill, 1971.

[4] V.S. Arpaci, Conduction Heat Transfer, Addison-Wesley, 1966.

[5] COMSOL, www.comsol.com.

[6] Solidworks, www.solidworks.com.

[7] ANSYS, www.ansys.com.

[8] CD-Adapco (now Siemens), mdx.plm.automation.siemens.com.

[9] R.V. Andrews, Solving conductive heat transfer problems with electrical-analogue shape factors, Chem. Eng. Prog. 51 (2) (1955) 67.

[10] J.E. Sunderland, K.R. Johnson, Shape factors for heat conduction through bodies with isothermal or convective boundary conditions, Trans. ASHRAE 10 (1964) 237–241.

[11] E. Hahne, U. Gringull, Formfaktor und Formwiderstand der stationaren mehr- dimensionalen Warmeleitung, Int. J. Heat Mass Transf. 18 (1975) 751–767.

[12] K.A. Gardner, Efficiency of extended surfaces, Trans. ASME 67 (1945) 621–631.

[13] W.B. Harper, D.R. Brown, Mathematical equations for heat conduction in the fins of air-cooled engines, NACA Report 158 (1922).

[14] A.D. Kraus, A. Aziz, J. Welty, Extended Surface Heat Transfer, John Wiley, 2001.

# Unsteady conduction

## Chapter outline

## 4.1 Introduction

In general, temperatures and heat flows in a body vary with both location and time. Chapter 3 discussed situations in which the heat transfer was steady, i.e., had no variation with time. In this chapter, we consider situations in which temperatures and heat flows vary with time.

We will first discuss lumped systems. For these systems, all locations in the body have essentially the same temperature at a given time. That is, there is no spatial variation of temperature. This is the

Heat Transfer Principles and Applications. https://doi.org/10.1016/B978-0-12-802296-2.00004-4

simplest case of unsteady conduction. It is valid when a body is sufficiently small, has high thermal conductivity, and/or is sufficiently insulated from its surroundings.

If a heat transfer situation does not meet the criteria for the lumped system analysis, then we must look at spatial variation of temperatures and heat flows in a body. We will first discuss conduction in large plane plates, long cylinders, and spheres. We will then show how these results can be applied to other geometries such as short cylinders and rectangular solids.

Finally, we discuss heat transfer in semiinfinite solids. This topic is of considerable practical interest. For example, it can be applied to determine the needed depth of burial of water pipes in the ground to prevent freezing and the needed depth of foundation footings to prevent frost heave.

## 4.2 Lumped systems (no spatial variation)

In this section, we consider systems in which all points in the body are at the same temperature at a given time. Although temperatures in the body change with time, there is no spatial variation at a given time. That is, the temperature in the body is only a function of time or

$$T = T(t) \tag{4.1}$$

### 4.2.1 Lumped systems analysis

Consider the body shown in Fig. 4.1. Let us say the body is initially at temperature $T_i$ and it is suddenly immersed in a fluid at a lower temperature $T_\infty$. Heat will flow by convection from the body to the fluid until the temperature of the body reaches the fluid temperature $T_\infty$.

We can write a word equation for this process as follows:

Rate of decrease in internal energy of the body = Rate of heat flow from the body to the fluid

$$\tag{4.2}$$

Mathematically, this equation is

$$-\rho c V \frac{dT}{dt} = hA(T - T_\infty) \tag{4.3}$$

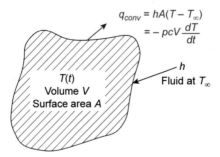

**FIGURE 4.1**

Body with convection at boundary.

where $\rho =$ density of body $(kg/m^3)$

$c =$ specific heat of body $(J / kg\ C)$
$V =$ volume of body $(m^3)$
$t =$ time $(s)$
$T =$ temperature of body at time $t$ $(C)$
$h =$ convective coefficient $(W / m^2\ C)$
$A =$ surface area of body in contact with fluid $(m^2)$
$T_\infty =$ fluid temperature $(C)$

All parameters in Eq. (4.3) are constants except temperature $T$ and time $t$.
As $T_\infty$ is constant, $\frac{dT}{dt} = \frac{d(T-T_\infty)}{dt}$, and Eq. (4.3) can be rearranged to

$$\frac{d(T - T_\infty)}{T - T_\infty} = -\frac{hA}{\rho c V}\ dt \tag{4.4}$$

The body is initially at uniform temperature $T_i$. That is, $T = T_i$ at $t = 0$. Integrating Eq. (4.4) from $t = 0$ to $t = t$, we have

$$\int_{T_i - T_\infty}^{T - T_\infty} \frac{d(T - T_\infty)}{T - T_\infty} = -\frac{hA}{\rho c V} \int_0^t dt \tag{4.5}$$

$$\ln\left(\frac{T - T_\infty}{T_i - T_\infty}\right) = -\frac{hA}{\rho c V}t \tag{4.6}$$

And, finally, we have

$$T - T_\infty = (T_i - T_\infty)e^{-\left(\frac{hA}{\rho c V}\right)t} \tag{4.7}$$

$$\text{or } T(t) = T_\infty + (T_i - T_\infty)\ e^{-\left(\frac{hA}{\rho c V}\right)t} \tag{4.8}$$

The instantaneous rate of heat flow at any time during the process is

$$q = -\rho c V \frac{dT}{dt} = hA(T - T_\infty) \tag{4.9}$$

Sometimes we are interested in the quantity of heat $Q$ transferred between the body and the fluid during the process. To determine this, we can integrate the heat flow rate $q$ over time, as follows:

$$Q = \int_0^t q\ dt \tag{4.10}$$

Using the expressions for $q$ in Eq. (4.9), we can determine $Q$ from either

$$Q = -\rho c V \int_{T_i}^{T} dT = -\rho c V (T - T_i) \qquad (4.11)$$

$$\text{or } Q = hA \int_{0}^{t} (T - T_\infty) dt \qquad (4.12)$$

We will get the same result for $Q$ from either Eq. (4.11) or Eq. (4.12). Using Eq. (4.11) is a little easier as integration of the exponential function is not needed. Continuing, therefore, with Eq. (4.11) and the function for $T$ in Eq. (4.8), we have

$$Q = -\rho c V (T - T_i) = -\rho c V \left( T_\infty + (T_i - T_\infty) e^{-\left(\frac{hA}{\rho c V}\right) t} - T_i \right)$$

Reducing this, we get

$$Q = \rho c V (T_i - T_\infty) \left[ 1 - e^{-\left(\frac{hA}{\rho c V}\right) t} \right] \qquad (4.13)$$

## 4.2.2 Application criterion

A body can be modeled as a lumped system when its internal resistance to heat conduction is small compared with the convective resistance at its surface. This requirement is best met by small bodies of high conductivity having a small convective coefficient at the surface.

A dimensionless parameter, the Biot number ($Bi_{lumped}$), is used to determine the applicability of lumped system analysis. This parameter is

$$Bi_{lumped} = \frac{hL}{k}$$

where $h$ is the convective coefficient at the surface of the body, L is a characteristic length of the body, usually taken as the volume V of the body divided by the surface area $A$ of the body, and $k$ is the thermal conductivity of the body.

Let us briefly discuss why we have a subscript of "lumped" on the Biot number: In this consideration of whether a lumped analysis is appropriate, the characteristic length $L$ is the volume V of the body divided by the surface area $A$. The length parameter in the Biot number is defined differently in other places in this text. As an example, let us consider a sphere of radius $r_o$. $Bi_{lumped}$ would be

$$Bi_{lumped} = \frac{h\left(\dfrac{V}{A}\right)}{k} = \frac{h\left(\dfrac{4/3\,\pi r_o^3}{4\pi r_o^2}\right)}{k} = \frac{hr_o}{3k}$$

In other sections of this text, the Biot number $Bi$ for a sphere is given as $hr_o/k$. Thus, there is a factor of three difference between $Bi_{lumped}$ and the other Biot number. In short, there are really two different Biot numbers: one for use in this section and one for use in other sections of this text. Hence, to avoid confusion, we have decided to use different nomenclatures for the two numbers rather than use the same $Bi$ for both.

Let us move on to the criterion for use of the lumped analysis. This criterion is

If $Bi_{lumped}$ is less than 0.1, then the spatial variation of temperature in the body is small, and the lumped system analysis is deemed appropriate.

$$\text{If} \quad Bi_{lumped} \quad < \quad 0.1 \quad \text{Lumped} - \text{System Analysis is applicable}$$

Some words about the criterion: First, the value 0.1 is a rule of thumb and is not necessarily a firm value. We discussed earlier the uncertainty in convective coefficient $h$. This coefficient is in the Biot number, which leads to some uncertainty in the 0.1 value. If $Bi_{lumped}$ is, say, 0.004, then we can be quite confident that the spatial variation of temperature in the body is negligible and the lumped analysis is appropriate. On the other hand, if $Bi_{lumped}$ is a bit greater than 0.1, say 0.13, we should not a priori assume that there is significant spatial variation of temperature in the body. Perhaps a closer investigation is needed.

Other factors may also enter into the decision to use a lumped analysis. Maybe only an estimate is needed of the temperature in the body at a given time, and the details of the spatial variation are not important. Or, maybe a quick answer is needed as to the temperature in the body. One could get a quick result from the lumped analysis and then, if necessary, later supplement this analysis with analyses that consider spatial variation.

---

### Example 4.1
### Cooling of a copper cube
*Problem*
The copper cube ($k = 400$ W/m C, $\rho = 8930$ kg/m³, $c = 390$ J/kg C) shown below is 2 cm by 2 cm by 2 cm. It is initially at 200 C in an oil bath. The cube is taken out of the bath and placed on an insulating board to cool by convection to the surrounding air. Assume the surrounding air is at 20 C and the convective coefficient is 10 W/m² C.
**(a)** What is the temperature of the center of the cube after 45 min?
**(b)** How much heat (J) flows to the surrounding air during the 45 min?
**(c)** If the cooling continues, how long will it take for the cube to reach the air temperature?

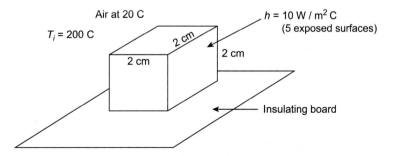

*Solution*
We will first determine the Biot number to determine if the lumped analysis is appropriate.

$$Bi_{lumped} = \frac{h\left(\dfrac{V}{A}\right)}{k} = \frac{(10)\dfrac{(0.02)^3}{(5)(0.02)^2}}{400} = 0.0001$$

(Please note that for the surface area, we have only included five of the six faces of the cube. One face is on the insulating board. This face does not convect to the air and we assume that there is no conduction to the board.)
The Biot number is much less that 0.1, so we can use the lumped analysis.
**(a)** Using Eq. (4.8), we have

$$T(t) = T_\infty + (T_i - T_\infty)\, e^{-\left(\frac{hA}{\rho cV}\right)t} \tag{4.8}$$

For this problem, $T_\infty = 20$ C.
$T_i = 200$ C.
$h = 10$ W/m$^2$ C.
$A = 5\,(0.02)^2 = 0.002$ m$^2$
$V = (0.02)^3 = 0.000008$ m$^3$
$\rho = 8930$ kg/m$^3$
$c = 390$ J/kg C.
$t = 45$ min $\times$ 60 s/min $= 2700$ s

$$\left(\frac{hA}{\rho cV}t\right) = \left(\frac{10(002)}{8930(390)(0.000008)}(2700)\right) = 1.938$$

So, from Eq. (4.8), we have

$$T(45\ \text{min}) = 20 + (200 - 20)e^{-1.938} = 45.9\ \text{C}$$

This is the temperature of the cube after 45 min. As we have a lumped system, all points in the cube will be at that temperature, including the center.
**(b)** Eq. (4.13) gives the heat transferred during time $t$.

$$Q = \rho cV(T_i - T_\infty)\left[1 - e^{-\left(\frac{hA}{\rho cV}\right)t}\right] \tag{4.13}$$

So, from Eq. (4.13), we have

$$Q = (8930)(390)(0.000008)(200 - 20)\left[1 - e^{-1.938}\right] = 4293\ \text{J}$$

Therefore, 4293 J of heat will be transferred from the cube to the surrounding air during the 45 min process.
**(c)** Eq. (4.8) gives the temperature of the cube after any amount of time. If we use this equation to see how long it will take to have the cube reach $T_\infty$, the result is that it will take an infinite amount of time. How can this possibly be true? Let us revisit this problem after we have discussed the time constant in the next section.

## 4.2.3 The time constant

The lumped system is a first-order system as it obeys Eq. (4.3), which is a first-order differential equation:

$$-\rho cV\frac{dT}{dt} = hA(T - T_\infty) \tag{4.3}$$

And the system receives an abrupt change at time zero, that is, it is suddenly subjected to convection at its surface. For such a first-order system with step input, it is useful to define a time constant τ.

Let's look at the response of our system, which is given by Eq. (4.8):

$$T(t) = T_\infty + (T_i - T_\infty)\ e^{-\left(\frac{hA}{\rho c V}\right) t} \tag{4.8}$$

The quantity $\rho c V / hA$ in the exponential term has the units of time and is called the time constant.

$$\tau = \frac{\rho c V}{hA} = \text{Time Constant}$$

Using this definition, Eq. (4.8) becomes

$$T(t) = T_\infty + (T_i - T_\infty)\ e^{-t/\tau} \tag{4.14}$$

Let us consider the process for the system:

At time zero, the system is at $T_i$. Convection begins and, assuming that $T_i$ is greater than $T_\infty$, the system starts to cool toward the fluid temperature $T_\infty$. After a long period of time (actually infinite time according to the equation), the system will reach the fluid temperature $T_\infty$. At any time $t$ during the process, the temperature of the system is $T$.

By time $t$, the system has cooled from its initial temperature $T_i$ to temperature $T$. For the complete process, the system cools from $T_i$ to $T_\infty$.

Therefore, at time $t$, the system has completed $\left[\frac{T_i - T}{T_i - T_\infty} \times 100\right]$ percent of the complete process. It can be shown that the process is

63.2% complete after one time constant.

86.5% complete after two time constants.

95.0% complete after three time constants.

98.2% complete after four time constants, and.

99.3% complete after five time constants.

It is usually said that the process is complete after five time constants. That is, the system's temperature has reached $T_\infty$ in five time constants.

Let us continue our discussion of item (c) of Example 4.1:

The time constant for Example 4.1 is $\tau = \frac{\rho c V}{hA} = \frac{(8930)(390)(0.000008)}{(10)(0.002)} = 1393\ s$

The process is essentially complete after five time constants, so the answer to item (c) of Example 4.1 is that the cube reaches the air temperature of 20 C in about (5) (1393) = 6965 s or 116 min.

## 4.3 Systems with spatial variation (large plate, long cylinder, sphere)
### 4.3.1 Overview

If a body does not meet the criterion for treatment as a lumped system, i.e., if the Biot number is not small enough, then we must consider the spatial distribution of temperature in the body. In this section, we consider the transient temperature distributions in large plane plates, long cylinders, and spheres due to convection boundary conditions.

What is "large" for the plates and "long" for the cylinders? For the plates, we are theoretically looking at plates that have a finite thickness of $2L$ and infinite lengths in the other two directions. We, of course, do not really have infinite plates. The plates just have to be large enough so that their spatial temperature distribution is only a function of one direction, say "$x$," whose origin is at the center plane of the plate. This is shown in Fig. 4.2.

Theoretically, the cylinders are infinitely long. Actually they just have to be long enough so that the spatial temperature distribution is only a function of radial distance $r$, whose origin is on the centerline of the cylinder. The temperature is not a function of the axial distance or angle. This is shown in Fig. 4.3.

And, for the sphere, the spatial temperature distribution is only a function of radial distance $r$, whose origin is at the center of the sphere. The temperature is not a function of angle. This is shown in Fig. 4.4.

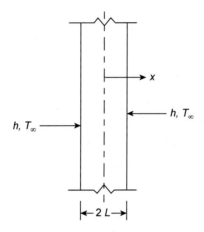

**FIGURE 4.2**

Large plane plate.

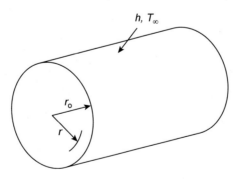

**FIGURE 4.3**

Long cylinder.

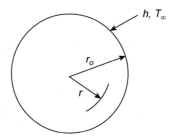

**FIGURE 4.4**

Sphere.

In 1947, Heisler [1] solved the governing differential equations for transient temperature distributions in these three geometries and provided the solutions in graphical form. These solutions are called the "Heisler Charts" Later, in 1961, Gröber, Erk, and Grigull [2] extended Heisler's work and provided charts giving the total energy transfer during a heating or cooling process. We are not providing the Heisler and Gröber charts in this text. Rather, we are providing the solutions for the three geometries in equation form. Section 4.2 dealt with the **lumped method.** We will call the method of this Section 4.3 the **Heisler method.**

### 4.3.2 Large plane plates

From Chapter 2, the general heat conduction equation in rectangular coordinates is

$$\frac{\partial}{\partial x}\left(k\frac{\partial T}{\partial x}\right) + \frac{\partial}{\partial y}\left(k\frac{\partial T}{\partial y}\right) + \frac{\partial}{\partial z}\left(k\frac{\partial T}{\partial z}\right) + q_{gen} = \rho\, c\, \frac{\partial T}{\partial t} \tag{4.15}$$

Consider the plate of Fig. 4.2 above. The plate has no internal heat generation and the conductivity is constant. The temperature distribution at a given time depends only on $x$. Eq. (4.15) then reduces to the one-dimensional transient equation

$$\frac{\partial^2 T}{\partial x^2} = \frac{1}{\alpha}\frac{\partial T}{\partial t} \tag{4.16}$$

where $\alpha$ = thermal diffusivity = $\frac{k}{\rho c}$

All points in the plate are initially at temperature $T_i$. Then, at time zero, the sides of the plate at $x = -L$ and $x = +L$ are suddenly subjected to convection to the environment at $T_\infty$ with a convective coefficient $h$.

The temperature solution is symmetrical about $x = 0$. Hence, the boundary condition equation at the center plane of the plate is

$$\text{At } x = 0, \quad \frac{\partial T}{\partial x} = 0 \tag{4.17}$$

At the surface of the plate, the convective boundary condition equation is

$$\text{At } x=L, \quad -k\frac{\partial T}{\partial x} = h \ (T - T_\infty) \tag{4.18}$$

The initial condition is: At $t=0$, $T(x) = T_i$ (4.19)

The solution of Eq. (4.16) with conditions (4.17) through (4.19) can be obtained by the separation of variables technique. The solution $T(x,t)$ is an infinite series:

$$T(x,t) = T_\infty + (T_i - T_\infty) \sum_{n=1}^{\infty} C_n e^{-\frac{\lambda_n^2 \alpha t}{L^2}} \cos(\lambda_n x / L) \tag{4.20}$$

This equation gives the temperature at any point in the plate at any time. This solution given by Eq. (4.20) is often written in dimensionless form:

$$\frac{\theta}{\theta_i} = \frac{T(x,t) - T_\infty}{T_i - T_\infty} = \sum_{n=1}^{\infty} C_n e^{-\lambda_n^2 Fo} \cos(\lambda_n x / L) \tag{4.21}$$

$$Fo = \frac{\alpha t}{L^2} = \text{Fourier Number}$$

The eigenvalues $\lambda_n$ are the positive roots of the transcendental equation

$$\lambda_n \tan \lambda_n = Bi \tag{4.22}$$

where $Bi$ is the Biot number. For plane plates, $Bi = \frac{hL}{k}$. (As we will see later, the Biot number for cylinders and spheres of radius $r_o$ is $Bi = \frac{h\, r_o}{k}$.) Among other methods, Eq. (4.22) may be solved for the $\lambda_n$ through use of the *Goal Seek* or *Solver* features of Excel or the *fzero* function of Matlab. See Appendix J for details.

Once the $\lambda_n$ values are determined, the $C_n$ values in Eqs. (4.20) and (4.21) may be obtained from

$$C_n = \frac{4 \sin \lambda_n}{2\lambda_n + \sin(2\lambda_n)} \tag{4.23}$$

Note that the trigonometric arguments in this section are in units of *radians*.

The series in Eqs. (4.20) and (4.21) converge rapidly with increasing time due to the exponential term. If the Fourier number is greater than 0.2, excellent results are obtained using just the first term of the series. This is called the **one-term approximation**. Keeping just the first term, Eq. (4.21) becomes

$$\frac{\theta}{\theta_i} = \frac{T(x,t) - T_\infty}{T_i - T_\infty} = C_1 e^{-\lambda_1^2 Fo} \cos(\lambda_1 x / L) \quad \text{for} \quad Fo > 0.2 \tag{4.24}$$

From Eq. (4.24), the temperature at the center plane of the plate, i.e., at $x = 0$, is

$$\frac{\theta_o}{\theta_i} = \frac{T(0,t) - T_\infty}{T_i - T_\infty} = C_1 e^{-\lambda_1^2 Fo} \quad \text{for} \quad Fo > 0.2 \tag{4.25}$$

If we are interested in an earlier time of the system's response, that is, when $Fo = \frac{\alpha t}{L^2} \leq 0.2$, we can just use more terms of the infinite series for the temperature.

**Table 4.1 Coefficients for the one-term approximation.**

| Bi | Plate $\lambda_1$ | Plate $C_1$ | Cylinder $\lambda_1$ | Cylinder $C_1$ | Sphere $\lambda_1$ | Sphere $C_1$ |
|---|---|---|---|---|---|---|
| 0.01 | 0.0998 | 1.0017 | 0.1412$'$ | 1.0025 | 0.1730 | 1.0030 |
| 0.02 | 0.1410 | 1.0033 | 0.1995 | 1.0050 | 0.2445 | 1.0060 |
| 0.04 | 0.1987 | 1.0066 | 0.2814 | 1.0099 | 0.3450 | 1.0120 |
| 0.06 | 0.2425 | 1.0098 | 0.3438 | 1.0148 | 0.4217 | 1.0179 |
| 0.08 | 0.2791 | 1.0130 | 0.3960 | 1.0197 | 0.4860 | 1.0239 |
| 0.1 | 0.3111 | 1.0161 | 0.4417 | 1.0246 | 0.5423 | 1.0298 |
| 0.2 | 0.4328 | 1.0311 | 0.6170 | 1.0483 | 0.7593 | 1.0592 |
| 0.3 | 0.5218 | 1.0450 | 0.7465 | 1.0712 | 0.9208 | 1.0880 |
| 0.4 | 0.5932 | 1.0580 | 0.8516 | 1.0931 | 1.0528 | 1.1164 |
| 0.5 | 0.6533 | 1.0701 | 0.9408 | 1.1143 | 1.1656 | 1.1441 |
| 0.6 | 0.7051 | 1.0814 | 1.0184 | 1.1345 | 1.2644 | 1.1713 |
| 0.7 | 0.7506 | 1.0918 | 1.0873 | 1.1539 | 1.3525 | 1.1978 |
| 0.8 | 0.7910 | 1.1016 | 1.1490 | 1.1724 | 1.4320 | 1.2236 |
| 0.9 | 0.8274 | 1.1107 | 1.2048 | 1.1902 | 1.5044 | 1.2488 |
| 1.0 | 0.8603 | 1.1191 | 1.2558 | 1.2071 | 1.5708 | 1.2732 |
| 1.5 | 0.9883 | 1.1537 | 1.4570 | 1.2807 | 1.8364 | 1.3849 |
| 2.0 | 1.0769 | 1.1785 | 1.5995 | 1.3384 | 2.0288 | 1.4793 |
| 3.0 | 1.1925 | 1.2102 | 1.7887 | 1.4191 | 2.2889 | 1.6227 |
| 4.0 | 1.2646 | 1.2287 | 1.9081 | 1.4698 | 2.4556 | 1.7202 |
| 5.0 | 1.3138 | 1.2403 | 1.9898 | 1.5029 | 2.5704 | 1.7870 |
| 6.0 | 1.3496 | 1.2479 | 2.0490 | 1.5253 | 2.6537 | 1.8338 |
| 7.0 | 1.3766 | 1.2532 | 2.0937 | 1.5411 | 2.7165 | 1.8673 |
| 8.0 | 1.3978 | 1.2570 | 2.1286 | 1.5526 | 2.7654 | 1.8920 |
| 9.0 | 1.4149 | 1.2598 | 2.1566 | 1.5611 | 2.8044 | 1.9106 |
| 10.0 | 1.4289 | 1.2620 | 2.1795 | 1.5677 | 2.8363 | 1.9249 |
| 15.0 | 1.4729 | 1.2676 | 2.2509 | 1.5800 | 2.9349 | 1.9630 |
| 20.0 | 1.4961 | 1.2699 | 2.2880 | 1.5919 | 2.9857 | 1.9781 |
| 30.0 | 1.5202 | 1.2717 | 2.3261 | 1.5973 | 3.0372 | 1.9898 |
| 40.0 | 1.5325 | 1.2723 | 2.3455 | 1.5993 | 3.0632 | 1.9942 |
| 50.0 | 1.5400 | 1.2727 | 2.3572 | 1.6002 | 3.0788 | 1.9962 |
| 100.0 | 1.5552 | 1.2731 | 2.3809 | 1.6015 | 3.1102 | 1.9990 |
| $\infty$ | 1.5708 | 1.2732 | 2.4048 | 1.6021 | 3.1416 | 2.0000 |

The $\lambda_1$ and $C_1$ values are given for different Biot numbers in Table 4.1. If more than one term is used in the series solution, then the $\lambda_n$ can be obtained through solution of Eq. (4.22) and the $C_n$ can then be obtained from Eq. (4.23).

Before continuing, let us look at Table 4.1. It is seen that there are entries for a Biot number of $\infty$. As $Bi$ is $hL/k$ for a plate, and $Bi$ is $hr_o/k$ for a cylinder or sphere, an infinite Biot number means that the convective coefficient is infinite. Infinite $h$ means there is zero convective resistance at the surface. So, the $\infty$ entry in the table corresponds to a temperature boundary condition. At time zero, a constant temperature is imposed on the two surfaces of the plate (or the surface of the cylinder or the surface of the sphere). This temperature is used in place of $T_\infty$ in the above equations.

We are often interested in the amount of energy that is transferred from (or to) the surroundings during a time period of the process. If a plate is initially uniform at $T_i$ and the process is fully completed, i.e., the plate is finally uniform at $T_\infty$, the decrease in internal energy of the plate is

$$Q_{max} = \rho c V (T_i - T_\infty) \tag{4.26}$$

where

$\rho =$ density of the plate.
$c =$ specific heat of the plate.
$V =$ volume of the plate.
If the process is only partially completed, the decrease in internal energy is

$$Q = \rho c \int_V (T_i - T) \, dV \tag{4.27}$$

The decrease in internal energy given in Eq. (4.27) is equal to the energy going into the surrounding medium by convection.

For the plane plate, $dV = A \, dx$, where $A$ is the cross sectional area of the plate. Therefore, Eq. (4.27) becomes

$$Q = \rho c A \int_{-L}^{L} [T_i - T(x,t)] \, dx \tag{4.28}$$

As the temperature distribution is symmetrical about the center plane, Eq. (4.28) is equivalent to

$$Q = 2\rho c A \int_{0}^{L} [T_i - T(x,t)] \, dx \tag{4.29}$$

The temperature in the plate at time $t$ is given by Eq. (4.20). Putting this into Eq. (4.29), and remembering that $Fo = \alpha t / L^2$, we have

$$Q = 2\rho c A \int_{0}^{L} \left[ T_i - T_\infty - (T_i - T_\infty) \sum_{n=1}^{\infty} C_n e^{-\lambda_n^2 Fo} \cos(\lambda_n x / L) \right] dx \tag{4.30}$$

Rearranging and recognizing that we are integrating only over $x$ and not $t$, Eq. (4.30) becomes

$$Q = 2\rho c A \left[ \int_{0}^{L} (T_i - T_\infty) \, dx - (T_i - T_\infty) \sum_{n=1}^{\infty} C_n \, e^{-\lambda_n^2 \, Fo} \int_{0}^{L} \cos(\lambda_n x / L) \, dx \right] \tag{4.31}$$

Performing the integrations and factoring out $(T_i - T_\infty)$, we get

$$Q = 2\rho c A L \ (T_i - T_\infty) \left[ 1 - \sum_{n=1}^{\infty} C_n \ e^{-\lambda_n^2 \ Fo} \ \frac{\sin \lambda_n}{\lambda_n} \right] \tag{4.32}$$

If we divide this result by $Q_{max}$ and recognize that the volume of the plate is $2LA$, we get the final dimensionless result:

$$\frac{Q}{Q_{max}} = 1 - \sum_{n=1}^{\infty} C_n \ e^{-\lambda_n^2 \ Fo} \ \frac{\sin \lambda_n}{\lambda_n} \tag{4.33}$$

where $Q_{max} = 2\rho c L A \ (T_i - T_\infty)$

For the one-term approximation, we have

$$\frac{Q}{Q_{max}} = 1 - C_1 e^{-\lambda_1^2 Fo} \frac{\sin \lambda_1}{\lambda_1} \tag{4.34}$$

---

## Example 4.2

### Cooling of a large plate

*Problem*

The large stainless steel plate ($\rho = 7900$ kg/m³, $c = 580$ J/kg C, $k = 23$ W/m C) shown below is 30 cm thick. It is initially in an oven and at a uniform temperature of 800 C. It is taken out of the oven, and it starts cooling to the 25 C room air. The convective coefficient is 75 W/m² C.

**(a)** Can the lumped method be used for this problem?
**(b)** How long does it take for the center of the plate to reach 300 C?
**(c)** What is the surface temperature of the plate at that time?
**(d)** During the cooling, how much heat goes into the room air?

If your answer is "no" to item (a), you may use the one-term approximation method for items (b), (c), and (d).

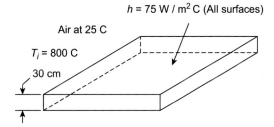

$h = 75$ W / m² C (All surfaces)

Air at 25 C

$T_i = 800$ C

30 cm

*Solution*

**(a)** We first calculate $Bi_{lumped}$ to see if we can use the lumped method.

$$Bi_{lumped} = \frac{h(V/A)}{k} = \frac{(75)(0.15)}{23} = 0.489$$

As $Bi_{lumped} > 0.1$ the lumped method is inappropriate. We will consider the spatial distribution of temperatures in the plate and use the one-term approximation method.

**(b)** For the center temperature, we use Eq. (4.25):

$$\frac{\theta_o}{\theta_i} = \frac{T(0,t) - T_\infty}{T_i - T_\infty} = C_1 e^{-\lambda_1^2 Fo} \text{ for } Fo > 0.2 \qquad (4.25)$$

$$Bi = \frac{hL}{k} = \frac{(75)(0.15)}{23} = 0.489$$

For $Bi = 0.489$, linear interpolation of entries in Table 4.1 gives $\lambda_1 = 0.6467$ and $C_1 = 1.0688$.

$$Fo = \frac{\alpha t}{L^2} \text{ and } \alpha = \frac{k}{\rho c}$$

Therefore, $\alpha = \frac{23}{(7900)(580)} = 5.02 \times 10^{-6} \text{ m}^2/\text{s}$ and $Fo = \frac{(5.02 \times 10^{-6})t}{(0.15)^2} = 2.231 \times 10^{-4} t$.
Putting the values in Eq. (4.25), we have

$$\frac{300 - 25}{800 - 25} = (1.0688) e^{-0.6467^2 (2.231 \times 10^{-4}) t}$$

or $e^{-9.3305 \times 10^{-5} t} = 0.3320$
Taking the natural log of both sides, we get $-9.3305 \times 10^{-5} t = \ln(0.3320) = -1.1026$ and arrive at the answer $t$ = **11817 s**.
Let us just check $Fo$ and make sure that $Fo$ is greater than 0.2, which is the criterion for use of the one-term approximation method:

$$Fo = 2.231 \times 10^{-4} t = 2.231 \times 10^{-4} (11817) = 2.636$$

As $Fo > 0.2$, our use of the one-term approximation was fine.

**(c)** To get the surface temperature, we use Eq. (4.24):

$$\frac{\theta}{\theta_i} = \frac{T(x,t) - T_\infty}{T_i - T_\infty} = C_1 e^{-\lambda_1^2 Fo} \cos(\lambda_1 x / L) \text{ for } Fo > 0.2 \qquad (4.24)$$

$$\frac{T(L = 0.15\text{m}, 11817\text{s}) - 25}{800 - 25} = 1.0688 \ e^{-0.6467^2 (2.636)} \cos(0.6467)$$

$$T = 244.5 \text{ C}$$

The surface temperature after 11817 s is **244.5 C**.

**(d)** Using Eq. (4.34), we have

$$\frac{Q}{Q_{max}} = 1 - C_1 e^{-\lambda_1^2 Fo} \frac{\sin \lambda_1}{\lambda_1} \qquad (4.34)$$

$$\frac{Q}{Q_{max}} = 1 - 1.0688 e^{-0.6467^2 (2.636)} \frac{\sin(0.6467)}{0.6467} = 0.6693$$

$$Q_{max} = \rho c V (T_i - T_\infty) = \rho c (2LA)(T_i - T_\infty) = (7900)(580)(0.3)(800 - 25)A = 1.065 \times 10^9 A$$

where $A$ is the cross-sectional area of the plate.
Therefore, $Q/A = 0.6693(1.065 \times 109) = 7.13 \times 10^8 \text{J/m}^2$.
During the process, **7.13 × 10⁸ J** goes into the surrounding fluid per square meter of cross-sectional area.

## 4.3.3 Long cylinders

From Chapter 2, the general heat conduction equation in cylindrical coordinates is

$$\frac{1}{r} \frac{\partial}{\partial r} \left( k \, r \frac{\partial T}{\partial r} \right) + \frac{1}{r^2} \frac{\partial}{\partial \varphi} \left( k \frac{\partial T}{\partial \varphi} \right) + \frac{\partial}{\partial z} \left( k \frac{\partial T}{\partial z} \right) + q_{gen} = \rho \, c \, \frac{\partial T}{\partial t} \qquad (4.35)$$

Consider the cylinder of Fig. 4.3 above. The cylinder has no internal heat generation and the conductivity is constant. The temperature distribution depends only on $r$. Eq. (4.35) then reduces to the one-dimensional transient equation.

$$\frac{\partial^2 T}{\partial r^2} + \frac{1}{r}\frac{\partial T}{\partial r} = \frac{1}{\alpha}\frac{\partial T}{\partial t} \tag{4.36}$$

where $\alpha$ = thermal diffusivity = $\frac{k}{\rho c}$

All points in the cylinder are initially at temperature $T_i$. Then, at time zero, the surface of the cylinder at $r = r_o$ is suddenly subjected to convection to the environment at $T_\infty$ with a convective coefficient $h$.

The temperature solution is symmetrical about $r = 0$. Hence, the boundary condition equation at the centerline of the cylinder is

$$\text{At } r = 0, \quad \frac{\partial T}{\partial r} = 0 \tag{4.37}$$

At the surface of the cylinder, the convective boundary condition equation is

$$\text{At } r = r_o, \quad -k\frac{\partial T}{\partial r} = h(T - T_\infty) \tag{4.38}$$

$$\text{The initial condition is: At } t = 0, \quad T(r) = T_i \tag{4.39}$$

The solution of Eq. (4.36) with conditions (4.37) through (4.39) is an infinite series:

$$T(r,t) = T_\infty + (T_i - T_\infty)\sum_{n=1}^{\infty} C_n e^{-\frac{\lambda_n^2 \alpha t}{r_o^2}} J_0(\lambda_n r / r_o) \tag{4.40}$$

This equation gives the temperature at any point in the cylinder at any time.
This solution given by Eq. (4.40) is often written in dimensionless form:

$$\frac{\theta}{\theta_i} = \frac{T(r,t) - T_\infty}{T_i - T_\infty} = \sum_{n=1}^{\infty} C_n e^{-\lambda_n^2 Fo} J_0(\lambda_n r / r_o) \tag{4.41}$$

$$Fo = \frac{\alpha t}{r_o^2} = \text{Fourier Number}$$

The eigenvalues $\lambda_n$ are the positive roots of the transcendental equation

$$\lambda_n \frac{J_1(\lambda_n)}{J_0(\lambda_n)} = Bi \tag{4.42}$$

where $Bi$ is the Biot number. For cylinders, $Bi = \frac{hr_o}{k}$.

Note: $J_0$ and $J_1$ are Bessel functions of the first kind. The functions are available in software such as Excel and Matlab, and a table of the functions is in Appendix G.

Among other methods, Eq. (4.42) may be solved for the $\lambda_n$ through use of the *Goal Seek* or *Solver* features of Excel or the *fzero* function of Matlab. See Appendix J for details.

Once the $\lambda_n$ values are determined, the $C_n$ values in Eqs. (4.40) and (4.41) may be obtained from

$$C_n = \frac{2}{\lambda_n} \frac{J_1(\lambda_n)}{J_0^2(\lambda_n) + J_1^2(\lambda_n)} \tag{4.43}$$

The series in Eqs. (4.40) and (4.41) converges rapidly with increasing time due to the exponential term. If the Fourier number is greater than 0.2, excellent results are obtained using just the first term of the series. This is called the **one-term approximation.** Keeping just the first term, Eq. (4.41) becomes

$$\frac{\theta}{\theta_i} \frac{T(r,t) - T_\infty}{T_i - T_\infty} = C_1 e^{-\lambda_1^2 Fo} \; J_0(\lambda_1 r / r_o) \text{ for } Fo > 0.2 \tag{4.44}$$

From Eq. (4.44), the temperature on the centerline of the cylinder, i.e., at $r = 0$ is

$$\frac{\theta_o}{\theta_i} = \frac{T(0,t) - T_\infty}{T_i - T_\infty} = C_1 e^{-\lambda_1^2 Fo} \text{ for } Fo > 0.2 \tag{4.45}$$

If we are interested in an earlier time of the system's response, that is, when $Fo = \frac{\alpha t}{r_o^2} \le 0.2$, we can just use more terms of the infinite series for the temperature.

The $\lambda_1$ and $C_1$ values are given for different Biot numbers in Table 4.1. If more than one term is used in the series solution, then the $\lambda_n$ can be obtained through solution of Eq. (4.42) and the $C_n$ can then be obtained from Eq. (4.43).

We discussed above the amount of energy transferred from a plate to the surroundings during a time period of the process. We will now do this for a cylinder: If the cylinder is initially uniform at $T_i$ and the process is fully completed, i.e., the cylinder is finally uniform at $T_\infty$, the decrease in internal energy of the cylinder is

$$Q_{max} = \rho c V (T_i - T_\infty) \tag{4.46}$$

where
$\rho$ = density of the cylinder.
$c$ = specific heat of the cylinder.
$V$ = volume of the cylinder = $\pi r_o^2 L$, where $L$ is the length of the cylinder.
If the process is only partially completed, the decrease in internal energy is

$$Q = \rho c \int_V (T_i - T) dV \tag{4.47}$$

The decrease in internal energy given in Eq. (4.47) is equal to the energy going into the surrounding medium by convection.

For the cylinder, $dV = 2\pi r L \; dr$, where $L$ is the length of the cylinder. Therefore, Eq. (4.47) becomes

$$Q = 2\pi \rho c L \int_0^{r_o} [T_i - T(r,t)] r dr \tag{4.48}$$

The temperature in the cylinder at time $t$ is given by Eq. (4.40). Putting this into Eq. (4.48), and remembering that $Fo = \alpha t / r_o^2$, we have

$$Q = 2\pi\rho cL \int_0^{r_o} \left[ T_i - T_\infty - (T_i - T_\infty) \sum_{n=1}^{\infty} C_n e^{-\lambda_n^2 Fo} J_0(\lambda_n r / r_o) \right] r dr \qquad (4.49)$$

Rearranging and recognizing that we are integrating only over $r$ and not $t$, Eq. (4.49) becomes

$$Q = 2\pi\rho cL \left[ \int_0^{r_o} (T_i - T_\infty) r dr - (T_i - T_\infty) \sum_{n=1}^{\infty} C_n e^{-\lambda_n^2 Fo} \int_0^{r_o} J_0(\lambda_n r / r_o) \, r dr \right] \qquad (4.50)$$

Performing the integrations and then factoring out $(T_i - T_\infty)$ and $r_o^2$, we get

$$Q = \rho cL\pi r_o^2 (T_i - T_\infty) \left[ 1 - 2 \sum_{n=1}^{\infty} C_n e^{-\lambda_n^2 Fo} \frac{J_1(\lambda_n)}{\lambda_n} \right] \qquad (4.51)$$

If we divide this result by $Q_{max}$ from Eq. (4.46) and recognize that the volume of the cylinder is $\pi r_o^2 L$, we get the final dimensionless result:

$$\frac{Q}{Q_{max}} = 1 - 2 \sum_{n=1}^{\infty} C_n e^{-\lambda_n^2 Fo} \frac{J_1(\lambda_n)}{\lambda_n} \qquad (4.52)$$

For the one-term approximation, we have

$$\frac{Q}{Q_{max}} = 1 - 2C_1 e^{-\lambda_1^2 Fo} \frac{J_1(\lambda_1)}{\lambda_1} \qquad (4.53)$$

where $Q_{max} = \rho c\pi r_o^2 L(T_i - T_\infty)$.

### 4.3.4 Spheres

From Chapter 2, the general heat conduction equation in spherical coordinates is

$$\frac{1}{r^2} \frac{\partial}{\partial r} \left( kr^2 \frac{\partial T}{\partial r} \right) + \frac{1}{r^2 \sin^2 \theta} \frac{\partial}{\partial \varphi} \left( k \frac{\partial T}{\partial \varphi} \right) + \frac{1}{r^2 \sin \theta} \frac{\partial}{\partial \theta} \left( k \sin \theta \frac{\partial T}{\partial \theta} \right) + q_{gen} = \rho c \frac{\partial T}{\partial t} \qquad (4.54)$$

Consider the sphere of Fig. 4.4 above. The sphere has no internal heat generation and the conductivity is constant. The temperature distribution depends only on $r$. Eq. (4.54) then reduces to the one-dimensional transient equation

$$\frac{\partial^2 T}{\partial r^2} + \frac{2}{r} \frac{\partial T}{\partial r} = \frac{1}{\alpha} \frac{\partial T}{\partial t} \qquad (4.55)$$

where $\alpha$ = thermal diffusivity = $\frac{k}{\rho c}$.

All points in the sphere are initially at temperature $T_i$. Then, at time zero, the surface of the sphere at $r = r_o$ is suddenly subjected to convection to the environment at $T_\infty$ with a convective coefficient $h$.

The temperature solution is symmetrical about $r = 0$. Hence, the boundary condition equation at the center of the sphere is:

$$\text{At } r = 0, \quad \frac{\partial T}{\partial r} = 0 \tag{4.56}$$

At the surface of the sphere, the convective boundary condition equation is

$$\text{At } r = r_o, \quad -k\frac{\partial T}{\partial r} = h(T - T_\infty) \tag{4.57}$$

$$\text{The initial condition is: } \text{At } t = 0, \quad T(r) = T_i \tag{4.58}$$

The solution of Eq. (4.55) with conditions (4.56) through (4.58) is an infinite series:

$$T(r,t) = T_\infty + (T_i - T_\infty)\sum_{n=1}^{\infty} C_n e^{-\frac{\lambda_n^2 \alpha t}{r_o^2}}\frac{\sin(\lambda_n r/r_o)}{\lambda_n r/r_o} \tag{4.59}$$

This equation gives the temperature at any point in the sphere at any time.
This solution given by Eq. (4.59) is often written in dimensionless form:

$$\frac{\theta}{\theta_i} = \frac{T(r,t) - T_\infty}{T_i - T_\infty} = \sum_{n=1}^{\infty} C_n e^{-\lambda_n^2 Fo}\frac{\sin(\lambda_n r/r_o)}{\lambda_n r/r_o} \tag{4.60}$$

$$Fo = \frac{\alpha t}{r_o^2} = \text{Fourier Number}$$

The eigenvalues $\lambda_n$ are the positive roots of the transcendental equation

$$1 - \lambda_n \cot\lambda_n = Bi \tag{4.61}$$

where $Bi$ is the Biot number. For spheres, $Bi = \frac{h\,r_o}{k}$. Among other methods, Eq. (4.61) may be solved for the $\lambda_n$ through use of the *Goal Seek* or *Solver* features of Excel or the *fzero* function of Matlab. See Appendix J for details.

Once the $\lambda_n$ values are determined, the $C_n$ values in Eqs. (4.59) and (4.60) may be obtained from

$$C_n = \frac{4(\sin \lambda_n - \lambda_n \cos \lambda_n)}{2\lambda_n - \sin(2\lambda_n)} \tag{4.62}$$

Note: The arguments of the trig functions are in *radians*.

The series in Eqs. (4.59) and (4.60) converge rapidly with increasing time due to the exponential term. If the Fourier number is greater than 0.2, excellent results are obtained using just the first term of the series. This is called the **one-term approximation.** Keeping just the first term, Eq. (4.60) becomes

$$\frac{\theta}{\theta_i} = \frac{T(r,t) - T_\infty}{T_i - T_\infty} = C_1 e^{-\lambda_1^2 Fo}\frac{\sin(\lambda_1 r/r_o)}{\lambda_1 r/r_o} \quad \text{for } Fo > 0.2 \tag{4.63}$$

As $r$ goes to 0, $\frac{\sin(\lambda_1 r/r_o)}{\lambda_1 r/r_o}$ goes to 1. Therefore, the temperature at the center of the sphere is

$$\frac{\theta_o}{\theta_i} = \frac{T(0,t) - T_\infty}{T_i - T_\infty} = C_1 e^{-\lambda_1^2 Fo} \quad \text{for } Fo > 0.2 \tag{4.64}$$

If we are interested in an earlier time of the system's response, that is, when $Fo = \frac{\alpha t}{r_o^2} \le 0.2$, we can just use more terms of the infinite series for the temperature.

The $\lambda_1$ and $C_1$ values are given for different Biot numbers in Table 4.1. If more than one term is used in the series solution, then the $\lambda_n$ can be obtained through solution of Eq. (4.61) and the $C_n$ can then be obtained from Eq. (4.62).

Regarding the amount of energy transferred from a sphere to the surroundings during a time period of the process: If the sphere is initially uniform at $T_i$ and the process is fully completed, i.e., the sphere is finally uniform at $T_\infty$, the decrease in internal energy of the sphere is

$$Q_{max} = \rho c V(T_i - T_\infty) \tag{4.65}$$

where

$\rho$ = density of the sphere.
$c$ = specific heat of the sphere.
$V$ = volume of the sphere = $\frac{4}{3}\pi r_o^3$.
If the process is only partially completed, the decrease in internal energy is

$$Q = \rho c \int_V (T_i - T)dV \tag{4.66}$$

The decrease in internal energy given in Eq. (4.66) is equal to the energy going into the surrounding medium by convection.

For the sphere, $dV = 4\pi r^2 \, dr$. Therefore, Eq. (4.66) becomes

$$Q = 4\pi\rho c \int_0^{r_o} [T_i - T(r,t)] \, r^2 dr \tag{4.67}$$

The temperature in the sphere at time $t$ is given by Eq. (4.59). Putting this into Eq. (4.67), and remembering that $Fo = \alpha t/r_o^2$, we have

$$Q = 4\pi\rho c \int_0^{r_o} \left[ T_i - T_\infty - (T_i - T_\infty) \sum_{n=1}^{\infty} C_n e^{-\lambda_n^2 Fo} \frac{\sin(\lambda_n r/r_o)}{\lambda_n r/r_o} \right] r^2 dr \tag{4.68}$$

Rearranging and recognizing that we are integrating only over $r$ and not $t$, Eq. (4.68) becomes

$$Q = 4\pi\rho c \left[ \int_0^{r_o} (T_i - T_\infty)r^2 dr - (T_i - T_\infty) \sum_{n=1}^{\infty} C_n e^{-\lambda_n^2 Fo} \int_0^{r_o} \frac{\sin(\lambda_n r/r_o)}{\lambda_n r/r_o} r^2 dr \right] \tag{4.69}$$

Performing the integrations and then rearranging the equation, we get

$$Q = \rho c \frac{4}{3}\pi r_o^3 (T_i - T_\infty) \left[ 1 - 3 \sum_{n=1}^{\infty} C_n e^{-\lambda_n^2 Fo} \frac{\sin \lambda_n - \lambda_n \cos \lambda_n}{\lambda_n^3} \right] \tag{4.70}$$

If we divide this result by $Q_{max}$ from Eq. (4.65) and recognize that the volume of the sphere is $\frac{4}{3}\pi r_o^3$, we get the final dimensionless result:

$$\frac{Q}{Q_{max}} = 1 - 3\sum_{n=1}^{\infty} C_n e^{-\lambda_n^2 Fo}\frac{\sin \lambda_n - \lambda_n \cos \lambda_n}{\lambda_n^3} \tag{4.71}$$

For the one-term approximation, we have

$$\frac{Q}{Q_{max}} = 1 - 3C_1 e^{-\lambda_1^2 Fo}\frac{\sin \lambda_1 - \lambda_1 \cos \lambda_1}{\lambda_1^3} \tag{4.72}$$

where $Q_{max} = \rho c\, \frac{4}{3}\pi r_o^3\, (T_i - T_\infty)$.

---

## Example 4.3

### Cooling of a glass sphere

*Problem*

A glass sphere ($\rho = 2800$ kg/m³, $c = 800$ J/kg C, $k = 0.8$ W/m C) is shown in the figure below. It has a diameter of 2 cm and is initially at a uniform temperature of 200 C. The sphere then begins cooling to the surrounding fluid that is at 20 C. The convective coefficient is 20 W/m² C.

**(a)** What is the temperature of the center of the sphere after 10 min?

**(b)** What is the temperature of the surface of the sphere after 10 min?

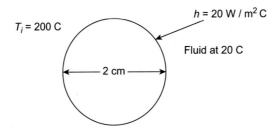

*Solution*

The first thing to do is to calculate $Bi_{lumped}$ to see if the lumped method can be used.

$$Bi_{lumped} = \frac{h(V/A)}{k} = \frac{h\left(\frac{\frac{4}{3}\pi r_o^3}{4\pi r_o^2}\right)}{k} = \frac{hr_o}{3k} = \frac{(20)(0.01)}{3(0.8)} = 0.0833$$

As $Bi_{lumped}$ is less than 0.1, the criterion for use of the lumped method is met. However, we are quite close to 0.1, so let us solve the problem in both ways—with the lumped method and also with the Heisler method. Doing this, we will be able to see the amount of spatial variation of temperature in the sphere.

**Lumped method.** We will use Eq. (4.8):

$$T(t) = T_\infty + (T_i - T_\infty)e^{-\left(\frac{hA}{\rho cV}\right)t} \tag{4.8}$$

$$t = 10 \text{ minutes} = 600\text{s}, \quad \frac{A}{V} = \frac{4\pi r_o^2}{\frac{4}{3}\pi r_o^3} = \frac{3}{r_o} = \frac{3}{0.01} = 300$$

so, Eq. (4.8) becomes $T = 20 + 180\, e^{-\frac{(20)(300)(600)}{(2800)(800)}} = 20 + 180\, e^{-1.6071} = 56.1$ C.

Using the lumped method, the temperature of the sphere after 600 s is **56.1 C.**

### Heisler method

**(a)** We will first check the Fourier number to see if the one-term approximation can be used.

$$Fo = \frac{\alpha t}{r_o^2} \quad \alpha = \frac{k}{\rho c} = \frac{0.8}{(2800)(800)} = 3.571 \times 10^{-7} \text{ m}^2/\text{s}$$

$Fo = \frac{3.571 \times 10^{-7}(600)}{(0.01)^2} = 2.143 > 0.2,$ so the one-term approximation is OK.

Eq. (4.64) gives the temperature at the center of the sphere:

$$\frac{\theta_o}{\theta_i} = \frac{T(0,t) - T_\infty}{T_i - T_\infty} = C_1 e^{-\lambda_1^2 Fo} \tag{4.64}$$

Eigenvalue $\lambda_1$ can be obtained from Table 4.1 or by solving the transcendental equation given in Eq. (4.61). Both methods require the Biot number.

$$Bi = \frac{hr_o}{k} = \frac{(20)(0.01)}{0.8} = 0.25$$

From linear interpolation of Table 4.1 for this Biot number, $\lambda_1 = 0.8401$ and $C_1 = 1.0736$.
Putting values in Eq. (4.64), we get

$$T(0, 600 \text{ s}) = 20 + (200 - 20)(1.0736)e^{-0.8401^2(2.143)} = 62.6 \text{ C}.$$

The temperature of the center of the sphere after 600 s is **61.9 C.**

**(b)** Eq. (4.63) is used to get the surface temperature.

$$\frac{\theta}{\theta_i} = \frac{T(r,t) - T_\infty}{T_i - T_\infty} = C_1 e^{-\lambda_1^2 Fo} \frac{\sin(\lambda_1 r/r_o)}{\lambda_1 r/r_o} \tag{4.63}$$

At the surface, $r = r_o$. Putting values into Eq. (4.63), we have

$$T(r_o, \ 600 \text{ s}) = 20 + (200 - 20)(1.0736)e^{-0.8401^2(2.143)} \frac{\sin(0.8401)}{0.8401} = \textbf{57.8 C}.$$

The temperature of the surface of the sphere after 600 s is **57.8 C.**

For this problem, the value of $Bi_{lumped}$ met the 0.1 criterion for use of the lumped method. However, we see that there is still some variation of temperature within the sphere. We conclude that in cases where the $Bi_{lumped}$ meets the 0.1 criterion for use of the lumped method, but where $Bi_{lumped}$ is fairly close to 0.1, it is prudent to also perform a Heisler method solution. We will now do a problem in which the Fourier number is less than 0.2 and multiple terms of the series solutions must be included for accurate results.

## Example 4.4

### Cooling of a large sphere

*Problem*

Note: This example is the same as Example 4.3 except the sphere is considerably larger. Its diameter is 10 cm rather than 2 cm.

A glass sphere ($\rho = 2800$ kg/m$^3$, $c = 800$ J/kg C, $k = 0.8$ W/m C) has a diameter of 10 cm and is initially at a uniform temperature of 200 C. The sphere then begins cooling to the surrounding fluid, which is at 20 C. The convective coefficient is 20 W/m$^2$ C.

**(a)** What is the temperature of the center of the sphere after 10 min?

**(b)** What is the temperature of the surface of the sphere after 10 min?

*Solution*

We first check $Bi_{lumped}$ to see if the lumped method can be used.

$$Bi_{lumped} = \frac{hr_o}{3k} = \frac{(20)(0.05)}{3(0.8)} = 0.417$$

This is considerably greater than 0.1 so the lumped method is inappropriate. We will use the Heisler method.

Let us check the Fourier number to see if the one-term approximation can be used. (From Example 4.3, $\alpha = 3.571 \times 10^{-7}$ m / s$^2$.)

$$Fo = \frac{\alpha t}{r_o^2} = \frac{(3.571 \times 10^{-7})(600)}{0.05^2} = 0.0857$$

As $Fo < 0.2$, we are too early in the process to use the one-term approximation. We should use more than one term in the infinite series.

**(a)** The solution for the temperatures in the sphere is given by Eq. (4.59)

$$T(r,t) = T_\infty + (T_i - T_\infty) \sum_{n=1}^{\infty} C_n e^{-\frac{\lambda_n^2 \alpha t}{r_o^2}} \frac{\sin(\lambda_n r/r_o)}{\lambda_n r/r_o} \tag{4.73}$$

As $r$ goes to 0, $\frac{\sin(\lambda_n r/r_o)}{\lambda_n r/r_o}$ goes to 1. Therefore, for the center temperature, we have

$$T(0,t) = T_\infty + (T_i - T_\infty) \sum_{n=1}^{\infty} C_n e^{-\frac{\lambda_n^2 \alpha t}{r_o^2}} \tag{4.74}$$

Let us arbitrarily keep four terms of the series. Also, $Fo = \frac{\alpha t}{r_o^2}$. Then Eq. (4.74) becomes

$$T(0,t) = T_\infty + (T_i - T_\infty) \left[ C_1 e^{-\lambda_1^2 Fo} + C_2 e^{-\lambda_2^2 Fo} + C_3 e^{-\lambda_3^2 Fo} + C_4 e^{-\lambda_4^2 Fo} \right] \tag{4.75}$$

The values of $\lambda_n$ are from the solution of the transcendental Eq. (4.61). Once we have the $\lambda_n$, we get the $C_n$ from Eq. (4.62). The Biot number is $Bi = \frac{hr_o}{k} = \frac{(20)(0.05)}{0.8} = 1.25$. For this Biot number, the $\lambda_n$ values are 1.7155, 4.7648, 7.8857, and 11.0183. The corresponding $C_n$ values are 1.3313, $-0.5183$, 0.3157, and $-0.2265$. Putting values in Eq. (4.75), we get

$$T(0,t) = 20 + 180 \left[ 1.3313 e^{-1.7155^2(0.0857)} - 0.5183 e^{-4.7648^2(0.0857)} + 0.3157 e^{-7.8857^2(0.0857)} - 0.2265 e^{-11.0183^2(0.0857)} \right]$$

$$T(0,t) = 20 + 180 \; (1.0345 - 0.0741 + 0.0015 - 6.9 \times 10^{-6}) = \textbf{193.1 C.}$$

The center temperature of the sphere is **193.1 C** after 600 s of cooling.

It is seen that the series converges very quickly. If we had only used the first two terms in the series, the result for the center temperature would have been 192.9 C.

We showed above that the one-term approximation was inappropriate for this problem as the Fourier number was too small. Indeed, if we had only used the first term of the series, the center temperature after 600 s would have been $T(0,t) = 20 + 180 \; (1.0345) = 206.2$ C. This is obviously wrong as the sphere is being cooled and it starts out with an initial temperature of 200 C.

**(b)** Eq. (4.59) gives the temperature at any location in the sphere.

$$T(r,t) = T_\infty + (T_i - T_\infty) \sum_{n=1}^{\infty} C_n e^{-\frac{\lambda_n^2 \alpha t}{r_o^2}} \frac{\sin(\lambda_n r/r_o)}{\lambda_n r/r_o} \tag{4.76}$$

At the surface of the sphere, $r = r_o$. And, like Part (a) above, we will keep the first four terms of the series. Eq. (4.76) then becomes

$$T(r_o,t) = T_\infty + (T_i - T_\infty) \left[ C_1 e^{-\lambda_1^2 Fo} \frac{\sin \lambda_1}{\lambda_1} + C_2 e^{-\lambda_2^2 Fo} \frac{\sin \lambda_2}{\lambda_2} + C_3 e^{-\lambda_3^2 Fo} \frac{\sin \lambda_3}{\lambda_3} + C_4 e^{-\lambda_4^2 Fo} \frac{\sin \lambda_4}{\lambda_4} \right]$$

Putting values into this equation, we get

$$T(r_o,t) = 20 + 180 (0.5967 - 0.0155 + 0.0002 - 6.3 \times 10^{-7}) = \textbf{124.7 C.}$$

The surface temperature after 600 s of cooling is **124.7 C.**

Like Part (a), it is seen that the series converges quickly and only the first two terms are significant.

## 4.4 Multidimensional systems with spatial variation
### 4.4.1 Overview

Section 4.3 dealt with three geometries: Large (actually infinite) plates of thickness $2L$, long (actually infinitely long) cylinders of radius $r_o$, and spheres of radius $r_o$. These geometries are one-dimensional. In all three cases, a location in the body can be described by a single parameter: "$x$" for the plates and "$r$" for the cylinders and spheres.

The Heisler method of Section 4.3 can be extended to two- and three-dimensional bodies. Let us start by visualizing how we can get multidimensional objects from the one-dimensional plates and cylinders.

First, let us consider two large plates. If they intersect each other at right angles, the intersected volume is a long (actually infinitely long) bar. If the two plates have the same thickness $2L$, then the cross section of the bar is square. If the two plates have different thicknesses $2L_1$ and $2L_2$, then the cross section of the bar is rectangular. This is shown in Fig. 4.5.

A second object is a short cylinder of length $2L$ and radius $r_o$. As shown in Fig. 4.6, this object is created by intersecting a long cylinder of radius $r_o$ with a large plate of thickness $2L$.

These two examples are two-dimensional objects as a location in the object can be described by two parameters. In the long bar, a location in the cross section is defined by the "$x$" locations in the two

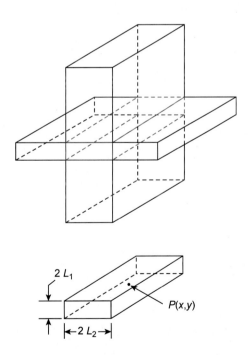

**FIGURE 4.5**

Long bar created by two large plates.

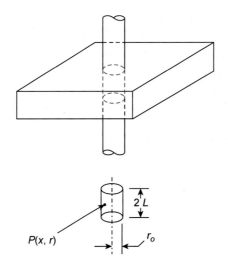

**FIGURE 4.6**

Short cylinder created by a long cylinder and a large plate.

intersecting large plates. For the short cylinder, a location in the cylinder is defined by the "$x$" location in the plate and the "$r$" location in the long cylinder.

An example of a three-dimensional object is the rectangular solid. This object is formed by the intersection of three large plates at right angles to each other. Consider Fig. 4.5 above and envision an additional vertical plate at right angles to the shown vertical plate. If this second vertical plate cuts through both the shown vertical plate and the horizontal plate, the intersected volume is a rectangular solid. Said another way, the rectangular solid is formed by the intersection of three large plates at right angles to each other. If all three plates have the same thickness, then the object is a cube of side $2L$. If the plates are not all the same thickness, then we have a rectangular solid of dimensions $2L_1$ by $2L_2$ by $2L_3$. A location in the solid is defined by the "$x$" locations for the three intersecting large plates. See Fig. 4.7. The rectangular solid is formed by two vertical plates of thickness $2L_1$ and $2L_2$, respectively, and a horizontal plate of thickness $2L_3$.

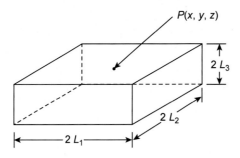

**FIGURE 4.7**

Rectangular solid created by three large plates.

It turns out that we can use the Heisler method, discussed in Section 4.3 for one-dimensional objects, to solve problems involving two- and three-dimensional objects, which are formed by the intersection of the one-dimensional objects. We simply solve the one-dimensional problems separately and then multiply these results to get the solutions for the two- and three-dimensional objects. That is, the product of the solutions for the one-dimensional objects is the solution for the multidimensional object. The theory behind this technique is discussed in Refs. [3−5].

The following equations outline the solutions for the three multidimensional objects described above:

For the long bar,

$$\left(\frac{T(x,y,t)-T_\infty}{T_i-T_\infty}\right)_{\text{long bar}} = \left(\frac{T(x,t)-T_\infty}{T_i-T_\infty}\right)_{\text{plate 1}} \left(\frac{T(y,t)-T_\infty}{T_i-T_\infty}\right)_{\text{plate 2}} \tag{4.77}$$

For the one-dimensional plates on the right side of Eq. (4.77): The "$x$" and "$y$" are the "$x$" locations for Plate 1 and Plate 2, respectively.

Eq. (4.77) could also be written as

$$\left(\frac{\theta}{\theta_i}\right)_{\text{long bar}} = \left(\frac{\theta}{\theta_i}\right)_{\text{plate 1}} \left(\frac{\theta}{\theta_i}\right)_{\text{plate 2}} \tag{4.78}$$

For the short cylinder,

$$\left(\frac{T(x,r,t)-T_\infty}{T_i-T_\infty}\right)_{\text{short cylinder}} = \left(\frac{T(r,t)-T_\infty}{T_i-T_\infty}\right)_{\text{long cylinder}} \left(\frac{T(x,t)-T_\infty}{T_i-T_\infty}\right)_{\text{plate}} \tag{4.79}$$

Eq. (4.79) could also be written as

$$\left(\frac{\theta}{\theta_i}\right)_{\text{short cylinder}} = \left(\frac{\theta}{\theta_i}\right)_{\text{long cylinder}} \left(\frac{\theta}{\theta_i}\right)_{\text{plate}} \tag{4.80}$$

For the rectangular solid,

$$\left(\frac{T(x,y,z,t)-T_\infty}{T_i-T_\infty}\right)_{\text{rect. solid}} = \left(\frac{T(x,t)-T_\infty}{T_i-T_\infty}\right)_{\text{plate 1}} \left(\frac{T(y,t)-T_\infty}{T_i-T_\infty}\right)_{\text{plate 2}} \left(\frac{T(z,t)-T_\infty}{T_i-T_\infty}\right)_{\text{plate 3}} \tag{4.81}$$

For the one-dimensional plates on the right side of Eq. (4.81): The "$x$," "$y$," and "$z$" are the "$x$" locations for Plates 1, 2 and 3, respectively.

Eq. (4.81) could also be written as

$$\left(\frac{\theta}{\theta_i}\right)_{\text{rect. solid}} = \left(\frac{\theta}{\theta_i}\right)_{\text{plate 1}} \left(\frac{\theta}{\theta_i}\right)_{\text{plate 2}} \left(\frac{\theta}{\theta_i}\right)_{\text{plate 3}} \tag{4.82}$$

For the product solution of multidimensional objects, the fluid temperature $T_\infty$ for the object must be the same as that for all the one-dimensional objects which were combined to form the object. The convective coefficients, however, need not be the same on all surfaces of the multidimensional object. However, they must be the same for the different surfaces of each one-dimensional object forming the multidimensional object. To clarify this: If we have a short cylinder, the $h$ value on the ends of the cylinder can be different from the $h$ value on the side of the cylinder. However, both ends of the cylinder must have the same $h$ value. For the rectangular solid, we can have a maximum of three different $h$ values. The two surfaces of each intersecting plate must have the same $h$ value, but the different plates can have different $h$ values.

Regarding the heat transfer from a multidimensional object, Langston [6] showed how the results from the one-dimensional objects could be combined to get the heat transfer results for the two- or three-dimensional object formed from the intersection of the one-dimensional objects, as follows:

For a two-dimensional object formed by the intersection of one-dimensional objects "1" and "2":

$$\left(\frac{Q}{Q_{max}}\right)_{2-D \text{ object}} = \left(\frac{Q}{Q_{max}}\right)_1 + \left(\frac{Q}{Q_{max}}\right)_2 \left[1 - \left(\frac{Q}{Q_{max}}\right)_1\right] \tag{4.83}$$

For a three-dimensional object formed by the intersection of one-dimensional objects "1", "2", and "3":

$$\left(\frac{Q}{Q_{max}}\right)_{3-D \text{ object}} = \left(\frac{Q}{Q_{max}}\right)_1 + \left(\frac{Q}{Q_{max}}\right)_2 \left[1 - \left(\frac{Q}{Q_{max}}\right)_1\right] + \left(\frac{Q}{Q_{max}}\right)_3 \left[1 - \left(\frac{Q}{Q_{max}}\right)_1\right]$$

$$\times \left[1 - \left(\frac{Q}{Q_{max}}\right)_2\right] \tag{4.84}$$

The following three sections (Sections 4.4.2−4.4.4) include problems illustrating the use of these equations for the multidimensional objects discussed above, i.e., a long bar, a short cylinder, and a rectangular solid. The application of the Heisler method to additional geometries may be found in Refs. [7−10].

## 4.4.2 Long bar

As shown in Fig. 4.5 above, a long bar is the intersection of two large plates. Example 4.5 illustrates the application of the Heisler method to this geometry.

### Example 4.5
#### Extrusion of a PVC bar
*Problem*
A continuous PVC bar $(\rho = 1400 \text{ kg} / \text{m}^3,\ c = 900 \text{ J} / \text{kg C},\ k = 0.19 \text{ W} / \text{m C})$ is being produced by extrusion. The bar has a rectangular cross section of 2 cm by 4 cm. It leaves the extrusion machine and enters a water bath for cooling. The water bath is at 35 C and the bar is 140 C when it enters the water. The bar is in the bath for 10 min. It may be assumed that the convective coefficient while the bar is in the water bath is 200 W/m² C.

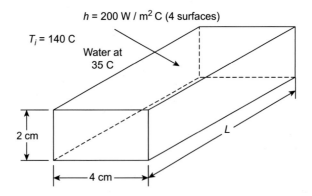

(a) What is the center temperature of the extruded bar as it leaves the water bath?

(b) What is the temperature on the surface of the bar, at the center of one of the 2-cm-wide sides as the bar leaves the bath?

(c) How much heat is transferred to the water bath per meter length of bar passing through it?

*Solution*

We first check $Bi_{lumped}$ to see if the lumped method can be used.

$$Bi_{lumped} = \frac{h(V/A)}{k} = \frac{(200)\left(\frac{(0.02)(0.04)L}{[2(02) + 2(04)]L}\right)}{0.19} = 7.018$$

(Note: In this calculation, "$L$" is the length of the bar. In the calculations below, "$L$" is one-half the thickness of the particular intersecting plate.)

As $Bi_{lumped}$ is greater than 0.1, the lumped method is inappropriate. We will use the Heisler method applied to a two-dimensional problem. The bar is the intersection of a 2-cm-thick large plate and a 4-cm-thick large plate.

$$\alpha = \frac{k}{\rho c} = \frac{0.19}{(1400)(900)} = 1.508 \times 10^{-7} \text{ m}^2/\text{s}$$

Let us calculate the Fourier numbers to see if the one-term approximation can be used.

$$Fo = \frac{\alpha t}{L^2} \quad t = 10 \text{ minutes} = 600\text{s}.$$

$$Fo \ (2 \text{ cm plate}) = \frac{1.508 \times 10^{-7}(600)}{(0.01)^2} = 0.905$$

$$Fo \ (4 \text{ cm plate}) = \frac{1.508 \times 10^{-7}(600)}{(0.02)^2} = 0.226$$

For both plates, the Fourier number is greater than 0.2, so the one-term approximation is appropriate. The product solution gives

$$\left(\frac{\theta}{\theta_i}\right)_{\text{bar}} = \left(\frac{\theta}{\theta_i}\right)_{2 \text{ cm plate}} \left(\frac{\theta}{\theta_i}\right)_{4 \text{ cm plate}} \tag{4.85}$$

(a) For this part of the problem, the location of interest, the center of the bar, is on the center plane of both plates. For the 2 cm plate ($L = 0.01$ m),

$$Bi = \frac{hL}{k} = \frac{(200)(0.01)}{0.19} = 10.526$$

From Equation (4.25), $\dfrac{\theta_o}{\theta_i} = \dfrac{T(0,t) - T_\infty}{T_i - T_\infty} = C_1 e^{-\lambda_1^2 Fo}$ (4.86)

$\lambda_1$ is from Eq. (4.22). Once we have $\lambda_1$ we can get $C_1$ from Eq. (4.23).

For the above Biot number, using Table 4.1 or from software mentioned above, we find that $\lambda_1 = 1.435$ and $C_1 = 1.263$. Putting these values into Eq. (4.86),

$$\left(\frac{\theta_o}{\theta_i}\right)_{2 \text{ cm plate}} = 1.263e^{-(1.435)^2(0.905)} = 0.1959 \tag{4.87}$$

For the 4 cm plate ($L = 0.02$ m),

$$Bi = \frac{hL}{k} = \frac{(200)(0.02)}{0.19} = 21.053$$

$$\frac{\theta_o}{\theta_i} = \frac{T(0,t) - T_\infty}{T_i - T_\infty} = C_1 e^{-\lambda_1^2 Fo} \tag{4.88}$$

$\lambda_1$ is from Eq. (4.22). Once we have $\lambda_1$ we can get $C_1$ from Eq. (4.23).

For the above Biot number, using Table 4.1 or from software mentioned above, we find that $\lambda_1 = 1.500$ and $C_1 = 1.270$.

Putting these values into Eq. (4.88),

$$\left(\frac{\theta_o}{\theta_i}\right)_{4 \text{ cm plate}} = 1.270 e^{-(1.500)^2(0.226)} = 0.7638 \tag{4.89}$$

For the bar,

$$\left(\frac{\theta_o}{\theta_i}\right)_{\text{bar}} = \frac{T_o - T_\infty}{T_i - T_\infty} = \left(\frac{\theta_o}{\theta_i}\right)_{2 \text{ cm plate}} \left(\frac{\theta_o}{\theta_i}\right)_{4 \text{ cm plate}} = (0.1959)(0.7638) = 0.1496 \tag{4.90}$$

$$T_o = 35 + (140 - 35)(0.1496) = 50.7 \text{ C}.$$

The temperature on the centerline of the bar as it leaves the water bath is **50.7 C.**

**(b)** The location of interest is at the center of the 2 cm plate and the surface of the 4 cm plate. Using values calculated in Part (a), and from Eq. (4.24), we have at the surface of the 4 cm plate

$$\left(\frac{\theta}{\theta_i}\right)_{4 \text{ cm plate}} = \frac{T(x = L, t) - T_\infty}{T_i - T_\infty} = C_1 e^{-\lambda_1^2 Fo} \cos \lambda_1 = 0.7638 \cos(1.500) = 0.05403 \tag{4.91}$$

So, for the bar, we have

$$\left(\frac{\theta}{\theta_i}\right)_{\text{bar}} = \frac{T - T_\infty}{T_i - T_\infty} = \left(\frac{\theta}{\theta_i}\right)_{2 \text{ cm plate}} \left(\frac{\theta}{\theta_i}\right)_{4 \text{ cm plate}} = (0.1959)(0.05403) = 0.01058 \tag{4.92}$$

$$T = T_\infty + (T_i - T_\infty)(0.01058) = 35 + (140 - 35)(0.01058) = \mathbf{36.1 \text{ C}.}$$

The temperature at the center of one of the 2-cm-wide sides is **36.1 C** as the bar leaves the water bath.

**(c)** From Eq. (4.83), we have

$$\left(\frac{Q}{Q_{max}}\right)_{\text{bar}} = \left(\frac{Q}{Q_{max}}\right)_{2 \text{ cm plate}} + \left(\frac{Q}{Q_{max}}\right)_{4 \text{ cm plate}} \left[1 - \left(\frac{Q}{Q_{max}}\right)_{2 \text{ cm plate}}\right] \tag{4.93}$$

For the 2 cm plate using Eq. (4.34),

$$\frac{Q}{Q_{max}} = 1 - C_1 e^{-\lambda_1^2 Fo} \frac{\sin \lambda_1}{\lambda_1} = 1 - 0.1959 \frac{\sin(1.435)}{1.435} = 0.8647 \tag{4.94}$$

For the 4 cm plate using Eq. (4.34),

$$\frac{Q}{Q_{max}} = 1 - C_1 e^{-\lambda_1^2 Fo} \frac{\sin \lambda_1}{\lambda_1} = 1 - 0.7638 \frac{\sin(1.500)}{1.500} = 0.4921 \tag{4.95}$$

So, from Equation (4.93), $\left(\frac{Q}{Q_{max}}\right)_{\text{bar}} = 0.8647 + (0.4921)(1 - 0.8647) = 0.9313$ \hfill (4.96)

$$(Q_{max})_{\text{bar}} = \rho c V (T_i - T_\infty) = (1400)(900)(0.02)(0.04)L(140 - 35) = 1.0584 \times 10^5 L \tag{4.97}$$

From Eqs. (4.96) and (4.97), $\frac{Q}{L} = (0.9313)(1.0584 \times 10^5) = 9.86 \times 10^4$ J/m length

The heat transfer to the water bath is **9.86 × 10⁴ J** per meter length of bar passing through the bath.

## 4.4.3 Short cylinder

As shown in Fig. 4.6 above, a short cylinder is the intersection of a long cylinder and a large plate. Example 4.6 illustrates the application of the Heisler method to this geometry.

---

## Example 4.6

### Cooling of a short cylinder
*Problem*

A stainless steel disk $(\rho = 7900 \text{ kg / m}^3, c = 480 \text{ J / kg C}, k = 15 \text{ W / m C})$ is 10 cm diameter and 5 cm thick. It goes into an oven where it is heated to a uniform temperature of 1200 C. It is then taken out of the oven and cooled by convection to a fluid, which is at 150 C. The convective coefficient on the curved surface of the disk is 100 W/m² C, and the convective coefficient on the two flat surfaces is 250 W/m² C.

**(a)** How long does it take for the center of the disk to reach 500 C?

**(b)** At the time obtained in Part (a): What is the temperature at a point located 3 cm from the centerline and 1.5 cm from one end?

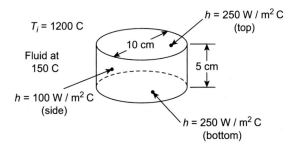

*Solution*

**(a)** Let us first calculate $Bi_{lumped}$. To be conservative, we will use the larger $h$ value that will give the larger Biot number.

$$Bi_{lumped} = \frac{h(V/A)}{k} = \frac{(250)\left(\dfrac{\pi r_o^2 L}{2\pi r_o L + 2\pi r_o^2}\right)}{k} = \frac{(250)\left(\dfrac{\pi (0.05)^2 (0.05)}{2\pi (0.05)(0.05) + 2\pi (0.05)^2}\right)}{15} = 0.208$$

As $Bi_{lumped}$ is greater than $0.1$, we should not use the lumped method. Therefore, we will use the Heisler method for multi-dimensional objects.

The disk is the intersection of a long cylinder of 10 cm diameter and a large plate of 5 cm thickness. We can use the product solution

$$\left(\frac{\theta}{\theta_i}\right)_{disk} = \left(\frac{\theta}{\theta_i}\right)_{cylinder} \left(\frac{\theta}{\theta_i}\right)_{plate} \tag{4.98}$$

The point of interest is the center of the disk, this is, at the center plane of the plate and the centerline of the cylinder.

$$\alpha = \frac{k}{\rho c} = \frac{15}{(7900)(480)} = 3.956 \times 10^{-6} \text{ m}^2/\text{s}$$

For the cylinder ($r_o = 5$ cm),

$$Fo = \frac{\alpha t}{r_o^2} = \frac{3.956 \times 10^{-6} t}{(0.05)^2} = 0.001582 t$$

From Eq. (4.45), the one-term approximation solution is

$$\frac{\theta_o}{\theta_i} = \frac{T(0,t) - T_\infty}{T_i - T_\infty} = C_1 e^{-\lambda_1^2 Fo} \tag{4.99}$$

$\lambda_1$ is from Eq. (4.42). Once we have $\lambda_1$ we can get $C_1$ from Eq. (4.43).

$$Bi = \frac{h r_o}{k} = \frac{(100)(0.05)}{15} = 0.3333$$

For this Biot number, using Table 4.1 or from software mentioned above, we find that $\lambda_1 = 0.7833$ and $C_1 = 1.0785$. Putting values into Eq. (4.98),

$$\left(\frac{\theta_o}{\theta_i}\right)_{cylinder} = 1.0785 e^{-(0.7833)^2 (0.0.001582 t)} = 1.0785 e^{-9.707 \times 10^{-4} t} \tag{4.100}$$

For the plate ($L = 2.5$ cm),

$$Fo = \frac{\alpha t}{L^2} = \frac{3.956 \times 10^{-6} t}{(0.025)^2} = 0.006330 t$$

From Eq. (4.25), the one-term approximation solution is

$$\frac{\theta_o}{\theta_i} = \frac{T(0,t) - T_\infty}{T_i - T_\infty} = C_1 e^{-\lambda_1^2 Fo} \tag{4.101}$$

$\lambda_1$ is from Eq. (4.22). Once we have $\lambda_1$, we can get $C_1$ from Eq. (4.23).

$$Bi = \frac{hL}{k} = \frac{(250)(0.025)}{15} = 0.4167$$

For this Biot number, using Table 4.1 or from software mentioned above, we find that $\lambda_1 = 0.6040$ and $C_1 = 1.0601$.
Putting values into Eq. (4.101),

$$\left(\frac{\theta_o}{\theta_i}\right)_{plate} = 1.0601 e^{-(0.6040)^2(0.0.00633t)} = 1.0601 e^{-2.309 \times 10^{-3} t}$$

Now we can put everything together. From Equation (4. 98) above,

$$\left(\frac{\theta}{\theta_i}\right)_{disk} = \left(\frac{\theta}{\theta_i}\right)_{cylinder} \left(\frac{\theta}{\theta_i}\right)_{plate} \tag{4.102}$$

For the disk, $\theta_o = T_o - T_\infty = 500 - 150 = 350$   $\theta_i = T_i - T_\infty = 1200 - 150 = 1050$
Continuing with Eq. (4.102),

$$\left(\frac{\theta_o}{\theta_i}\right)_{disk} = \frac{350}{1050} = 0.3333 = \left[1.0785 e^{-9.707 \times 10^{-4} t}\right]\left[1.0601 e^{-2.309 \times 10^{-3} t}\right]$$

$$0.3333 = 1.1433 e^{-0.003280 t}$$

Rearranging and taking natural logs, we have

$$e^{-0.003280\ t} = 0.2915$$

$$-0.003280t = \ln(0.2915) = -1.2327$$

$$t = \mathbf{376\ s} = \mathbf{6.26}\ \text{minutes}$$

The center of the disk cools to 500 C after **6.26 min.**
To complete the solution, we should check the Fourier numbers to make sure they are greater than 0.2. This would confirm that use of the one-term approximation method was appropriate.
For the cylinder, $Fo = 0.001582\ t = (0.001582)(376) = 0.595$
For the plate, $Fo = 0.00633\ t = (0.00633)(376) = 2.380$
Both numbers are greater than 0.2, so use of the one-term approximation method was fine.
**(b)** The point of interest for this part of the problem is at $r = 3$ cm for the cylinder and $x = 1$ cm for the plate. We will use calculations from Part (a) as much as possible.
From Eq. (4.98),

$$\left(\frac{\theta}{\theta_i}\right)_{disk} = \left(\frac{\theta}{\theta_i}\right)_{cylinder} \left(\frac{\theta}{\theta_i}\right)_{plate} \tag{4.103}$$

For the cylinder,
From Eq. (4.44),

$$\left(\frac{\theta}{\theta_i}\right)_{cylinder} = \frac{T(0.03\ m,\ 376\ s) - T_\infty}{T_i - T_\infty} = C_1 e^{-\lambda_1^2 Fo} J_0(\lambda_1 r / r_o)$$

$$\left(\frac{\theta}{\theta_i}\right)_{cylinder} = 1.0785 e^{-(0.7833)^2(0.595)} J_0\left((0.7833)\frac{0.03}{0.05}\right) = 0.7079$$

For the plate,
From Eq. (4.24),

$$\left(\frac{\theta}{\theta_i}\right)_{plate} = \frac{T(0.01\ m, 376\ s) - T_\infty}{T_i - T_\infty} = C_1 e^{-\lambda_1^2 Fo} \cos(\lambda_1 x / L)\ \text{for}\ Fo > 0.2$$

$$\left(\frac{\theta}{\theta_i}\right)_{\text{plate}} = 1.0601 e^{-(0.6040)^2(2.380)} \cos\left((0.6040)\frac{0.01}{0.025}\right) = 0.4320$$

From Eq. (4.103),

$$\left(\frac{\theta}{\theta_i}\right)_{\text{disk}} = \frac{T - T_\infty}{T_i - T_\infty} = \left(\frac{\theta}{\theta_i}\right)_{\text{cylinder}}\left(\frac{\theta}{\theta_i}\right)_{\text{plate}} = (0.7079)(0.4320) = 0.3058$$

and

$$T = 150 + (1200 - 150)(0.3058) = \mathbf{471.1\ C}$$

The temperature in the disk at a location 3 cm from the centerline and 1.5 cm from one end after 376 s of cooling is **471.1 C**.

### 4.4.4 Rectangular solid

A rectangular solid is the intersection of three large plates. This is shown in Fig. 4.7 above. Example 4.7 illustrates the application of the Heisler method to a rectangular solid.

---

## Example 4.7
### Cooling of a cube
*Problem*

A steel cube $(\rho = 7800\ \text{kg}\ /\ \text{m}^3, c = 470\ \text{J}\ /\ \text{kg C}, k = 40\ \text{W}\ /\ \text{m C})$ is shown in the figure below. It is 8 cm by 8 cm by 8 cm. The cube is initially at a uniform temperature of 500 C, and it is suddenly immersed in a fluid bath that is at 150 C. The convective coefficient is 200 W/m$^2$ C. The cube stays in the fluid for 2 min.

**(a)** What is the temperature of the center of the cube when it leaves the fluid bath?

**(b)** What is the temperature of the center of one face of the cube when it leaves the bath?

**(c)** What is the temperature of a corner of the cube when it leaves the bath?

**(d)** If 100 cubes are cooled per hour, what size cooling system is needed to keep the bath at 150 C?

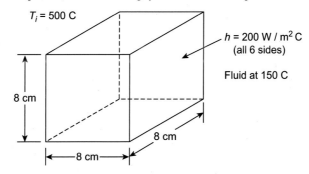

*Solution*

The cube is the intersection of three large plates of 8 cm thickness ($L = 4$ cm). We can use the product solution

$$\left(\frac{\theta}{\theta_i}\right)_{\text{cube}} = \left(\frac{\theta}{\theta_i}\right)_{\text{plate 1}}\left(\frac{\theta}{\theta_i}\right)_{\text{plate 2}}\left(\frac{\theta}{\theta_i}\right)_{\text{plate 3}} \qquad (4.104)$$

(a) The center of the cube is at the center of all three plates. Let us look at Plate 1. The results for Plates 2 and 3 will be the same.

$$\alpha = \frac{k}{\rho c} = \frac{40}{(7800)(470)} = 1.091 \times 10^{-5} \ \text{m}^2/\text{s}$$

$$t = 2 \text{ minutes} = 120 \text{ s}$$

$$Fo = \frac{\alpha t}{L^2} = \frac{1.091 \times 10^{-5}(120)}{(0.04)^2} = 0.8183$$

We will use the one-term approximation as $Fo > 0.2$.
Using Eq. (4.25),

$$\left(\frac{\theta_o}{\theta_i}\right)_{\text{plate 1}} = \frac{T(0,t) - T_\infty}{T_i - T_\infty} = C_1 e^{-\lambda_1^2 Fo} \tag{4.105}$$

$$Bi = \frac{hL}{k} = \frac{(200)(0.04)}{40} = 0.2$$

From Table 4.1, for $Bi = 0.2$, $\lambda_1 = 0.4328$ and $C_1 = 1.0311$. We could have alternatively obtained $\lambda_1$ and $C_1$ from Eqs. (4.22) and (4.23).
Putting values into Eq. (4.105), we have

$$\left(\frac{\theta_o}{\theta_i}\right)_{\text{plate 1}} = 1.0311 e^{-0.4328^2(0.8183)} = 0.8846$$

As all three plates are the same, from Eq. (4.104), we get

$$\left(\frac{\theta_o}{\theta_i}\right)_{\text{cube}} = (0.8846)^3 = 0.6922$$

So, $\left(\frac{T-T_\infty}{T_i-T_\infty}\right)_{\text{cube}} = 0.6922$ and

$$T = T_\infty + (T_i - T_\infty)(0.6922) = 150 + (500 - 150)(0.6922) = \textbf{392 C.}$$

The temperature at the center of the cube is **392 C** after 2 minutes of immersion.

(b) For the location at the center of one face of the cube: This location is on the center plane of Plates 1 and 2 and on the surface of Plate 3.
At the surface of Plate 3, i.e., at $x = L$, we have from Eq. (4.24):

$$\left(\frac{\theta}{\theta_i}\right)_{\text{plate 3}} = C_1 e^{-\lambda_1^2 Fo} \cos \lambda_1 \tag{4.106}$$

From Part (a): $Fo = 0.8183$, $Bi = 0.2$, $\lambda_1 = 0.4328$ and $C_1 = 1.0311$.
Using the results of calculations in Part (a), we have

$$\left(\frac{\theta}{\theta_i}\right)_{\text{plate 3}} = 0.8846 \cos \lambda_1 = 0.8846 \cos(0.4328) = 0.8030$$

So, $\left(\frac{\theta}{\theta_i}\right)_{\text{cube}} = \left(\frac{\theta}{\theta_i}\right)_{\text{plate 1}} \left(\frac{\theta}{\theta_i}\right)_{\text{plate 2}} \left(\frac{\theta}{\theta_i}\right)_{\text{plate 3}} = (0.8846)(0.8846)(0.8030) = 0.6284.$
And $T = T_\infty + (T_i - T_\infty)(0.6284) = 150 + (500 - 150)(0.6284) = \textbf{370 C.}$
The temperature of the center of one face of the cube after 2 minutes of immersion is **370 C.**

(c) For the location at a corner of the cube: This location is on the surface of all three plates. From the calculations in

Parts (a) and (b), we have $\left(\frac{\theta}{\theta_i}\right)_{\text{cube}} = (0.8030)^3 = 0.5178$

And $T = T_\infty + (T_i - T_\infty)(0.5178) = 150 + (500 - 150)(0.5178) = \textbf{331 C.}$
The temperature of a corner of the cube after 2 minutes of immersion is **331 C.**

Looking at the results of Parts (a), (b), and (c): The corner cools quicker than the center of one face, and the center of a face cools quicker than the center of the cube. Makes sense!

(d) From Eq. (4.84),

$$\left(\frac{Q}{Q_{max}}\right)_{cube} = \left(\frac{Q}{Q_{max}}\right)_{plate\ 1} + \left(\frac{Q}{Q_{max}}\right)_{plate\ 2}\left[1 - \left(\frac{Q}{Q_{max}}\right)_{plate\ 1}\right]$$
$$+ \left(\frac{Q}{Q_{max}}\right)_{plate\ 3}\left[1 - \left(\frac{Q}{Q_{max}}\right)_{plate\ 1}\right]\left[1 - \left(\frac{Q}{Q_{max}}\right)_{plate\ 2}\right] \tag{4.107}$$

All three plates are the same ($L = 4$ cm). From Eq. (4.34), for each plate,

$$\left(\frac{Q}{Q_{max}}\right) = 1 - C_1 e^{-\lambda_1^2 Fo}\frac{\sin \lambda_1}{\lambda_1}$$

Using the calculations in Parts (a) and (b), we have

$$\left(\frac{Q}{Q_{max}}\right)_{plates\ 1,\ 2,\ and\ 3} = 1 - (0.8846)\frac{\sin(0.4328)}{(0.4328)} = 0.1428$$

Therefore, from Eq. (4.107),

$$\left(\frac{Q}{Q_{max}}\right)_{cube} = 0.1428 + 0.1428(1 - 0.1428) + 0.1428(1 - 0.1428)^2 = 0.3701$$

For the cube, $Q_{max} = \rho c V(T_i - T_\infty) = (7800)(470)(0.08^3)(500 - 150) = 6.569 \times 10^5$ J.

So, $Q_{cube} = 0.3701\ Q_{max} = (0.3701)(6.569 \times 10^5) = 2.431 \times 10^5$ J.

Each cube puts $2.431 \times 10^5$ J of heat into the fluid bath. If 100 cubes per hour are processed, the rate of heat into the bath is

$$(100\ cubes/hour) \times (2.431 \times 10^5\ J/cube) \times (1\ hour/3600\ s) = 6750\ J/s = 6.75\ kW$$

The needed size of the cooling system is **6.75 kW.**

As a final item, let us calculate $Bi_{lumped}$. (Actually, we probably should have done this at the beginning of the problem.)

$$Bi = \frac{h(V/A)}{k} = \frac{(200)\left(\dfrac{(0.08)^3}{6(0.08)^2}\right)}{40} = 0.06667$$

This is less than 0.1, so let us see what results would be obtained from the lumped method.

From Eq. (4.8),

$$T(2\ minutes) = T_\infty + (T_i - T_\infty)e^{-\left(\frac{hA}{\rho c V}\right)t} = 150 + (500 - 150)e^{-\left(\frac{200(6)(0.08)^2}{7800(470)(0.08)^3}(120)\right)} = 364\ C.$$

Regarding heat flow: From Eq. (4.13),

$$Q = \rho c V(T_i - T_\infty)\left[1 - e^{-\left(\frac{hA}{\rho c V}\right)t}\right] = (7800)(470)(0.08)^3(500 - 150)\left[1 - e^{-0.4910}\right]$$

$$= 2.55 \times 10^5\ Joules\ per\ cube$$

It was shown above that the temperatures in cube do indeed vary somewhat with location, and it was good that we performed a Heisler analysis even though the Biot number was less than 0.1. However, the lumped method gives results that are very good. The temperature result is well within the range of temperatures found by the Heisler method. And, the heat flow result differs only 4.8% from the result obtained by the Heisler method.

# 4.5 Semi-infinite solid

## 4.5.1 Overview

In this section, we consider a semi-infinite solid, also called a semi-infinite slab. This geometry has one physical boundary (a plane surface) and extends internally to infinity in all directions. The semi-infinite solid is shown in Fig. 4.8. The distance from the boundary to a location in the solid is "$x$."

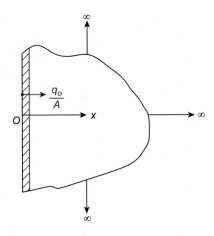

**FIGURE 4.8**

The semi-infinite solid.

A variety of conditions can be imposed at the boundary. We will consider temperature, heat flux, and convection boundary conditions.

The semi-infinite geometry is particularly applicable to large, thick bodies and to the earth's surface and ground. In addition, bodies can often be modeled as semi-infinite solids during short periods of time after a change has been made at a surface of a body. During this time period, the change is felt in the region near the affected surface but has not yet been felt by the other surfaces of the body.

In the analysis that follows, the semi-infinite solid has an initial uniform temperature $T_i$; the imposed boundary condition remains constant during its application; and all thermal and physical properties are constant.

### 4.5.2 Temperature boundary condition

The semi-infinite solid is initially at uniform temperature $T_i$. At time $t = 0$, the temperature of the surface is suddenly changed to $T_o$. The temperature at a depth $x$ at time $t$ is

$$\frac{T(x,t) - T_i}{T_o - T_i} = \text{erfc}\left(\frac{x}{2\sqrt{\alpha t}}\right) \tag{4.108}$$

where $\alpha$ = thermal diffusivity = $k/\rho c$ and erfc = complementary error function. This function is available on the internet and in software, e.g., Excel and Matlab. A table of the function is given in Appendix I.

The heat flux $\frac{q_o}{A}$ at the surface at time $t$ is

$$\frac{q_o}{A}(t) = \frac{k(T_o - T_i)}{\sqrt{\pi \alpha t}} \tag{4.109}$$

### 4.5.3 Heat flux boundary condition

The semi-infinite solid is initially at uniform temperature $T_i$. At time $t = 0$, a constant heat flux $\frac{q_o}{A}$ is imposed on the surface. The temperature at a depth $x$ at time $t$ is

$$T(x, t) = T_i + \frac{(q_o/A)}{k} \left[ \sqrt{\frac{4\alpha t}{\pi}} e^{-\frac{x^2}{4\alpha t}} - x \operatorname{erfc}\left(\frac{x}{2\sqrt{\alpha t}}\right) \right] \qquad (4.110)$$

### 4.5.4 Convection boundary condition

The semi-infinite solid is initially at uniform temperature $T_i$. At time $t = 0$, convection to a fluid at $T_\infty$ with a convective coefficient $h$ occurs at the surface. The temperature at a depth $x$ from the surface at time $t$ is

$$\frac{T(x, t) - T_i}{T_\infty - T_i} = \operatorname{erfc}\left(\frac{x}{2\sqrt{\alpha t}}\right) - e^{\left(\frac{hx}{k} + \frac{h^2 \alpha t}{k^2}\right)} \operatorname{erfc}\left(\frac{x}{2\sqrt{\alpha t}} + \frac{h\sqrt{\alpha t}}{k}\right) \qquad (4.111)$$

The temperature response is also shown in Fig. 4.9. Note that there is a line in the figure for $\frac{h\sqrt{\alpha t}}{k} = \infty$. Infinite $h$ corresponds to the case where the temperature at the surface equals the temperature of the fluid. Therefore, this line actually shows the response due to the sudden imposition of temperature $T_o$ at the surface at $t = 0$. On the vertical axis label, just change $T_\infty$ to $T_o$ for this case.

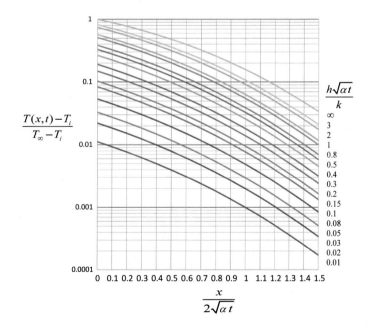

**FIGURE 4.9**

Temperature response of semi-infinite solid with convection at boundary.

## Example 4.8

### Burial depth of a water pipe

*Problem*

In a town in the Northern United States, the air temperature can be as low as $-18$ C for as long as 4 weeks. Assume that the ground is initially at a uniform temperature of 10 C. How deep should water pipes be buried to prevent freezing? Assume that the combined convective and radiative coefficient at the ground's surface is an average 30 W/m$^2$ C. The ground has a thermal conductivity of 2 W/m C, a density of 1800 kg/m$^3$, and a specific heat of 2100 J/kg C.

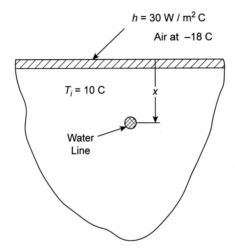

$h = 30$ W / m$^2$ C

Air at $-18$ C

$T_i = 10$ C

$x$

Water Line

*Solution*

We want to find the depth at which the temperature is the freezing temperature of water (0 C).

$$\alpha = \frac{k}{\rho c} = \frac{2}{(1800)(2100)} = 5.291 \times 10^{-7} \text{ m}^2/\text{s}$$

where $t = 4$ weeks $\times$ 7 days/week $\times$ 24 h/day $\times$ 3600 s/h $= 2.42 \times 10^6$ s.

We will use Eq. (4.111):

$$\frac{T(x,t) - T_i}{T_\infty - T_i} = \text{erfc}\left(\frac{x}{2\sqrt{\alpha t}}\right) - e^{\left(\frac{hx}{k} + \frac{h^2 \alpha t}{k^2}\right)} \text{erfc}\left(\frac{x}{2\sqrt{\alpha t}} + \frac{h\sqrt{\alpha t}}{k}\right) \tag{4.111}$$

$$\alpha t = (5.291 \times 10^{-7})(2.420 \times 10^6) = 1.280$$

Putting values into Eq. (4.111), we have

$$\frac{0 - 10}{-18 - 10} = \text{erfc}\left(\frac{x}{2\sqrt{1.280}}\right) - e^{\left(\frac{30x}{2} + \frac{30^2(1.280)}{2^2}\right)} \text{erfc}\left(\frac{x}{2\sqrt{1.280}} + \frac{30\sqrt{1.280}}{2}\right)$$

$$0.3571 = \text{erfc}(0.4419x) - e^{(15x+288)} \text{erfc}(0.4419x + 16.971)$$

This equation can be solved by trial and error, that is, guessing a value of x, calculating the right side of the equation, and seeing if it equals 0.3571. If not, choose another x and try again. Of course, it is much more efficient to use software. Solving the equation for x using Excel's *Goal Seek* gives x = 1.474 m = **4.84 ft**.

Being conservative, we will round this answer upwards and conclude that the water pipe should be buried at a depth of **5 feet or greater.**

Another way to solve this problem is to use Fig. 4.9. The left axis is $\frac{T(x,t)-T_i}{T_\infty -T_i} = 0.3571$ and the lines on the figure are for different $\frac{h\sqrt{\alpha t}}{k} = 16.971$. For our problem, the left axis is 0.3571 and the line should be the one for 16.971, which is close to the one for $\infty$. Reading the figure, we get

$$\eta = \frac{x}{2\sqrt{\alpha t}} = 0.64 \text{ and } x = (0.64)(2)\sqrt{1.280} = 1.45 \text{ m} = \textbf{4.76 ft}$$

This result agrees with the above result from Eq. (4.111).

## 4.6 Chapter summary and final remarks

In this chapter, we discussed unsteady conduction where temperatures and heat flows in a body vary with time. The first step in an analysis is to calculate the Biot number, $Bi_{lumped}$, to see if the body can be considered to be a lumped system. If so, all locations in the body at a given time have essentially the same temperature and a simplified analysis, the lumped method, can be performed. If the body does not meet the criterion for lumped system analysis, then we have to consider spatial variation of temperature in the body. We discussed the Heisler method that provides solutions for spatial variation in basic geometric shapes such as plates, cylinders, and spheres. We also considered temperature distributions in semi-infinite solids due to different boundary conditions at the exposed surface of the solid.

Some bodies have complex shapes that do not fall within the basic geometric categories of this chapter. In addition, some situations do not meet the requirements of the lumped or Heisler methods. For example, thermophysical properties may vary with location or time, and the body may not have a uniform temperature at the beginning of the process. For these cases, problems can often be solved numerically. This is the topic of the next chapter.

## 4.7 Problems

Notes:

- If needed information is not given in the problem statement, then use the Appendix for material properties and other information. If the Appendix does not have sufficient information, then use the Internet or other reference sources. Some good reference sources are mentioned at the end of Chapter 1. If you use the Internet, double-check the validity of the information by using more than one source.
- For this chapter, it is often helpful at the beginning of a solution to calculate $Bi_{lumped}$ to see if the lumped method can be used. If you use the Heisler method, you should calculate the Fourier number to see if the one-term approximation can be used or if more than one term is needed. Of course, you have to pay close attention to the geometry of the problem to see what sections of the chapter are relevant.
- Your solutions should include a sketch of the problem.

**4-1** A large plate has thickness $L$ and cross-sectional area $A$. One side of the plate is perfectly insulated. The other side gets a radiation input $q = \varepsilon \, \sigma A \left( T_{surr}^4 - T^4 \right)$, where $T$ is the temperature of the plate at time $t$. Assume that the lumped method can be used. The initial uniform temperature of the plate is $T_i$.

**(a)** Write the differential equation for the temperature $T(t)$ of the plate.

**(b)** Solve the differential equation for $T(t)$.

**4-2** A sphere of radius $r_o$ is initially at a uniform temperature $T_i$. It cools by radiation and convection to the surroundings at $T_{surr}$ and the adjacent fluid at $T_\infty$. The convective coefficient is $h$ and the emissivity of the surface of the sphere is $\varepsilon$. Assume the lumped method can be used. Write the differential equation for the temperature $T(t)$ of the sphere.

**4-3** A wire has a radius $r_o$ and is initially at the same temperature as the surroundings $(T_\infty)$ Current suddenly starts flowing through the wire, resulting in a uniform and steady volumetric heat generation rate of $q_{gen}$. The surface of the wire convects to the surrounding fluid at $T_\infty$ with a convective coefficient $h$. Assuming the lumped method is appropriate.

**(a)** Write the differential equation for the temperature $T(t)$ of the wire.

**(b)** Solve the differential equation for $T(t)$.

**4-4** An aluminum sphere $(\rho = 2700 \text{ kg } / \text{ m}^3, \quad c = 900 \text{ J } / \text{ kg C}, \quad k = 240 \text{ W } / \text{ m C})$ of 3 cm diameter is initially at a uniform temperature of 300 C. It is suddenly placed in a 75 C air stream. The convective coefficient at the sphere's surface is 50 W/m$^2$ C.

**(a)** How long does it take for the sphere to cool to 150 C?

**(b)** How much heat goes into the air during the cooling?

**4-5** A copper cylinder $(\rho = 8900 \text{ kg}/\text{m}^3, \quad c = 380 \text{ J}/\text{kg C}, \quad k = 400 \text{ W}/\text{m C})$ has a diameter of 5 cm and a length of 3 cm. It is initially at a uniform temperature of 250 C. It is suddenly placed in a fluid whose temperature is 100 C. The convective coefficient at the cylinder's surface is 150 W/m$^2$ C

**(a)** How long does it take for the center of the cylinder to reach 200 C?

**(b)** How much heat goes into the fluid during the cooling?

**4-6** A large iron plate $(\rho = 7800 \text{ kg}/\text{m}^3, \quad c = 450 \text{ J}/\text{kg C}, \quad k = 80 \text{ W}/\text{m C})$ is being heated in an oven. The plate is 1 cm thick and has a cross-sectional area of 3 m$^2$. The plate is initially at a uniform temperature of 20 C. The oven is at 800 C and the convective coefficient at the plate's surfaces is 150 W/m$^2$ C. The plate stays in the oven for 2 min.

What is the temperature of the plate when it leaves the oven?

**4-7** A large copper plate $(\rho = 8900 \text{ kg}/\text{m}^3, \quad c = 380 \text{ J}/\text{kg C}, \quad k = 400 \text{ W}/\text{m C})$ is 2 cm thick. The plate is initially at a uniform temperature of 30 C. It is then subjected to convective cooling on both of its sides. The adjacent fluids on both sides of the plate are at 90 C and the convective coefficients are 110 W/m$^2$ C on both sides.

How long will it take for the plate to reach 55 C?

**4-8** A large copper plate $(\rho = 8900 \text{ kg}/\text{m}^3, \quad c = 380 \text{ J}/\text{kg C}, \quad k = 400 \text{ W}/\text{m C})$ is 2 cm thick. The plate is initially at a uniform temperature of 30 C. It is then subjected to convective cooling on both of its sides. On one side of the plate, the adjacent fluid is at 70 C and the convective coefficient is 45 W/m$^2$ C. On the other side of the plate, the adjacent fluid is at 90 C and the convective coefficient is 110 W/m$^2$ C.

How long will it take for the plate to reach 55 C?

**4-9** A mercury/glass thermometer initially at 20 C is suddenly immersed into oil that is kept at 120 C. After 4 s, the thermometer reads 90 C.

What is the time constant of the thermometer?

**4-10** In a lab experiment, a large horizontal flat plate is heated to a constant temperature by electrical heating tapes under the plate. It is desired to see how uniform the temperature is on the surface of the plate. A handheld surface probe is used to measure 20 different points on the plate. The probe has a time constant of 5 s, and temperature readings are taken at the different points every 5 s. The results are reviewed, and they do not make any sense.
**(a)** What is going wrong?
**(b)** What are two possible ways to remedy the problem?

**4-11** An aluminum cube $(\rho = 2700 \text{ kg/m}^3, \ c = 900 \text{ J/kg C}, \ k = 240 \text{ W/m C})$ is 2 cm by 2 cm by 2 cm. It is initially at a temperature of 200 C. It is suddenly put into an air stream that is at 40 C. In 1 minute, the temperature of the cube drops to 80 C.
**(a)** What is the convective coefficient at the surface of the cube?
**(b)** Using the results of Part (a), how long does it take for the cube to reach the 40 C temperature of the air stream?

**4-12** A temperature sensor has a time constant of 1 second and performs as a first-order system. The sensor is initially at 50 C and is then immersed in a fluid that is at 85 C.
What temperature does the sensor indicate 10 s after it is immersed?

**4-13** A long aluminum cylinder $(\rho = 170 \text{ lbm/ft}^3, \ c = 0.22 \text{ Btu/lbm F}, \ k = 140 \text{ Btu/h ft F})$ has a diameter of 3 inches. It initially has a uniform temperature of 600 F. It is suddenly quenched in a water bath for 2 min. The water bath is at 120 F and the convective coefficient is 100 Btu/h ft$^2$ F.
What is the temperature of the cylinder when it leaves the water bath?

**4-14** An orange can be considered to be a sphere of 8 cm diameter. The orange is initially at a uniform temperature of 10 C when the air temperature quickly falls to −5 C and remains at that temperature. Assume that the orange has the properties of water and that the convective coefficient at the surface of the orange is 15 W/m$^2$ C.
**(a)** How long will it take for the surface of the orange to freeze, i.e., reach 0 C ?
**(b)** How long will it take for the center of the orange to freeze, i.e., reach 0 C ?

**4-15** A long hot dog $(\rho = 950 \text{ kg/m}^3, \ c = 3400 \text{ J/kg C}, \ k = 0.41 \text{ W/m C})$ has a diameter of 2.5 cm. It is initially at 5 C and is dropped into boiling water that is at 100 C. Assume that $h = 2000 \text{ W/m}^2$ C.
How long will it take for the center of the hot dog to reach a safe temperature of 65 C? Compare your answer to the recommendations of a cookbook.

**4-16** A ham $(\rho = 1030 \text{ kg/m}^3, \ c = 3500 \text{ J/kg C}, \ k = 0.48 \text{ W/m C})$ is in the shape of a short cylinder: 20 cm diameter and 15 cm long. The ham is initially at 5 C and is placed in a 190 C oven. The convective coefficient for the oven is 30 W/m$^2$ C.
**(a)** How long will it take for the center of the ham to reach 65 C?
Compare your answer to the recommendations of a cookbook.
**(b)** If the ham had been modeled as a sphere having the same volume as the cylinder, what would have been the answer to Part (a)? Compare the two results.

**4-17** A 3 kg roast beef $(\rho = 950 \text{ kg/m}^3, \ c = 3400 \text{ J/kg C}, \ k = 0.45 \text{ W/m C})$ is initially at a temperature of 4 C. It is being cooked in a 200 C oven where the convective coefficient is 25 W/m$^2$ C. The roast beef can be considered to be a sphere.
**(a)** How long will it take for the center of the beef to reach 75 C?
**(b)** What is the surface temperature at the time found in Part (a)?

**4-18** A furniture company wants to char the surface of some wood rods by passing them through an oven. The rods are 6 cm diameter and 5 m long. Their properties are $\rho = 420 \text{ kg/m}^3$, $c = 2700 \text{ J/kg C}$, $k = 0.15 \text{ W/m C}$. The rods are to stay in the oven for 30 min. The rods will char if their surface temperature reaches 300 C. The rods are initially at 20 C before they enter the oven, and the convective coefficient is 40 W/m$^2$ C.
What should the oven temperature be?

**4-19** Same problem as 4–18 except the company wants to speed up the process and keep the rods in the oven for only 10 min.
What should the oven temperature be?

**4-20** A large wood sheet ($\rho = 420 \text{ kg/m}^3$, $c = 2700 \text{ J/kg C}$, $k = 0.15 \text{ W/m C}$) is 10 cm thick. It is at 20 C when it enters an oven. The convective coefficient on both sides of the wood sheet is 40 W/m$^2$ C. The wood will ignite if its surface temperature reaches 350 C.
**(a)** If the oven is at 360 C, how long will it take for the wood to ignite?
**(b)** If the oven is at 450 C, how long will it take for the wood to ignite?

**4-21** A thick slab of wood ($\rho = 400 \text{ kg/m}^3$, $c = 2800 \text{ J/kg C}$, $k = 0.15 \text{ W/m C}$) is heated on its surface by hot gases in an oven. The gases are at 600 C, and the convective coefficient at the surface of the wood is 40 W/m$^2$ C. The wood is at 20 C when it enters the oven.
How long will it take for the wood to ignite if the ignition temperature is 380 C?

**4-22** A large steel plate ($\rho = 7800 \text{ kg/m}^3$, $c = 480 \text{ J/kg C}$, $k = 50 \text{ W/m C}$) is 15 cm thick and initially at a uniform temperature of 500 C. The plate is horizontal and on an insulating pad so there is no heat flow through its bottom surface. An air stream at 20 C is blown across the top surface to cool the plate. The convective coefficient at this top surface is 150 W/m$^2$ C.
**(a)** What is the temperature at the top surface of the plate after 10 min of cooling?
**(b)** What is the temperature at the bottom insulated surface after 10 min of cooling?
**(c)** What is the temperature at the midpoint of the plate after 10 min of cooling?
(Hint: Take a close look at the geometry and boundary conditions. This can be solved with the Heisler method for a plate.)

**4-23** A long stainless steel rod $\left(\rho = 7900 \text{ kg/m}^3, \ c = 480 \text{ J/kg C}, \ k = 15 \text{ W/m C}\right)$ of 12 cm diameter is initially at 450 C and is being cooled by a 30 C air stream. The convective coefficient at the rod's surface is 200 W/m$^2$ C.
**(a)** How long does it take for the center of the rod to reach 100 C?
**(b)** How much heat goes into the air during the process?

**4-24** Table 4.1 gives eigenvalue $\lambda_1$ and constant $C_1$ for a large plate. Determine the next three eigenvalues and constants for a large plate with a Biot number of 3.5.

**4-25** Table 4.1 gives eigenvalue $\lambda_1$ and constant $C_1$ for a long cylinder. Determine the next three eigenvalues and constants for a long cylinder with a Biot number of 3.5.

**4-26** Table 4.1 gives eigenvalue $\lambda_1$ and constant $C_1$ for a sphere. Determine the next three eigenvalues and constants for a sphere with a Biot number of 3.5.

**4-27** An aluminum cylinder $\left(\rho = 2700 \text{ kg/m}^3, \ c = 900 \text{ J/kg C}, \ k = 240 \text{ W/m C}\right)$ has a 40 cm diameter and is 40 cm long. It is heated in an oven until its temperature is uniform at 900 C. It is then taken out of the oven and cooled by convection to the 25 C room air. The convective coefficient between the cylinder and the room air is 20 W/m$^2$ C.
What is the temperature of the center of the cylinder 15 min after it is taken out of the oven?

**4-28** A long concrete column $\left(\rho = 2100\ \text{kg}/\text{m}^3,\ c = 910\ \text{J}/\text{kg C},\ k = 1.4\ \text{W}/\text{m C}\right)$ of square cross section (10 cm by 10 cm) is initially at a uniform temperature of 220 C. Cooling suddenly starts to the 25 C surroundings with a convective coefficient of 45 W/m$^2$ C.
(a) What is the temperature of the center of the column 40 min after the cooling begins?
(b) What is the temperature at an edge of the column 40 min after the cooling begins?
(c) How much heat leaves the column per meter length during the cooling period?

**4-29** A thick concrete slab $\left(\rho = 2100\ \text{kg}/\text{m}^3,\ c = 910\ \text{J}/\text{kg C},\ k = 1.4\ \text{W}/\text{m C}\right)$ is at a uniform initial temperature of 280 C. It suddenly starts cooling from its surface to the room air. The air is at 20 C and the convective coefficient is 25 W/m$^2$ C.
(a) What is the temperature at the surface of the slab after 30 min of cooling?
(b) What is the temperature at a depth of 5 cm from the surface after 30 min of cooling?

**4-30** The surface of a thick block of wood is being subjected to a heat flux of 3000 W/m$^2$. The properties of the wood are $\rho = 400\ \text{kg}/\text{m}^3,\ c = 2500\ \text{J}/\text{kg C},\ k = 0.12\ \text{W}/\text{m C}$. The wood is initially at a uniform temperature of 25 C. The wood will ignite if its temperature reaches 400 C.
Will the wood ignite if it is subjected to the surface heat flux for 20 min?

**4-31** The surface of a thick block of wood is subjected to a heat flux of 3500 W/m$^2$ for 15 min. The wood has properties: $\rho = 400\ \text{kg}/\text{m}^3,\ c = 2500\ \text{J}/\text{kg C},\ k = 0.12\ \text{W}/\text{m C}$.
The wood is initially at a uniform temperature of 20 C.
What is the temperature at a depth of 1 cm from the surface at the end of the heating?

**4-32** The surface of a thick slab of iron $\left(\rho = 8000\ \text{kg}/\text{m}^3,\ c = 450\ \text{J}/\text{kg C},\ k = 80\ \text{W}/\text{m C}\right)$ is subjected to a heat flux of 10,000 W/m$^2$ for 1 hour. The slab is initially at a uniform temperature of 30 C.
At the end of the heating process, at what depth is the temperature 50 C?

**4-33** A stainless steel cube $\left(\rho = 7900\ \text{kg}/\text{m}^3,\ c = 500\ \text{J}/\text{kg C},\ k = 15\ \text{W}/\text{m C}\right)$ is 12 cm by 12 cm by 12 cm. It initially is at 50 C and is heated by hot gases at 900 C with a convective coefficient of 150 W/m$^2$ C on all six sides.
(a) What is the temperature at the center of the cube after 30 min of heating?
(b) What is the temperature at a corner of the cube after 30 min of heating?
(c) How much heat goes into the cube during the 30-minute heating process?

**4-34** A stainless steel rectangular solid $\left(\rho = 7900\ \text{kg}/\text{m}^3,\ c = 500\ \text{J}/\text{kg C},\ k = 15\ \text{W}/\text{m C}\right)$ is 6 cm by 15 cm by 20 cm. It initially is at 550 C and cools by convection to the room air, which is at 25 C. The convective coefficient on the two 6 cm by 15 cm sides is 20 W/m$^2$C. The convective coefficient on the other four sides is 80 W/m$^2$ C.
(a) What is the center temperature of the solid after 40 min of cooling?
(b) What is the temperature at the center of a 15 cm by 20 cm side after 40 min of cooling?

**4-35** A thick concrete slab $\left(\rho = 2100\ \text{kg}/\text{m}^3,\ c = 910\ \text{J}/\text{kg C},\ k = 1.4\ \text{W}/\text{m C}\right)$ is initially at a uniform temperature of 50 C. The temperature of the surface of the slab is suddenly lowered to 10 C.
(a) What is the temperature at a depth of 5 cm from the surface 45 min after the surface temperature was lowered?
(b) What is the heat flux at the surface 45 min after the surface temperature was lowered?

**4-36** A thick steel slab $(\rho = 7800\,\text{kg/m}^3,\ c = 480\,\text{J/kg C},\ k = 45\,\text{W/m C})$ is initially at 20 C. It is suddenly put into boiling water at 100 C. The convective coefficient at the surface of the slab is 2000 W/m$^2$ C.

What is the temperature at a depth 1 cm from the surface 5 min after the slab was put into the water?

**4-37** A thick steel slab $(\rho = 490\,\text{lbm/ft}^3,\ c = 0.1\,\text{Btu/lbm F},\ k = 35\,\text{Btu/h ft F})$ is initially at a uniform temperature of 500 F. Its surface temperature is suddenly lowered to 50 F.

**(a)** How long will it take for the temperature at a depth of 1 inch from the surface to reach 250 F?

**(b)** How much heat per square foot of surface area leaves the slab during this process?

## References

[1] M.P. Heisler, Temperature charts for induction and constant temperature heating, ASME Trans. 69 (1947) 227–236.

[2] H. Gröber, S. Erk, U. Grigull, Fundamentals of Heat Transfer, McGraw-Hill, 1961.

[3] H.S. Carslaw, J.C. Jaeger, Conduction of Heat in Solids, Oxford University Press, 1959 (Section 1.15).

[4] V.S. Arpaci, Conduction Heat Transfer, Addison-Wesley, 1966 (Section 5.2).

[5] J. Sucec, Heat Transfer, W. C. Brown, 1985 (Section 3.2e).

[6] L.S. Langston, Heat transfer from multidimensional objects using one-dimensional solutions for heat loss, Int. J. Heat Mass Transf. 25 (1982) 149–150.

[7] A.F. Mills, Heat Transfer, second ed., Prentice Hall, 1999 (Section 3.4.4 and Table 3.6).

[8] F. Kreith, W.Z. Black, Basic Heat Transfer, Harper and Row, 1980 (Tables 3-2 and 3-3).

[9] J.P. Holman, Heat Transfer, tenth ed., McGraw-Hill, 2010 (Section 4-5, Figure 4-18).

[10] Y.A. Çengel, A.J. Ghajar, Heat and Mass Transfer, fourth ed., McGraw-Hill, 2011 (Table 4-5).

# Numerical methods (steady and unsteady)

## 5.1 Introduction

Almost all of the material in previous chapters dealt with bodies having simple geometries, i.e., plates, cylinders, and spheres, and constant and uniform thermophysical properties. For situations involving convection, the convective coefficient $h$ and the fluid temperature $T_\infty$ were constant. They did not vary with time. If radiation was involved, the emissivity $\varepsilon$ and the temperature of the surroundings $T_{surr}$ were also time invariant. Additional constraints were also present in the Lumped and Heisler Methods of Chapter 4. The bodies were required to have uniform temperatures at the beginning of heat transfer processes. Solutions were for convective, not radiative, boundary conditions. For the flat plate of the Heisler Method, the convective coefficients on both surfaces had to be the same.

In this chapter, we discuss numerical solution of heat transfer problems. Through the use of numerical methods, the restrictions mentioned above can be removed. Numerical methods can be applied to complex, as well as basic geometries, and can be used for situations in which properties and boundary conditions are nonuniform and varying with time.

Numerical methods for heat transfer problems have been used since at least the 1940s [1−3]. With the widespread availability of digital computers in the 1960s and 1970s, researchers and students became able to solve problems that are very difficult or impossible to be solved analytically.

There are several types of numerical methods used in heat transfer, including finite differences, finite elements, boundary elements, and control volumes [4−9]. We discuss the finite-difference method in depth and show its application to both steady and unsteady problems. Numerical solutions to several problems will be presented. Some of these problems were solved earlier using other techniques. For those problems, comparison will be made of the numerical results and the previously obtained results. The chapter concludes with a brief discussion of the finite element method.

After four chapters of differential equations and relatively complex mathematics, many students will find this chapter to be refreshing and a welcome change, although temporary! The chapter only uses algebra and the three basic equations for the modes of heat transfer:

$$q_{cond} = \frac{kA}{L}(T_1 - T_2) \tag{5.1}$$

$$q_{conv} = hA(T_s - T_\infty) \tag{5.2}$$

$$q_{rad} = \varepsilon\sigma A\left(T_s^4 - T_{surr}^4\right) \tag{5.3}$$

In all three equations, $A$ is the area through which heat flows. Eq. (5.1) is for conduction through a wall of thickness $L$ with surface temperatures $T_1$ and $T_2$. Eq. (5.2) is for convection from a surface at $T_s$ to a fluid at $T_\infty$ with convective coefficient $h$. And, in Eq. (5.3) for radiation, $\varepsilon$ is the emissivity of the surface, $\sigma$ is the Stefan–Boltzmann constant, $T_s$ is the surface temperature, and $T_{surr}$ is the temperature of the surroundings. We also have to remember that, for radiation, temperatures must be in absolute temperature units—Rankine or Kelvin.

## 5.2 Finite-difference method

Let us first consider the meaning of "finite difference". In general, the temperature in a body is a function of location and time. For example, in rectangular Cartesian coordinates, $T = T(x, y, z, t)$. The parameters $x$, $y$, $z$, and $t$ are continuous variables over their respective ranges. In the finite-difference method, these four parameters are not continuous. They are only defined at specific locations or times: **nodes** for the spatial parameters and **time steps** for the time parameter.

The first step in a finite-difference analysis is to determine the grid and nodal points for the object. Let us say that we have a large wall of thickness $L$. The surface temperatures of the wall are, given, and we wish to determine the interior temperatures. We can divide the thickness into any number of parts. For example, if we decide that there should be 10 equally sized parts, the nodal points for the wall will be as shown in Fig. 5.1.

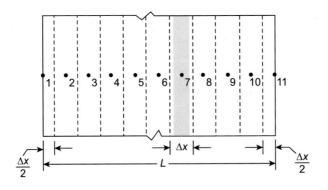

**FIGURE 5.1**

Nodes for wall of thickness $L$.

As the surface temperatures are of primary importance, the two wall surfaces have been selected as nodal points. The other nodal points are spaced equally across the thickness. The size of increment $\Delta x$ is given by

$$\Delta x = \frac{\text{total length}}{\text{number of divisions}} = \frac{L}{10} \tag{5.4}$$

Let us look at Fig. 5.1. Nodes 1 and 11 are on the boundary and hence are called "boundary nodes" The other nodes, nodes 2 through 10, are "interior nodes" The dashed lines in Fig. 5.1 are the boundaries between the different nodes. In our finite-difference analysis, we will be looking at heat flows into and out of the various nodes. The cross-hatched area in Fig. 5.1 is the extent of node 7. The node has node 6 on its left and node 8 on its right. Note that the interior nodes 2−10 have widths of $\Delta x$ and the boundary nodes 1 and 11 have widths of $\Delta x/2$.

Fig. 5.2 shows a rectangular plate that has been given a rectangular grid. There are 16 nodes, and the increments $\Delta x$ and $\Delta y$ are given by

$$\Delta x = \frac{L_1}{3} \quad \text{and} \quad \Delta y = \frac{L_2}{3}. \tag{5.5}$$

There are three cross-hatched areas in Fig. 5.2, which show the extents of corner node 1, interior node 7, and edge node 15. (A hint: To determine the extent of a given node, just go halfway to all adjacent nodes and box-in the area.)

For the rectangular plate of Fig. 5.2, $L_2 = 2L_1$. Therefore, we could have used a square grid, as shown in Fig. 5.3.

The new grid has 28 nodes instead of the original 16 nodes. Increment $\Delta x$ is still $L_1/3$, but $\Delta y$ is now $L_2/6$. When the temperatures at these 28 nodes are determined, we will have a better knowledge

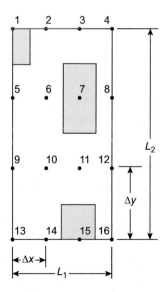

**FIGURE 5.2**

Nodes for a rectangular plate.

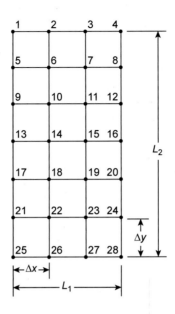

**FIGURE 5.3**

Square grid for the rectangular plate.

of the temperature distribution in the plate than we would have obtained with the original 16 nodes. Actually, we could have used an even larger number of smaller nodes to get an even more-detailed indication of the temperature distribution in the plate.

In the next section, we will be formulating finite difference equations to solve for the nodal temperatures. Such equations can certainly be obtained for a rectangular grid. However, we will see that use of a square grid provides some simplifications.

Grids and nodes can also be established for other geometries. For example, consider Fig. 5.4 that shows a circular plate.

There are five equally spaced nodes in Fig. 5.4. The distance between nodal points is $\Delta r$. If the outer radius of the plate is $r_o$, then the increment $\Delta r = r_o/4$. The dashed lines in Fig. 5.4 show the boundaries of the various nodes. The center node, node 1, is a circular area. The other four nodes are ring areas. As an example, the cross-hatched area is node 4. Note that the interior nodes 2, 3, and 4 have widths of $\Delta r$. Center node 1 is circular with a radius of $\Delta r/2$. Boundary node 5 is a ring of width $\Delta r/2$.

## 5.2.1 Steady state

In this section, we will develop finite difference equations for steady-state heat transfer in one and two-dimensional objects.

Fig. 5.5 shows some nodes in a one-dimensional object.

For steady state, temperatures and heat flows are constant; they do not change with time. Let us look at the possible heat flows for node i. There is conduction between node i and its two adjacent nodes i−1 and i+1. And, there may be convection and/or radiation at the upper, lower, front, and/or

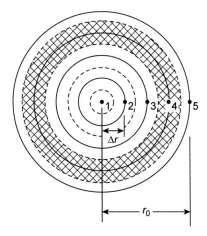

**FIGURE 5.4**

Nodes for a circular plate.

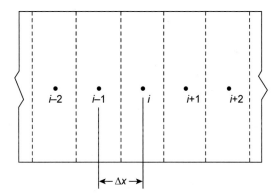

**FIGURE 5.5**

Steady one-dimensional nodes.

back boundaries of node i. (There could also be imposed heat fluxes at these boundaries or the boundaries could be insulated.) Finally, there may also be internal heat generation within node i at a volumetric strength of $q_{gen}$.

For steady state, from conservation of energy, the total rate of heat flow **into** node i equals the total rate of heat flow **out of** node i. Fig. 5.6 shows the heat flows for node i.

We have assumed that the incoming flows are the conduction from node i−1 and the internal heat generation and that the outgoing flows are the convection to the adjacent fluid, the radiation to the surroundings, and the conduction to node i+1. We actually do not know the directions of the conductive, convective, or radiative flows until we solve the finite-difference equations and determine the temperatures in the object. However, we need to assume directions so we can write the energy

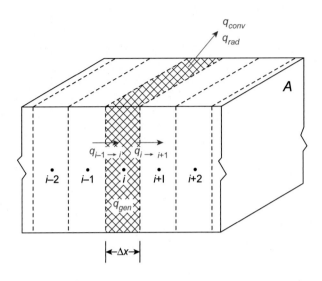

**FIGURE 5.6**

Heat flows for node i.

conservation equation. (Remember that the direction of heat flow is from the higher temperature to the lower temperature.)

The energy conservation equation is: heat flow in = heat flow out. For node i we have

$$q_{(i-1)\to i} + q_{generated} = q_{rad} + q_{conv} + q_{i\to(i+1)} \tag{5.6}$$

Using Eqs. (5.1), (5.2), and (5.3) for the various heat flows, Eq. (5.6) becomes

$$\frac{kA}{\Delta x}(T_{i-1} - T_i) + q_{gen}A\ \Delta x = \frac{kA}{\Delta x}(T_i - T_{i+1}) + hA(T_i - T_\infty) + \varepsilon\sigma A\left(T_i^4 - T_{surr}^4\right) \tag{5.7}$$

The direction of the internal heat generation is, of course, **into** node i. But, we have assumed the directions for the other heat flows and have written the terms in Eq. (5.7) accordingly. An alternative way to write the energy conservation equation is to assume that **all** the heat flows are **into** the node. Then, rather than heat flow in = heat flow out, the energy conservation equation for steady-state would be: total heat flow in = 0. Let us do that. We will assume that all heat flows are **into** node i. The energy conservation equation is then

$$\frac{kA}{\Delta x}(T_{i-1} - T_i) + \frac{kA}{\Delta x}(T_{i+1} - T_i) + hA(T_\infty - T_i) + \varepsilon\sigma A\left(T_{surr}^4 - T_i^4\right) + q_{gen}A\ \Delta x = 0 \tag{5.8}$$

Comparing Eqs. (5.7) and (5.8), it is seen that the equations are identical.

To summarize, we can assume the directions for the different heat flows associated with a node and write an equation using heat flow in = heat flow out. Alternatively, we can assume that all heat flows are **into** the node and write an equation using total heat flow in = 0. The author prefers the second approach, and that is the way we will determine the finite-difference equations for nodes.

Eq. (5.8) is the finite-difference equation for node i. Similar equations can be developed for all the other nodes in the object. The equations may then be solved simultaneously for the nodal temperatures.

Continuing our discussion of one-dimensional steady-state numerical analysis, let us now look again at the pin fin of Example 3.11. The heat flow for the fin was determined analytically in Chapter 3. Let us determine the heat flow numerically and compare the two answers.

## Example 5.1
### Heat transfer from a pin fin

*Problem*
A pin fin ($k = 250$ W/m C) is attached to a wall. The fin's diameter is 1.5 cm, and it is 10 cm long. The wall is at 200 C. The surrounding air is at 15 C and the convective coefficient is 9 W/m² C. Determine, numerically, the rate of heat transfer from the fin to the surrounding air.

*Solution*
As discussed in Chapter 3, fins are attached to surfaces to enhance the heat transfer. This problem was solved analytically as Example 3.11. We now solve it numerically and compare the result with that of the analytical solution.
The first step is to determine the grid for the problem. We decided to have the grid shown in Fig. 5.7. There are 11 nodes. Node 1 is at the wall and node 11 is at the end of the fin. These are boundary nodes. The other nodes (nodes 2 through 10) are interior nodes. The spacing between nodes is uniform with $\Delta x = 1$ cm $= 0.01$ m. We wanted nodal points at the wall and at the end of the fin, so nodes 1 and 11 have widths of $\Delta x/2$. The other nodes have widths of $\Delta x$.
The temperature of node 1 is known. It is the wall temperature, i.e., 200 C. The other nodal temperatures are determined through simultaneous solution of the 10 finite-difference equations for nodes 2 through 11.
Let us determine the equation for node 2. The node gets heat by conduction from adjacent nodes 1 and 3. It also gets heat by convection from the surrounding fluid. As discussed above, we will assume that all heat flows are **into** node 2. We have

$$q_{1 \to 2} + q_{3 \to 2} + q_{\infty \to 2} = 0 \tag{5.9}$$

Putting in the expressions for the three heat flows, Eq. (5.9) becomes

$$\frac{kA}{\Delta x}(T_1 - T_2) + \frac{kA}{\Delta x}(T_3 - T_2) + hP\Delta x(T_\infty - T_2) = 0 \tag{5.10}$$

In Eq. (5.10), $P$ is the perimeter $\pi D$ of the fin, and $A$ is the cross section area $\pi D^2/4$ of the fin. $D$ is the diameter of the fin. Equations similar to Eq. (5.10) can be written for the other eight interior nodes, i.e., nodes 3–10. The only thing that changes is the subscripts on the temperatures. That is.
For $i = 3$ to 10,

$$\frac{kA}{\Delta x}(T_{i-1} - T_i) + \frac{kA}{\Delta x}(T_{i+1} - T_i) + hP\Delta x(T_\infty - T_i) = 0 \tag{5.11}$$

Node 11 is a special node. It is a boundary node and gets conduction from node 10 and convection from the side and end of the fin. Its nodal equation is

$$\frac{kA}{\Delta x}(T_{10} - T_{11}) + hP\frac{\Delta x}{2}(T_\infty - T_{11}) + hA(T_\infty - T_{11}) = 0 \tag{5.12}$$

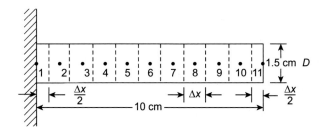

**FIGURE 5.7**

Grid for example 5.1.

Eqs. (5.10), (5.11), and (5.12) can be solved simultaneously for the 10 unknown temperatures $T_2$ through $T_{11}$. (Remember $T_1$ is a given. It is 200 C)

For this problem, $k = 250$ W/m C, $h = 9$ W/m$^2$ C, $D = 1.5$ cm $= 0.015$ m, $\Delta x = 1$ cm $= 0.01$ m, $P = \pi D = 0.047124$ m, $A = \pi D^2/4 = 0.00017672$ m$^2$, $T_\infty = 15$ C, and $T_1 = 200$.

Putting these values into Eqs. (5.10), (5.11), and (5.12), we have the following 10 equations to solve for the unknown temperatures $T_2$ through $T_{11}$:

$$-8.840T_2 + 4.418T_3 = -883.66 \tag{5.13}$$

$$4.418T_2 - 8.840T_3 + 4.418T_4 = -0.06362 \tag{5.14}$$

$$4.418T_3 - 8.840T_4 + 4.418T_5 = -0.06362 \tag{5.15}$$

$$4.418T_4 - 8.840T_5 + 4.418T_6 = -0.06362 \tag{5.16}$$

$$4.418T_5 - 8.840T_6 + 4.418T_7 = -0.06362 \tag{5.17}$$

$$4.418T_6 - 8.840T_7 + 4.418T_8 = -0.06362 \tag{5.18}$$

$$4.418T_7 - 8.840T_8 + 4.418T_9 = -0.06362 \tag{5.19}$$

$$4.418T_8 - 8.840T_9 + 4.418T_{10} = -0.06362 \tag{5.20}$$

$$4.418T_9 - 8.840T_{10} + 4.418T_{11} = -0.06362 \tag{5.21}$$

$$4.418T_{10} - 4.4217T_{11} = -0.05567 \tag{5.22}$$

It is seen that these equations are linear equations with constant coefficients. Several methods are available to solve these simultaneous equations. We will discuss their solution by matrix inversion, the Gauss—Seidel method, and the Excel add-in *Solver*.

Matrix Inversion:

Equations (5.13) through (5.22) can be written in the matrix form

$$AT = B \tag{5.23}$$

where **A** is the $10 \times 10$ matrix of the coefficients, T is the column vector of the 10 unknown temperatures, and **B** is the column vector of the constants on the right-hand sides of the equations. That is

**A =**

| −8.840 | 4.418 | 0 | 0 | 0 | 0 | 0 | 0 | 0 | 0 |
|---|---|---|---|---|---|---|---|---|---|
| 4.418 | −8.840 | 4.418 | 0 | 0 | 0 | 0 | 0 | 0 | 0 |
| 0 | 4.418 | −8.840 | 4.418 | 0 | 0 | 0 | 0 | 0 | 0 |
| 0 | 0 | 4.418 | −8.840 | 4.418 | 0 | 0 | 0 | 0 | 0 |
| 0 | 0 | 0 | 4.418 | −8.840 | 4.418 | 0 | 0 | 0 | 0 |
| 0 | 0 | 0 | 0 | 4.418 | −8.840 | 4.418 | 0 | 0 | 0 |
| 0 | 0 | 0 | 0 | 0 | 4.418 | −8.840 | 4.418 | 0 | 0 |
| 0 | 0 | 0 | 0 | 0 | 0 | 4.418 | −8.840 | 4.418 | 0 |
| 0 | 0 | 0 | 0 | 0 | 0 | 0 | 4.418 | −8.840 | 4.418 |
| 0 | 0 | 0 | 0 | 0 | 0 | 0 | 0 | 4.418 | −4.4217 |

| **T =** | **B =** |
|---|---|
| $T_2$ | −883.66 |
| $T_3$ | −0.06362 |
| $T_4$ | −0.06362 |
| $T_5$ | −0.06362 |
| $T_6$ | −0.06362 |
| $T_7$ | −0.06362 |
| $T_8$ | −0.06362 |
| $T_9$ | −0.06362 |
| $T_{10}$ | −0.06362 |
| $T_{11}$ | −0.05567 |

To obtain the temperatures of vector **T**, we first determine the inverse $\mathbf{A}^{-1}$ of the coefficients matrix **A**. The inverse is then multiplied by vector **B** to get the temperatures. That is

$$\mathbf{T} = \mathbf{A}^{-1}\mathbf{B} \tag{5.24}$$

The following outlines the procedures for solution using Excel and Matlab. Detailed instructions are included for Excel, and the reader is encouraged to run the program to get experience in solving equations by matrix inversion using Excel. Procedures are also given below for Matlab solution.

In the Excel spreadsheet, we choose locations for the following four areas: a $10 \times 10$ block for matrix **A**, a $10 \times 10$ block for the inverse $\mathbf{A}^{-1}$ of **A**, a 10-element column for **B**, and a 10-element column where we want to have the temperatures. For example, let us decide to put matrix **A** in block A1:J10, matrix $\mathbf{A}^{-1}$ in block A12:J21, vector **B** in column A23:A32, and vector **T** in column A34:A43. We enter the values of the elements for **A** and **B** and then proceed as follows:

To get the inverse matrix $\mathbf{A}^{-1}$, highlight the block A12:J21 and then enter the formula = MINVERSE(A1:J10) in cell A12. Press Ctrl+Shift+Enter. You will see the elements of the $\mathbf{A}^{-1}$ matrix appear. As shown in Eq. (5.24), we then have to multiply $\mathbf{A}^{-1}$ by **B** to get the temperatures. To do this, we highlight the column A34:A43 and then enter the formula = MMULT(A12:J21,A23:A32) in cell A34. Press Ctrl+Shift+Enter and the elements of vector **T** will appear. These elements are the desired values of $T_2$ through $T_{11}$.

For Matlab, the solution may be achieved interactively or through use of an m-file. Matrix A and vector B are defined by assignment statements, and the temperatures $T_2$ through $T_{11}$ are obtained by the statement T = inv(A)*B or T = A\B. The m-file for the solution is given in Appendix K.

The results using matrix inversion were the same for both Matlab and Excel. The $T_2$ through $T_{11}$ temperatures were 198.40, 196.96, 195.69, 194.58, 193.64, 192.85, 192.23, 191.76, 191.45, and 191.31 C. From the problem statement, $T_1 = 200$ C.

It should be noted that the matrix inversion method is only appropriate for the solution of **linear** simultaneous equations. If some equations are nonlinear (e.g., if we have radiative heat transfer and there are $T_4$ terms), then matrix inversion cannot be used. We now discuss the Gauss–Seidel method and the Excel *Solver* add-in. These methods can be used for both linear and nonlinear equations.

Gauss–Seidel Method:

This is a very useful method that can be used to solve both linear and nonlinear simultaneous equations. Let us return to the 10 equations of this problem, i.e., Eqs. (5.13)–(5.22). Eq. (5.13) was for node 2, Eq. (5.14) was for node 3, etc. Let us rearrange the equations for solution by the Gauss–Seidel Method:

Eq. (5.13), for node two is

$$-8.840T_2 + 4.418T_3 = -883.66 \tag{5.13}$$

Rearrange the equation so that $T_2$ is the only item on the left side of the equal sign. Doing this, we get

$$T_2 = 0.4998T_3 + 99.962 \tag{5.25}$$

The equation for node 3 is

$$4.418T_2 - 8.840T_3 + 4.418T_4 = -0.06362 \tag{5.14}$$

Rearranging this so that $T_3$ is the only item left of the equal sign, we get

$$T_3 = 0.4998(T_2 + T_4) + 0.007197 \tag{5.26}$$

Continuing this for the other eight equations, we have

$$T_4 = 0.4998(T_3 + T_5) + 0.007197 \tag{5.27}$$

$$T_5 = 0.4998(T_4 + T_6) + 0.007197 \tag{5.28}$$

$$T_6 = 0.4998(T_5 + T_7) + 0.007197 \tag{5.29}$$

$$T_7 = 0.4998(T_6 + T_8) + 0.007197 \tag{5.30}$$

$$T_8 = 0.4998(T_7 + T_9) + 0.007197 \tag{5.31}$$

$$T_9 = 0.4998(T_8 + T_{10}) + 0.007197 \tag{5.32}$$

$$T_{10} = 0.4998(T_9 + T_{11}) + 0.007197 \tag{5.33}$$

$$T_{11} = 0.9992T_{10} + 0.01259 \tag{5.34}$$

Using Matlab or similar software, the solution of these 10 equations (Eqs. 5.25 through 5.34) is obtained by successive looping through all the equations, one equation after another. Each time an equation is executed, the right side of the equation is calculated and the result becomes a new value for the temperature on the left side of the equation. After each complete loop through all the equations, we have new values for all the temperatures. Before starting the process, we have to give starting values for the unknown temperatures. For this particular problem, the base temperature is 200 C and the fluid temperature is 15 C. Initial guesses in this temperature range would be reasonable. When a loop is performed for all 10 equations, all the temperatures have been given new values. If the Gauss–Seidel Method is successful, successive looping will result in convergence where the temperatures do not change with additional looping. For our equations, the method is usually successful. As mentioned above, this method can be used for nonlinear, as well as linear, sets of equations. If the equations are nonlinear, convergence may be affected by the initial guesses for the temperatures. If the temperatures do not converge, try other starting temperatures. The Matlab program for this problem is given in Appendix K.

The Excel *Solver* add-in program:

Excel *Solver*, which comes with the Excel software, is a very useful program for solving both linear and nonlinear simultaneous equations. We will use it to solve for the temperatures in this Example 5.1, as follows:

We will solve the nodal equations, Eqs. (5.25) through (5.34). Alternatively, we could solve Eqs. (5.13) through (5.22) as they are the same equations as (5.25) through (5.34).

Proceeding with *Solver*:

Starting with Eq. (5.25), rearrange the equation so that everything is to the left of the equal sign.

$$T_2 = 0.4998T_3 + 99.962 \qquad (5.25)$$

Eq. (5.25) then becomes

$$T_2 - 0.4998T_3 - 99.962 = 0 \qquad (5.35)$$

Do this with the other nine nodal equations. Eqs. (5.26) through (5.34) then become

$$T_3 - 0.4998(T_2 + T_4) - 0.007197 = 0 \qquad (5.36)$$

$$T_4 - 0.4998(T_3 + T_5) - 0.007197 = 0 \qquad (5.37)$$

$$T_5 - 0.4998(T_4 + T_6) - 0.007197 = 0 \qquad (5.38)$$

$$T_6 - 0.4998(T_5 + T_7) - 0.007197 = 0 \qquad (5.39)$$

$$T_7 - 0.4998(T_6 + T_8) - 0.007197 = 0 \qquad (5.40)$$

$$T_8 - 0.4998(T_7 + T_9) - 0.007197 = 0 \qquad (5.41)$$

$$T_9 - 0.4998(T_8 + T_{10}) - 0.007197 = 0 \qquad (5.42)$$

$$T_{10} - 0.4998(T_9 + T_{11}) - 0.007197 = 0 \qquad (5.43)$$

$$T_{11} - 0.9992T_{10} - 0.01259 = 0 \qquad (5.44)$$

Looking at these 10 equations, we see that the left sides of the equations are all functions of the unknown temperatures $T_2$ through $T_{11}$. Let us call these functions $f_2$ through $f_{11}$, respectively.

For example, from Eq. (5.35), $f_2 = T_2 - 0.4998T_3 - 99.962$. From Eq. (5.36), $f_3 = T_3 - 0.4998(T_2 + T_4) - 0.007197$, etc.

So, if we find temperatures such that all the functions $f_2$ through $f_{11}$ are equal to zero, then we have solved the nodal equations. Now comes the main part to get ready for using *Solver*. The requirement that all 10 functions be zero can be reduced to a single requirement; namely, that the sum of the squares of all the functions be zero.

That is, we have the solution to the nodal equations if

$$\sum_{i=2}^{11} f_i^2 = 0 \qquad (5.45)$$

Now on to *Solver*:

Open a new Excel spreadsheet. We decided to put the temperatures in the column A2 through A11. We need starting values (i.e., initial guesses) for the temperatures, so we put 200 in these 10 temperature cells. We entered the functions $f_2$ through $f_{11}$, one-by-one, into the column C2 through C11. For example, cell C2 for function $f_2$ is $= A2 - 0.4998 * A3 - 99.962$. We then entered the squares of the functions, $f_2^2$ through $f_{11}^2$, one-by-one, into the column E2 through E11. Finally, we entered =SUM(E2:E11) into cell G1.

Calling up the *Solver* program, we instructed the program to make G1 zero by changing cells A2 through A11. We pressed *Solve*. The temperature values changed, but we got a statement that a solution could not be found. This often happens if the program does not get a result close enough to the required value of zero. In our case, the value in G1 was quite small, 0.00124, but not small enough for the program to declare a success. We reran the program asking it to make G1 a minimum rather than zero and got the same answers as before, but now the program declared that it was successful. Even though G1 was not zero, it was quite small, and the values for all the functions were also small. Hence, we decided to accept the temperature values that were, from $T_2$ through $T_{11}$: 198.38, 196.92, 195.62, 194.48, 193.49, 192.67, 192.01, 191.53, 191.21, and 191.06 C. From the problem statement, $T_1 = 200$ C.

To summarize, we solved the nodal equations using the following methods: matrix inversion with Excel and Matlab, Gauss–Seidel with Matlab, and Excel *Solver*. The two matrix inversion solutions gave the same results for the temperatures. The Gauss–Seidel and *Solver* runs gave results slightly different from each other and slightly different from the matrix inversion results. This was expected, as the Gauss–Seidel and *Solver* computational methods are different and are iterative. But the bottom line is that the results from all methods were close to each other and all were deemed acceptable.

We will use the temperature results to finally solve this problem, namely, to determine the heat flow from the fin. Recalling the discussion on fins in Chapter 3, the heat flow can be determined by looking at the heat flow into the fin from the base or by considering the heat flow by convection into the fluid. It turns out that looking at the convection to the fluid is the better alternative, for the following reason:

The heat flow into the fin at the base is $q = -kA \left( \frac{\partial T}{\partial x} \right)_{x=0}$. In our finite-difference model, the partial derivative is approximated by $\frac{(T_2 - T_1)}{\Delta x}$. As the grid is quite coarse, this approximation does not accurately represent the partial derivative. If we had taken additional smaller nodes near the base, we would have obtained a better result for $(\partial T / \partial x)$ and a more accurate result for heat flow $q$.

So, let us sum up the convective heat flows from the different nodes to the fluid. Looking at Fig. 5.7, we have

$$q = hP(\Delta x / 2)(T_1 - T_\infty) + \sum_{i=2}^{10} hP(\Delta x)(T_i - T_\infty) + hP(\Delta x / 2)(T_{11} - T_\infty) + hA(T_{11} - T_\infty) \quad (5.46)$$

Eq. (5.46) reduces to

$$q = hP(\Delta x)\left[ \frac{1}{2}(T_1 + T_{11}) + \sum_{i=2}^{10} T_i - 10T_\infty \right] + hA(T_{11} - T_\infty) \quad (5.47)$$

Putting, into Eq. (5.47), the above-noted temperatures for the matrix inversion solution and the values of the other parameters in the problem, it is found that the heat flow from the fin to the fluid is **7.89 W**. This is only a 0.25% difference from the 7.87 W result obtained for the identical Example 3.11 of Chapter 3 which was done analytically.

After this long, but needed, discussion of solution methods, let us move on to derive nodal equations for two-dimensional objects.

Fig. 5.8 shows a flat plate with a rectangular grid. Nodes are of size $\Delta x$ by $\Delta y$. Let us derive the steady-state nodal equation for node $(i, j)$, which is an internal node for the plate, i.e., the node is not on a boundary.

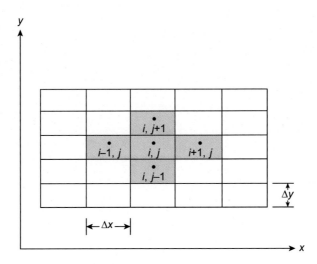

**FIGURE 5.8**

Two-dimensional plate with rectangular grid.

Node $(i, j)$ is surrounded by nodes $(i-1, j)$, $(i+1, j)$, $(i, j-1)$, and $(i, j+1)$. Let us assume that all heat flows are **into** node $(i, j)$ and that there could also be internal heat generation at volumetric strength $q_{gen}$. Node i gets conductive heat flow from the four surrounding nodes and also heat from the internal generation. To better visualize the energy balance, let us give the plate a thickness "$b$." This thickness will ultimately cancel out of the equation.

Looking at the five heat flows into $(i, j)$, we have

$$\frac{kb\Delta y}{\Delta x}\left(T_{i-1,j} - T_{i,j}\right) + \frac{kb\Delta y}{\Delta x}\left(T_{i+1,j} - T_{i,j}\right) + \frac{kb\Delta x}{\Delta y}\left(T_{i,j-1} - T_{i,j}\right) + \frac{kb\Delta x}{\Delta y}\left(T_{i,j+1} - T_{i,j}\right) + q_{gen}b\Delta x\Delta y$$
$$= 0$$

$$(5.48)$$

This can be reduced to

$$\frac{kb\Delta y}{\Delta x}\left(T_{i-1,j} + T_{i+1,j} - 2T_{i,j}\right) + \frac{kb\Delta x}{\Delta y}\left(T_{i,j-1} + T_{i,j+1} - 2T_{i,j}\right) + q_{gen}b\Delta x\Delta y = 0 \qquad (5.49)$$

If there is no internal heat generation, Eq. (5.49) becomes

$$\frac{kb\Delta y}{\Delta x}\left(T_{i-1,j} + T_{i+1,j} - 2T_{i,j}\right) + \frac{kb\Delta x}{\Delta y}\left(T_{i,j-1} + T_{i,j+1} - 2T_{i,j}\right) = 0 \qquad (5.50)$$

Finally, if conductivity $k$ is uniform throughout the body and we have a square grid, i.e., $\Delta x = \Delta y$, Eq. (5.50) becomes

$$T_{i-1,j} + T_{i+1,j} + T_{i,j-1} + T_{i,j+1} - 4T_{i,j} = 0 \tag{5.51}$$

Rearranging this equation, we get

$$T_{i,j} = \frac{T_{i-1,j} + T_{i+1,j} + T_{i,j-1} + T_{i,j+1}}{4} \tag{5.52}$$

Eq. (5.52) says that the temperature of a node is the average of the temperatures of the four surrounding nodes.

The above nodal equations are for an interior node. Similar equations can be written for the other interior nodes of the object.

Eqs. (5.49) through (5.52) were derived for an interior node $(i, j)$ by considering an energy balance on the node. They could have alternatively been obtained through converting the appropriate differential equation to a numerical form, as follows:

The general heat conduction equation was presented in Chapter 2:

$$\frac{\partial}{\partial x}\left(k\frac{\partial T}{\partial x}\right) + \frac{\partial}{\partial y}\left(k\frac{\partial T}{\partial y}\right) + \frac{\partial}{\partial z}\left(k\frac{\partial T}{\partial z}\right) + q_{gen} = \rho c \frac{\partial T}{\partial t} \tag{2.3}$$

If we have steady-state, constant conductivity, and two-dimensional conduction, then Eq. (2.3) reduces to

$$\frac{\partial^2 T}{\partial x^2} + \frac{\partial^2 T}{\partial y^2} + \frac{q_{gen}}{k} = 0 \tag{a}$$

Consider the five nodes shown in Fig. 5.8. We will convert the partial derivatives in Eq. (a) to their finite-difference form:

$$\left.\frac{\partial T}{\partial x}\right|_{i+\frac{1}{2},j} \approx \frac{T_{i+1,j} - T_{i,j}}{\Delta x} \tag{b}$$

$$\left.\frac{\partial T}{\partial x}\right|_{i-\frac{1}{2},j} \approx \frac{T_{i,j} - T_{i-1,j}}{\Delta x} \tag{c}$$

$$\left.\frac{\partial T}{\partial y}\right|_{i,j+\frac{1}{2}} \approx \frac{T_{i,j+1} - T_{i,j}}{\Delta y} \tag{d}$$

$$\left.\frac{\partial T}{\partial y}\right|_{i,j-\frac{1}{2}} \approx \frac{T_{i,j} - T_{i,j-1}}{\Delta y} \tag{e}$$

$$\left.\frac{\partial^2 T}{\partial x^2}\right|_{i,j} \approx \frac{\left.\frac{\partial T}{\partial x}\right|_{i+\frac{1}{2},j} - \left.\frac{\partial T}{\partial x}\right|_{i-\frac{1}{2},j}}{\Delta x} \tag{f}$$

$$\left.\frac{\partial^2 T}{\partial x^2}\right|_{i,j} \approx \left(\frac{T_{i+1,j} - T_{i,j}}{(\Delta x)^2}\right) - \left(\frac{T_{i,j} - T_{i-1,j}}{(\Delta x)^2}\right) \tag{g}$$

$$\left.\frac{\partial^2 T}{\partial x^2}\right|_{i,j} \approx \frac{T_{i+1,j} - 2T_{i,j} + T_{i-1,j}}{(\Delta x)^2} \tag{h}$$

Similarly, for y, $\left.\dfrac{\partial^2 T}{\partial y^2}\right|_{i,j} \approx \dfrac{T_{i,j+1} - 2T_{i,j} + T_{i,j-1}}{(\Delta y)^2}$ (i)

Putting Eqs. (h) and (i) into Eq. (a), we have the finite-difference form of the differential equation:

$$\frac{T_{i+1,j} + T_{i-1,j} - 2T_{i,j}}{(\Delta x)^2} + \frac{T_{i,j+1} + T_{i,j-1} - 2T_{i,j}}{(\Delta y)^2} + \frac{q_{gen}}{k} = 0 \tag{j}$$

If there is no internal heat generation, Eq. (j) becomes

$$\frac{T_{i+1,j} + T_{i-1,j} - 2T_{i,j}}{(\Delta x)^2} + \frac{T_{i,j+1} + T_{i,j-1} - 2T_{i,j}}{(\Delta y)^2} = 0 \tag{k}$$

Finally, if the grid is square with $\Delta x = \Delta y$, Eq. (k) becomes

$$T_{i+1,j} + T_{i-1,j} + T_{1,j+1} + T_{i,j-1} - 4T_{i,j} = 0 \tag{l}$$

Eq. (l), obtained from the differential equation, is the same as Eq. (5.51), which was obtained from an energy-balance approach.

Let us now consider nodes on the boundary of an object. Fig. 5.9 shows node $(i, j)$ and its surrounding nodes.

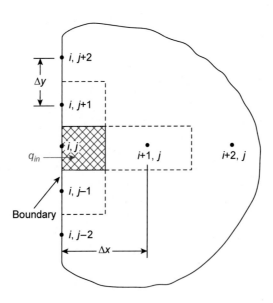

**FIGURE 5.9**

Node $(i, j)$ on a boundary.

The boundary node $(i, j)$ gets heat by conduction from its three surrounding nodes, possible heat input $q_{in}$ at the boundary, and possible heat input by internal generation.

We will sum all the heat flows into $(i, j)$ and set the result equal to zero as follows:

$$q_{in} + \frac{kb(\Delta x/2)}{\Delta y}\left(T_{i,j+1} - T_{i,j}\right) + \frac{kb(\Delta x/2)}{\Delta y}\left(T_{i,j-1} - T_{i,j}\right) + \frac{kb\Delta y}{\Delta x}\left(T_{i+1,j} - T_{i,j}\right)$$

$$+ q_{gen}b(\Delta x / 2)(\Delta y) = 0 \tag{5.53}$$

If we have convection at the boundary, then

$$q_{in} = hb\Delta y\left(T_\infty - T_{i,j}\right) \tag{5.54}$$

If we have radiation at the boundary, then

$$q_{in} = \varepsilon \sigma b\Delta y\left(T_{surr}^4 - T_{i,j}^4\right) \tag{5.55}$$

If the boundary is perfectly insulated, then

$$q_{in} = 0 \tag{5.56}$$

For example, if we have a square grid $(\Delta x = \Delta y)$, convection at the boundary, and no internal heat generation, then Eq. (5.53) reduces to the following equation for node $(i, j)$:

$$T_{i,j} = \frac{T_{i+1,j} + \dfrac{1}{2}\left(T_{i,j+1} + T_{i,j-1}\right) + \dfrac{h\Delta x}{k}T_\infty}{2 + \dfrac{h\Delta x}{k}} \tag{5.57}$$

Another boundary situation is shown in Fig. 5.10, where we have node $(i, j)$ as an exterior corner.

The exterior corner node $(i, j)$ gets heat by conduction from its two adjacent nodes, possible heat input $q_{in}$ at the boundary, and possible heat input by internal generation.

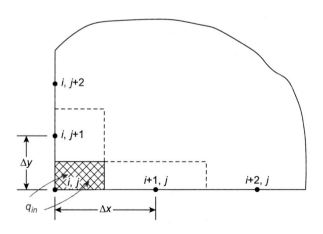

**FIGURE 5.10**

Node $(i, j)$ at exterior corner.

Summing all the heat flows into $(i, j)$ and setting the result equal to zero, we have

$$q_{in} + \frac{kb(\Delta x/2)}{\Delta y}\left(T_{i,j+1} - T_{i,j}\right) + \frac{kb(\Delta y/2)}{\Delta x}\left(T_{i+1,j} - T_{i,j}\right) + q_{gen}b(\Delta x / 2)(\Delta y / 2) = 0 \qquad (5.58)$$

If we have convection at the boundary, then

$$q_{in} = hb\left(\frac{\Delta x}{2} + \frac{\Delta y}{2}\right)\left(T_\infty - T_{i,j}\right) \qquad (5.59)$$

If we have radiation at the boundary, then

$$q_{in} = \varepsilon\sigma b\left(\frac{\Delta x}{2} + \frac{\Delta y}{2}\right)\left(T^4_{surr} - T^4_{i,j}\right) \qquad (5.60)$$

If the boundary is perfectly insulated, then

$$q_{in} = 0 \qquad (5.61)$$

For example, if we have a square grid $(\Delta x = \Delta y)$, convection at the boundary, and no internal heat generation, then Eq. (5.58) reduces to the following equation for node $(i, j)$:

$$T_{i,j} = \frac{\frac{1}{2}\left(T_{i,j+1} + T_{i+1,j}\right) + \frac{h\Delta x}{k}T_\infty}{1 + \frac{h\Delta x}{k}} \qquad (5.62)$$

As a final example of a boundary node, let us look at Fig. 5.11, where we have a node on an interior corner.

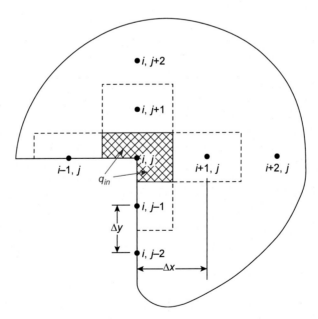

**FIGURE 5.11**

Node $(i, j)$ at interior corner.

The interior corner node $(i, j)$ gets heat by conduction from its four adjacent nodes, possible heat input $q_{in}$ at the boundary, and possible heat input by internal generation.

Summing all the heat flows into node $(i, j)$ and setting the result to zero, we have

$$q_{in} + \frac{kb(\Delta x/2)}{\Delta y}\left(T_{i,j-1} - T_{i,j}\right) + \frac{kb(\Delta y/2)}{\Delta x}\left(T_{i-1,j} - T_{i,j}\right) + \frac{kb\Delta x}{\Delta y}\left(T_{i,j+1} - T_{i,j}\right) + \frac{kb\Delta y}{\Delta x}\left(T_{i+1,j} - T_{i,j}\right)$$

$$+ q_{gen}b\left(\frac{3}{4}\Delta x\Delta y\right)$$

$$= 0$$

$$\text{(5.63)}$$

If we have convection at the boundary, then

$$q_{in} = hb\left(\frac{\Delta x}{2} + \frac{\Delta y}{2}\right)\left(T_\infty - T_{i,j}\right) \tag{5.64}$$

If we have radiation at the boundary, then

$$q_{in} = \varepsilon\sigma b\left(\frac{\Delta x}{2} + \frac{\Delta y}{2}\right)\left(T_{surr}^4 - T_{i,j}^4\right) \tag{5.65}$$

If the boundary is perfectly insulated, then

$$q_{in} = 0 \tag{5.66}$$

For example, if we have a square grid $(\Delta x = \Delta y)$, convection at the boundary, and no internal heat generation, then Eq. (5.63) reduces to the following equation for node $(i, j)$:

$$T_{i,j} = \frac{T_{i,j+1} + T_{i+1,j} + \frac{1}{2}\left(T_{i,j-1} + T_{i-1,j}\right) + \frac{h\Delta x}{k}T_\infty}{3 + \frac{h\Delta x}{k}} \tag{5.67}$$

---

## Example 5.2
### Rectangular plate with temperature boundary conditions

*Problem*

The rectangular plate shown in Fig. 5.12 has temperature boundary conditions. Determine, numerically, the temperatures at the four internal nodes for steady conditions.

*Solution*

The nodal spacing is $\Delta x = 20$ cm and $\Delta y = 10$ cm. There is no internal heat generation. The four nodes are interior nodes, so we can use Eq. (5.50).

$$\frac{kb\Delta y}{\Delta x}\left(T_{i-1,j} + T_{i+1,j} - 2T_{i,j}\right) + \frac{kb\Delta x}{\Delta y}\left(T_{i,j-1} + T_{i,j+1} - 2T_{i,j}\right) = 0 \tag{5.50}$$

Applying this equation to node 1, we have

$$0.5(300 + T_2 - 2T_1) + 2(T_3 + 100 - 2T_1) = 0 \tag{5.68}$$

Rearranging Eq. (5.68), the equation for temperature $T_1$ becomes

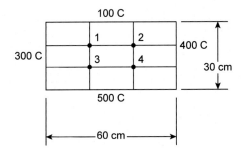

**FIGURE 5.12**

Plate for Example 5.2.

$$T_1 = \frac{0.5T_2 + 2T_3 + 350}{5} \qquad (5.69)$$

Applying Eq. (5.50) to node 2, we have

$$0.5(T_1 + 400 - 2T_2) + 2(T_4 + 100 - 2T_2) = 0 \qquad (5.70)$$

$$T_2 = \frac{0.5T_1 + 2T_4 + 400}{5} \qquad (5.71)$$

Continuing the same procedure for nodes 3 and 4, we get

$$T_3 = \frac{0.5T_4 + 2T_1 + 1150}{5} \qquad (5.72)$$

$$\text{and} \quad T_4 = \frac{0.5T_3 + 2T_2 + 1200}{5} \qquad (5.73)$$

Solving Eqs. (5.69), (5.71), (5.72), and (5.73) by Gauss—Seidel, we got the results

$$T_1 = 241.3, \quad T_2 = 255.6, \quad T_3 = 364.4, \quad T_4 = 378.7 \; C$$

An alternative way of solving this problem is to change the grid from a rectangular grid to a square grid, as shown in Fig. 5.13.

There are more nodes and equations (10 vs. 4), but it is easier to write the equations as the ratio $\Delta y/\Delta x$ becomes 1. Eq. (5.52) shows that, for an interior node and a square grid, a nodal temperature is simply the average of the temperatures of the four surrounding nodes. The equations for the 10 nodes of this revised grid can be quickly written by inspection. For example, the temperature of node 1 is $(300 + 100 + T_2 + T_6)/4$; the temperature of node 2 is $(T_1 + 100 + T_3 + T_7)/4$, etc. Solving the 10 equations by Gauss—Seidel, we got the below results:

$$T_1 = 248.0, \quad T_2 = 238.4, \quad T_3 = 240.1, \quad T_4 = 250.9, \quad T_5 = 285.5,$$

$$T_6 = 353.4, \quad T_7 = 365.7, \quad T_8 = 371.0, \quad T_9 = 378.2, \quad T_{10} = 390.9 \; C$$

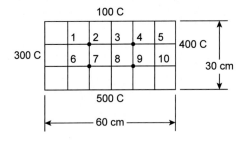

**FIGURE 5.13**

Revised grid for Example 5.2.

In this revised, finer grid, node 2 is the same as the original node 1; node 4 is the same as the original node 2; node 7 is the same as the original node 3, and node 9 is the same as the original node 4. Using the labeling of the original grid, the result is

$$T_1 = 238.4, \quad T_2 = 250.9, \quad T_3 = 365.7, \quad T_4 = 378.2 \ C \quad \text{(10 node grid)}$$

$$T_1 = 241.3, \quad T_2 = 255.6, \quad T_3 = 364.4, \quad T_4 = 378.7 \ C \quad \text{(4 node grid)}$$

It is seen that there is a small difference between the results of the two models. In general, a finer grid gives more accurate results than a coarser grid unless the grid is so fine that round-off error becomes significant.

The Matlab programs for both the 4-node and 10-node models are in Appendix K.

(Note: Although this problem was for a plate, it could have been for a long column of rectangular cross section having different temperatures on its four sides.)

Example 5.2 only had interior nodes. The next example has boundary, as well as interior, nodes. It also has internal heat generation.

## Example 5.3
### Plate with a variety of boundary conditions and heat generation

*Problem*

The plate shown in Fig. 5.14 has a variety of boundary conditions. The left surface is isothermal at 500 C. The bottom surface is isothermal at 100 C. The top and right surfaces are insulated, and the inside corner on the left has convection to a fluid. The convective coefficient is 100 W/m² C, and the fluid temperature is 30 C. The plate is steel and has a conductivity of 43 W/m C. The grid is square with $\Delta x = \Delta y = 8$ cm.

**(a)** If there is no internal heat generation, determine the nodal temperatures for steady state.

**(b)** If the plate has internal heat generation at a strength of $10^6$ W/m³, determine the nodal temperatures for steady state.

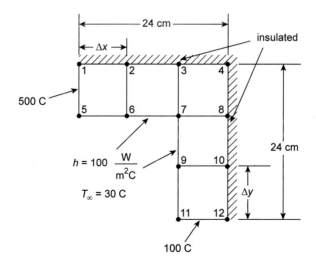

**FIGURE 5.14**

Plate for Example 5.3.

*Solution*

Let us start the solution by deriving the nodal equations. For a particular node, we assume that all heat flows go into the node and we sum up the heat flows to zero. Again, we will use "*b*" as the thickness of the plate. As Part (b) has heat generation, we will include the heat generation term in the nodal equations.

Insulated nodes (2, 3, 4, 8, 10):

Each node gets heat flow by conduction from its adjacent nodes. For node 2, we have conduction from nodes 1, 3, and 6. The nodal equation is

$$\frac{kb(\Delta y/2)}{\Delta x}(T_1 - T_2) + \frac{kb(\Delta y/2)}{\Delta x}(T_3 - T_2) + \frac{kb\Delta x}{\Delta y}(T_6 - T_2) + q_{gen}b\frac{\Delta y}{2}\Delta x = 0 \tag{5.74}$$

For node 3, we have conduction from nodes 2, 4, and 7, and the nodal equation is

$$\frac{kb(\Delta y/2)}{\Delta x}(T_2 - T_3) + \frac{kb(\Delta y/2)}{\Delta x}(T_4 - T_3) + \frac{kb\Delta x}{\Delta y}(T_7 - T_3) + q_{gen}b\frac{\Delta y}{2}\Delta x = 0 \tag{5.75}$$

Similarly for nodes 4, 8, and 10:

Node 4

$$\frac{kb(\Delta y/2)}{\Delta x}(T_3 - T_4) + \frac{kb(\Delta x/2)}{\Delta y}(T_8 - T_4) + q_{gen}b\frac{\Delta y}{2}\frac{\Delta x}{2} = 0 \tag{5.76}$$

Node 8

$$\frac{kb(\Delta x/2)}{\Delta y}(T_4 - T_8) + \frac{kb(\Delta x/2)}{\Delta y}(T_{10} - T_8) + \frac{kb\Delta y}{\Delta x}(T_7 - T_8) + q_{gen}b\frac{\Delta x}{2}\Delta y = 0 \tag{5.77}$$

Node 10

$$\frac{kb(\Delta x/2)}{\Delta y}(T_8 - T_{10}) + \frac{kb(\Delta x/2)}{\Delta y}(T_{12} - T_{10}) + \frac{kb\Delta y}{\Delta x}(T_9 - T_{10}) + q_{gen}b\frac{\Delta x}{2}\Delta y = 0 \tag{5.78}$$

Convection nodes (6, 7, 9):

These nodes get conduction from adjacent nodes and convection from the adjacent fluid.

For node 6, we have conduction from nodes 2, 5, and 7 and convection from the fluid. The nodal equation is

$$\frac{kb(\Delta y/2)}{\Delta x}(T_5 - T_6) + \frac{kb(\Delta y/2)}{\Delta x}(T_7 - T_6) + \frac{kb\Delta x}{\Delta y}(T_2 - T_6) + hb\Delta x(T_\infty - T_6) + q_{gen}b\frac{\Delta y}{2}\Delta x = 0 \tag{5.79}$$

Similarly for nodes 7 and 9:

Node 7

$$\frac{kb(\Delta y/2)}{\Delta x}(T_6 - T_7) + \frac{kb(\Delta x/2)}{\Delta y}(T_9 - T_7) + \frac{kb\Delta y}{\Delta x}(T_8 - T_7) + \frac{kb\Delta x}{\Delta y}(T_3 - T_7)$$
$$+ hb((\Delta x/2) + (\Delta y/2))(T_\infty - T_7) + q_{gen}b\left(\frac{3}{4}\Delta x\Delta y\right) = 0 \tag{5.80}$$

Node 9

$$\frac{kb(\Delta x/2)}{\Delta y}(T_7 - T_9) + \frac{kb(\Delta x/2)}{\Delta y}(T_{11} - T_9) + \frac{kb\Delta y}{\Delta x}(T_{10} - T_9) + hb\Delta y(T_\infty - T_9) + q_{gen}b\frac{\Delta x}{2}\Delta y = 0 \tag{5.81}$$

Isothermal nodes (1, 5, 11, 12):

$$T_1 = T_5 = 500\ C \text{ and } T_{11} = T_{12} = 100\ C \tag{5.82}$$

We will solve the nodal equations by matrix inversion. We have a square grid ($\Delta x = \Delta y$), which simplifies the equations considerably. To use matrix inversion, we need to rearrange the terms in Eqs. (5.74)–(5.82) to facilitate creation of a coefficients matrix and a constants vector. Doing this, and incorporating the square matrix feature, the equations become

$$\text{For Node 1: } T_1 = 500 \tag{5.83}$$

$$\text{For Node 2: } -\frac{1}{2}T_1 + 2T_2 - \frac{1}{2}T_3 - T_6 = \frac{q_{gen}(\Delta x)^2}{2k} \tag{5.84}$$

$$\text{For Node 3: } -\frac{1}{2}T_2 + 2T_3 - \frac{1}{2}T_4 - T_7 = \frac{q_{gen}(\Delta x)^2}{2k} \tag{5.85}$$

$$\text{For Node 4: } -\frac{1}{2}T_3 + T_4 - \frac{1}{2}T_8 = \frac{q_{gen}(\Delta x)^2}{4k} \tag{5.86}$$

$$\text{For Node 5: } T_5 = 500 \tag{5.87}$$

$$\text{For Node 6: } -T_2 - \frac{1}{2}T_5 + \left(2 + \frac{h\Delta x}{k}\right)T_6 - \frac{1}{2}T_7 = \frac{q_{gen}(\Delta x)^2}{2k} + \frac{h\Delta x}{k}T_\infty \tag{5.88}$$

$$\text{For Node 7: } -T_3 - \frac{1}{2}T_6 + \left(3 + \frac{h\Delta x}{k}\right)T_7 - T_8 - \frac{1}{2}T_9 = \frac{3}{4}\frac{q_{gen}(\Delta x)^2}{k} + \frac{h\Delta x}{k}T_\infty \tag{5.89}$$

$$\text{For Node 8: } -\frac{1}{2}T_4 - T_7 + 2T_8 - \frac{1}{2}T_{10} = \frac{q_{gen}(\Delta x)^2}{2k} \tag{5.90}$$

$$\text{For Node 9: } -\frac{1}{2}T_7 + \left(2 + \frac{h\Delta x}{k}\right)T_9 - T_{10} - \frac{1}{2}T_{11} = \frac{q_{gen}(\Delta x)^2}{2k} + \frac{h\Delta x}{k}T_\infty \tag{5.91}$$

$$\text{For Node 10: } -\frac{1}{2}T_8 - T_9 + 2T_{10} - \frac{1}{2}T_{12} = \frac{q_{gen}(\Delta x)^2}{2k} \tag{5.92}$$

$$\text{For Node 11: } T_{11} = 100 \tag{5.93}$$

$$\text{For Node 12: } T_{12} = 100 \tag{5.94}$$

For our problem, $k = 43$ W/mC, $h = 100$ W/m² C, $T_\infty = 30$ C, and $\Delta x = 8$ cm $= 0.08$ m.
We put these values into Eqs. (5.83)–(5.94) and created the coefficients matrix and the constants vector. For Part (a) of the problem, $q_{gen} = 0$. For Part (b) of the problem, $q_{gen} = 10^6$ W/m³. The problem was solved using matrix inversion and Matlab, with the following results:

| | (a) $q_{gen} = 0$ | (b) $q_{gen} = 10^6$ W/m³ |
|---|---|---|
| $T_1$ | 500 C | 500 C |
| $T_2$ | 353.9 | 589.8 |
| $T_3$ | 256.7 | 617.0 |
| $T_4$ | 231.0 | 628.5 |
| $T_5$ | 500 | 500 |
| $T_6$ | 329.3 | 546.6 |
| $T_7$ | 221.0 | 550.4 |
| $T_8$ | 205.3 | 565.6 |
| $T_9$ | 143.8 | 361.1 |
| $T_{10}$ | 148.2 | 384.1 |
| $T_{11}$ | 100 | 100 |
| $T_{12}$ | 100 | 100 |

The Matlab program for this solution is in Appendix K.
(Note: Although this problem was for a plate, it could have been for a long L-shaped structural support having different conditions on its various surfaces.)

Let us now move on to a discussion of numerical solution of unsteady problems.

## 5.2.2 Unsteady state

If we have unsteady heat flow in an object, we have to consider time, as well as location. For steady state, we had spatial increments such as $\Delta x$ and $\Delta y$. For unsteady state, we have a time increment, or "step," $\Delta t$ in addition to the spatial increments. As shown in Fig. 5.15, the time scale progresses from the origin at time zero, to one time step ($\Delta t$), two time steps ($2\Delta t$), three time steps ($3\Delta t$), etc.

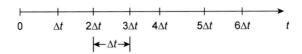

**FIGURE 5.15**

The time scale.

In general, temperatures in an object are known at the beginning of a process, i.e., at time zero. Problem solution proceeds from one time step to the next, with temperatures being determined for each successive time step.

Let us now determine an energy conservation equation for unsteady state situations. Fig. 5.16 shows an interior node i in a one-dimensional object.

Nodes i−1 and i+1 are the two nodes adjacent to node i. There is conductive heat flow between these nodes and node i. There may also be internal heat generation. Like steady state, we will consider all heat flows to be **into** a node. A positive net heat flow into node i will increase the node's internal energy. A negative net heat flow into node i will decrease the node's internal energy. Let us look at the change in internal energy of node i during a time increment $\Delta t$. In words, we have

$$\text{The net heat flow into Node i} = \text{Change in internal energy of material}$$

$$\text{during time } \Delta t \qquad\qquad \text{in node i during time } \Delta t \qquad\qquad (5.95)$$

Mathematically, this is

$$\frac{kA}{\Delta x}\left(T_{i-1}^{b} - T_{i}^{b}\right)\Delta t + \frac{kA}{\Delta x}\left(T_{i+1}^{b} - T_{i}^{b}\right)\Delta t + q_{gen}A\Delta x\ \Delta t = \rho c A\Delta x\left(T_{i}^{e} - T_{i}^{b}\right) \qquad (5.96)$$

where $k =$ thermal conductivity of the material

$\rho =$ density of the material
$c =$ specific heat of the material

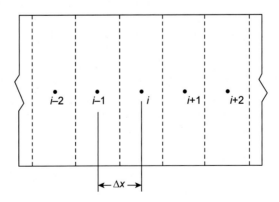

**FIGURE 5.16**

Interior nodes for one-dimensional object.

$\Delta x$ = spatial increment

$\Delta t$ = time increment

$A$ = cross-sectional area

$q_{gen}$ = volumetric heat generation rate

Eq. (5.96) describes the change in internal energy during one time step $\Delta t$. The terms on the left are the heat flows into node i during the time step. We have the heat flow rates times $\Delta t$ to give the heat quantity into node i during the time step. The change in internal energy of the material in node i during the time step is on the right side of Eq. (5.96). This internal energy change causes a change in temperature of the node. At the beginning of the time step, the temperature of node i is $T_i^b$. At the end of the time step, the temperature is $T_i^e$. Let us talk a little more about these temperatures.

For numerical finite differences, time is not continuous. Rather, it is a series of successive, discrete time steps. Temperatures are defined at the beginning and end of these time steps. In Eq. (5.96), temperature superscript "$b$" designates temperatures at the beginning of the time step, and superscript "$e$" designates temperatures at the end of the time step. Looking at the left side of Eq. (5.96), it is seen that the heat flows into the node are based on temperatures at the beginning of the time step. We need temperatures at both the beginning and the end of the time step on the right side of Eq. (5.96) as we need the temperature *change* for the step.

Equations like Eq. (5.96) may be written for all interior nodes in an object. Let us rearrange the equation so that $T_i^e$ is on the left side of the equation. Doing this, we have

$$T_i^e = T_i^b \left(1 - 2\frac{\alpha\Delta t}{(\Delta x)^2}\right) + \frac{\alpha\Delta t}{(\Delta x)^2}\left(T_{i-1}^b + T_{i+1}^b\right) + \frac{q_{gen}\Delta t}{\rho c} \tag{5.97}$$

where $\alpha$ = thermal diffusivity = $k/\rho\, c$.

A two-dimensional plate is shown in Fig. 5.17.

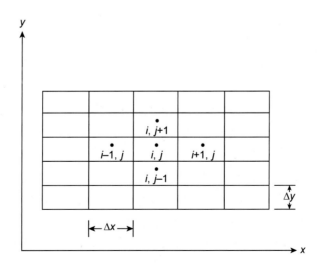

**FIGURE 5.17**

Interior nodes for two-dimensional plate.

Interior node $(i,j)$ gets conduction from surrounding nodes $(i-1,j)$, $(i+1,j)$, $(i,j-1)$, and $(i,j+1)$. It also may get heat from internal heat generation. Like the one-dimensional case, a positive net heat flow into node $(i,j)$ will raise the internal energy of the node. The energy conservation equation for node $(i, j)$ is

$$\frac{kb\Delta y}{\Delta x}\left(T_{i-1,j}^b - T_{i,j}^b\right)\Delta t + \frac{kb\Delta y}{\Delta x}\left(T_{i+1,j}^b - T_{i,j}^b\right)\Delta t + \frac{kb\Delta x}{\Delta y}\left(T_{i,j-1}^b - T_{i,j}^b\right)\Delta t + \frac{kb\Delta x}{\Delta y}\left(T_{i,j+1}^b - T_{i,j}^b\right)\Delta t$$

$$+ q_{gen}b\Delta x\Delta y\Delta t = \rho c b\Delta x\Delta y\left(T_{i,j}^e - T_{i,j}^b\right)$$

(5.98)

Rearranging this equation so that $T_{i,j}^e$ is on the left side of the equation, we get

$$T_{i,j}^e = T_{i,j}^b\left(1 - 2\frac{\alpha\Delta t}{(\Delta x)^2} - 2\frac{\alpha\Delta t}{(\Delta y)^2}\right) + \frac{\alpha\Delta t}{(\Delta x)^2}\left(T_{i-1,j}^b + T_{i+1,j}^b\right) + \frac{\alpha\Delta t}{(\Delta y)^2}\left(T_{i,j-1}^b + T_{i,j+1}^b\right) + \frac{q_{gen}\Delta t}{\rho c}$$

(5.99)

If we have a square grid ($\Delta x = \Delta y$) and no heat generation, Eq. (5.99) reduces to

$$T_{i,j}^e = T_{i,j}^b\left(1 - 4\frac{\alpha\Delta t}{(\Delta x)^2}\right) + \frac{\alpha\Delta t}{(\Delta x)^2}\left(T_{i-1,j}^b + T_{i+1,j}^b + T_{i,j-1}^b + T_{i,j+1}^b\right)$$

(5.100)

Furthermore, if we also pick the time step $\Delta t$ such that $1 - 4\frac{\alpha\Delta t}{(\Delta x)^2} = 0$. That is, if $\Delta t = \frac{(\Delta x)^2}{4\alpha}$, then Eq. (5.100) becomes

$$T_{i,j}^e = \frac{1}{4}\left(T_{i-1,j}^b + T_{i+1,j}^b + T_{i,j-1}^b + T_{i,j+1}^b\right)$$

(5.101)

Eq. (5.101) states that the temperature of node $(i, j)$ at the end of the time step is equal to the average of the temperatures of the four surrounding nodes at the beginning of the time step.

Energy conservation equations may also be written for boundary nodes. Consider, for example, the corner in Fig. 5.18, which has convection from the adjacent fluid and conduction from nodes $(i, j+1)$ and $(i+1, j)$.

The energy equation for corner node $(i, j)$ is

$$\frac{kb(\Delta x/2)}{\Delta y}\left(T_{i,j+1}^b - T_{i,j}^b\right)\Delta t + \frac{kb(\Delta y/2)}{\Delta x}\left(T_{i+1,j}^b - T_{i,j}^b\right)\Delta t + hb\left(\frac{\Delta x}{2} + \frac{\Delta y}{2}\right)\left(T_\infty - T_{i,j}^b\right)\Delta t$$

$$+ q_{gen}b(\Delta x/2)\ (\Delta y/2)\Delta t = \rho c b(\Delta x/2)\ (\Delta y/2)\left(T_{i,j}^e - T_{i,j}^b\right)$$

(5.102)

Rearranging this equation so that $T_{i,j}^e$ is on the left side, we get

$$T_{i,j}^e = T_{i,j}^b\left(1 - 2\frac{\alpha\Delta t}{(\Delta x)^2} - 2\frac{\alpha\Delta t}{(\Delta y)^2} - 2\frac{h\Delta t(\Delta x + \Delta y)}{\rho c(\Delta x)(\Delta y)}\right) + 2\frac{\alpha\Delta t}{(\Delta x)^2}T_{i+1,j}^b + 2\frac{\alpha\Delta t}{(\Delta y)^2}T_{i,j+1}^b$$

(5.103)

$$+ 2\frac{h\Delta t(\Delta x + \Delta y)}{\rho c(\Delta x)(\Delta y)}T_\infty + \frac{q_{gen}\Delta t}{\rho c}$$

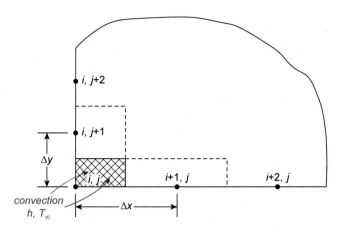

**FIGURE 5.18**

Exterior corner with convection.

If the grid is square ($\Delta x = \Delta y$) and there is no heat generation, Eq. (5.103) reduces to

$$T_{i,j}^e = T_{i,j}^b \left(1 - 4\frac{\alpha \Delta t}{(\Delta x)^2} - 4\frac{h \Delta t}{\rho c (\Delta x)}\right) + 2\frac{\alpha \Delta t}{(\Delta x)^2}\left(T_{i+1,j}^b + T_{i,j+1}^b\right) + 4\frac{h \Delta t}{\rho c (\Delta x)}T_\infty \qquad (5.104)$$

We can write nodal equations similar to Eqs. (5.97), (5.99), (5.100), (5.101), (5.103), and (5.104) for all the interior and boundary nodes of a problem. If we look at these equations, it is seen from the superscripts that the temperature on the left side of the equation is the temperature at the end of a time step and all the temperatures on the right side are temperatures at the beginning of a time step. In solving the equations, we work forward in time. For each time step, we know the temperatures at the beginning of the time step. We calculate the right sides of the equations, and the result becomes the temperatures on the left side of the equation. Then we continue with the next time step. The temperatures we got on the left side of the equations now become the beginning temperatures for the next time step. We use these temperatures on the right side of the equations to calculate end temperatures for the time step. And, the process continues for as many time steps as we desire. For steady-state problems we had to simultaneously solve the nodal equations. For unsteady problems, we simply list the equations, one after another, and go through the equations successively. We calculate the right-hand sides of the equations using temperatures that we already know and we get new temperatures on the left sides of the equations that are temperatures at the end of that time step. These temperatures are now used on the right sides of the equations to determine the temperatures at the end of the next time step. This procedure can be easily done using loops in Matlab or iterations in Excel.

We will give a few examples of the procedure, but let us first discuss the determination of time increment $\Delta t$.

When we solve a problem, we first decide on the spatial increments such as $\Delta x$, $\Delta y$, $\Delta z$, $\Delta r$, etc. We then decide on what time increment $\Delta t$ to use. The above nodal equations were developed using the **forward-difference** or **explicit approach**. As described above, we know the temperatures at the beginning of the time step and we get the temperatures at the end of the time step using the nodal equations. We are working *forward* in time. It turns out that selection of the spatial and time increments are not independent with this forward-difference approach. If we decide on too large a $\Delta t$, the procedure will be unstable and not result in the correct solution.

What is "too large" for $\Delta t$ ? Let us look at Eq. (5.97), which is the equation for an interior node ($i$) in a one-dimensional object, and Eq. (5.99), which is the equation for an interior node ($i, j$) in a two-dimensional object. The coefficients for nodes ($i$) and ($i, j$) in these equations must be zero or positive for a stable solution. If $\Delta t$ is too large, the coefficients become negative and the solution is unstable. That is, for the forward-difference approach,

$$1 - 2\frac{\alpha \Delta t}{(\Delta x)^2} \geq 0 \text{ and } \Delta t \leq \frac{(\Delta x)^2}{2\alpha} \text{ (One − Dimensional)} \tag{5.105}$$

$$1 - 2\frac{\alpha \Delta t}{(\Delta x)^2} - 2\frac{\alpha \Delta t}{(\Delta y)^2} \geq 0 \text{ and } \Delta t \leq \frac{1}{\left(\dfrac{2\alpha}{(\Delta x)^2} + \dfrac{2\alpha}{(\Delta y)^2}\right)} \text{ (Two − Dimensional)} \tag{5.106}$$

These restrictions on the size of $\Delta t$ are based on interior points in a body. If there are boundary nodes, $\Delta t$ may have to be a bit smaller. See, for example, Eq. (5.103), which is for a two-dimensional boundary node with convection. The coefficient for node ($i, j$) needs to be zero or positive, and the convection term makes the acceptable $\Delta t$ smaller than that given by Eq. (5.106).

As a practical matter, these restrictions on $\Delta t$ usually have no impact. Most of the time we are able to use a $\Delta t$ meeting the criteria. However, if we really need to use a larger $\Delta t$, then we can use the **backward-difference** or **implicit approach.** With this approach, the spatial and time increments are independent, and there is no restriction on the size of $\Delta t$. However, as we will see, much more computation is needed for the backward-difference approach than for the forward-difference approach. Let us look at this backward-difference approach. The previous equation we had for an interior node in a two-dimensional object was

$$\frac{kb\Delta y}{\Delta x}\left(T_{i-1,j}^b - T_{i,j}^b\right)\Delta t + \frac{kb\Delta y}{\Delta x}\left(T_{i+1,j}^b - T_{i,j}^b\right)\Delta t + \frac{kb\Delta x}{\Delta y}\left(T_{i,j-1}^b - T_{i,j}^b\right)\Delta t + \frac{kb\Delta x}{\Delta y}\left(T_{i,j+1}^b - T_{i,j}^b\right)\Delta t$$

$$+ q_{gen}b\Delta x\Delta y\Delta t$$

$$= \rho cb\Delta x\Delta y\left(T_{i,j}^e - T_{i,j}^b\right) \tag{5.98}$$

The temperature differences on the left side of the equation have superscript "$b$," which means they are the differences at the *beginning* of the time step. Let us instead make these differences the differences at the *end* of the time step. That is, we will give them the superscript "$e$."

$$\frac{kb\Delta y}{\Delta x}\left(T_{i-1,j}^e - T_{i,j}^e\right)\Delta t + \frac{kb\Delta y}{\Delta x}\left(T_{i+1,j}^e - T_{i,j}^e\right)\Delta t + \frac{kb\Delta x}{\Delta y}\left(T_{i,j-1}^e - T_{i,j}^e\right)\Delta t + \frac{kb\Delta x}{\Delta y}\left(T_{i,j+1}^e - T_{i,j}^e\right)\Delta t$$

$$+ q_{gen}b\Delta x\Delta y\Delta t = \rho cb\Delta x\Delta y\left(T_{i,j}^e - T_{i,j}^b\right) \tag{5.107}$$

Rearranging this equation, we get

$$T^b_{i,j} = T^e_{i,j}\left(1 + 2\frac{\alpha\Delta t}{(\Delta x)^2} + 2\frac{\alpha\Delta t}{(\Delta y)^2}\right) - \frac{\alpha\Delta t}{(\Delta x)^2}\left(T^e_{i-1,j} + T^e_{i+1,j}\right) - \frac{\alpha\Delta t}{(\Delta y)^2}\left(T^e_{i,j-1} + T^e_{i,j+1}\right) - \frac{q_{gen}\Delta t}{\rho c}$$

(5.108)

All items in the coefficient for $T_{i,j}$ in Eq. (5.108) are now positive, and there is no restriction on the size of $\Delta t$. Similar equations may be written for all nodes in an object, giving us a set of equations to solve for the temperatures at the end of a time step. Unfortunately, there are many unknowns in each equation: namely, the temperatures of the various nodes at the end of the time step. We have to solve the set of equations simultaneously to get the temperatures at the end of each time step before we can move forward to the next time step. The computational effort is much more than we had with the forward-difference approach. In using the backward-difference approach, we have eliminated the size restriction on $\Delta t$, but we have increased the computational effort greatly. With the forward-difference approach, we merely looped through the nodal equations, getting new values for the temperatures and working forward one time step per loop. With the backward-difference approach, we have to solve simultaneous equations for the temperatures for each time step as we work forward in time.

We will now give some examples illustrating solution of unsteady state problems.

## Example 5.4
### Rectangular plate with unsteady conduction
*Problem*
The stainless steel ($k = 15$ W/m C, $\rho = 7900$ kg/m$^3$, $c = 477$ J/kg C) plate shown in Fig. 5.19 has an initial uniform temperature of 30 C. At time zero, the boundary temperatures shown in the figure are imposed on the plate.
**(a)** How long does it take for node 1 to reach 150 C and what are the temperatures of the other nodes at that time?
**(b)** What are the steady-state temperatures for the nodes?
**(c)** How long does it take for the plate to achieve steady state?

*Solution*
The four nodes of the plate are interior nodes. Each node is surrounded by four other nodes. If we call the node we are writing the equation for "node $(i, j)$," then the surrounding nodes are nodes $(i-1, j)$, $(i+1, j)$, $(i, j-1)$, and $(i, j+1)$. There is no internal heat generation. Therefore, Eq. (5.99), with $q_{gen} = 0$, can be used to obtain the nodal equations. This equation is

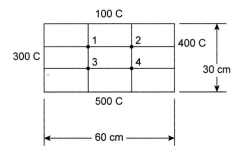

**FIGURE 5.19**

Plate for Example 5.4.

$$T_{i,j}^e = T_{i,j}^b\left(1 - 2\frac{\alpha\Delta t}{(\Delta x)^2} - 2\frac{\alpha\Delta t}{(\Delta y)^2}\right) + \frac{\alpha\Delta t}{(\Delta x)^2}\left(T_{i-1,j}^b + T_{i+1,j}^b\right) + \frac{\alpha\Delta t}{(\Delta y)^2}\left(T_{i,j-1}^b + T_{i,j+1}^b\right) \qquad (5.109)$$

Applying Eq. (5.109) to the four nodes, we get the following nodal equations:

$$T_1^e = T_1^b\left(1 - 2\frac{\alpha\Delta t}{(\Delta x)^2} - 2\frac{\alpha\Delta t}{(\Delta y)^2}\right) + \frac{\alpha\Delta t}{(\Delta x)^2}\left(300 + T_2^b\right) + \frac{\alpha\Delta t}{(\Delta y)^2}\left(100 + T_3^b\right) \qquad (5.110)$$

$$T_2^e = T_2^b\left(1 - 2\frac{\alpha\Delta t}{(\Delta x)^2} - 2\frac{\alpha\Delta t}{(\Delta y)^2}\right) + \frac{\alpha\Delta t}{(\Delta x)^2}\left(T_1^b + 400\right) + \frac{\alpha\Delta t}{(\Delta y)^2}\left(T_4^b + 100\right) \qquad (5.111)$$

$$T_3^e = T_3^b\left(1 - 2\frac{\alpha\Delta t}{(\Delta x)^2} - 2\frac{\alpha\Delta t}{(\Delta y)^2}\right) + \frac{\alpha\Delta t}{(\Delta x)^2}\left(300 + T_4^b\right) + \frac{\alpha\Delta t}{(\Delta y)^2}\left(500 + T_1^b\right) \qquad (5.112)$$

$$T_4^e = T_4^b\left(1 - 2\frac{\alpha\Delta t}{(\Delta x)^2} - 2\frac{\alpha\Delta t}{(\Delta y)^2}\right) + \frac{\alpha\Delta t}{(\Delta x)^2}\left(T_3^b + 400\right) + \frac{\alpha\Delta t}{(\Delta y)^2}\left(T_2^b + 500\right) \qquad (5.113)$$

One of the first things to do is decide on a time step. For the forward-difference approach, the bracketed expression $\left(1 - 2\frac{\alpha\Delta t}{(\Delta x)^2} - 2\frac{\alpha\Delta t}{(\Delta y)^2}\right)$ cannot be negative. This limits the size of $\Delta t$. For this problem, $\Delta x = 0.2$ m and $\Delta y = 0.1$ m and $\alpha = \frac{k}{\rho c} = \frac{15}{(7900)(477)} = 3.981 \times 10^{-6}$ m$^2$/s.

Putting these values into the bracketed expression, it is found that $\Delta t$ must be less than or equal to 1004 s. We used a $\Delta t$ of 1 s, certainly well within the limit. A small $\Delta t$ will give more-accurate results. A time step of 10 s gave very good results and a time step of 100 s gave good results. Above that time step, the results were not accurate.

(a) The Matlab program for this problem is in Appendix K. The four nodes are given initial values of 30 C and then the program loops through Eq. (5.110)–(5.113) successively. Each loop (or pass) moves the time ahead one $\Delta t$. When node 1 reaches 150 C, the program stops and prints out all the nodal temperatures.

  The program took 2134 loops for node 1 to reach 150 C. With $\Delta t = 1$ s, this means node 1 reached 150 C in **2134 s = 35.6 min.** At this time, $T_2 = $ **161.1 C**, $T_3 = $ **265.3 C**, and $T_4 = $ **276.4 C.**

(b) We ran the program without the $T_1 = 150$ breakpoint to see what final temperatures would be reached. For 19,300 loops and greater, the temperatures did not change. Steady state had been reached. Final temperatures were $T_1 = $ **241.3 C,** $T_2 = $ **255.6 C,** $T_3 = $ **364.4 C,** and $T_4 = $ **378.7 C.** These temperatures are the same as we got for the Gauss–Seidel solution to Example 5.2, which was the steady-state problem for this geometry.

(c) As mentioned in Part (b): By running the program to steady-state conditions, we found that temperatures stopped changing and steady state was reached in 19,300 loops, or 19,300 s = **5.4 h.** We may wish to define steady state a bit less stringently, as follows: Earlier in this text, we mentioned that, for a first-order system, a process is often considered to be complete after five time constants. At this point, the process is actually 99.33% complete. Let us consider node 1. For its process, the node goes from an initial 30 C to a final 241.3 C. The process is therefore a temperature change of 241.3–30 or 211.3 C. The 99.33% change for the process is $211.3 \times 0.9933 = 209.9$ C, which gives a final temperature of $30 + 209.9 = 239.9$ C. Running the Matlab program again, it is found that node 1 reaches 239.9 C in 10,610 s = 2.95 h. Similar results are found for the other three nodes. So, if we use this "five time constant" criterion, we could say that steady state is essentially reached after **3 h.**

(Note: Although this problem was for a plate, it could have been for a long column of rectangular cross section having different temperatures on its four sides.)

We now look at an example in cylindrical coordinates, which have both convection and radiation boundary conditions.

## Example 5.5

### Disk with convection and radiation

*Problem*

The stainless steel disk ($k = 15$ W/m C, $\rho = 7900$ kg/m$^3$, $c = 477$ J/kg C) shown in Fig. 5.20 has a diameter of 8 cm and a thickness of 1 cm. It is perfectly insulated on the bottom and has convection and radiation from the top. The top surface has

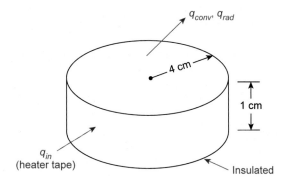

**FIGURE 5.20**

Disk for Example 5.5.

an emissivity of 0.5 and a convective coefficient of 100 W/m² C. Radiation is to the surroundings that are at 20 C, and convection is to a fluid also at 20 C. There is an electric heater tape on the side of the disk that inputs 300 W into the disk. The disk is initially uniform at 20 C. The heater tape is then turned on. Assume that the temperature in the disk varies only radially. There is no axial variation.

**(a)** Assume there are five nodes as shown in the top view of the disk in Fig. 5.21. What are the nodal temperatures 4 minutes after the heater is turned on?

**(b)** What are the nodal temperatures of the disk at steady state?

*Solution*

Let us first get the areas and volumes for the five nodes. Looking at Fig. 5.21, the areas on the top surface in contact with the surroundings are

$$A_1 = \pi(\Delta r/2)^2 \quad A_2 = \pi\left[(3\Delta r/2)^2 - (\Delta r/2)^2\right] \quad A_3 = \pi\left[(5\Delta r/2)^2 - (3\Delta r/2)^2\right]$$
$$A_4 = \pi\left[(7\Delta r/2)^2 - (5\Delta r/2)^2\right] \quad A_5 = \pi\left[(4\Delta r)^2 - (7\Delta r/2)^2\right]$$

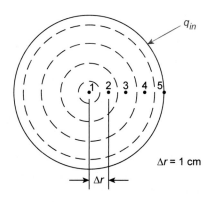

**FIGURE 5.21**

Top view of disk showing nodes.

Each node gets conduction from adjacent nodes. The thickness of the disk is "$b$." The areas are the cross-sectional areas for the radial conduction. The subscripts on the areas denote the two nodes sharing conduction. For example, $A_{12}$ is for conduction between nodes 1 and 2.

$$A_{12} = b[2\pi(\Delta r/2)] \quad A_{23} = b[2\pi(3\Delta r/2)] \quad A_{34} = b[2\pi(5\Delta r/2)] \quad A_{45} = b[2\pi(7\Delta r/2)]$$

The volumes of the nodes are

$$V_1 = bA_1 \quad V_2 = bA_2 \quad V_3 = bA_3 \quad V_4 = bA_4 \quad V_5 = bA_5$$

In deriving the nodal equations, we use the basic equations for conduction, convection, and radiation. That is, $q_{cond} = \frac{kA}{L}(\Delta T) \quad q_{conv} = hA(T - T_\infty) \quad q_{rad} = \varepsilon\sigma A(T^4 - T_{surr}^4)$

For the conduction, $A$ is the cross-sectional area through which the heat flows and $L$ is the distance between nodes. For convection and radiation, $A$ is the surface area of the particular node.

As we have done previously, we assume that all heat flows *into* the node we are considering.

Node 1 gets heat by conduction from node 2, convection from the adjacent fluid, and radiation from the surroundings. The rate of heat flow into node 1 is

$$q_1 = \frac{kA_{12}}{\Delta r}(T_2 - T_1) + hA_1(T_\infty - T_1) + \varepsilon\sigma A_1(T_{surr}^4 - T_1^4) \tag{5.114}$$

The rate of heat flows into the other nodes 2, 3, and 4 are

$$q_2 = \frac{kA_{12}}{\Delta r}(T_1 - T_2) + \frac{kA_{23}}{\Delta r}(T_3 - T_2) + hA_2(T_\infty - T_2) + \varepsilon\sigma A_2(T_{surr}^4 - T_2^4) \tag{5.115}$$

$$q_3 = \frac{kA_{23}}{\Delta r}(T_2 - T_3) + \frac{kA_{34}}{\Delta r}(T_4 - T_3) + hA_3(T_\infty - T_3) + \varepsilon\sigma A_3(T_{surr}^4 - T_3^4) \tag{5.116}$$

$$q_4 = \frac{kA_{34}}{\Delta r}(T_3 - T_4) + \frac{kA_{45}}{\Delta r}(T_5 - T_4) + hA_4(T_\infty - T_4) + \varepsilon\sigma A_4(T_{surr}^4 - T_4^4) \tag{5.117}$$

Node 5 gets conduction from node 4, convection from the adjacent fluid, radiation from the surroundings, and heat input $q_{in}$ from the heater tape.

$$q_5 = \frac{kA_{45}}{\Delta r}(T_4 - T_5) + hA_5(T_\infty - T_5) + \varepsilon\sigma A_5(T_{surr}^4 - T_5^4) + q_{in} \tag{5.118}$$

If we multiply the heat flow rates in Equations (5.114) through (5.118) by $\Delta t$, we get the heat going into a node during the time step $\Delta t$. This heat input is equal to the increase in the internal energy of the node during the time step. For example, for node 1, from Eq. (5.114), we have

$$q_1\Delta t = \left(\frac{kA_{12}}{\Delta r}\left(T_2^b - T_1^b\right) + hA_1\left(T_\infty - T_1^b\right) + \varepsilon\sigma A_1\left[T_{surr}^4 - \left(T_1^b\right)^4\right]\right)\Delta t = \rho c V_1\left(T_1^e - T_1^b\right) \tag{5.119}$$

Similarly, for the other four nodes we have

$$q_2\Delta t = \left(\frac{kA_{12}}{\Delta r}\left(T_1^b - T_2^b\right) + \frac{kA_{23}}{\Delta r}\left(T_3^b - T_2^b\right) + hA_2\left(T_\infty - T_2^b\right) + \varepsilon\sigma A_2\left[T_{surr}^4 - \left(T_2^b\right)^4\right]\right)\Delta t = \rho c V_2\left(T_2^e - T_2^b\right) \tag{5.120}$$

$$q_3\Delta t = \left(\frac{kA_{23}}{\Delta r}\left(T_2^b - T_3^b\right) + \frac{kA_{34}}{\Delta r}\left(T_4^b - T_3^b\right) + hA_3\left(T_\infty - T_3^b\right) + \varepsilon\sigma A_3\left[T_{surr}^4 - \left(T_3^b\right)^4\right]\right)\Delta t = \rho c V_3\left(T_3^e - T_3^b\right) \tag{5.121}$$

$$q_4\Delta t = \left(\frac{kA_{34}}{\Delta r}\left(T_3^b - T_4^b\right) + \frac{kA_{45}}{\Delta r}\left(T_5^b - T_4^b\right) + hA_4\left(T_\infty - T_4^b\right) + \varepsilon\sigma A_4\left[T_{surr}^4 - \left(T_4^b\right)^4\right]\right)\Delta t = \rho c V_4\left(T_4^e - T_4^b\right) \tag{5.122}$$

$$q_5\Delta t = \left(\frac{kA_{45}}{\Delta r}\left(T_4^b - T_5^b\right) + hA_5\left(T_\infty - T_5^b\right) + \varepsilon\sigma A_4\left[T_{surr}^4 - \left(T_5^b\right)^4\right] + q_{in}\right)\Delta t = \rho c V_5\left(T_5^e - T_5^b\right) \tag{5.123}$$

Rearranging Equations (5.119) through (5.123) so that the temperatures at the **end** of the time step are on the left side of the equations, we get

$$T_1^e = T_1^b\left(1 - \frac{A_{12}\alpha\Delta t}{(\Delta r)V_1} - \frac{hA_1\Delta t}{\rho c V_1} - \frac{\varepsilon\sigma A_1\Delta t}{\rho c V_1}\left(T_1^b\right)^3\right) + \frac{A_{12}\alpha\Delta t}{(\Delta r)V_1}T_2^b + \frac{hA_1\Delta t}{\rho c V_1}T_\infty + \frac{\varepsilon\sigma A_1\Delta t}{\rho c V_1}T_{surr}^4 \tag{5.124}$$

$$T_2^e = T_2^b \left(1 - \frac{A_{12}\alpha\Delta t}{(\Delta r)V_2} - \frac{A_{23}\alpha\Delta t}{(\Delta r)V_2} - \frac{hA_2\Delta t}{\rho c V_2} - \frac{\varepsilon\sigma A_2\Delta t}{\rho c V_2}\left(T_2^b\right)^3\right) + \frac{A_{12}\alpha\Delta t}{(\Delta r)V_2}T_1^b + \frac{A_{23}\alpha\Delta t}{(\Delta r)V_2}T_3^b$$
$$+ \frac{hA_2\Delta t}{\rho c V_2}T_\infty + \frac{\varepsilon\sigma A_2\Delta t}{\rho c V_2}T_{surr}^4$$

(5.125)

$$T_3^e = T_3^b \left(1 - \frac{A_{23}\alpha\Delta t}{(\Delta r)V_3} - \frac{A_{34}\alpha\Delta t}{(\Delta r)V_3} - \frac{hA_3\Delta t}{\rho c V_3} - \frac{\varepsilon\sigma A_3\Delta t}{\rho c V_3}\left(T_3^b\right)^3\right) + \frac{A_{23}\alpha\Delta t}{(\Delta r)V_3}T_2^b + \frac{A_{34}\alpha\Delta t}{(\Delta r)V_3}T_4^b$$
$$+ \frac{hA_3\Delta t}{\rho c V_3}T_\infty + \frac{\varepsilon\sigma A_3\Delta t}{\rho c V_3}T_{surr}^4$$

(5.126)

$$T_4^e = T_4^b \left(1 - \frac{A_{34}\alpha\Delta t}{(\Delta r)V_4} - \frac{A_{45}\alpha\Delta t}{(\Delta r)V_4} - \frac{hA_4\Delta t}{\rho c V_4} - \frac{\varepsilon\sigma A_4\Delta t}{\rho c V_4}\left(T_4^b\right)^3\right) + \frac{A_{34}\alpha\Delta t}{(\Delta r)V_4}T_3^b + \frac{A_{45}\alpha\Delta t}{(\Delta r)V_4}T_5^b$$
$$+ \frac{hA_4\Delta t}{\rho c V_4}T_\infty + \frac{\varepsilon\sigma A_4\Delta t}{\rho c V_4}T_{surr}^4$$

(5.127)

$$T_5^e = T_5^b \left(1 - \frac{A_{45}\alpha\Delta t}{(\Delta r)V_5} - \frac{hA_5\Delta t}{\rho c V_5} - \frac{\varepsilon\sigma A_5\Delta t}{\rho c V_5}\left(T_5^b\right)^3\right) + \frac{A_{45}\alpha\Delta t}{(\Delta r)V_5}T_4^b + \frac{hA_5\Delta t}{\rho c V_5}T_\infty$$
$$+ \frac{\varepsilon\sigma A_5\Delta t}{\rho c V_5}T_{surr}^4 + \frac{q_{in}\Delta t}{\rho c V_5}$$

(5.128)

(a) A Matlab program was written to solve the problem by the forward-difference approach. The program is in Appendix K. As radiation is present, we needed to use the absolute temperature scale, i.e., Kelvin. The five nodes were given the initial temperature of 20 C = 293.15 K. Eqs. (5.124) through (5.128) were arranged successively in a loop. We picked a time increment $\Delta t$ of 1 second. As we wanted the nodal temperatures after 4 min = 240 s, the program performed 240 loops and then output the nodal temperatures. The temperatures after 4 min were found to be

$$T_1 = 494.5 \text{ K} = 221.4 \text{ C} \quad T_2 = 503.0 \text{ K} = 229.9 \text{ C} \quad T_3 = 529.0 \text{ K} = 255.9 \text{ C}$$

$$T_4 = 573.9 \text{ K} = 300.8 \text{ C} \quad T_5 = 640.4 \text{ K} = 367.3 \text{ C}$$

(b) To determine the steady-state temperatures: Instead of only doing 240 loops (4 min), we continued the looping of Eqs. (5.124) through (5.128) until the temperatures stopped changing with successive loops. The steady-state temperatures were found to be

$$T_1 = 711.7 \text{ K} = 438.6 \text{ C} \quad T_2 = 719.9 \text{ K} = 446.7 \text{ C} \quad T_3 = 744.9 \text{ K} = 471.7 \text{ C}$$

$$T_4 = 788.5 \text{ K} = 515.3 \text{ C} \quad T_5 = 854.1 \text{ K} = 580.9 \text{ C}$$

We also determined the steady-state temperatures using Excel *Solver*. If we set Eqs. (5.114)–(5.118) all equal to zero, we have the energy equations for steady state. And, the expressions in the equations are the functions for use by *Solver*. There was excellent agreement between the *Solver* and finite-difference results; only a maximum of 0.4 C difference in temperatures for all five nodes.

To close this section, we will use numerical forward-difference to solve two problems that were previously solved analytically. The first problem, Example 5.6, is the same as Example 4.8, which determined the acceptable burial depth of a water line using semiinfinite slab theory.

## Example 5.6

### Burial depth of a water pipe

*Problem*

In a town in the Northern United States, the air temperature can be as low as −18 C for as long as 4 weeks. Assume that the ground is initially at a uniform temperature of 10 C. How deep should water pipes be buried to prevent freezing? Assume that the combined convective and radiative coefficient at the ground's surface is an average 30 W/m² C. The ground has a thermal conductivity of 2 W/m C, a density of 1800 kg/m³, and a specific heat of 2100 J/kg C. Fig. 5.22.

*Solution*

This is a one-dimensional problem. The first step is to determine the grid. We decided to make node 1 at the ground surface. This is a boundary node with convection. We need another boundary at some depth within the ground. This node, node "n", has to be at a depth sufficiently greater than that of the water pipe (More about this later). We decided to make this node an insulated boundary node. The grid is shown in Fig. 5.23.

Let us now develop the energy equations. Like we have done previously, we will assume that all heat flows *into* a node. And, the heat input during time step $\Delta t$ is equal to the increase in the internal energy of the node during $\Delta t$.

We have two boundary nodes (nodes 1 and n) and (n−2) interior nodes (nodes 2 through n−1). Let us consider a cross-sectional area $A$ for the heat flow. (This area $A$ eventually cancels out of the equations.) The two boundary nodes have a thickness of $\Delta x/2$. The interior nodes have a thickness of $\Delta x$. Therefore, the volume of a boundary node is $(A\,\Delta x/2)$, and the volume of an interior node is $(A\,\Delta x)$.

Node 1 has heat input by convection from the adjacent fluid and conduction from node 2. The energy equation is

$$\left[hA\left(T_\infty - T_1^b\right) + \frac{kA}{\Delta x}\left(T_2^b - T_1^b\right)\right]\Delta t = \rho cA(\Delta x / 2)\left(T_1^e - T_1^b\right) \tag{5.129}$$

Node n only has conduction from node n−1. Its equation is

$$\left[\frac{kA}{\Delta x}\left(T_{n-1}^b - T_n^b\right)\right]\Delta t = \rho cA(\Delta x / 2)\left(T_n^e - T_n^b\right) \tag{5.130}$$

Each interior node has conduction from its two adjacent nodes. The nodal equations for the interior nodes are as follow: For nodes $i = 2$ to $i = $n−1,

$$\left[\frac{kA}{\Delta x}\left(T_{i-1}^b - T_i^b\right) + \frac{kA}{\Delta x}\left(T_{i+1}^b - T_i^b\right)\right]\Delta t = \rho cA(\Delta x)\left(T_i^e - T_i^b\right) \tag{5.131}$$

Rearranging these equations so that the temperatures at the end of a time step are on the left side of the equations and the temperatures at the beginning of a time step are on the right side, we have

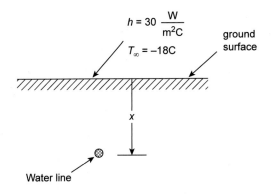

**FIGURE 5.22**

Buried water line.

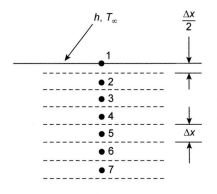

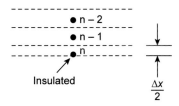

**FIGURE 5.23**

Nodes for Example 5.6.

For node 1:

$$T_1^e = T_1^b\left(1 - \frac{2\alpha\Delta t}{(\Delta x)^2} - \frac{2h\Delta t}{\rho c(\Delta x)}\right) + \frac{2\alpha\Delta t}{(\Delta x)^2}T_2^b + \frac{2h\Delta t}{\rho c(\Delta x)}T_\infty \tag{5.132}$$

For node $n$:

$$T_n^e = T_n^b\left(1 - \frac{2\alpha\Delta t}{(\Delta x)^2}\right) + \frac{2\alpha\Delta t}{(\Delta x)^2}T_{n-1}^b \tag{5.133}$$

For nodes $i = 2$ to $i = n-1$:

$$T_i^e = T_i^b\left(1 - \frac{2\alpha\Delta t}{(\Delta x)^2}\right) + \frac{\alpha\Delta t}{(\Delta x)^2}\left(T_{i-1}^b + T_{i+1}^b\right) \tag{5.134}$$

Water pipes in the northern states of the USA are typically buried at depths of four feet or greater to prevent freezing. We decided to use a nodal increment $\Delta x$ of 1 inch = 0.0254 m. As we are solving this problem using forward differences, we have to make sure that the chosen time step meets the stability criterion. For our problem, the limiting criterion is that the bracketed expression in Eq. (5.132) may not be negative. That is, for stability,

$$\left(1 - \frac{2\alpha\Delta t}{(\Delta x)^2} - \frac{2h\Delta t}{\rho c(\Delta x)}\right) \geq 0 \tag{5.135}$$

Putting in the values of the parameters in Eq. (5.135), it is found that $\Delta t$ must be equal to or less than 441 s. We decided to make $\Delta t = 5$ min = 300 s. From the problem statement, we want nodal temperatures after 4 weeks. This is equivalent to $2.4192 \times 10^6$ s. With a time step of 300 s, we need $2.4192 \times 10^6$ s/300 s = 8064 time steps.

The Matlab program used for the solution is in Appendix K. All nodes were given initial temperatures of 10 C and then we had 8064 successive loops of Equations (5.132) through (5.134). Each loop updated the nodal temperatures for a time step. At the end of the 8064 loops, we had the nodal temperatures after 4 weeks. We then had the program determine the first node at which the temperature was above the freezing point of water, 0 C. The location of that node was the required burial depth.

Finally, as mentioned above, node n has to be sufficiently deeper than the water pipe. We ran the program for several depths, i.e., several different numbers of nodes. We found that if node n was 10 feet or more from the surface, the results of the program did not change. As we had chosen $\Delta x$ to be 1 inch, 10 feet is equivalent to 120 $\Delta x$. Hence, we used 121 nodes in the program.

It was found that the water pipe had to be buried at a depth of at least **4.75 feet** to prevent freezing. This result by numerical forward differences is in excellent agreement with the **4.84 feet** result of Example 4.8, which was solved using the semi-infinite slab equations.

Our final example of numerical solution is the same as Example 4.6, which used the Multidimensional Heisler Method to determine the required cooling time for a short cylinder and the temperature of a point in the cylinder at that time.

## Example 5.7

### Cooling of a short cylinder

*Problem*

The stainless steel disk $\left(\rho = 7900 \text{ kg/m}^3, \ c = 480 \text{ J/kg C}, \ k = 15 \text{ W/mC}\right)$ shown in Fig. 5.24 is 10 cm in diameter and 5 cm thick. It goes into an oven where it is heated to a uniform temperature of 1200 C. It is then taken out of the oven and cooled by convection to a fluid that is at 150 C. The convective coefficient on the curved surface of the disk is 100 W/$m^2$ C and the convective coefficient on the two flat surfaces is 250 W/$m^2$ C.

**(a)** How long does it take for the center of the disk to reach 500 C?

**(b)** At the time obtained in Part (a): What is the temperature at a point located 3 cm from the centerline and 1.5 cm from one end?

*Solution*

The nodal labeling is shown in Figs. 5.25 and 5.26. Fig. 5.25 shows the radial numbering, and Fig. 5.26A and B show the axial numbering. Because of symmetry, there is no need to write nodal equations for axial nodes 7 through 11 of

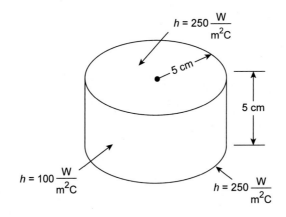

**FIGURE 5.24**

Cylinder for Example 5.7.

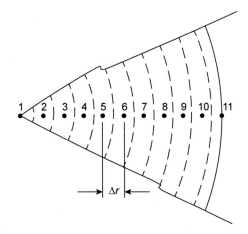

**FIGURE 5.25**

Radial numbering of nodes.

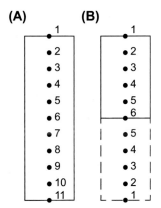

**FIGURE 5.26**

(A) Original axial numbering of nodes. (B) Revised axial numbering of nodes.

Fig. 5.26A. The temperature at node 7 is the same as that of node 5; the temperature of node 8 is the same as that of node 4, etc. Hence, we can use the revised axial grid of Fig. 5.26B and have fewer nodal equations.

Nodal identification is by two subscripts $(i, j)$. For example, the temperature $T_{i,j}$ denotes the node at radial location "$i$" and axial location "$j$." Nodes on the axis of the cylinder (that is, $i = 1$) are cylinders of radius $\Delta r/2$ with thickness $\Delta x$ for interior nodes and $\Delta x/2$ for the boundary node ($j = 1$). The other nodes are rings of width $\Delta r$ except for the boundary node ($i = 11$), which has a width of $\Delta r/2$. The ring nodes have a thickness $\Delta x$ except for the boundary nodes ($j = 1$), which have a thickness $\Delta x/2$.

From the dimensions of the cylinder and the nodal grid, we have $\Delta r = 0.5$ cm $= 0.005$ m and $\Delta x = 0.5$ cm $= 0.005$ m. Writing the nodal equations is straightforward but very tedious. There are many nodes, and one must be very careful in the correct assignment of subscripts and heat flows for a node. We will derive the equations for two nodes—one on the axis of the cylinder and the other on the side of the cylinder. The other equations are in the Matlab solution, which is in Appendix K.

**Node (1, 4)** is an interior node on the axis of the cylinder. It is a cylinder of radius $\Delta r/2$ and thickness $\Delta x$. The node gets conduction from nodes (1, 3), (1, 5), and (2, 4). The heat input during a time step raises the internal energy of the material in the node. The energy equation for the node is

$$\frac{kA_1}{\Delta x}\left(T_{1,3}^b - T_{1,4}^b\right)\Delta t + \frac{kA_1}{\Delta x}\left(T_{1,5}^b - T_{1,4}^b\right)\Delta t + \frac{kA_2}{\Delta r}\left(T_{2,4}^b - T_{1,4}^b\right)\Delta t = \rho c V_{1,4}\left(T_{1,4}^e - T_{1,4}^b\right) \tag{5.136}$$

where $A_1 = \pi(\Delta r/2)^2$ $A_2 = 2\pi(\Delta r/2)(\Delta x)$ and $V_{1,4} = \pi(\Delta r/2)^2(\Delta x)$.
Rearranging Eq. (5.136) for forward-difference solution, we have

$$T_{1,4}^e = T_{1,4}^b\left[1 - 2\frac{A_1\alpha\Delta t}{V_{1,4}(\Delta x)} - \frac{A_2\alpha\Delta t}{V_{1,4}(\Delta r)}\right] + \frac{A_1\alpha\Delta t}{V_{1,4}(\Delta x)}\left(T_{1,3}^b + T_{1,5}^b\right) + \frac{A_2\alpha\Delta t}{V_{1,4}(\Delta r)}T_{2,4}^b \tag{5.137}$$

**Node (11, 3)** is on the side of the cylinder. It is a ring of width $\Delta r/2$ and thickness $\Delta x$. Its inner radius is $\left(\frac{19}{2}\Delta r\right)$ and its outer radius is $(10\ \Delta r)$. The node gets convection on its outer surface and conduction from nodes (11, 2), (11, 4), and (10, 3). The energy equation for the node is

$$\frac{kA_3}{\Delta x}\left(T_{11,2}^b - T_{11,3}^b\right)\Delta t + \frac{kA_3}{\Delta x}\left(T_{11,4}^b - T_{11,3}^b\right)\Delta t + \frac{kA_4}{\Delta r}\left(T_{10,3}^b - T_{11,3}^b\right)\Delta t + h_{side}A_5\left(T_\infty - T_{11,3}^b\right)\Delta t$$

$$= \rho c V_{11,3}\left(T_{11,3}^e - T_{11,3}^b\right) \tag{5.138}$$

where

$$A_3 = \pi\left[(10\ \Delta r)^2 - \left(\frac{19}{2}\ \Delta r\right)^2\right]$$

$$A_4 = 2\pi\left(\frac{19}{2}\Delta r\right)(\Delta x)$$

$$A_5 = 2\pi(10\ \Delta r)(\Delta x)$$

$$V_{11,3} = A_3(\Delta x)$$

Rearranging Eq. (5.138) for forward-difference solution, we have

$$T_{11,3}^e = T_{11,3}^b\left[1 - 2\frac{A_3\alpha\Delta t}{V_{11,3}(\Delta x)} - \frac{A_4\alpha\Delta t}{V_{11,3}(\Delta r)} - \frac{h_{side}A_5\Delta t}{\rho c V_{11,3}}\right] + \frac{A_3\alpha\Delta t}{V_{11,3}(\Delta x)}\left(T_{11,2}^b + T_{11,4}^b\right) + \frac{A_4\alpha\Delta t}{V_{11,3}(\Delta r)}T_{10,3}^b + \frac{h_{side}A_5\Delta t}{\rho c V_{11,3}}T_\infty \tag{5.139}$$

In the Matlab program, all the nodes are given the initial temperature of 1200 C. The program then loops through these two equations and the other nodal equations. With each pass through the equations, the time advances another time step.

**(a)** We used a time step of $\Delta t = 1$ s, and it took 376 loops for the center node (node (1,6)) to reach 500 C. So, the center node reached 500 C in **376 s**.

**(b)** Node (7,4) is the node that is 3 cm from the centerline of the cylinder and 1.5 cm from an end. After 376 s, the temperature of this node was **470.6 C**.

These results are in excellent agreement with the results of the Multidimensional Heisler solution of Example 4.6, which were 376 s and 471.1 C.

The Matlab program also determines the amount of heat flow into the adjacent fluid during the 376 s of the process. The result was $Q = 1.108 \times 10^6$ J. The Heisler solution for the heat flow was $1.110 \times 10^6$ J. Again, there was excellent agreement between the finite-difference and Heisler solutions.

# 5.3 Finite element method

We have discussed the finite-difference method in detail. There are also other methods in general use today. These include the finite element method, the boundary element method, and the control volume approach.

In the finite element method, an object is divided into spatial regions, i.e., finite elements. Triangles are commonly used for two-dimensional objects and tetrahedrons for three-dimensional objects. The elements are typically larger than the nodal volumes for finite differences, leading to a lesser number of volumes. And, the finite element method gives a better approximation of irregularly and complex-shaped geometries and complicated boundary conditions. However, the finite element method is harder to implement than the finite-difference method. References at the end of this chapter discuss the finite element method in detail. There are also many other books available on the subject.

The finite element method has historically been the predominant method used for structural mechanics and stress calculations. Determination of the temperature distribution in an object is a necessary precursor to determination of thermal stresses. Hence, several companies have developed finite element software for determining both temperatures and stresses. Some of these companies provide low-cost software packages for colleges and students. Major software providers include ANSYS [11], COMSOL [12], SolidWorks [13], Autodesk [14], and Siemens (formerly, CD-adapco) [15]. This list is by no means inclusive. There are several other providers, both open source and commercial, of software applicable to the heat transfer field.

## 5.4 Chapter summary and final remarks

In this chapter, we discussed the finite-difference method for solving both steady and unsteady heat transfer problems. For steady state, we learned how to divide an object into nodal volumes and develop energy equations for both interior and boundary nodes. We then gave much attention to different ways of solving these simultaneous equations for the nodal temperatures. For unsteady state, we discussed the nodal grid and the development of the nodal equations. We discussed the forward-difference approach in detail and mentioned the stability criteria associated with it. We also outlined the backward-difference approach that eliminates stability concerns. Finally, we briefly discussed the finite element method, which is one alternative to the finite-difference method.

Several examples were included in the chapter to demonstrate the application of numerical methods to both steady-state and unsteady-state problems. Some of these examples had previously been solved using the advanced mathematics of differential equations, semi-infinite slab theory, and the Heisler Method. It was found that the numerical methods, using only algebra, gave results in excellent agreement with the results obtained through the higher-level mathematics.

Moving on to the next chapter: Up to now, if a problem involved convection, we provided the value of the convective coefficient $h$. This is very unrealistic. For a problem in the "real world", $h$ will not normally be a given. One has to either make a good assumption based on experience or determine $h$ using information from heat transfer researchers. A limited amount of this information comes from analytical solutions. A major part of the information is in the form of correlations obtained by experimentation. The next two chapters discuss convection and the determination of convective coefficients for various geometries and situations.

## 5.5 Problems

Notes:

- If needed information is not given in the problem statement, then use the Appendix for material properties and other information. If the Appendix does not have sufficient information, then use the internet, but double-check the validity of the information by using more than one source.

- Some problems specify the nodal spacing and time increment (if an unsteady problem). If these are not given, you should choose the values yourself and clearly note them in your solution.
- Your solutions should include a sketch of the problem. Node points should be numbered and clearly indicated on the sketch. Solutions should include the nodal equations and indicate the method used for solving them.
- Caution: Make sure that you use absolute temperatures (Kelvin or Rankine) if the problem involves radiation.

**Problems 5-1 through 5-15 are steady; problems 5-16 through 5-30 are unsteady.**

**5-1**  A long square bar ($k = 10$ W/m C) has a cross section of 20 cm by 20 cm and has boundary conditions as shown in Fig. P 5.1.

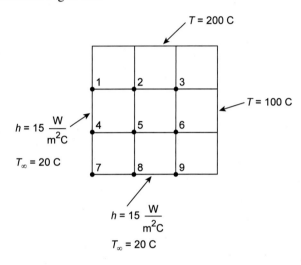

**FIGURE P 5.1**

(a) What are the steady-state temperatures for the nodes?
(b) What is the rate of heat flow to the fluid per meter length of the bar?

**5-2**  A square steel plate (k = 45 W/m C) has temperatures on its four sides as shown in Fig. P 5.2. What are the steady-state nodal temperatures?

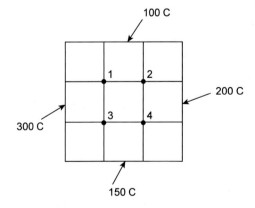

**FIGURE P 5.2**

**5-3** An aluminum conical fin ($k = 240$ W/m C) is attached to a wall that is at 250 C. The diameter of the fin at the point of attachment is 0.5 cm and the fin is 20 cm long. The surrounding fluid is at 20 C, and the convective coefficient is 15 W/m² C.

**(a)** Using a nodal spacing of $\Delta x = 1$ cm, determine the rate of heat transfer from the fin to the surrounding fluid.

**(b)** Do the problem using Section 3.8.4.3 and compare the result with that of Part (a).

**5-4** For the pin fin of Example 5.1: Replace the convection boundary condition with a radiation condition. The surroundings are at 15 C, and the emissivity of the fin surface is 0.8. Determine the rate of heat transfer from the fin to the surroundings.

**5-5** The straight triangular fin ($k = 150$ W/m C) shown in Fig. P 5.5 has a length of 8 cm and a height at the base of 0.5 cm. The temperature of the base is 250 C, and there is convection from the two sides of the fin to the surrounding 50 C fluid with a convective coefficient of 40 W/m² C.

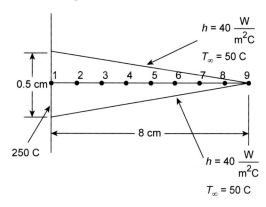

**FIGURE P 5.5**

**(a)** What are the steady-state nodal temperatures?

**(b)** What is the rate of heat transfer to the fluid?

**(c)** Determine the rate of heat transfer to the fluid using the information in Section 3.8.4.2 and compare the result with that of Part (b).

**5-6** The cross section of a long carbon steel I-beam ($k = 62$ W/m C) is shown in Fig. P 5.6. The top surface of the beam is at 50 C, and the bottom surface is at 10 C. All other surfaces are perfectly insulated. Determine the rate of heat flow through the beam (top to bottom) per meter length. Use a nodal spacing of $\Delta x = \Delta y = 0.5$ cm.

**5-7** The square plate ($k = 1.2$ W/m C) of Fig. P 5.7 has temperature boundary conditions on the top and right side, a convective boundary condition on the bottom, and is insulated on the left side. The plate is 6 cm by 6 cm. Determine the steady-state nodal temperatures.

**5-8** The rectangular cross section of a long bar ($k = 5$ W/m C) is shown in Fig. P 5.8. The bar has convection on the top surface, a temperature boundary condition on the right side, radiation on the bottom, and is insulated on the left side.

**(a)** What are the steady-state nodal temperatures?

**(b)** What is the rate of heat flow to the fluid from the top surface per meter length of the bar?

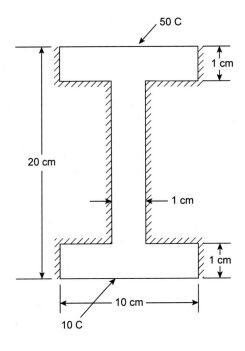

**FIGURE P 5.6**

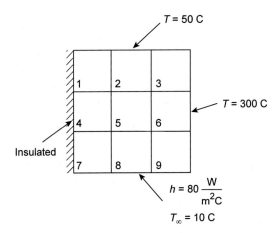

**FIGURE P 5.7**

(c) What is the rate of heat flow to the surroundings from the bottom surface per meter length of the bar?

**5-9** The plate ($k = 3$ W/m C) of Fig. P 5.9 has temperature, convection, and insulated boundary conditions as shown. The thickness of the plate is 0.5 cm.

(a) What are the steady-state nodal temperatures?

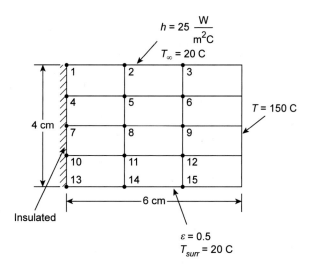

**FIGURE P 5.8**

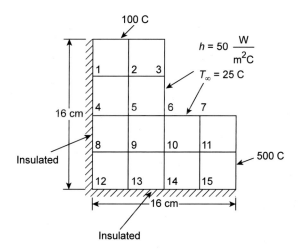

**FIGURE P 5.9**

**(b)** What is the rate of heat flow to the fluid from the sides of the plate that have convection?

**5-10** The nichrome wire ($k = 11.3$ W/m C) shown in Fig. P 5.10 has a diameter of 1.626 mm and a length of 20 cm. An electrical current is flowing through the wire, causing a uniform heat generation of $2 \times 10^8$ W/m³. The two ends of the wire are thermally anchored to sinks at 20 C. The surface of the wire radiates with an emissivity of 0.7 to the surroundings at 20 C. As shown in the figure, use a uniform nodal spacing of $\Delta x = 1$ cm. Node 1 is at one end of the wire, and node 21 is at the other end.

What is the steady-state temperature of node 11, which is at the center of the wire?

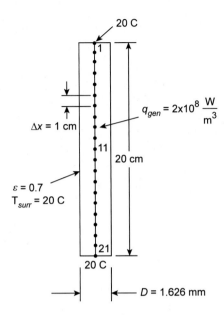

20 C

$q_{gen} = 2 \times 10^8 \; \dfrac{W}{m^3}$

$\Delta x = 1$ cm

20 cm

$\varepsilon = 0.7$

$T_{surr} = 20$ C

20 C

$D = 1.626$ mm

**FIGURE P 5.10**

**5-11** A copper circumferential fin ($k = 250$ W/m C) is attached to a tube that has an outer surface temperature of 80 C. The fin has an inside radius of 1.5 cm, an outside radius of 6 cm, and a thickness of 0.05 cm. Let there be 11 equally spaced nodes between the inner and outer surfaces of the fin (i.e., $\Delta r = 0.45$ cm). Node 1 is at the inner surface of the fin, and node 11 is at the outer surface. There is convection from the top, bottom, and edge of the fin to the 18 C surrounding air with a convective coefficient of 7.5 W/m$^2$ C (Fig. P 5.11).
  **(a)** Determine the steady-state temperatures of nodes 2 through 11 (node 1 is 80 C)
  **(b)** What is the rate of heat flow from the fin to the surrounding air?
  **(c)** What is the fin efficiency?
  **(d)** Determine the fin efficiency using Fig. 3.30 and compare the result with the numerical result from Part (c).
**5-12** Do Problem 5-7, but let the plate also have uniform internal heat generation of $2 \times 10^6$ W/m$^3$.
**5-13** A chimney has a cross section as shown in Fig. P 5.13. If the inner surface is at 800 C and the outer surface is at 30 C, what are the steady-state nodal temperatures? (Note: You can use the symmetry of the problem to reduce the number of nodes from 20 to 6.)
**5-14** A chimney ($k = 0.7$ W/m C) is 8 m high and has a cross section as shown in Fig. P 5.14. The inner surface is at 800 C, and the outer surface convects to the 10 C ambient air with a convective coefficient of 25 W/m$^2$ C.
  **(a)** What are the steady-state nodal temperatures?
  **(b)** What is the rate of heat flow to the ambient air?
**5-15** As shown in Fig. P 5-15, two layers of material are attached to a surface. Both layers have the same thickness. Layer A is firmly attached to the surface with a very high melting point adhesive. However, the glue between layers A and B will melt if its temperature is greater than

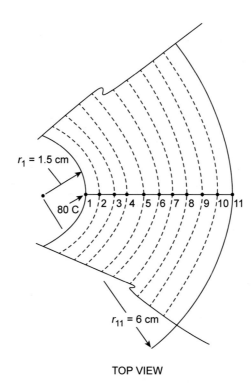

$r_1 = 1.5$ cm

80 C  1 2 3 4 5 6 7 8 9 10 11

$r_{11} = 6$ cm

TOP VIEW

**FIGURE P 5.11**

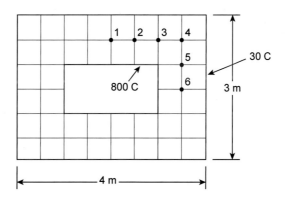

1  2  3  4

5

800 C    6

30 C

3 m

4 m

**FIGURE P 5.13**

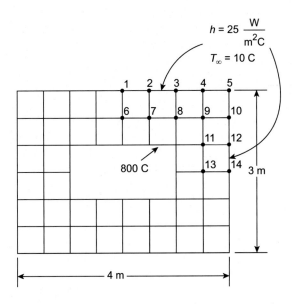

**FIGURE P 5.14**

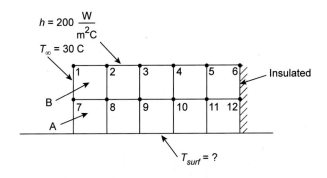

**FIGURE P 5.15**

80 C and the layers will separate from each other. There is convection on the left and top surfaces with the fluid being 30 C and the convective coefficient being 200 W/m² C. The right surface is perfectly insulated. Layer A has a conductivity of 30 W/m C and layer B has a conductivity of 5 W/m C. The grid is square, with $\Delta x = \Delta y = 1.5$ cm.

How high can the temperature of the bottom surface, $T_{surf}$, be without layers A and B separating?

**5-16** The nichrome wire ($k = 11.3$ W/m C, $\rho = 8400$ kg/m³, $c = 450$ J/kg C) of Problem 5–10 initially has a uniform temperature of 20 C. The electric current is then turned-on. How long does it take for the center of the wire (node 11) to reach 700 C?

**5-17** A concrete wall ($k = 0.8$ Btu/h ft F, $\rho = 140$ lbm/ft³, $c = 0.18$ Btu/lbm F) is 8 inches thick and is initially at a uniform temperature of 65 F. The outside surface of the wall is suddenly subjected to sunlight that inputs a heat flux of 120 Btu/h ft$^2$ to the wall. The inside surface of the wall has convection to the 70 F room air with a convective coefficient of 2.5 Btu/h ft$^2$ F.

(a) How long does it take for the inside surface of the wall to reach 75 F?

(b) What is the steady-state temperature of the inside surface?

**5-18** A thick steel plate ($k = 45$ W/m C, $\rho = 7800$ kg/m$^3$, $c = 480$ J/kg C) is initially at a uniform temperature of 50 C. A surface of the plate is suddenly lowered to 10 C.

(a) What is the temperature 5 cm from the surface after 2 min? Use a nodal spacing of 0.1 cm and assume that the furthest node from the surface is an insulated node at a distance of 20 cm from the surface.

(b) Do the problem using the semiinfinite slab theory of Section 4.5 and compare the result with the Part (a) result.

**5-19** A large steel plate ($k = 45$ W/m C, $\rho = 7800$ kg/m$^3$, $c = 480$ J/kg C) is 30 cm thick and initially at a uniform temperature of 500 C. The plate is horizontal. Air at 20 C is blown across the top and bottom surfaces of the plate to cool it. The convective coefficient is 150 W/m$^2$ C for both surfaces.

(a) How long does it take for the center of the plate to reach 350 C?

(b) What is the temperature of the top surface of the plate at the time found in Part (a)?

(c) Do this problem using the Heisler Method of Chapter 4 and compare the results with the numerical results of Parts (a) and (b).

**5-20** A large steel plate ($k = 45$ W/m C, $\rho = 7800$ kg/m$^3$, $c = 480$ J/kg C) is 30 cm thick and initially at a uniform temperature of 500 C. The plate is horizontal. Air at 20 C is blown across the top and bottom surfaces of the plate to cool it. The convective coefficient on the top surface is 200 W/m$^2$ C and the coefficient on the bottom surface is 50 W/m$^2$ C.

(a) How long does it take for the center of the plate to reach 350 C?

(b) What are the temperatures of the two surfaces of the plate at the time found in Part (a)?

**5-21** A long aluminum cylinder ($k = 240$ W/m C, $\rho = 2700$ kg/m$^3$, $c = 900$ J/kg C) has a diameter of 40 cm. It is heated in an oven until its temperature is uniform at 400 C. It is then taken out of the oven and cooled by convection and radiation to the 25 C room air and surroundings. The convective coefficient at the cylinder's surface is 500 W/m$^2$ C and the emissivity of the cylinder's surface is 0.3.

What is the temperature of the centerline of the cylinder 5 min after it is taken out of the oven? Use $\Delta r = 2$ cm. Node 1 is at the centerline of the cylinder and node 11 is at the surface.

**5-22** A long stainless steel cylinder ($k = 15$ W/m C, $\rho = 7900$ kg/m$^3$, $c = 480$ J/kg C) has a diameter of 12 cm. It is initially at a uniform temperature of 450 C. The surface of the cylinder is suddenly subjected to a 30 C air stream with a convective coefficient of 200 W/m$^2$ C.

(a) How long does it take for the center of the cylinder to reach 100 C?

(b) How much heat goes into the air during the process?

**5-23** An apple is modeled as a sphere of 10 cm diameter. It is initially at a uniform temperature of 10 C when the air temperature suddenly falls to $-5$ C and remains at that temperature. Assume that the apple has the properties of water and that the convective coefficient at the apple's surface is 15 W/m$^2$ C.

(a) How long will it take for the surface of the apple to freeze, i.e., reach 0 C?

(b) What is the temperature of the center of the apple at the time found in Part (a)?

**5-24** A large iron plate ($k = 80$ W/m C, $\rho = 7800$ kg/m³, $c = 450$ J/kg C) is being heated in an oven. The plate is 1 cm thick and is initially at a uniform temperature of 20 C. The oven is at 800 C and the convective coefficient at both of the plate's surfaces is 150 W/m² C. The plate is in the oven for 2 min.

(a) What is the temperature of the center of the plate when it leaves the oven?

(b) What is the temperature of the surface of the plate when it leaves the oven?

Use a nodal spacing of $\Delta x = 0.1$ cm.

**5-25** A stainless steel sphere ($k = 15$ W/m C, $\rho = 7900$ kg/m³, $c = 480$ J/kg C) has a diameter of 16 cm and is initially at a uniform temperature of 750 C. It is suddenly placed in a large vacuum chamber where it radiates to the walls of the chamber. The walls are at 20 C and the emissivity of the surface of the sphere is 0.9. How long does it take for the center of the sphere to reach 500 C? (Use a nodal spacing of $\Delta r = 2$ cm. Node 1 is at the center of the sphere and node 5 is at the surface.).

**5-26** A thick piece of wood ($k = 0.15$ W/m C, $\rho = 700$ kg/m³, $c = 1300$ J/kg C) is initially at a uniform temperature of 20 C. The surface of the wood is suddenly exposed to 500 C combustion gases. The convective coefficient at the surface is 35 W/m² C.

(a) How long will it take for the surface to reach 250 C?

(b) Do the problem using the semiinfinite solid theory of Section 4.5 and compare the result with the numerical result of Part (a).

**5-27** As shown below, a man is soldering an elbow fitting on the end of a long 1/2-inch Type L copper tube. He is using a propane torch that heats the fitting at a rate of about 200 W. The solder melts at 240 C. The man is holding the torch but does not have a third hand to hold the tube. Therefore, he asks his wife to hold the tube about 25 cm from the end that is being heated. He assures her that she will not be burned. The wife is greatly concerned but helps out and holds the tube. The copper elbow has a mass of 0.018 kg. Before the torch is applied, the tube and the fitting are at 20 C. There is convection from the outer surface of the tube to the room air. The air is at 20 C, and the convective coefficient is 15 W/m² C. Neglect any heat transfer at the inner surface of the tube or in the air inside the tube (Fig. P 5.27).

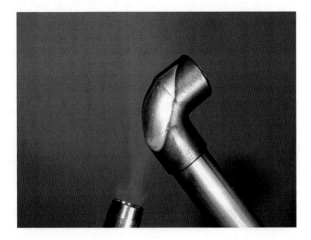

**FIGURE P 5.27**

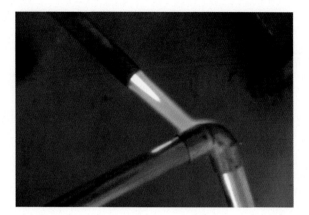

**FIGURE P 5.28**

**(a)** How long does it take for the fitting to reach 240 C and melt the solder?
**(b)** What is the temperature at the wife's hand at the time found in Part (a)?
**(c)** Was the man correct, or did the wife have to drop the tube to avoid getting burned?

**5-28** A man is soldering an elbow fitting on the end of two long 1/2-inch Type L copper tubes as shown below. He is using a propane torch that heats the fitting at a rate of about 200 W. The solder melts at 240 C. The copper elbow has a mass of 0.018 kg. Before the torch is applied, the fitting and the tubes are at 20 C. There is convection from the outer surface of the tubes to the room air. The air is at 20 C and the convective coefficient is 15 W/m² C. Neglect any heat transfer at the inner surfaces of the tubes or in the air inside the tubes (Fig. P 5.28).

**(a)** How long does it take for the fitting to reach 240 C and melt the solder?
**(b)** What is the temperature of one of the tubes at a location 10 cm from the elbow at the time found in Part (a)?

**5-29** The plate of Problem 5–7 ($k = 1.2$ W/m C, $\rho = 2500$ kg/m³, $c = 850$ J/kg C) has an initial uniform temperature of 10 C. The boundary conditions are suddenly applied at time zero. Plot the temperature of node 3 versus time until the node reaches steady state.

**5-30** A long steel cylinder ($k = 40$ W/m C, $\rho = 7800$ kg/m³, $c = 470$ J/kg C) has a diameter of 12 cm. It has an initial uniform temperature of 300 C. The cylinder is suddenly quenched in an oil bath that is at 40 C. The center of the cylinder cools to 75 C in 300 s. What is the average convective coefficient during the cooling?

# References

[1] R.V. Southwell, Relaxation Methods in Engineering Science, Oxford University Press, 1940.
[2] H.W. Emmons, The numerical solution of heat-conduction problems, Trans. ASME 65 (1943) 607–612.
[3] G.M. Dusinberre, Numerical Analysis of Heat Flow, McGraw-Hill, 1949.
[4] E.L. Wilson, R.E. Nickell, Application of the finite element method to heat conduction analysis, Nucl. Eng. Des. 4 (1966) 276–286.
[5] Y. Jaluria, Computer Methods for Engineering, Allyn and Bacon, 1988.

[6] G.E. Myers, Analytical Methods in Conduction Heat Transfer, McGraw-Hill, 1971 (Chapter 8, Finite Differences; Chapter 9, Finite Elements).

[7] W.J. Minkowycz, E.M. Sparrow, G.E. Schneider, R.H. Pletcher (Eds.), Handbook of Numerical Heat Transfer, John Wiley & Sons, 1988.

[8] C.S. Desai, Elementary Finite Element Method, Prentice-Hall, 1979.

[9] L.C. Thomas, Heat Transfer – Professional Version, Capstone Publishing, 1999 (Appendix F - Finite Element Method).

[10] W.M. Rohsenow, J.P. Hartnett (Eds.), Handbook of Heat Transfer, McGraw- Hill, 1973 (Section 4 by P. Razelos).

[11] www.ansys.com.

[12] www.comsol.com.

[13] www.solidworks.com.

[14] www.autodesk.com.

[15] www.plm.automation.siemens.com.

# Forced convection

## 6.1  Introduction

Up to now, the convective coefficient, $h$, has been a given in all problems involving convective heat transfer. This is very unrealistic. Indeed, the coefficient is usually not a given. Rather, one has to determine the coefficient either by making an informed estimate based on previous experience or by using available analytical or experimental results. The extent of analytical solutions is limited, and $h$-values are usually obtained from correlations of experimental data.

There are three major categories of convection: forced convection; natural (or free) convection; and boiling, condensation, and evaporation. Forced convection is covered in this chapter. Natural convection is in Chapter 7. Evaporation is in Chapter 11, and boiling and condensation are in Chapter 12.

## 6.2 Basic considerations

Let us consider convective heat transfer between a surface and a fluid. In forced convection, the fluid is moving over the surface at a significant velocity. This velocity is often created by a pump (for liquids) or a fan or blower (for gases). Other examples of forced convection are wind blowing across the roof or exterior walls of a building and a liquid flowing by gravity over an inclined plate.

Experimental data for forced convection are correlated through the use of three dimensionless parameters: the Reynolds number, the Prandtl number, and the Nusselt number. These parameters are as follows:

$$\text{Reynolds number} = \text{Re} = \frac{\rho V L}{\mu} \tag{6.1}$$

where $\rho$ = density of the fluid.

$V$ = velocity of the fluid

$L$ = a characteristic length of the geometry. (For example, $L$ may be the length of a plate or the inside diameter of a pipe.)

$\mu$ = absolute viscosity of the fluid

The kinematic viscosity, $v$, is related to the absolute viscosity by $v = \mu/\rho$. Hence, Eq. (6.1) may be written as

$$\text{Reynolds number} = \text{Re} = \frac{V L}{v} \tag{6.2}$$

$$\text{Prandtl number} = \text{Pr} = \frac{c_p \mu}{k} \tag{6.3}$$

where $c_p$ = specific heat at constant pressure of the fluid

$\mu$ = absolute viscosity of the fluid

$k$ = thermal conductivity of the fluid

$$\text{Nusselt number} = \text{Nu} = \frac{h L}{k} \tag{6.4}$$

where $h$ = convective coefficient

$L$ = a characteristic length of the geometry

$k$ = thermal conductivity of the fluid

Correlation of experimental data gives the Nusselt number as a function of the two other numbers. That is

$$Nu = f \ (Re, Pr) \tag{6.5}$$

The functional relationship is often of the form $f = C \ Re^a \ Pr^b$, where C, a, and b are constants.

After the Nusselt number has been determined for a problem, the convective coefficient, $h$, can be obtained from the definition of Nu given in Eq. (6.4).

Fluid motion and velocities for natural (or free) convection are much smaller than those of forced convection. There typically is no equipment moving the fluid over the surface, and the fluid motion occurs through density gradients in the fluid created by the temperature difference between the surface and the fluid. Because velocities are small, the Reynolds number, which has the fluid velocity in it, is no longer significant and relevant. It is replaced by the dimensionless Grashof number, which has the coefficient of thermal expansion, $\beta$, and the acceleration of gravity, $g$, in its definition.

$$Grashof \ Number = Gr = \frac{g\beta(T_s - T_\infty)L^3}{v^2} \tag{6.6}$$

where $g$ = acceleration of gravity

$\beta$ = coefficient of thermal expansion of the fluid

$L$ = a characteristic length of the geometry

$v$ = kinematic viscosity of the fluid

$T_s$ and $T_\infty$ are the temperatures of the surface and fluid, respectively

For natural convection, experimental data are correlated using the Grashof, Prandtl, and Nusselt numbers:

$$Nu = \frac{hL}{k} = f \ (Gr, Pr) \tag{6.7}$$

The functional relationship is often of the form $f = C \ (Gr \ Pr)^a$, where C and a are constants.

The product of the Grashof and Prandtl numbers is the Rayleigh number, Ra.

Like forced convection, once the Nusselt number is determined for a problem, the convective coefficient, $h$, is obtained from the definition of Nu, i.e., from Eq. (6.4).

We will now continue with our discussion of forced convection. We first look at flow over flat plates. Following that, we consider flow over cylinders and spheres and flow through tube banks. Finally, we look at flow through tubes, pipes, and ducts.

## 6.3 External flow

This section considers external flow over surfaces. We first look at flow over a flat plate; then flow over cylinders and spheres; and finally, flow through tube banks.

### 6.3.1 Flow over a flat plate

Consider two-dimensional fluid flow over a flat plate, as shown in Fig. 6.1.

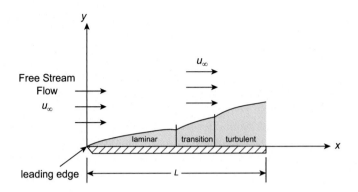

**FIGURE 6.1**

Flow over a flat plate.

Flow velocity in the $x$-direction is $u$ and flow velocity in the $y$ direction is $v$. There is no flow in the $z$-direction. The incoming flow is solely in the $x$-direction at a velocity $u_\infty$. This is called the "free-stream" velocity. The fluid encounters the plate at $x = 0$, which is called the "leading edge" of the plate. Flow of the fluid over the plate causes a "boundary layer" to form on the surface of the plate. At the plate surface ($y = 0$), the $x$ and $y$ components of the fluid velocity are zero. That is, the fluid sticks to the plate. Within the boundary layer, the fluid has both $u$ and $v$ components. However, $u$ is much larger than $v$. Outside the boundary layer, the fluid is moving in the $x$-direction at the free-stream velocity $u_\infty$.

If the plate is short enough in the direction of flow, there is only a laminar boundary layer region. Longer plates will have transition and turbulent boundary layer regions. The transitional region is short, and the flow over a plate is often modeled as a laminar boundary layer immediately followed by a turbulent boundary layer, as shown in Fig. 6.2.

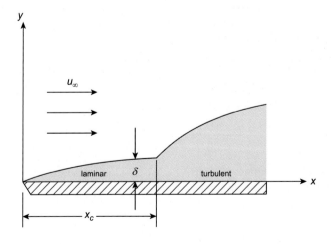

**FIGURE 6.2**

Laminar and turbulent boundary layers on a flat plate.

The Reynolds number was defined in Eqs. (6.1) and (6.2). For flow over a flat plate, the flow velocity in the Reynolds number is the free-stream velocity $u_\infty$ and the characteristic length is the distance $x$ downstream of the leading edge. As shown in Fig. 6.2, the flow changes from laminar to turbulent at a distance $x_c$ downstream of the leading edge. This distance, the "critical distance" is determined from the critical Reynolds number.

$$\text{Critical Reynolds number} = \text{Re}_{x_c} = \frac{\rho u_\infty x_c}{\mu} \tag{6.8}$$

The critical Reynolds number varies, depending on the quality of the flow as it passes the leading edge. If the flow is quiet and very smooth at the leading edge, the Critical Reynolds number is higher than if the flow is noisy and rough. The value typically used is $5 \times 10^5$. With this value, the critical distance $x_c$, from Eq. (6.8), is

$$x_c = \frac{\left(5 \times 10^5\right)\mu}{\rho u_\infty} = \frac{\left(5 \times 10^5\right)v}{u_\infty} \tag{6.9}$$

The thickness of the boundary layer, $\delta$, varies with $x$. It is usually defined as the distance from the plate at which the $x$-component of velocity is 99% of the free-stream velocity. That is,

$$\text{At } y = \delta, \quad \frac{u}{u_\infty} = 0.99 \tag{6.10}$$

The boundary layer thickness, $\delta$, does not increase as rapidly with downstream distance $x$ as shown in Figs. 6.1 and 6.2. For example, let us consider air flowing across a plate at a velocity $u_\infty = 2$ m/s. If the plate temperature is 100 C and the free-stream air temperature is 20 C, then the boundary layer thickness at a location 1 m downstream of the leading edge is only 1.5 cm.

### 6.3.1.1 Laminar boundary layer

In this section, we discuss the fluid mechanics and heat transfer relevant to the laminar boundary layer. Let us first obtain the continuity and momentum equations for an infinitesimal control volume in the boundary layer. This elemental control volume is shown in Fig. 6.3. Also shown in the figure are the

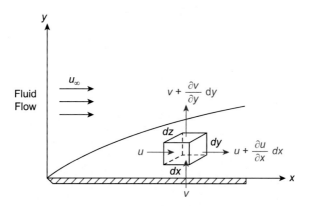

**FIGURE 6.3**

Control volume for laminar boundary layer.

flow velocities entering and leaving the control volume. The flow is two-dimensional. There is no flow in the $z$-direction. Solution of the continuity and momentum equations will give us the velocity components $u(x, y)$ and $v(x, y)$ in the boundary layer.

### 6.3.1.1.1 Continuity equation

The incoming mass flows are $\rho u\ dydz$ and $\rho v\ dxdz$; the outgoing flows are $\rho\left(u +\frac{\partial u}{\partial x}dx\right)dydz$ and $\rho\left(v +\frac{\partial v}{\partial y}dy\right)dxdz$. We will assume incompressible, steady flow.

Then, from mass conservation, we have

$$\text{Mass flow rate in} = \text{Mass flow rate out} \tag{6.11}$$

$$\rho u\ dydz + \rho v\ dxdz = \rho\left(u + \frac{\partial u}{\partial x}dx\right)dydz + \rho\left(v + \frac{\partial v}{\partial y}dy\right)dxdz \tag{6.12}$$

$$\text{This simplifies to } \frac{\partial u}{\partial x} + \frac{\partial v}{\partial y} = 0 \tag{6.13}$$

### 6.3.1.1.2 Momentum equation

We now look at the momentum in the $x$-direction. From Newton's Second Law,

$$\sum F_x = ma_x \tag{6.14}$$

where $\sum F_x$ = sum of forces on the element in the $x$-direction.
$m$ = mass inside the element = $\rho dxdydz$
$a_x$ = acceleration of the element in the $x$-direction = $\frac{du}{dt}$

Let us look at Fig. 6.3 and consider the forces acting on the fluid element. There are shear forces on the top and bottom of the element and pressure forces on the left and right faces of the element. The shear forces are due to the velocity gradient in the $y$ direction, $\partial u/\partial y$, and the fluid viscosity $\mu$. The shear stress in the $x$-direction in the fluid is $\tau = \mu\frac{\partial u}{\partial y}$, and the shear force is the shear stress times the area it acts on. Because of the fluid motion, the fluid at the bottom of the element is moving faster in the $x$-direction than the fluid below the element. Hence, the fluid below the element is exerting a shear force on the element in the $-x$ direction. At the top face of the element, the shear force on the element is in the $+x$ direction as the fluid above the element is moving faster than the fluid at the top face of the element.

The pressure forces on the left and right faces of the element act inward to the element. The pressure on the left face is $P$ and that on the right face is $P + \frac{\partial P}{\partial x}dx$.

Summing the forces on the element, we have

$$\sum F_x = \mu\left[\frac{\partial u}{\partial y} + \frac{\partial}{\partial y}\left(\frac{\partial u}{\partial y}\right)dy\right]dxdz - \mu\frac{\partial u}{\partial y}dxdz + Pdydz - \left(P + \frac{\partial P}{\partial x}dx\right)dydz$$

Simplifying, we get

$$\sum F_x = \left(\mu\frac{\partial^2 u}{\partial y^2} - \frac{\partial P}{\partial x}\right)dxdydz \tag{6.15}$$

From Eqs. (6.14) and (6.15), we get

$$\mu \frac{\partial^2 u}{\partial y^2} - \frac{\partial P}{\partial x} = \rho \frac{du}{dt} \tag{6.16}$$

As $u = u(x, y)$, we have, by the chain rule, $\frac{du}{dt} = \frac{\partial u}{\partial x}\frac{dx}{dt} + \frac{\partial u}{\partial y}\frac{dy}{dt}$
And, as $u = \frac{dx}{dt}$ and $v = \frac{dy}{dt}$, Eq. (6.16) becomes

$$\mu \frac{\partial^2 u}{\partial y^2} - \frac{\partial P}{dx} = \rho \left( u \frac{\partial u}{\partial x} + v \frac{\partial u}{\partial y} \right) \tag{6.17}$$

For a flat plate, $\partial P / dx = 0$, so we have the x $-$ momentum equation

$$u \frac{\partial u}{\partial x} + v \frac{\partial u}{\partial y} = v \frac{\partial^2 u}{\partial y^2} \tag{6.18}$$

From before, we had for continuity

$$\frac{\partial u}{\partial x} + \frac{\partial v}{\partial y} = 0 \tag{6.19}$$

Eqs. (6.18) and (6.19) can be solved simultaneously for the velocity components $u(x, y)$ and $v(x, y)$ within the laminar boundary layer. This was first done by Blasius [1] in 1908. He changed the two partial differential equations into the ordinary differential equation

$$f \frac{d^2 f}{d\eta^2} + 2 \frac{d^3 f}{d\eta^3} = 0 \tag{6.20}$$

This was done by defining a similarity variable $\eta$, which is a function of both $x$ and $y$, and the streamline function $\psi(x, y)$. For the streamline function $\psi$,

$$u = \frac{\partial \psi}{\partial y} \quad \text{and} \quad v = -\frac{\partial \psi}{\partial x} \tag{6.21}$$

If we apply Eq. (6.21) to Eq. (6.19), the continuity equation is satisfied identically, so we only have to consider the momentum equation, Eq. (6.18).

The variable $\eta(x, y)$ is defined as

$$\eta(x, y) = y \sqrt{\frac{u_\infty}{vx}} \tag{6.22}$$

The streamline function $\psi$ is related to the function $f(\eta)$ by

$$\psi(x, y) = \sqrt{u_\infty vx} f(\eta) \tag{6.23}$$

Eq. (6.20) was obtained by applying the definitions of Eqs. (6.22) and (6.23) to the terms of the momentum equation, Eq. (6.18).

Using Eqs. (6.21)–(6.23), it can be shown that the velocity components $u$ and $v$ are

$$u = u_\infty \frac{df}{d\eta} \quad \text{and} \quad v = \frac{1}{2} \sqrt{\frac{u_\infty v}{x}} \left( \eta \frac{df}{d\eta} - f \right) \tag{6.24}$$

Boundary conditions are needed for the solution of Eq. (6.20). These are as follows:

- At the plate surface $(y = 0)$, $u = 0$ and $v = 0$. From Eq. (6.22) and Eq. (6.24), these convert to

$$\text{At } \eta = 0, \quad \frac{df}{d\eta} = 0 \text{ and } f = 0$$

- As $y \to \infty$, $u \to u_\infty$ converts to

$$\text{As } \eta \to \infty, \quad \frac{df}{d\eta} \to 1$$

Solution of Eq. (6.20) is difficult due to its nonlinearity. Blasius solved it using a power series. More-recent solutions have been obtained through numerical methods. Table 6.1 gives the function $f(\eta)$ and its first two derivatives. The velocity components $u$ and $v$ may be obtained from this table by using Eq. (6.24). The velocity components are also available from Fig. 6.4: component $u$ from the left axis and component $v$ from the right axis. Note the straight line on the figure. This is the tangent line to the $u/u_\infty$ versus $\eta$ curve at $\eta = 0$. The slope of the line is 0.332, which is consistent with the 0.332 entry in Table 6.1 for $d^2f/d\eta^2$ at $\eta = 0$. The "0.332" factor will be discussed later in more detail.

**Table 6.1 Blasius function and its derivatives for laminar boundary layer.**

| $\eta$ | $f$ | $\frac{df}{d\eta}$ | $\frac{d^2f}{d\eta^2}$ |
|---|---|---|---|
| 0.0 | 0.0000 | 0.0000 | 0.3321 |
| 0.2 | 0.0066 | 0.0664 | 0.3320 |
| 0.4 | 0.0266 | 0.1328 | 0.3315 |
| 0.6 | 0.0597 | 0.1989 | 0.3301 |
| 0.8 | 0.1061 | 0.2647 | 0.3274 |
| 1.0 | 0.1656 | 0.3298 | 0.3230 |
| 1.2 | 0.2379 | 0.3938 | 0.3166 |
| 1.4 | 0.3230 | 0.4563 | 0.3079 |
| 1.6 | 0.4203 | 0.5168 | 0.2967 |
| 1.8 | 0.5295 | 0.5748 | 0.2829 |
| 2.0 | 0.6500 | 0.6298 | 0.2668 |
| 2.2 | 0.7812 | 0.6813 | 0.2484 |
| 2.4 | 0.9223 | 0.7290 | 0.2281 |
| 2.6 | 1.0725 | 0.7725 | 0.2065 |
| 2.8 | 1.2310 | 0.8115 | 0.1840 |
| 3.0 | 1.3968 | 0.8460 | 0.1614 |
| 3.2 | 1.5691 | 0.8761 | 0.1391 |
| 3.4 | 1.7470 | 0.9018 | 0.1179 |
| 3.6 | 1.9300 | 0.9233 | 0.0981 |

| $\eta$ | $f$ | $\frac{df}{d\eta}$ | $\frac{d^2f}{d\eta^2}$ |
|---|---|---|---|
| | **Table 6.1 Blasius function and its derivatives for laminar boundary layer.—cont'd** | | |
| 3.8 | 2.1160 | 0.9411 | 0.0801 |
| 4.0 | 2.3057 | 0.9555 | 0.0642 |
| 4.2 | 2.4980 | 0.9670 | 0.0505 |
| 4.4 | 2.6924 | 0.9759 | 0.0390 |
| 4.6 | 2.8882 | 0.9827 | 0.0295 |
| 4.8 | 3.0853 | 0.9878 | 0.0219 |
| 5.0 | 3.2833 | 0.9915 | 0.0159 |
| 5.2 | 3.4819 | 0.9942 | 0.0113 |
| 5.4 | 3.6809 | 0.9962 | 0.0079 |
| 5.6 | 3.8803 | 0.9975 | 0.0054 |
| 5.8 | 4.0800 | 0.9984 | 0.0036 |
| 6.0 | 4.2796 | 0.9990 | 0.0024 |

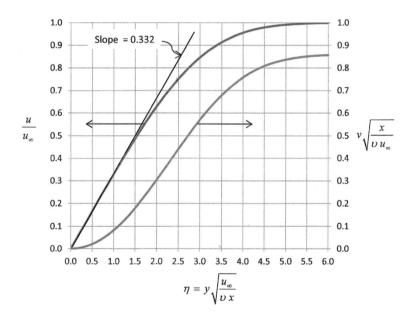

**FIGURE 6.4**

Velocity components for laminar boundary layer.

We discussed the boundary layer thickness, $\delta$, earlier. It is the distance $y$ from the plate surface at which the $x$-component of velocity is 99% of the free-stream velocity $u_\infty$. Looking at Table 6.1, we see that $\frac{u}{u_\infty} = \frac{df}{d\eta} = 0.99$ at about $\eta = 5$ (actually $\eta = 4.92$). From the definition of $\eta$, the boundary layer thickness at location $x$ from the leading edge is

$$\delta = 4.92\sqrt{\frac{\nu x}{u_\infty}} \tag{6.25}$$

In terms of the local Reynolds number $Re_x = \frac{u_\infty x}{v}$, the boundary layer thickness is

$$\delta = \frac{4.92x}{\sqrt{Re_x}} \tag{6.26}$$

For external flows such as flow over a plate, fluid properties are evaluated at the so-called *film temperature*, which is the average of the surface temperature $T_s$ and the fluid temperature $T_\infty$.

## Example 6.1
### Flow over a flat plate

*Problem*
Air at 20 C flows at a velocity of 2 m/s across a rectangular flat plate. The plate is 25 cm long in the direction of flow and 5 cm wide. It is at a temperature of 160 C.
**(a)** What types of boundary layers are on the plate, laminar or turbulent?
**(b)** What is the boundary layer thickness at a distance 15 cm downstream of the leading edge?
**(c)** What are the velocity components $u$ and $v$ at a location 15 cm downstream of the leading edge and at the middle of the boundary layer?

*Solution*
The first thing to do in this, and similar, problems is to the determine the Reynolds number at the far end of the plate, i.e., at $x = L$. For this problem, $L = 25$ cm.
The fluid properties are taken at the film temperature $T_f = \frac{T_s + T_\infty}{2} = \frac{160 + 20}{2} = 90$ C.
For this problem, we only need the kinematic viscosity of air. At 90 C, $v = 2.20 \times 10^{-5}$ m$^2$/s.
**(a)** At the end of the plate, $Re_L = \frac{u_\infty L}{v} = \frac{2(0.25)}{2.20 \times 10^{-5}} = 2.27 \times 10^4$. As $Re_L < 5 \times 10^5$, the boundary layer is laminar at the end of the plate and has a laminar boundary layer on it. Hence, we can use the equations of this section for determining $\delta$, $u$, and $v$.
**(b)** From Eq. (6.26), $\delta = \frac{4.92x}{\sqrt{Re_x}}$.

$$Re_x = u_\infty x/v = (2)(0.15)/2.20 \times 10^{-5} = 1.364 \times 10^4$$

$$\delta = \frac{4.92(0.15)}{\sqrt{1.364 \times 10^4}} = 0.00632 \text{ m} = 6.32 \text{ mm}.$$

**(c)** We want $u$ and $v$ at $x = 0.15$ m and $y = 0.00316$ m. For these values,

$$\eta = y\sqrt{\frac{u_\infty}{vx}} = 0.00316\sqrt{\frac{2}{(2.20 \times 10^{-5})(0.15)}} = 2.46$$

From Table 6.1, for this $\eta$, $f = 0.9674$; $\frac{df}{d\eta} = 0.7421$; $\frac{d^2f}{d\eta^2} = 0.2216$.

Using Eq. (6.24), $u = u_\infty \frac{df}{d\eta}$ and $v = \frac{1}{2}\sqrt{\frac{u_\infty v}{x}}\left(\eta\frac{df}{d\eta} - f\right)$

$$u = 2(0.7421) = 1.484 \text{ m/s} \text{ and } v = 0.5\sqrt{\frac{2(2.20 \times 10^{-5})}{0.15}}(2.46(0.7421) - 0.9674) = 0.00735 \text{ m/s}$$

As expected, it is seen that $u \gg v$.
Alternatively, Fig. 6.4 could have been used to solve this part. From the figure, for $\eta = 2.46$, $\frac{u}{u_\infty} \approx 0.74$ and $v\sqrt{\frac{x}{vu_\infty}} \approx 0.43$.
The $u$ and $v$ results are the same as the above results obtained from Table 6.1.

### 6.3.1.1.3 Drag force

Flow of a fluid across the flat plate exerts a drag force on the plate. The shear stress at the plate surface is

$$\tau = \mu \left(\frac{\partial u}{\partial y}\right)_{y=0\text{(plate surface)}} \tag{6.27}$$

Using the definition of $\eta$, $\left(\frac{\partial u}{\partial y}\right)_{y=0} = \sqrt{\frac{u_\infty}{vx}}\left(\frac{\partial u}{\partial \eta}\right)_{\eta=0}$ and Eq. (6.27) may be written as

$$\tau_x = \mu\sqrt{\frac{u_\infty}{vx}}\left(\frac{\partial u}{\partial \eta}\right)_{\eta=0} \tag{6.28}$$

Note that the subscript $x$ has been added to $\tau$. The shear stress at the surface of the plate varies with $x$, and $\tau_x$ is the local shear stress at location $x$ downstream of the leading edge.

As $u = u_\infty \frac{df}{d\eta}$, $\left(\frac{\partial u}{\partial \eta}\right)_{\eta=0} = u_\infty \left(\frac{d^2f}{d\eta^2}\right)_{\eta=0}$ and Eq. (6.28) becomes

$$\tau_x = \mu u_\infty \sqrt{\frac{u_\infty}{vx}}\left(\frac{d^2f}{d\eta^2}\right)_{\eta=0} \tag{6.29}$$

From Table 6.1, for $\eta = 0$, $\left(\frac{d^2f}{d\eta^2}\right) = 0.332$ so Eq. (6.29) becomes

$$\tau_x = 0.332\ \mu u_\infty \sqrt{\frac{u_\infty}{vx}} \tag{6.30}$$

The local Reynolds number is $\text{Re}_x = \frac{\rho u_\infty x}{\mu} = \frac{u_\infty x}{v}$. Using this definition, Eq. (6.30) may be written as

$$\tau_x = 0.332\frac{\mu u_\infty \sqrt{\text{Re}_x}}{x} \tag{6.31}$$

The local drag or friction coefficient $C_{fx}$ is defined by

$$\tau_x = C_{fx}\frac{\rho u_\infty^2}{2} \tag{6.32}$$

From Eqs. (6.31) and (6.32),

$$C_{fx} = \frac{0.664}{\sqrt{\text{Re}_x}} \tag{6.33}$$

By integrating the local friction coefficient over the length $L$ of the plate, we get the average drag coefficient $C_f$ for the plate.

$$C_f = \frac{1}{L}\int_0^L C_{fx}dx = \frac{1.328}{\sqrt{\text{Re}_L}} \tag{6.34}$$

$\text{Re}_L$ is the local Reynolds number at the end of the plate, that is, at $x = L$. Therefore, the average friction coefficient for the whole plate is twice the value of the local friction coefficient at $x = L$.

The average shear stress on the plate's surface is $\tau = C_f \frac{\rho u_\infty^2}{2}$. If the area of the plate in contact with the fluid is $A$, then the drag force $DF$ on the plate is

$$DF = \tau A = C_f A \frac{\rho u_\infty^2}{2} \tag{6.35}$$

---

## Example 6.2
### Drag force on a flat plate

*Problem*
Air at 20 C flows at a velocity of 2 m/s across a rectangular flat plate. The plate is 25 cm long in the direction of flow and 5 cm wide. It is at a temperature of 160 C.
What is the drag force exerted by the fluid on the plate?

*Solution*
We must first determine the Reynolds number at the far end of the plate, i.e., at $x = L$, to determine which types of boundary layers are on the plate, i.e., laminar or turbulent, or both. For this problem, $L = 25$ cm.
The fluid properties are taken at the film temperature $T_f = \frac{T_s + T_\infty}{2} = \frac{160 + 20}{2} = 90$ C.
At 90 C, $v = 2.20 \times 10^{-5}$ m$^2$/s, $\mu = 2.14 \times 10^{-5}$ kg/ms, and $\rho = 0.972$ kg/m$^3$.
At the end of the plate, $Re_L = \frac{u_\infty L}{v} = \frac{2(0.25)}{2.20 \times 10^{-5}} = 2.27 \times 10^4$. Hence, as $Re_L < 5 \times 10^5$, the equations for laminar flow on a flat plate may be used.
From Eq. (6.34), $C_f = \frac{1.328}{\sqrt{Re_L}} = \frac{1.328}{\sqrt{2.27 \times 10^4}} = 0.008814$.
From Eq. (6.35),

$$DF = \tau A = C_f A \frac{\rho u_\infty^2}{2} = 0.008814(0.25 \times 0.05)(0.972)(2^2)/2 = 2.14 \times 10^{-4}\,\text{N}$$

The drag force on the plate is **$2.14 \times 10^{-4}$ N**.

## 6.3.1.1.4 Energy equation
Let us now look at the thermal aspects of the laminar boundary layer. As we saw above, the flow of fluid across a plate creates a hydrodynamic boundary layer. If the plate is at a different temperature than the fluid, a thermal boundary layer is also created. Let us say that the plate temperature is $T_s$ and the free-stream fluid temperature is $T_\infty$. The temperature of the thermal boundary layer at the plate's surface is $T_s$. At the outer surface of the boundary layer, the fluid temperature is essentially at $T_\infty$. Being more specific, the thermal boundary layer thickness is normally defined as the distance from the plate at which the difference between the fluid temperature $T$ and the plate temperature $T_s$ is 99% of the difference between the free-stream fluid temperature $T_\infty$ and the plate temperature $T_s$.

Fig. 6.5 shows an infinitesimal control volume within the laminar thermal boundary layer.

Let us look at the energy transfers into and out of the control volume. We assume that the fluid is incompressible and the flow is steady. There is also no internal heat generation in the fluid due to viscous shearing and no changes in potential and kinetic energies. The flow in the boundary layer is two-dimensional. There is no flow in the $z$-direction. Finally, fluid properties are constant.

Energy transfers to and from the control volume are by conduction and flow of enthalpy. Enthalpy is given the symbol $h_{en}$ to distinguish it from the convective coefficient $h$.

The energy flows into and out of the control volume are shown in Fig. 6.5. They are

$$(\text{Energy in})_{\text{left}} = -kdydz\frac{\partial T}{\partial x} + \rho u h_{en}dydz \tag{6.36}$$

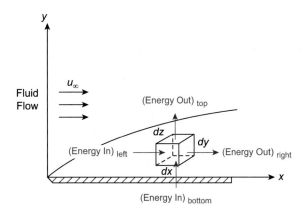

**FIGURE 6.5**

Control volume in the laminar thermal boundary layer.

$$(\text{Energy in})_{\text{bottom}} = -kdxdz\frac{\partial T}{\partial y} + \rho v h_{en}dxdz \tag{6.37}$$

$$(\text{Energy out})_{\text{right}} = -kdydz\left[\frac{\partial T}{\partial x} + \frac{\partial}{\partial x}\left(\frac{\partial T}{\partial x}\right)dx\right] + \rho\left[\left(u+\frac{\partial u}{\partial x}\right)dx\left(h_{en}+\frac{\partial h_{en}}{\partial x}\right)dx\right]dydz \tag{6.38}$$

$$(\text{Energy out})_{\text{top}} = -kdxdz\left[\frac{\partial T}{\partial y} + \frac{\partial}{\partial y}\left(\frac{\partial T}{\partial y}\right)dy\right] + \rho\left[\left(v+\frac{\partial v}{\partial y}\right)dy\left(h_{en}+\frac{\partial h_{en}}{\partial y}\right)dy\right]dxdz \tag{6.39}$$

For an incompressible fluid,

$$dh_{en} = c_p dT \tag{6.40}$$

where $c_p$ = specific heat at constant pressure.

And, as there is no internal heat generation and conditions are steady, the energy conservation equation for the control volume is

$$(\text{Energy In})_{\text{left}} + (\text{Energy In})_{\text{bottom}} = (\text{Energy Out})_{\text{right}} + (\text{Energy Out})_{\text{top}} \tag{6.41}$$

Using Eqs. (6.36) through (6.40) and the continuity equation, Eq. (6.19), and also neglecting second-order terms, Eq. (6.41) becomes

$$u\frac{\partial T}{\partial x} + v\frac{\partial T}{\partial y} = \alpha\left(\frac{\partial^2 T}{\partial x^2} + \frac{\partial^2 T}{\partial y^2}\right) \tag{6.42}$$

where $\alpha = \frac{k}{\rho c_p}$ = thermal diffusivity.

In a boundary layer, $\frac{\partial^2 T}{\partial x^2} \ll \frac{\partial^2 T}{\partial y^2}$, so the final form of the energy equation is

$$u\frac{\partial T}{\partial x} + v\frac{\partial T}{\partial y} = \alpha\frac{\partial^2 T}{\partial y^2} \tag{6.43}$$

Let us define a dimensionless temperature parameter $\theta_1 = \frac{T-T_s}{T_\infty - T_s}$. Then, $\frac{\partial T}{\partial x} = (T_\infty - T_s)\frac{\partial \theta_1}{\partial x}$, $\frac{\partial T}{\partial y} = (T_\infty - T_s)\frac{\partial \theta_1}{\partial y}$ and $\frac{\partial^2 T}{\partial y^2} = (T_\infty - T_s)\frac{\partial^2 \theta_1}{\partial y^2}$. Using these expressions in Eq. (6.43), we get

$$u\frac{\partial \theta_1}{\partial x} + v\frac{\partial \theta_1}{\partial y} = \alpha\frac{\partial^2 \theta_1}{\partial y^2} \tag{6.44}$$

Let us now look at the momentum equation obtained earlier. The momentum equation is

$$u\frac{\partial u}{\partial x} + v\frac{\partial u}{\partial y} = v\frac{\partial^2 u}{\partial y^2} \tag{6.18}$$

Defining a dimensionless velocity parameter as $\theta_2 = \frac{u}{u_\infty}$, we get $\frac{\partial u}{\partial x} = u_\infty\frac{\partial \theta_2}{\partial x}$, $\frac{\partial u}{\partial y} = u_\infty\frac{\partial \theta_2}{\partial y}$, and $\frac{\partial^2 u}{\partial y^2} = u_\infty\frac{\partial^2 \theta_2}{\partial y^2}$. Using these expressions in Eq. (6.18), we get

$$u\frac{\partial \theta_2}{\partial x} + v\frac{\partial \theta_2}{\partial y} = v\frac{\partial^2 \theta_2}{\partial y^2} \tag{6.45}$$

Let us look at the boundary conditions for Eqs. (6.44) and (6.45). For Eq. (6.44), the boundary conditions are

- At the surface ($y = 0$), $T = T_s$ and $\theta_1 = 0$.
- As $y \to \infty$, $T \to T_\infty$ and $\theta_1 \to 1$.
- As $y \to \infty$, $\frac{\partial T}{\partial y} \to 0$ and $\frac{\partial \theta_1}{\partial y} \to 0$.

For Eq. (6.45), the boundary conditions are

- At the surface ($y = 0$), $u = 0$ and $\theta_2 = 0$.
- As $y \to \infty$, $u \to u_\infty$ and $\theta_2 \to 1$.
- As $y \to \infty$, $\frac{\partial u}{\partial y} \to 0$ and $\frac{\partial \theta_2}{\partial y} \to 0$.

Summarizing, it is seen that the boundary conditions for Eqs. (6.44) and (6.45) are identical. Hence, the solutions for $\theta_1$ in Eq. (6.44) and $\theta_2$ in Eq. (6.45) will be the same if the kinematic viscosity $v$ of the fluid is equal to the thermal diffusivity $\alpha$ of the fluid. However, the ratio $\frac{v}{\alpha}$ is the Prandtl number $\text{Pr} = \frac{c_p\mu}{k}$. So, the solutions for $\theta_1 = \frac{T-T_s}{T_\infty - T_s}$ and $\theta_2 = \frac{u}{u_\infty}$ will be the same if the Prandtl number of the fluid is 1.

Pohlhausen [2] provided the first solution to the energy equation in 1921. The dimensionless temperature distribution in the laminar boundary layer is shown in Fig. 6.6 for several different Prandtl numbers. As discussed above, the curve for $\text{Pr} = 1$ is identical to the $u/u_\infty$ curve of Fig. 6.4.

If the abscissa $y\sqrt{\frac{u_\infty}{vx}}$ of Fig. 6.6 is changed to $y\sqrt{\frac{u_\infty}{vx}}\text{Pr}^{1/3}$, then the curves of Fig. 6.6 converge to a single curve. This is shown in Fig. 6.7.

### 6.3.1.1.5 Thermal boundary layer thickness
Both the hydrodynamic and thermal boundary layers begin at the leading edge for flow over a heated plate. Eq. (6.26) gave the thickness $\delta$ of the hydrodynamic boundary layer:

$$\delta = \frac{4.92x}{\sqrt{\text{Re}_x}} \tag{6.26}$$

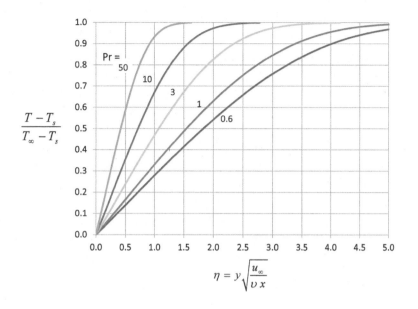

**FIGURE 6.6**

Temperature in laminar boundary layer.

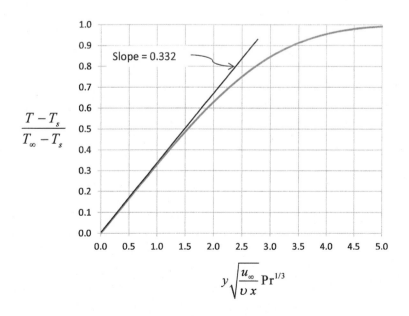

**FIGURE 6.7**

Temperature in laminar boundary layer (consolidated curve).

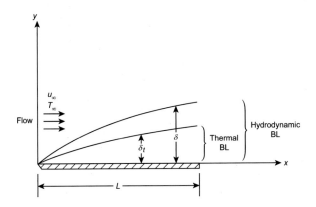

**FIGURE 6.8**

Boundary layers for laminar flow on flat plate.

For laminar flow over a flat plate, the thickness $\delta_t$ of the thermal boundary layer is

$$\delta_t = \delta Pr^{-1/3} = \frac{4.92x}{Pr^{1/3}\sqrt{Re_x}} \qquad (6.46)$$

From this equation, if the Prandtl number of the fluid is 1, then the hydrodynamic and thermal boundary layers are identical. If $Pr > 1$, then the thermal boundary layer thickness is less than the hydrodynamic boundary layer thickness at a given location. If $Pr < 1$, the thermal boundary layer thickness is greater than the hydrodynamic boundary layer thickness at a given location.

Fig. 6.8 shows the hydrodynamic and thermal boundary layers for laminar flow over a flat plate. Considering the relative thicknesses of the layers, the figure is for a fluid having $Pr > 1$.

### 6.3.1.1.6 Convective coefficient and Nusselt number

Let us first determine a relation for the convective coefficient at a location $x$ downstream of the leading edge. This is the **local** coefficient $h_x$. The coefficient we have previously been using, $h$, is the **average** coefficient for an entire surface.

As mentioned before, the fluid velocity is zero at the plate's surface, i.e., at $y = 0$. Hence, heat transfer from the plate's surface to the fluid is by conduction. If the area of the plate surface at location $x$ is $dA$, then the heat transfer to the fluid from area $dA$ is

$$q_{y=0} = -kdA\left(\frac{\partial T}{\partial y}\right)_{y=0} \qquad (6.47)$$

From the definition of the convective coefficient, we have

$$q_{y=0} = h_x dA(T_s - T_\infty) \qquad (6.48)$$

Equating the right sides of Eqs. (6.47) and (6.48) and rearranging, we get

$$h_x = -\frac{k}{T_s - T_\infty}\left(\frac{\partial T}{\partial y}\right)_{y=0} \qquad (6.49)$$

We can get $\left(\frac{\partial T}{\partial y}\right)_{y=0}$ from Fig. 6.7. The straight line on the figure is the tangent to the curve at $y = 0$. The line has a slope of 0.332. That is,

$$\text{At } y = 0, \frac{\partial\left(\dfrac{T - T_s}{T_\infty - T_s}\right)}{\partial\left(y\sqrt{\dfrac{u_\infty}{\upsilon x}}\text{Pr}^{1/3}\right)} = 0.332 \tag{6.50}$$

Everything in Eq. (6.50) is constant except $T$ and $y$, so we have

$$\left(\frac{\partial T}{\partial y}\right)_{y=0} = 0.332(T_\infty - T_s)\sqrt{\frac{u_\infty}{\upsilon x}}\text{Pr}^{1/3} \tag{6.51}$$

Using this in Eq. (6.49), the local convective coefficient is

$$h_x = 0.332k\sqrt{\frac{u_\infty}{\upsilon x}}\text{Pr}^{1/3} \tag{6.52}$$

And, as the local Reynolds number is $\text{Re}_x = \frac{u_\infty x}{\upsilon}$, Eq. (6.52) may also be written as

$$h_x = \frac{0.332k}{x}\sqrt{\text{Re}_x}\text{Pr}^{1/3} \tag{6.53}$$

Using the definition of the Nusselt number given in Eq. (6.4), the local Nusselt number is

$$\text{Nu}_x = \frac{h_x x}{k} = 0.332\sqrt{\text{Re}_x}\text{Pr}^{1/3} \tag{6.54}$$

To get the average convective coefficient and average Nusselt number for the entire plate, $h_x$ and $\text{Nu}_x$ from Eqs. (6.53) and (6.54) are integrated over the length of the plate. If the length of the plate in the direction of flow is $L$, the average values are

$$h = \frac{1}{L}\int_0^L h_x dx = \frac{0.664k}{L}\sqrt{\text{Re}_L}\text{Pr}^{1/3} \tag{6.55}$$

$$\text{Nu} = \frac{1}{L}\int_0^L \text{Nu}_x dx = 0.664\sqrt{\text{Re}_L}\text{Pr}^{1/3} \tag{6.56}$$

Looking at Eq. (6.53) through (6.56), it is seen that the average values of the convective coefficient and Nusselt number for the entire plate are twice the local values at the trailing edge, i.e., at $x = L$.

The above equations for laminar flow on a flat plate are valid for fluids with Prandtl numbers between about 0.6 and 50. These fluids include common gases (e.g., air, oxygen, nitrogen, hydrogen, carbon dioxide) and low viscosity liquids such as water, refrigerants, and propane. The equations are not valid for high viscosity fluids such as heavy oils and glycerin that have high Pr or for liquid metals that have low Pr.

Churchill and Ozoe [3] have correlated experimental data to arrive at Eq. (6.57), which is valid for all Prandtl numbers.

$$\text{Nu}_x = \frac{h_x x}{k} = \frac{0.3387 \ \text{Re}_x^{1/2} \text{Pr}^{1/3}}{\left[1 + \left(\dfrac{0.0468}{\text{Pr}}\right)^{2/3}\right]^{1/4}} \quad \text{for} \quad \text{Re}_x \text{Pr} > 100 \tag{6.57}$$

We saw above that the average Nusselt number and convective coefficient for the plate were twice the local values at the trailing edge. Churchill [4] recommends that this can be done for Eq. (6.57). That is, the average Nusselt number for the plate is $\text{Nu} = 2 \ \text{Nu}_L$. Using Eq. (6.57), we have

$$\text{Nu} = \frac{hL}{k} = \frac{0.6774 \text{Re}_L^{1/2} \text{Pr}^{1/3}}{\left[1 + \left(\dfrac{0.0468}{\text{Pr}}\right)^{2/3}\right]^{1/4}} \tag{6.58}$$

Eqs. (6.57) and (6.58) are valid for all values of Pr.

---

## Example 6.3
### Convection from an isothermal flat plate

*Problem*
Air at 15 C flows across a rectangular flat plate at a velocity of 3 m/s. The plate is 1.5 m wide and 2 m in the direction of flow. The plate is at a uniform temperature of 85 C. What is the rate of heat transfer from the plate to the air?

*Solution*
The first step is to determine the Reynolds number at the trailing edge to see if the boundary layer on the plate is entirely laminar or laminar followed by turbulent.
The film temperature is $T_f = \frac{T_s + T_\infty}{2} = \frac{85 + 15}{2} = 50$ C.
For air at 50 C, $k = 0.02735$ W/m C, $v = 1.798 \times 10^{-5}$ m$^2$/s, Pr $= 0.7228$

$$\text{Re}_L = \frac{u_\infty L}{v} = \frac{3(2)}{1.798 \times 10^{-5}} = 3.34 \times 10^5$$

As $\text{Re}_L < 5 \times 10^5$, there is a laminar boundary layer on the entire plate.
Using Eq. (6.55),

$$h = \frac{0.664k}{L} \sqrt{\text{Re}_L} \text{Pr}^{1/3} = \frac{0.664(0.02735)}{2} \sqrt{3.34 \times 10^5} (0.7228)^{1/3} = 4.71 \text{ W/m}^2 \text{ C}$$

The heat transfer is $q = hA(T_s - T_\infty) = (4.71)[(1.5)(2)](85 - 15) = \mathbf{989}$ W.
It is interesting to compare this result with the results obtained from Eq. (6.58). If we had used that equation, Nu = 338.5, $h = \frac{k \, \text{Nu}}{L} = 4.63$ W/m$^2$ C, and $q = 972$ W. This is less than a 2% difference from the Pohlhausen result.

---

## 6.3.1.1.7 Reynolds-Colburn analogy
The electric-heat analogy was discussed in Chapter 3. The analogy used electrical circuits to model heat transfer systems. The circuits were very useful in the visualization of the heat transfer processes and in the development of equations for solution of heat transfer problems.

In this section, we discuss another very useful analogy. The analogy relates the drag force for laminar flow over the surface of a flat plate to the convective coefficient for the surface. We will first develop the analogy and then discuss its applications.

The local drag coefficient and local Nusselt number for laminar flow over a flat plate are given by Eqs. (6.33) and (6.54):

$$C_{fx} = 0.664 \text{Re}_x^{-1/2} \tag{6.33}$$

$$\text{Nu}_x = \frac{h_x x}{k} = 0.332 \text{Re}^{1/2} \text{Pr}^{1/3} \tag{6.54}$$

The local **Stanton number**, $\text{St}_x$, is defined as

$$\text{St}_x = \frac{h_x}{\rho c_p u_\infty} \tag{6.59}$$

The Stanton number is dimensionless. It is a combination of three other dimensionless parameters:

$$\text{St}_x = \frac{\text{Nu}_x}{\text{Re}_x \text{Pr}} \tag{6.60}$$

From Eq. (6.33), we have

$$\frac{1}{2} C_{fx} = 0.332 \text{Re}_x^{-1/2} \tag{6.61}$$

From Eqs. (6.60) and (6.54), we have

$$\text{St}_x = \frac{0.332 \text{ Re}_x^{1/2} \text{ Pr}^{1/3}}{\text{Re}_x \text{Pr}} = 0.332 \text{ Re}_x^{-1/2} \text{ Pr}^{-2/3} \tag{6.62}$$

which can be changed to

$$\text{St}_x \text{Pr}^{2/3} = 0.332 \text{ Re}_x^{-1/2} \tag{6.63}$$

The right-hand sides of Eqs. (6.61) and (6.63) are the same. Therefore,

$$\text{St}_x \text{Pr}^{2/3} = \frac{1}{2} C_{fx} \tag{6.64}$$

Eq. (6.64) is the **Reynolds-Colburn analogy.** It relates the local Stanton number to the local drag coefficient.

Eq. (6.64) may also be written for the average Stanton number and the average drag coefficient:

$$\text{St} \text{ Pr}^{2/3} = \frac{1}{2} C_f \tag{6.65}$$

The convective coefficient is in the definition of the Stanton number. Hence, Eqs. (6.64) and (6.65) are relationships between the convective coefficient ($h_x$ or $h$) and the drag coefficient ($C_{fx}$ or $C_f$)

The Reynolds-Colburn analogy is very useful. Measurement of drag force for laminar flow on a flat plate can be used to predict the convective coefficient. That is, convective coefficients can be estimated from mechanical measurements without the performance of heat transfer experiments, which can often be very complex and time-consuming. Conversely, if the convective heat transfer for the plate is known, the drag force on the plate can be estimated without the performance of drag force experiments. Example 6.4 illustrates application of the Reynolds-Colburn analogy.

---

### Example 6.4
#### Reynolds-Colburn analogy

*Problem*

Nitrogen at 20 C flows at a speed of 5 m/s across a flat plate that has a uniform surface temperature of 80 C. The plate is 0.6 m long in the direction of flow and has a width of 0.3 m. The nitrogen exerts a drag force of 0.002 N on the plate. Estimate the average convective coefficient $h$ for the plate's surface.

*Solution*

$$T_{film} = \frac{T_s + T_\infty}{2} = \frac{80 + 20}{2} = 50 \text{ C}$$

For nitrogen at 50 C, $\rho = 1.049 \text{ kg/m}^3$, $c_p = 1041 \text{ J/kg C}$, $v = 17.97 \times 10^{-6} \text{ m}^2/\text{s}$, and Pr $= 0.714$.

At $x = L = 0.6$ m, $\text{Re}_L = \frac{u_\infty L}{v} = \frac{(5)(0.6)}{17.97 \times 10^{-6}} = 1.669 \times 10^5 \text{ (laminar since} < 5 \times 10^5)$

There is a laminar boundary layer over the entire surface.

Eq. (6.35): Drag Force $= C_f A \frac{\rho u_\infty^2}{2}$

Putting in values, we have $0.002 = \frac{C_f (0.6)(0.3)(1.049)(5)^2}{2}$

and $C_f = 8.474 \times 10^{-4}$

The Reynolds-Colburn analogy is Eq. (6.65):

$$\text{St Pr}^{2/3} = \frac{1}{2} C_f$$

Rearranging and putting in values, we have

$$\text{St} = \frac{\frac{1}{2} C_f}{\text{Pr}^{2/3}} = \frac{\frac{1}{2}(8.474 \times 10^{-4})}{(0.714)^{2/3}} = 5.304 \times 10^{-4}$$

From Eq. (6.59) modified for average coefficients rather than local coefficients:

$$\text{St} = \frac{h}{\rho c_p u_\infty}$$

Rearranging and putting in values, we have

$$h = \rho c_p u_\infty \text{ St} = (1.049)(1041)(5)(5.304 \times 10^{-4}) = 2.90 \text{ W/m}^2 \text{ C}$$

The estimated convective coefficient for the surface is 2.90 W/m² C.

## 6.3.1.1.8 Constant heat flux

The above discussion is for situations in which the plate has a uniform surface temperature $T_s$. However, some situations, e.g., electrically heated plates, have a constant heat flux boundary condition at the plate's surface rather than a constant temperature boundary condition.

Let us consider heat flow $q$ to the fluid from an area $A$ on the plate's surface. The heat flux $q/A$ is constant over the plate's surface. Although the heat flux is constant, the convective coefficient $h$ and the temperature difference $T_s - T_\infty$ vary with the $x$-location on the plate.

For the convective heat transfer, we have

$$\left(\frac{q}{A}\right) = h_x (T_s - T_\infty)_x \tag{6.66}$$

The local convective coefficient is

$$h_x = \frac{(q/A)}{(T_s - T_\infty)_x} \tag{6.67}$$

From this equation, the plate temperature $T_s$ at location $x$ is

$$T_s(x) = T_\infty + \frac{(q/A)}{h_x} \tag{6.68}$$

The local Nusselt number for constant heat flux is

$$\text{Nu}_x = \frac{h_x x}{k} = 0.453 \text{Re}_x^{1/2} \text{Pr}^{1/3} \tag{6.69}$$

As $\mathrm{Nu}_x = \frac{h_x x}{k}$, the local convective coefficient is

$$h_x = \frac{0.453k}{x} \mathrm{Re}_x^{1/2} \mathrm{Pr}^{1/3} \tag{6.70}$$

The average temperature difference $(T_s - T_\infty)_{\mathrm{avg}}$ for the entire plate of length $L$ in the direction of flow is

$$(T_s - T_\infty)_{\mathrm{avg}} = \frac{1}{L} \int_0^L (T_s - T_\infty)_x dx \tag{6.71}$$

Using Eqs. (6.68) and (6.70), and performing the integration of Eq. (6.71), we get

$$(T_s - T_\infty)_{\mathrm{avg}} = \left(\frac{q}{A}\right) \frac{L/k}{0.6795 \mathrm{Re}_L^{1/2} \mathrm{Pr}^{1/3}} \tag{6.72}$$

where $\mathrm{Re}_L$ is the local Reynolds number at $x = L$.

The average temperature difference is also

$$(T_s - T_\infty)_{\mathrm{avg}} = \frac{2}{3h_L}\left(\frac{q}{A}\right) \tag{6.73}$$

where $h_L$ is the local convective coefficient at $x = L$.

The material properties in Eqs. (6.69), (6.70), and (6.72) are evaluated at the "film" temperature, which is the average of the free-stream and surface temperatures. However, the surface temperature varies along the plate and is not known at the beginning of the solution. An iterative procedure must be used. This procedure is illustrated in Example 6.5.

## Example 6.5
### Heated plate with uniform heat flux

*Problem*
Air at 30 C flows at a speed of 2.4 m/s across the top of a thin, square electrically heated plate. The bottom of the plate is perfectly insulated. The plate is 25 cm on a side and inputs 150 W to the air.
**(a)** What is the average temperature of the plate?
**(b)** What is the maximum temperature of the plate?

*Solution*
Because of the small dimensions of the plate, we will assume that the plate only has a laminar boundary layer. We can check this assumption after plate temperatures are determined.
As mentioned above, we need an iterative technique to determine the surface temperature of the plate. One technique is as follows:
We first rearrange Eq. (6.73) to get $h_L$ on the left side:

$$h_L = \frac{2}{3\left[(T_s)_{\mathrm{avg}} - T_\infty\right]}\left(\frac{q}{A}\right) \tag{6.74}$$

From Eq. (6.70), the convective coefficient at $x = L$ is

$$h_L = \frac{0.453k}{L}\text{Re}_L^{1/2}\text{Pr}^{1/3} \tag{6.75}$$

Equating the right sides of Eqs. (6.74) and (6.75), we get

$$\frac{2}{3\left[(T_s)_{avg} - T_\infty\right]}\left(\frac{q}{A}\right) = \frac{0.453k}{L}\text{Re}_L^{1/2}\text{Pr}^{1/3} \tag{6.76}$$

Rearranging Eq. (6.73) for the average temperature of the surface $(T_s)_{avg}$, we get

$$(T_s)_{avg} = T_\infty + 1.4717\left(\frac{q}{A}\right)\frac{1}{k\text{Pr}^{1/3}}\sqrt{\frac{Lv}{u_\infty}} \tag{6.77}$$

For our problem, $T_\infty = 30$ C, $\left(\frac{q}{A}\right) = \frac{150}{(0.25)^2} = 2400$ W/m$^2$, $L = 0.25$ m, and $u_\infty = 2.4$ m/s, so Eq. (6.77) becomes

$$(T_s)_{avg} = 30 + 1140\frac{v^{1/2}}{k\text{Pr}^{1/3}} \tag{6.78}$$

We can now proceed with the iterative process: guess a value for the average surface temperature $(T_s)_{avg}$; determine the fluid properties at the film temperature $T_f = \frac{(T_s)_{avg}+T_\infty}{2}$; and calculate $(T_s)_{avg}$ using Eq. (6.78) and compare this result with the guessed surface temperature. If they agree, we have a solution. If not, pick another guess for $(T_s)_{avg}$ and do the process again and again until the guessed and calculated values of the surface temperature agree.

Following this procedure, let us pick an initial value for $(T_s)_{avg}$ of 200 C. Then, $T_f = (200+30)/2 = 115$ C. At 115 C, $k = 0.0320$ W/m C, $v = 2.468 \times 10^{-5}$ m$^2$/s, and Pr $= 0.7082$.

Calculating $(T_s)_{avg}$ using Eq. (6.78) gives $(T_s)_{avg} = 228.6$ C, which does not agree with our guess of 200C. Picking 230 C as our new guess, we have $T_f = (230+30)/2 = 130$ C.

At 130 C, $k = 0.03305$ W/m C, $v = 2.633 \times 10^{-5}$ m$^2$/s, and Pr $= 0.7057$. And, using Eq. (6.78), we get $(T_s)_{avg} = 228.8$ C, which agrees closely with our guess of 230 C. Iterating a final time with 229 C, we have agreement between our guess and the calculated value.

So, the solution to Part (a) is: **The average surface temperature of the plate is 229 C.**

From Eqs. (6.68) and (6.70), it can be seen that the maximum surface temperature is at the trailing edge $x = L$. From Eq. (6.70), we have

$$h_L = \frac{0.453k}{L}\text{Re}_L^{1/2}\text{Pr}^{1/3} = \frac{0.453(0.033)}{0.25}\left(\frac{2.4(0.25)}{2.6275 \times 10^{-5}}\right)^{1/2}(0.7058)^{1/3} = 8.05 \text{ W/m}^2\text{C}$$

From Eq. (6.68), we have

$$T_s(L) = T_\infty + \frac{(q/A)}{h_L} = 30 + \frac{150/(0.25)^2}{8.05} = 328 \text{ C}$$

So, the solution to Part (b) is: **The maximum surface temperature of the plate is 328 C.**

At the beginning, we assumed that the plate had only a laminar boundary layer. Let us check this by calculating the Reynolds number at $x = L$.

$$\text{Re}_L = \frac{u_\infty L}{v} = \frac{(2.4)(0.25)}{2.6275 \times 10^{-5}} = 2.284 \times 10^4$$

As $\text{Re}_L < 5 \times 10^5$, the boundary layer is laminar at $x = L$ and the plate only has a laminar boundary layer on it.

## 6.3.1.1.9 Unheated starting length

The above sections considered laminar flow over a flat plate where the entire plate was at a uniform surface temperature $T_s$. Let us now consider flow over a plate where the plate is insulated or unheated for a portion of the plate immediately downstream of the leading edge. This is shown in Fig. 6.9.

The plate is unheated or insulated for $0 \leq x \leq x_o$. For this portion, the plate surface is at the free-stream temperature $T_\infty$. For $x > x_o$, the plate surface is at uniform temperature $T_s$. The hydrodynamic boundary layer starts at the leading edge, but the thermal boundary layer starts at $x = x_o$. The boundary layers are laminar over the entire plate.

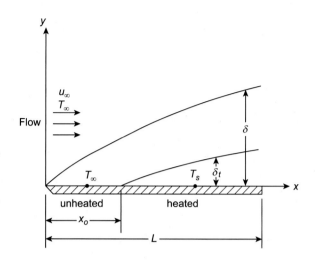

**FIGURE 6.9**

Plate with unheated or insulated starting length.

The local Nusselt number is [5]

$$\text{Nu}_x = \frac{h_x x}{k} = \frac{0.332 \text{Re}_x^{1/2} \text{Pr}^{1/3}}{\left[1 - \left(\frac{x_o}{x}\right)^{3/4}\right]^{1/3}} \tag{6.79}$$

The average convective coefficient for the heated portion of the plate $(x \geq x_o)$ is found by integrating $h_x$ over the heated part of the plate. It is

$$h = \frac{0.664 k}{(L - x_o)} \left[1 - \left(\frac{x_o}{L}\right)^{3/4}\right]^{2/3} \text{Re}_L^{1/2} \text{Pr}^{1/3} \tag{6.80}$$

Comparing Eqs. (6.79) and (6.80), it can be seen that the relationship between the average coefficient $h$ for the heated section and the local coefficient $h_L$ at the trailing edge $x = L$ is

$$h = 2 \frac{\left[1 - \left(\frac{x_o}{L}\right)^{3/4}\right]}{\left[1 - \left(\frac{x_o}{L}\right)\right]} h_L \tag{6.81}$$

Eqs. (6.79) and (6.80) are for fluids with Prandtl numbers in the range of 0.6–50. For fluids outside this range, Eq. (6.57) modified for an unheated starting length may be used. This is

$$\text{Nu}_x = \frac{h_x x}{k} = \frac{0.3387 \text{Re}_x^{1/2} \text{Pr}^{1/3}}{\left[1 - \left(\frac{x_o}{x}\right)^{3/4}\right]^{1/3} \left[1 + \left(\frac{0.0468}{\text{Pr}}\right)^{2/3}\right]^{1/4}} \tag{6.82}$$

Eq. (6.81) may be used to determine the average convective coefficient $h$ for the heated portion of the plate. From Eq. (6.82), the convective coefficient at the trailing edge of the plate is

$$h_L = \frac{0.3387 Re_L^{1/2} Pr^{1/3}}{\left[1 - \left(\frac{x_o}{L}\right)^{3/4}\right]^{1/3} \left[1 + \left(\frac{0.0468}{Pr}\right)^{2/3}\right]^{1/4}} \left(\frac{k}{L}\right) \qquad (6.83)$$

Using this $h_L$ in Eq. (6.81), we get the average convective coefficient:

$$h = \frac{0.6774 \left[1 - \left(\frac{x_o}{L}\right)^{3/4}\right]^{2/3} Re_L^{1/2} Pr^{1/3}}{\left[1 - \left(\frac{x_o}{L}\right)\right] \left[1 + \left(\frac{0.0468}{Pr}\right)^{2/3}\right]^{1/4}} \left(\frac{k}{L}\right) \qquad (6.84)$$

Summarizing: For fluids with Prandtl numbers in the 0.6–50 range, Eq. (6.80) should be used to determine the average convective coefficient. Eq. (6.84) should be used for fluids with Prandtl numbers outside this range.

### 6.3.1.2 Turbulent boundary layer

If a plate is long enough, there will generally be a laminar boundary layer followed by a turbulent boundary layer. If the laminar boundary layer portion is small or if the flow upstream of the plate is rough and agitated, then it is often reasonable to assume that the plate has a turbulent boundary layer on its entire length. This section presents equations for plates having a turbulent boundary layer region.

Eqs. (6.33) and (6.34) gave equations for the local friction coefficient $C_{fx}$ and average drag coefficient $C_f$ for a laminar boundary layer. If the flow is turbulent over the entire plate, the corresponding relations are

$$C_{fx} = 0.059 \, Re_x^{-1/5} \quad \text{and} \quad C_f = 0.074 \, Re_L^{-1/5} \qquad (6.85)$$

Eqs. (6.53) and (6.55) gave equations for the local convective coefficient $h_x$ and the average convective coefficient $h$ for a laminar boundary layer. If the flow is turbulent over the entire plate, the corresponding relations are

$$h_x = 0.0296 \, Re_x^{0.8} Pr^{1/3}(k/x) \quad \text{and} \quad h = 0.037 \, Re_L^{0.8} Pr^{1/3}(k/L) \qquad (6.86)$$

Eqs. (6.85) and (6.86) are for Reynolds numbers in the range $5 \times 10^5 < Re_L < 10^7$.

Regarding plates with an insulated or unheated starting length: Eqs. (6.82) and (6.84) gave relations for the local and average convective coefficients for laminar flow of fluids outside the $0.6 < Pr < 50$ range. The corresponding equations for turbulent flow are

$$Nu_x = \frac{h_x x}{k} = \frac{0.0296 Re_x^{0.8} Pr^{1/3}}{\left[1 - \left(\frac{x_o}{x}\right)^{9/10}\right]^{1/9}} \qquad (6.87)$$

$$\text{and } h = \frac{0.037\left[1 - \left(\frac{x_o}{L}\right)^{9/10}\right]^{8/9}\text{Re}_L^{0.8}\text{Pr}^{1/3}}{\left[1 - \left(\frac{x_o}{L}\right)\right]}\left(\frac{k}{L}\right) \tag{6.88}$$

If the plate has a laminar region followed by a turbulent region, then we can integrate the laminar and turbulent relations over their respective regions to get the average drag and convective coefficients for the entire plate. That is

$$C_f = \frac{1}{L}\left[\int_0^{x_c}(C_{fx})_{\text{laminar}}dx + \int_{x_c}^L(C_{fx})_{\text{turbulent}}dx\right] \tag{6.89}$$

and

$$h = \frac{1}{L}\left[\int_0^{x_c}(h_x)_{\text{laminar}}dx + \int_{x_c}^L(h_x)_{\text{turbulent}}dx\right] \tag{6.90}$$

The critical length $x_c$ is the distance from the leading edge of the plate where the flow changes from laminar to turbulent. This is taken to be where the Reynolds number is $5 \times 10^5$.

When the integrations of Eqs. (6.89) and (6.90) are performed using the $C_{fx}$ relations of Eqs. (6.33) and (6.85) and the $h_x$ relations of Eqs. (6.53) and (6.86), it is found that the average coefficients for a plate having a laminar boundary layer followed by a turbulent boundary layer are

$$C_f = 0.074\,\text{Re}_L^{-1/5} - \frac{1742}{\text{Re}_L} \tag{6.91}$$

and

$$h = \left(0.037\text{Re}_L^{0.8} - 871\right)\text{Pr}^{1/3}(k\,/\,L) \tag{6.92}$$

---

## Example 6.6
### Heated plate with laminar and turbulent boundary layers

*Problem*
Air at 20 C flows at a speed of 20 m/s over a flat plate. The plate is 20 cm wide and 1 m long in the direction of flow. The plate is at 80 C.
**(a)** What is the rate of convective heat transfer from the plate to the air?
**(b)** If the plate surface has an emissivity of 0.8 and the surroundings are at 20 C, what is the rate of radiative heat transfer from the plate to the surroundings?

*Solution*
**(a)** The film temperature is $T_f = \frac{T_s + T_\infty}{2} = \frac{80+20}{2} = 50$ C. Air at 50 C has the properties

$$k = 0.0274\text{ W/m C}, \quad v = 1.80 \times 10^{-5}\text{ m}^2/\text{s}, \quad \text{Pr} = 0.723$$

$$\text{Re}_L = \frac{u_\infty L}{v} = \frac{(20)(1)}{1.80 \times 10^{-5}} = 1.111 \times 10^6$$

As $Re_L > 5 \times 10^5$, the plate has both laminar and turbulent boundary layers, and Eq. (6.92) should be used for $h$.

$$h = (0.037Re_L^{0.8} - 871)Pr^{1/3}(k/L) = \left[(0.037)(1.111 \times 10^6)^{0.8} - 871\right](0.723)^{1/3}(0.0274/1)$$

$$h = 41.0 \text{ W/m}^2 \text{ C}$$

The heat flow by convection from the plate to the air is

$$q = hA(T_s - T_\infty) = (41.0)[(0.2)(1)](80 - 20) = \textbf{492 W}$$

**(b)** The radiative heat transfer from the plate to the surroundings is

$$q = \varepsilon\sigma A(T_s^4 - T_{surr}^4) = (0.8)(5.67 \times 10^{-8})[(0.2)(1)]\left[(80 + 273.15)^4 - (20 + 273.15)^4\right]$$

$$q = \textbf{74 W}$$

Note that we had to use absolute Kelvin temperatures for the radiative heat transfer and that the radiative heat transfer is small compared with the convective heat transfer.

As a final topic in the category of flow over a flat plate, let us look at how the local convective coefficient $h_x$ changes with distance $x$ from the leading edge. It will be seen that $h_x$ increases significantly as the flow changes from laminar to turbulent.

## Example 6.7
### Variation of local convective coefficient with location

*Problem*

Air at 30 C flows across a flat plate at a velocity of 8 m/s. The plate is at a uniform temperature of 90 C and is 2.5 m long in the direction of air flow. Determine how the local convective coefficient $h_x$ varies with distance $x$, the downstream distance from the leading edge. Plot $h_x$ versus $x$ from the leading edge ($x = 0$) to the downstream end of the plate ($x = L = 2.5$ m).

*Solution*

Let us first calculate the Reynolds number at $x = L$ and confirm that the plate has both laminar and turbulent boundary layers.

Properties of air at the film temperature of $(90 + 30)/2 = 60$ C are, from Appendix D:

$$k = 0.0281 \text{ W/m C}, \quad v = 18.97 \times 10^{-6} \text{ m}^2/\text{s}, \quad Pr = 0.72.$$

The Reynolds number at $x = L = 2.5$ m is $Re_L = \frac{u_\infty L}{v} = \frac{(8)(2.5)}{18.97 \times 10^{-6}} = 1.054 \times 10^6$. As this is greater than the value of the critical Reynolds number ($5 \times 10^5$), the flow at the downstream end of the plate is turbulent, and there are both laminar and turbulent regions on the plate.

The flow changes from laminar to turbulent at the critical distance $x_c$ where the Reynolds number is $5 \times 10^5$. From the definition of the Reynolds number, we have

$$Re_{x_c} = \frac{u_\infty x_c}{v} = \frac{8x_c}{18.97 \times 10^{-6}} = 5 \times 10^5, \quad \text{and} \quad x_c = 1.19 \text{ m}$$

From $x = 0$ to $x = 1.19$ m, the flow is laminar. From $x = 1.19$ m to $x = 2.5$ m, the flow is turbulent.

In the laminar region, the local convective coefficient is given by Eq. (6.54):

$$h_x = 0.332Re_x^{1/2}Pr^{1/3}(k/x) \tag{6.54}$$

When values of the various parameters are put into this equation, we have

$$h_x = 5.430x^{-0.5} \text{ (W/m}^2 \text{ C)} \tag{6.93}$$

In the turbulent region, the local convective coefficient is given by Eq. (6.86):

$$h_x = 0.0296Re_x^{0.8}Pr^{1/3}(k/x) \tag{6.86}$$

When values of the various parameters are put into this equation, we have

$$h_x = 23.58x^{-0.2} \ (\text{W} \, / \, \text{m}^2 \ \text{C}) \tag{6.94}$$

The following figure shows the variation of $h_x$ with $x$. Eq. (6.93) was used for the laminar region and Eq. (6.94) for the turbulent region.

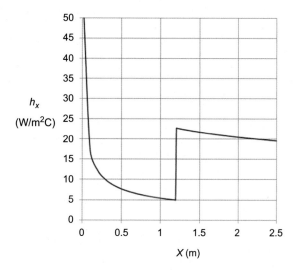

In the laminar region from $x = 0$ to $x = 1.19$ m, it is seen that $h_x$ decreases from a very large value near the leading edge ($x = 0$) to a value of about 5 at the end of the laminar region. Then, $h_x$ has a jump to about 22.5 at the transition location between laminar and turbulent. As $x$ continues to increase, $h_x$ gradually decreases to a value of about 20 at the end of the plate. It is seen that the rate of $h_x$ decrease in the laminar region is higher than that in the turbulent region. This is because, from Eqs. (6.93) and (6.94), $h_x$ is proportional to $x^{-0.5}$ in the laminar region and proportional to $x^{-0.2}$ in the turbulent region.

(Note: The figure shows an abrupt jump in $h_x$ where the flow changes from laminar to turbulent. In reality, there will be a transitional region between the laminar and turbulent regions and the change in $h_x$, although still very significant, will not be as abrupt.)

## 6.3.2 Flow over cylinders and spheres

Fig. 6.10 shows crossflow over cylinders and spheres. The free-stream velocity is $u_\infty$ and the free-stream temperature is $T_\infty$. The cylinder's and sphere's diameter is $D$ and its surface temperature is $T_s$.

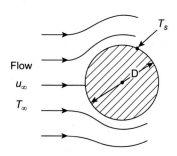

**FIGURE 6.10**

Cross flow over cylinders and spheres.

### 6.3.2.1 Cylinders

#### 6.3.2.1.1 Circular cylinders

Many experimental studies have been performed to determine the convective heat transfer for isothermal circular cylinders in crossflow. Some correlation equations from these studies are given below. For these equations, the characteristic length in the Reynolds number is the diameter $D$ of the cylinder. That is $\mathrm{Re}_D = \frac{u_\infty D}{v}$.

Zukauskas [6] performed experiments with air, water, and transformer oil and recommended the equation

$$\mathrm{Nu} = \frac{hD}{k} = C\mathrm{Re}_D^m \mathrm{Pr}^{0.37}\left(\frac{\mathrm{Pr}}{\mathrm{Pr}_s}\right)^{1/4} \tag{6.95}$$

All parameters in Eq. (6.95) are at $T_\infty$ except for $\mathrm{Pr}_s$, which is at $T_s$. The constants $C$ and $m$ are given in Table 6.2.

Churchill and Bernstein [7] recommended the equation

$$\mathrm{Nu} = \frac{hD}{k} = 0.3 + \frac{0.62\mathrm{Re}_D^{1/2}\mathrm{Pr}^{1/3}}{\left[1 + \left(\frac{0.4}{\mathrm{Pr}}\right)^{2/3}\right]^{1/4}}\left[1 + \left(\frac{\mathrm{Re}_D}{2.82 \times 10^5}\right)^{5/8}\right]^{4/5} \tag{6.96}$$

$$\text{for } \mathrm{Re}_D\mathrm{Pr} > 0.2$$

The fluid parameters in Eq. (6.96) are evaluated at the film temperature $T_f = (T_s + T_\infty)/2$. Sparrow, Abraham, and Tong [8] recommended the equation

$$\mathrm{Nu} = \frac{hD}{k} = 0.25 + \left(0.4\mathrm{Re}_D^{1/2} + 0.06\mathrm{Re}_D^{2/3}\right)\mathrm{Pr}^{0.37}\left(\frac{\mu}{\mu_s}\right)^{1/4} \tag{6.97}$$

$$\text{for } 1 \leq \mathrm{Re}_D \leq 10^5$$

The fluid parameters in Eq. (6.97) are at $T_\infty$ except for $\mu_s$, which is at $T_s$.

#### 6.3.2.1.2 Noncircular cylinders

Sparrow, Abraham, and Tong [8] reviewed the existing experimental data and correlations for air flow normal to blunt objects such as flat plates and noncircular cylinders. Their recommendations for flow normal to square and rectangular cylinders and normal to flat plates are given in Eq. (6.98)–(6.107) for the objects shown in Fig. 6.11.

**Table 6.2 Coefficients for Eq. (6.95).**

| $\mathrm{Re}_D$ | $C$ | $m$ |
|---|---|---|
| 1–40 | 0.75 | 0.4 |
| 40–1000 | 0.51 | 0.5 |
| $1000$–$2 \times 10^5$ | 0.26 | 0.6 |
| $2 \times 10^5$–$1 \times 10^6$ | 0.076 | 0.7 |

Shape                                               Correlations

Square Cylinder    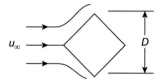                Equation (6.98)

Rotated
Square Cylinder                                        Equation (6.99)

Flat Plate     Front Surface   Equation (6.100)
                                                       Rear Surface    Equation (6.101)
                                                       Both Surfaces   Equation (6.102)

Rectangular Cylinder

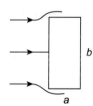                  Aspect Ratio $\dfrac{a}{b}$ = 0.2    Equation (6.103)

                                                       0.33    Equation (6.104)
                                                       0.67    Equation (6.105)
                                                       1.33    Equation (6.106)
                                                       1.5     Equation (6.107)

**FIGURE 6.11**

Air flow over noncircular cylinders and flat plates.

Correlations for objects in Fig. 6.11

Square cylinder   $\mathrm{Nu}_D = \dfrac{hD}{k} = 0.14\mathrm{Re}_D^{0.66}$   for   $5,000 \leq \mathrm{Re}_D \leq 60,000$ $\hspace{1cm}$ (6.98)

Rotated square cylinder   $\mathrm{Nu}_D = \dfrac{hD}{k} = 0.27\mathrm{Re}_D^{0.59}$   for   $6,000 \leq \mathrm{Re}_D \leq 60,000$ $\hspace{0.5cm}$ (6.99)

Plate (front surface)   $\mathrm{Nu}_D = \dfrac{hD}{k} = 0.592\mathrm{Re}_D^{1/2}$   for   $10,000 \leq \mathrm{Re}_D \leq 50,000$ $\hspace{0.5cm}$ (6.100)

Plate (rear surface)   $\mathrm{Nu}_D = \dfrac{hD}{k} = 0.17\mathrm{Re}_D^{2/3}$   for   $7,000 \leq \mathrm{Re}_D \leq 80,000$ $\hspace{0.5cm}$ (6.101)

Plate (both surfaces) $\quad Nu_D = \dfrac{hD}{k} = 0.25Re_D^{0.61}$ for $\quad 10,000 \le Re_D \le 50,000$ $\qquad$ (6.102)

Rectangular cylinder

$$a/b = 0.2 \quad Nu_b = \dfrac{hb}{k} = 0.26Re_b^{0.60} \quad \text{for} \quad 13,000 \le Re_b \le 77,000 \qquad (6.103)$$

$$a/b = 0.33 \quad Nu_b = \dfrac{hb}{k} = 0.25Re_b^{0.62} \quad \text{for} \quad 7,500 \le Re_b \le 37,500 \qquad (6.104)$$

$$a/b = 0.67 \quad Nu_b = \dfrac{hb}{k} = 0.163Re_b^{0.667} \quad \text{for} \quad 7,500 \le Re_b \le 37,500 \qquad (6.105)$$

$$a/b = 1.33 \quad Nu_b = \dfrac{h\,b}{k} = 0.127\ Re_b^{0.667} \quad \text{for} \quad 7,500 \le Re_b \le 37,500 \qquad (6.106)$$

$$a/b = 1.5 \quad Nu_b = \dfrac{hb}{k} = 0.116Re_b^{0.667} \quad \text{for} \quad 7,500 \le Re_b \le 37,500 \qquad (6.107)$$

## Example 6.8
### Flow over a square exhaust duct
*Problem*

A worker in a factory is cold. There is a sheet metal exhaust duct carrying hot air near his station, and he thinks he can get some warmth from the duct by directing a fan over its hot surface. Using an infrared thermometer, he finds that the temperature of the outer surface of the duct is 120 F. The room air is at 55 F. The duct has a square cross section 10 inches by 10 inches. Air from the fan moves at 8 ft/s.

**(a)** To get the greatest heating from the duct, should the fan be directed on a flat side of the duct or on a corner of the duct? (See Fig. 6.12 below)

**(b)** What is the expected convective heat flow from the duct surface to the room air per foot length of duct? Will this have a significant impact on the person's comfort?

**(c)** Estimate the duct's radiative heat flow to the room and compare it with the convective heat flow you found in Part (B).

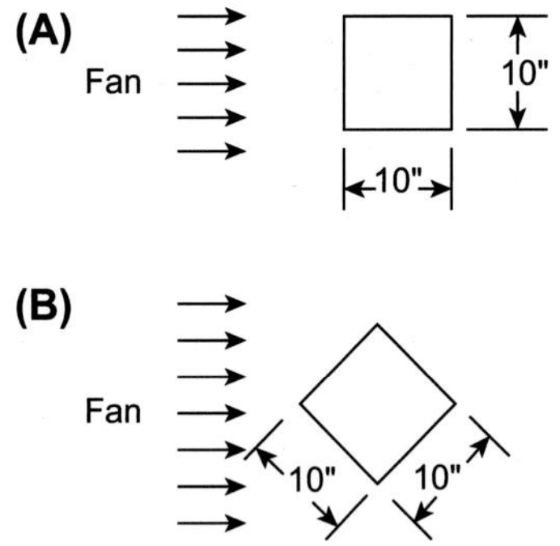

**FIGURE 6.12**

For Example 6.8: Which is preferred? A or B?

*Solution*

**(a)** The film temperature is $T_f = \frac{T_s + T_\infty}{2} = \frac{120 + 55}{2} = 87.5$ F. At this temperature, for air, we have
$k = 0.015$ Btu/h ft F and $v = 1.74 \times 10^{-4}$ ft$^2$/s.

For Orientation (A):

Looking at Fig. 6.11, D = 10" = 0.8333 ft.

$$\text{Then } Re_D = \frac{u_\infty D}{v} = \frac{(8)(0.8333)}{1.74 \times 10^{-4}} = 38300$$

For this orientation, the appropriate correlation is Eq. (6.98)

$$Nu_D = \frac{hD}{k} = 0.14 Re_D^{0.66} = 0.14(38300)^{0.66} = 148.3$$

$$h = \frac{k}{D} Nu_D = \frac{0.015}{0.8333}(148.3) = 2.67 \text{ Btu/h ft}^2\text{F}$$

For orientation (B):

Looking at Fig. 6.11, $D = \sqrt{10^2 + 10^2} = 14.14$ inch $= 1.178$ ft.

$$\text{Then } Re_D = \frac{u_\infty D}{v} = \frac{(8)(1.178)}{1.74 \times 10^{-4}} = 54160$$

For this orientation, the appropriate correlation is Eq. (6.99):

$$Nu_D = \frac{hD}{k} = 0.27 Re_D^{0.59} = 0.27(54160)^{0.59} = 167.6$$

$$h = \frac{k}{D} Nu_D = \frac{0.015}{1.178}(167.6) = 2.13 \text{ Btu/h ft}^2\text{F}$$

Conclusion: As Orientation (A) gives a larger convective coefficient, the air heating will be larger for that orientation.

**(b)** The surface area for a one-foot length of duct is $4 \times 0.8333 \times 1 = 3.333$ ft$^2$.

$$q_{conv} = hA\Delta T = (2.67)(3.333)(120 - 55) = 578 \text{ Btu/h per foot length of duct}$$

If there is, say, about 10 feet of duct near the person, then about 5800 Btu/h will be added to the air near the person. This should result in a positive impact on the person's comfort but perhaps not a *significant* impact.

**(c)** We estimate that the sheet metal duct has an emissivity of about 0.3.

For radiative heat transfer, $q_{rad} = \varepsilon \sigma A (T_s^4 - T_{surr}^4)$. In English units,

$$\sigma = 0.1714 \times 10^{-8} \text{ Btu/h ft}^2\text{R}^4$$

$T_s$ and $T_{surr}$ must be in absolute temperature units (Rankine)

$$\text{So, } q_{rad} = (0.3)(0.1714 \times 10^{-8})(3.333)\left((120 + 459.67)^4 - (55 + 459.67)^4\right)$$

$$q_{rad} = 73 \text{ Btu/h per foot length of duct.}$$

The radiative heat transfer is small compared with the convective heat transfer. (Note: Painting the duct black would increase the emissivity to about 0.9, which would triple the radiative heat transfer.)

### 6.3.2.2 Spheres

Many experimental studies have been performed to determine convective coefficients for spheres in crossflow. Although most of these studies are for heat transfer in air or water, there have been some studies for other fluids, such as oils. We mention a few forced convection correlations for spheres. There are many more in the literature.

Vliet and Leppert [9] recommended the following correlation for forced convection for spheres in water:

$$Nu = \frac{hD}{k} = \left(2.7 + 0.12 Re_D^{0.66}\right) Pr^{0.5} \left(\frac{\mu}{\mu_s}\right)^{1/4} \tag{6.108}$$

$$\text{for } 50 \leq Re_D \leq 300,000$$

They also recommended the following correlation for forced convection for spheres in fluids of a wide range of Prandtl numbers:

$$\text{Nu} = \frac{hD}{k} = \left(1.2 + 0.53\text{Re}_D^{0.54}\right)\text{Pr}^{0.3}\left(\frac{\mu}{\mu_s}\right)^{1/4} \tag{6.109}$$

$$\text{for } 1 \leq \text{Re}_D \leq 300,000, \quad 2 \leq \text{Pr} \leq 380$$

In Eqs. (6.108) and (6.109), all fluid properties are at $T_\infty$ except for $\mu_s$, which is at $T_s$.

Will, Kruyt, and Venner [10] studied forced convection for spheres in air and recommended the correlation

$$\text{Nu} = \frac{hD}{k} = A\text{Re}_D^{1/2} + B\text{Re}_D \tag{6.110}$$

where

$$A = 0.493 \pm 0.015$$
$$B = 0.0011 \times (1 \pm 0.035)$$
$$7.8 \times 10^3 \leq \text{Re}_D \leq 2.9 \times 10^5$$

In Eq. (6.110), the fluid properties are at the film temperature $T_f = \frac{T_s + T_\infty}{2}$.

The most often used correlation for forced convection for spheres is that by Whitaker [11]:

$$\text{Nu} = \frac{hD}{k} = 2 + \left(0.4\text{Re}_D^{1/2} + 0.06\text{Re}_D^{2/3}\right)\text{Pr}^{0.4}\left(\frac{\mu}{\mu_s}\right)^{1/4} \tag{6.111}$$

$$\text{for } 3.5 \leq \text{Re}_D \leq 7.6 \times 10^4, \quad 0.71 \leq \text{Pr} \leq 380, \quad \text{and } 1 \leq (\mu/\mu_s) \leq 3.2$$

In Eq. (6.111), all fluid properties are at $T_\infty$ except for $\mu_s$, which is at $T_s$.

### 6.3.3 Flow through tube banks

Heat exchangers are used to transfer heat between two fluids. The exchangers often have many parallel tubes arranged in a bundle or "tube bank" One fluid flows through the tubes and the other fluid flows through the tube bank, i.e., across the outer surfaces of the tubes. Convective coefficients for flow over a single cylinder (or tube) were discussed in Section 6.3.2.1. Flow over a tube in a tube bank is more complicated due to the impact on the flow by adjacent tubes in the tube bank. Such altering of the flow has a significant effect on the convective coefficient for the tube in question. This section discusses the convective coefficient for the outer surface of a tube in a tube bank. Section 6.4 covers the convective coefficient for the inner surface of the tube.

Two common arrangements of tubes in a tube bank are "inline" and "staggered" as shown in Fig. 6.13. The tube spacing in the flow direction is $S_L$ and the spacing in the transverse direction is $S_T$. A diagonal spacing $S_D$ is also shown for the staggered arrangement. In terms of $S_L$ and $S_T$,

$$S_D = \sqrt{S_L^2 + \left(\frac{S_T}{2}\right)^2} \tag{6.112}$$

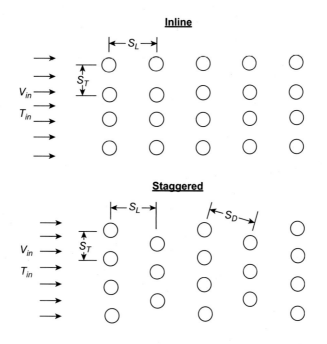

**FIGURE 6.13**

Tube arrangements for a tube bank.

Many researchers have determined correlations for the Nusselt number for flow through a tube bank. These correlations contain the Reynolds number, which is based on the maximum fluid velocity in the tube bank, and the outside diameter of the tubes. That is,

$$\text{Re}_D = \frac{\rho V_{\text{max}} D}{\mu} = \frac{V_{\text{max}} D}{\upsilon} \tag{6.113}$$

where $D$ is the outside diameter of the tube.

Let us look at the maximum velocity for the two arrangements: For both arrangements, fluid enters the tube bank from the left at uniform velocity $V_{in}$ and inlet temperature $T_{in}$. As the fluid passes the first row of tubes, it speeds up due to the narrowing of the flow area. From continuity, the velocity as it goes through the opening between two tubes in the first row is $\frac{S_T}{S_T - D} V_{in}$. For the inline arrangement, this is the maximum velocity in the tube bank. It is also the maximum velocity for the staggered arrangement unless the spacing of the rows is small. Looking at the geometry and mass flow conservation, it can be shown that if $S_D < \frac{S_T + D}{2}$, then the maximum velocity is through the diagonal passage of width $S_D - D$. To summarize:

$$\text{For the inline arrangement: } V_{\text{max}} = \frac{S_T}{S_T - D} V_{in} \tag{6.114}$$

$$\text{For the staggered arrangement: } V_{\text{max}} = \frac{S_T}{S_T - D} V_{in} \text{ if } S_D > \frac{S_T + D}{2} \tag{6.115}$$

**Table 6.3 Coefficients for Eq. (6.117).**

| | $Re_D$ | C | m | n |
|---|---|---|---|---|
| For inline | 1–100 | 0.9 | 0.4 | 0.36 |
| | 100–1000 | 0.52 | 0.5 | 0.36 |
| | 1000–$2 \times 10^5$ | 0.27 | 0.63 | 0.36 |
| | $2 \times 10^5$–$2 \times 10^6$ | 0.033 | 0.8 | 0.4 |
| For staggered | 1–500 | 1.04 | 0.4 | 0.36 |
| | 500–1000 | 0.71 | 0.5 | 0.36 |
| | 1000–$2 \times 10^5$ | $0.35(S_T/S_L)^{0.2}$ | 0.6 | 0.36 |
| | $2 \times 10^5$–$2 \times 10^6$ | $0.031(S_T/S_L)^{0.2}$ | 0.8 | 0.36 |

$$V_{max} = \frac{S_T}{2(S_D - D)}V_{in} \ \text{ if } \ S_D < \frac{S_T + D}{2} \tag{6.116}$$

Several researchers have performed experiments on tube banks and developed correlations for the Nusselt number. The recommendation of Zukauskas [12] is

$$Nu = \frac{hD}{k} = CRe_D^m Pr^n \left(\frac{Pr}{Pr_s}\right)^{1/4} F \tag{6.117}$$

Fluid parameters in Eq. (6.117) are at $T_\infty$ except for $Pr_s$, which is at $T_s$. The constants $C$, $m$, and $n$ are given in Table 6.3. Factor $F$ depends on the number of rows in the flow direction. If $Re_D > 1000$ and there are 16 or more rows, then $F$ is 1. For less than 16 rows, $F$ is less than 1. For 13, 10, 7, 5, and 4 rows, it was found that $F$ is, respectively, 0.99, 0.98, 0.96, 0.93, and 0.9. This data can be interpolated for other numbers of rows. These data fit very well to the fourth degree polynomial

$$F = -1.4894 \times 10^{-5}R^4 + 7.2245 \times 10^{-4}R^3 - 1.3009 \times 10^{-2}R^2 + 1.0674 \times 10^{-1}R + 0.63929$$

where $R$ is the number of rows in the flow direction.

## Example 6.9
### Flow through a tube bank
*Problem*
An inline tube bank consists of 30 tubes. There are five rows of tubes in the flow direction and six rows in the transverse direction. The tube spacing is shown in Fig. 6.14. The tubes are 1/2-inch Type M copper tubes, 0.5 m long. Air enters the tube bank at a velocity of 3 m/s and a temperature of 20 C. The outer surfaces of the tubes are at 110 C. What is the rate of heat transfer by forced convection to the air?

*Solution*
From an Internet search, the tubes have an outside diameter of 0.625 inch = 0.01588 m.
Looking at Fig. 6.14, we have

$$S_L = D + 5 \text{ mm} = 0.02088 \text{ m} \ \text{ and } \ S_T = D + 1 \text{ cm} = 0.02588 \text{ m}$$

From Eq. (6.114), $V_{max} = \frac{S_T}{S_T - D}V_{in} = \frac{0.02588}{0.02588 - 0.01588}(3) = 7.764 \text{ m/s}$.

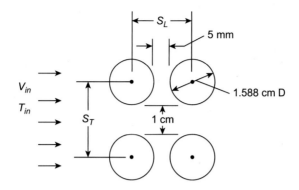

**FIGURE 6.14**

Tube spacing for example 6.9.

The Reynolds number is based on the maximum velocity and the outside diameter of the tubes. Fluid properties are to be taken at the average temperature of the incoming and outgoing flows from the tube bank. The air enters at 20 C. Let us assume that the air exits at 40 C (to be checked later). The properties at this average temperature 30 C are $k = 0.0259$ W/m C; $\rho = 1.164$ kg/m$^3$; $v = 1.608 \times 10^{-5}$ m$^2$/s; Pr $= 0.728$; and $c_p = 1007$ J/kg C. We also need Pr at the tube surface temperature of 110 C, that is 0.709.

$$\text{The Reynolds number is } Re_D = \frac{\rho V_{max} D}{\mu} = \frac{V_{max} D}{v} = \frac{(7.764)(0.01588)}{1.608 \times 10^{-5}} = 7670$$

From Eq. (6.117) and Table 6.3, the appropriate equation for this Reynolds number is

$$\text{Nu} = \frac{hD}{k} = 0.27 Re_D^{0.63} Pr^{0.36} \left(\frac{Pr}{Pr_s}\right)^{1/4} F$$

Putting values into this equation and recognizing that $F = 0.93$ as we only have five rows in the direction of flow, we get

$$\text{Nu} = \frac{h(0.01588)}{0.0259} = 0.27(7670)^{0.63}(0.728)^{0.36}\left(\frac{0.728}{0.709}\right)^{1/4}(0.93)$$

Solving this for the convective coefficient, we get $h = 103.0$ W/m$^2$ C.

The heat transfer to the air is $q = hN_{tubes}A_{tube}(T_s - T_\infty)$, where $N_{tubes}$ is the number of tubes, $A_{tube}$ is the surface area per tube and $T_\infty$ is the average fluid temperature. We assumed $T_\infty$ was 30 C.

So, we have $q = (103.0)(30)[\pi(0.01588)(0.5)](110 - 30) = 6166$ W.

The rate of heat transfer by forced convection to the air is 6166 W.

We assumed that the exit temperature of the air from the tube bank was 40 C. This gave us an average air temperature of 30 C for determination of the fluid properties. Now that we have the heat transfer amount, we can check whether our assumption of 40 C exit temperature was appropriate. The heat gain of the air is $q = (\rho_{in} V_{in} A_{in}) c_p (T_{out} - T_{in})$. The first parenthesis is the mass flow rate of the incoming air. Then we have the specific heat, and finally the change of temperature of the air as it goes through the tube bank. Putting values into this equation and using the calculated 6166 W for $q$, we get

$$q = 6166 = (1.172)(3)[(6)(0.02588)(0.5)](1007)(T_{out} - 20)$$

Solving for $T_{out}$, we get $T_{out} = 42.4$ C. This gives an average air temperature of $(20 + 42.4)/2 = 31.2$ C. Our assumption was very close, 30 C, and there is no need for iteration on the exit temperature and fluid properties to get a better value for $q$. Our calculated value is fine.

## 6.4 Internal flow

The previous section dealt with flow *over* objects. For example, we looked at flow over a cylinder and discussed correlations for the convective coefficient at the outer surface of the cylinder. This section deals with flow *through* cylinders, i.e., pipes and tubes, and ducts. We will discuss correlations for the convective coefficient at the inner surface of the cylinders and ducts.

Consider Fig. 6.15, which shows a fluid flowing through a circular tube.

The fluid enters the tube with a uniform velocity $V_{in}$. As the fluid goes through the tube, it is acted on by shear stresses at the wall which impact the velocity profile $u(x, r)$. The fluid sticks to the wall (zero velocity) and a boundary layer forms. If the flow is laminar and the tube is long enough, the velocity profile ultimately becomes parabolic. At this point, the flow is said to be "fully developed hydrodynamically" and the velocity profile stays the same as the fluid progresses further down the tube. If the flow is turbulent, rather than laminar, the ultimate velocity profile is not parabolic. Rather, it is flatter near the center of the tube, as shown in Fig. 6.16. The distance from the inlet to the fully developed region is $L_h$, called the "hydrodynamic entrance length" For $x > L_h$, $\frac{\partial u}{\partial x} = 0$.

We have used the terms "laminar" and "turbulent" In laminar flow, fluid particles glide through the tube. There is little cross mixing of the particles. They tend to stay in the same relative layer as they travel down the tube. In turbulent flow, the flow is rougher, with cross mixing of fluid particles. There are random variations of velocity and other fluid properties such as pressure and temperature. Whether

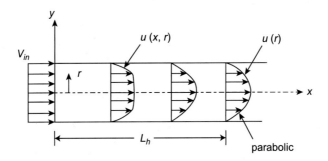

**FIGURE 6.15**

Laminar flow through a tube.

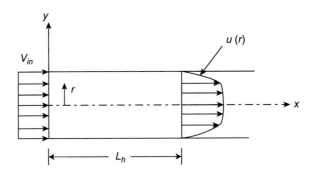

**FIGURE 6.16**

Fully developed velocity profile for turbulent flow.

a flow is laminar or turbulent depends a lot on the fluid velocity and viscosity. Indeed, the Reynolds number is used to determine whether the flow is laminar or turbulent. The Reynolds number for internal flow is

$$\text{Re}_D = \frac{\rho VD}{\mu} = \frac{VD}{v} \tag{6.118}$$

where $D$ is the inside diameter of the tube.

$V$ is the flow velocity

$\rho, \mu,$ and $v$ are the fluid density, absolute viscosity, and kinematic viscosity

From experiments, flow is usually laminar if $\text{Re}_D < 2300$ and turbulent if $\text{Re}_D > 4000$. (These are rough, not firm, values.) Between these values, the flow is "transitional."

Looking at the definition of the Reynolds number, it is seen that laminar flow generally occurs for fluids of high viscosity flowing at low velocity. Most flows encountered in engineering practice are of the turbulent category.

If we consider the thermal aspects of the flow, there is a thermal entrance length $L_t$ analogous to the hydrodynamic entrance length $L_h$ discussed above. For thermally fully developed flow, i.e., for $x > L_t$, $\frac{\partial}{\partial x} \left( \frac{T_s - T}{T_s - T_m} \right) = 0$. $T_s$ is the surface temperature, $T$ is the fluid temperature (which can vary with $y$), and $T_m$ is the mean or bulk fluid temperature, all at location $x$.

## 6.4.1 Entrance lengths

Experimentation has provided the following estimates of entrance lengths $L_h$ and $L_t$:

$$\text{For laminar flows, } L_h \approx 0.05\text{Re}_D D \tag{6.119}$$

$$L_t \approx 0.05\text{Re}_D D\text{Pr} \tag{6.120}$$

For turbulent flows, the entrance lengths $L_h$ and $L_t$ are about the same and they are essentially independent of Prandtl number Pr. It is often assumed that turbulent flows are fully developed, both hydrodynamically and thermally, about 10 diameters downstream of the entrance.

## 6.4.2 Mean velocity and mean temperature

The mass flow rate of the fluid is $\dot{m} = \rho A_c V_{avg}$, where $A_c$ is the cross-sectional area of the flow and $V_{avg}$ is the average flow velocity at a given location $x$. The fluid velocity varies over the cross section, and the average (or mean) velocity is obtained by integrating the velocity over the cross section:

$$V_{avg} = \frac{1}{A_c} \int_{A_c} u\, dA_c \tag{6.121}$$

The energy flowing in the fluid at a given location $x$ is $\dot{E} = \dot{m}c_p T_m = \int_{A_c} \rho u c_p T\, dA_c$. For constant density and specific heat at location $x$, the average or mean temperature at location $x$ is

$$T_m = \frac{1}{V_{avg} A_c} \int_{A_c} u T\, dA_c \tag{6.122}$$

$T_m$ is also called the "bulk" fluid temperature.

Later in this chapter, we will be providing correlation equations for the convective coefficient for internal flows. These equations often depend on the boundary conditions at the wall surface. The two common boundary conditions are constant heat flux and constant temperature, which we consider now.

### 6.4.3 Constant heat flux

Fig. 6.17 shows a differential control volume in a fluid flowing through a channel. There is a constant heat flux on the surface of the channel. Such a constant heat flux could have been applied by heating tapes wrapped around the tube or perhaps a radiant heat source beaming on the outer surface of the tube. We will do an energy balance on the control volume.

The heat flux $q'_s$ is constant, and the rate of heat flow into the control volume is $q'_s dA_s$, where $dA_s$ is the differential area at the surface of the tube. The heat flow into the volume raises the energy of the fluid in the volume a net amount $\dot{m}c_p dT_m$. The energy balance is

Net rate of heat in = net rate of energy increase of fluid.

$$dq_s = q'_s dA_s = \dot{m}c_p(T_m + dT_m) - \dot{m}c_p T_m \tag{6.123}$$

The differential surface area $dA_s$ is the perimeter $P$ times $dx$. Eq. (6.123) then becomes:

$$\frac{dT_m}{dx} = \frac{q'_s P}{\dot{m}c_p} = \text{constant} \tag{6.124}$$

It is seen that the mean fluid temperature varies linearly with $x$. Solving Eq. (6.124) for $T_m(x)$, between locations $x_1$ and $x_2$, we get

$$T_m(x_2) = T_m(x_1) + \frac{q'_s P}{\dot{m}c_p}(x_2 - x_1) \tag{6.125}$$

In Eqs. (6.124) and (6.125), the specific heat $c_p$ should be evaluated at the average of the mean temperatures at $x_1$ and $x_2$.

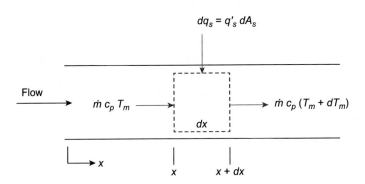

**FIGURE 6.17**

Constant heat flux.

## 6.4.4 Constant surface temperature

Fig. 6.18 shows a differential control volume in a fluid flowing through a channel. The channel has a constant surface temperature. Perhaps such a constant surface temperature occurred from a fluid condensing or vaporizing on the outer surface of the tube. We will do an energy balance on the control volume.

The surface temperature is $T_s$. The center of the volume is at location $x + dx/2$. At this location, the mean fluid temperature is $T_m + dT_m/2$. The rate of convective heat flow into the control volume is $dq_s = hdA_s[T_s - (T_m + dT_m/2)]$, where $h$ is the average convective coefficient for the tube section being considered. Area $dA_s$ is the differential area at the surface of the tube. The heat flow into the volume raises the energy of the fluid in the volume a net amount $\dot{m}c_p dT_m$. The energy balance is

$$\text{Net Rate of Heat in} = \text{Net Rate of Energy Increase of Fluid}$$

$$dq_s = hdA_s[T_s - (T_m + dT_m/2)] = \dot{m}c_p(T_m + dT_m) - \dot{m}c_p T_m \tag{6.126}$$

The differential surface area $dA_s$ is the perimeter $P$ times $dx$. And, we can eliminate the second-order term $dx\, dT_m$. Eq. (6.126) then becomes

$$dq_s = hPdx(T_s - T_m) = \dot{m}c_p dT_m \tag{6.127}$$

As $T_s$ is constant, $dT_m = -d(T_s - T_m)$. Eq. (6.127) can then be rearranged to

$$\frac{d(T_s - T_m)}{T_s - T_m} = -\frac{hP}{\dot{m}c_p}dx \tag{6.128}$$

Integrating Eq. (6.128) between locations $x_1$ and $x_2$, we get

$$[\ln(T_s - T_m)]_{x_1}^{x_2} = -\frac{hP}{\dot{m}c_p}(x_2 - x_1) \tag{6.129}$$

$$\ln\left[\frac{(T_s - T_m)_{x_2}}{(T_s - T_m)_{x_1}}\right] = -\frac{hP}{\dot{m}c_p}(x_2 - x_1) \tag{6.130}$$

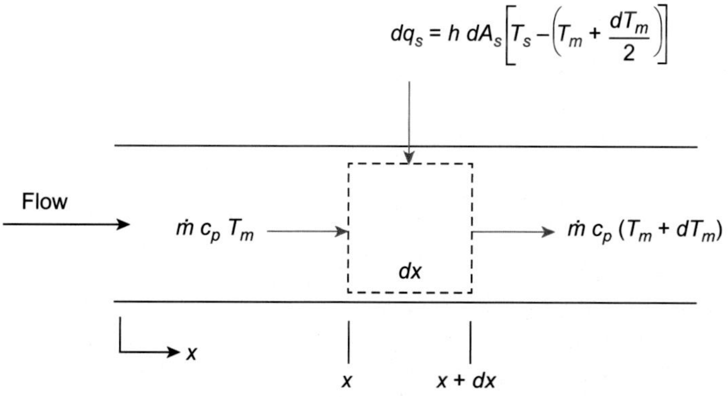

$$dq_s = h\, dA_s\left[T_s - \left(T_m + \frac{dT_m}{2}\right)\right]$$

**FIGURE 6.18**

Constant surface temperature.

And, finally

$$(T_s - T_m)_{x_2} = (T_s - T_m)_{x_1} e^{-\frac{hP}{\dot{m}c_p}(x_2 - x_1)} \tag{6.131}$$

Fluid properties in $h$, and the specific heat $c_p$, should be evaluated at the average of the mean temperatures at $x_1$ and $x_2$.

## 6.4.5 Equivalent diameter for flow through noncircular tubes

Many parameters, including the Reynolds number, Nusselt number, and friction factor, contain the inside diameter $D$ of the tube. If the tube has a noncircular area, then an equivalent diameter, the **hydraulic diameter,** should be used for the inside diameter. This diameter is defined by

$$D_h = \frac{4A_c}{P} \tag{6.132}$$

where $A_c$ is the flow area and $P$ is the wetted perimeter, i.e., the perimeter of the surface seeing fluid. Some examples are the following:

If the flow area is circular with diameter $D$, then the hydraulic diameter is the same as the actual diameter $D$ of the tube. That is, $D_h = \frac{4(\pi D^2/4)}{\pi D} = D$.

If the flow area is square, with a side $a$, then $D_h = \frac{4a^2}{4a} = a$.

If the flow area is rectangular, with sides of $a$ and $b$, then $D_h = \frac{4ab}{2a+2b} = \frac{2ab}{a+b}$.

If the flow area is an equilateral triangle of side $a$, then $D_h = \frac{4\left(\frac{\sqrt{3}}{4}a^2\right)}{3a} = \frac{\sqrt{3}}{3}a$.

If the flow area is an open channel of rectangular cross section, with two vertical sides $a$ and bottom $b$, then $D_h = \frac{4ab}{2a+b}$.

## 6.4.6 Correlations for the Nusselt number and convective coefficient

Internal flow has been extensively investigated and there are many correlations for the Nusselt number and convective coefficient. We will present some of them here. The correlations are grouped according to type of flow—laminar or turbulent.

### 6.4.6.1 Laminar flow; entrance region

- **Hausen [13]** (for constant surface temperature)

$$Nu_D = \frac{hD}{k} = 3.66 + \frac{0.0668(D/L)Re_D Pr}{1 + 0.04[(D/L)Re_D Pr]^{2/3}} \tag{6.133}$$

$$\text{for } Re_D Pr(D/L) < 100$$

Fluid properties are at the mean bulk temperature.

- **Sieder and Tate [14]** (for constant surface temperature)

$$Nu_D = \frac{hD}{k} = 1.86(Re_D Pr)^{1/3}(D/L)^{1/3}\left(\frac{\mu}{\mu_s}\right)^{0.14} \tag{6.134}$$

$$\text{for } 0.5 < Pr < 16,700 \text{ and } Re_D Pr(D/L) > 10$$

All properties are at mean bulk temperature except $\mu_s$, which is at the surface temperature.

### 6.4.6.2 Laminar flow; fully developed

Nusselt numbers for circular and noncircular cylinders are given in Table 6.4. There are entries for two boundary conditions: constant heat flux and constant surface temperature.

### 6.4.6.3 Turbulent flow; fully developed

In turbulent flow, the flow is usually fully developed within 10 or 20 diameters downstream of the entrance. As the entrance region is so short, we will only include correlation equations for the fully developed portion of the tube and will use these equations also for the entrance region as needed. In addition, the type of boundary condition (e.g., constant flux or constant temperature) has less impact on the convective coefficients for turbulent flow than for laminar flow. Therefore, the following equations may be used for either boundary condition.

- **Dittus and Boelter** [15]

$$\text{Nu}_D = \frac{hD}{k} = 0.023\text{Re}_D^{0.8}\text{Pr}^n \tag{6.135}$$

where $n = 0.4$ for heating of the fluid and 0.3 for cooling of the fluid and $0.6 < \text{Pr} < 100$.

Properties are at the mean bulk temperature.

If there is substantial temperature difference between the fluid and the surface of the tube, then the following equation may give better results:

- **Sieder and Tate** [14]

$$\text{Nu}_D = \frac{hD}{k} = 0.027\text{Re}_D^{0.8}\text{Pr}^{1/3}\left(\frac{\mu}{\mu_s}\right)^{0.14} \tag{6.136}$$

for $0.7 < \text{Pr} < 16,700$

**Table 6.4 Nusselt numbers for fully developed laminar flow.**

| Shape | | $\text{Nu} = \frac{hD_h}{k}$ | |
|---|---|---|---|
| | | Constant heat flux | Constant surface temperature |
| Circular cylinder | | 4.36 | 3.66 |
| Rectangular channel | a/b | | |
| | 1 | 3.61 | 2.98 |
| | 2 | 4.12 | 3.39 |
| | 4 | 5.33 | 4.44 |
| | 8 | 6.49 | 5.60 |
| | ∞ | 8.24 | 7.54 |
| Triangular channel | | 3.11 | 2.47 |

All properties are at mean bulk temperature except $\mu_s$, which is at the surface temperature.

Gnielinski performed an extensive review of correlation equations and recommended the following equation which includes friction factor $f$:

- **Gnielinski [16].**

$$Nu_D = \frac{hD}{k} = \frac{(f/8)(Re_D - 1000)Pr}{1 + 12.7(f/8)^{1/2}(Pr^{2/3} - 1)}\left[1 + \left(\frac{D}{L}\right)^{2/3}\right]\left(\frac{Pr}{Pr_s}\right)^{0.11} \tag{6.137}$$

$$\text{for }\ 0.6 < Pr < 10^5$$

$$2300 < Re < 10^6$$

All properties are at mean bulk temperature except $Pr_s$, which is at the surface temperature.

The friction factor $f$ for turbulent flow is a function of the Reynolds number and the relative roughness of the pipe surface. It may be obtained from the **Moody Diagram [17]** or the following **Colebrook Equation [18]**.

$$\frac{1}{\sqrt{f}} = -2\log_{10}\left(\frac{\varepsilon/D}{3.7} + \frac{2.51}{Re_D\sqrt{f}}\right) \tag{6.138}$$

The values for the pipe roughness $\varepsilon$ can vary greatly with manufacturing process and manufacturer. Typical values are

| | |
|---|---|
| Drawn tubing, PVC | 0.0015 mm |
| Commercial steel | 0.045 mm |
| Galvanized iron | 0.15 mm |
| Cast iron | 0.25 mm |

It is seen that Eq. (6.138) is not explicit for the friction factor $f$. It is on both sides of the equation. The equation can be solved for $f$ by manual trial-and-error or by using equation solving software. For example, it can be solved using *Goal Seek* or *Solver* of Excel or using the *fzero* function of Matlab.

---

## Example 6.10
### Flow through a circular tube

*Problem*

We want to cool engine oil from 100 to 80 C. To do this, we plan to pass the oil through a tube whose surface is at a uniform temperature of 40 C. The oil flows through the tube at a velocity of 0.03 m/s, and the tube has an inside diameter of 2 cm. How long does the tube have to be to achieve the desired cooling of the oil?

*Solution*

The mean bulk temperature of the oil is $(100 + 80)/2 = 90$ C. Properties of engine oil at 90 C are

$$\rho = 846 \text{ kg/m}^3; \quad c_p = 2176 \text{ J/kg C}; \quad k = 0.138 \text{ W/m C}$$
$$v = 2.81 \times 10^{-5} \text{ m}^2/\text{s}; \quad Pr = 360$$

The Reynolds number is $Re_D = \dfrac{VD}{v} = \dfrac{(0.03)(0.02)}{2.81 \times 10^{-5}} = 21.35$   Laminar since $< 2300$

Let us check if the flow is fully developed: From Eqs. (6.119) and (6.120),

$$L_h \approx 0.05 Re_D D = 0.05(21.35)(0.02) = 0.021 \text{ m}$$

$$L_t \approx 0.05 Re_D D \text{ Pr} = 0.021(360) = 7.6 \text{ m}$$

The flow is fully hydrodynamically developed, but probably thermally developing. Hence, we cannot say that the flow is fully developed, and we should use either Eq. (6.133) or (6.134) for the Nusselt number. We will use Eq. (6.133):

$$Nu_D = \frac{hD}{k} = 3.66 + \frac{0.0668(D/L)Re_D Pr}{1 + 0.04[(D/L)Re_D Pr]^{2/3}} \tag{6.139}$$

There are two unknowns in this equation, $h$ and $L$. We need a second equation with $h$ and $L$. As the temperature of the tube surface is uniform, we can use the energy equation for constant surface temperature, Eq. (6.130):

$$\ln\left[\frac{(T_s - T_m)_{x_2}}{(T_s - T_m)_{x_1}}\right] = -\frac{hP}{\dot{m}c_p}(x_2 - x_1) \tag{6.140}$$

Note: $L = x_2 - x_1$ and $P = \pi D$.

$$\dot{m} = \rho A V = \rho \left(\frac{\pi}{4}D^2\right)V = (846)(\pi/4)(0.02)^2(0.03) = 0.00797 \text{ kg/s}$$

Rearranging Eq. (6.139) for $h$, we have

$$h = \frac{k}{D}\left[3.66 + \frac{0.0668(D/L)Re_D Pr}{1 + 0.04[(D/L)Re_D Pr]^{2/3}}\right] \tag{6.141}$$

Rearranging Eq. (6.140) for $h$, we have

$$h = -\frac{\dot{m}c_p}{\pi DL}\ln\left[\frac{(T_s - T_m)_{x=L}}{(T_s - T_m)_{x=0}}\right] \tag{6.142}$$

Equating the right-hand sides of Eqs. (6.141) and (6.142), we have

$$\frac{k}{D}\left[3.66 + \frac{0.0668(D/L)Re_D Pr}{1 + 0.04[(D/L)Re_D Pr]^{2/3}}\right] = -\frac{\dot{m}c_p}{\pi DL}\ln\left[\frac{(T_s - T_m)_{x=L}}{(T_s - T_m)_{x=0}}\right] \tag{6.143}$$

The only unknown in Eq. (6.143) is the desired tube length $L$. So, if we solve Eq. (6.143) for $L$, we solve the problem. One way to do this is to use Excel. Moving the right side of Eq. (6.143) to the left of the equal sign, we have

$$\frac{k}{D}\left[3.66 + \frac{0.0668(D/L)Re_D Pr}{1 + 0.04[(D/L)Re_D Pr]^{2/3}}\right] + \frac{\dot{m}c_p}{\pi DL}\ln\left[\frac{(T_s - T_m)_{x=L}}{(T_s - T_m)_{x=0}}\right] = 0 \tag{6.144}$$

Putting values into Eq. (6.144), we have

$$\frac{0.138}{0.02}\left[3.66 + \frac{0.0668(0.02/L)(21.35)(360)}{1 + 0.04[(0.02/L)(21.35)(360)]^{2/3}}\right] + \frac{(0.00797)(2176)}{\pi(0.02)L}\ln\left[\frac{(40 - 80)}{(40 - 100)}\right] = 0$$

Reducing this, we have

$$\left[3.66 + \frac{(10.2685/L)}{1 + 0.04(153.72/L)^{2/3}} - \frac{16.220}{L} = 0\right] \tag{6.145}$$

Using Excel's *Goal Seek*, we got $L = 2.67$ m.

The tube has to be 2.67 m long for the desired cooling of the oil.

---

## Example 6.11

### Flow through a noncircular tube

*Problem*

Water at 20 C having a flow rate of 5 kg/s enters a tube of rectangular cross section. The tube cross section is 3 cm by 6 cm, and it is 15 m long. The inner surface of the tube is maintained at a constant temperature of 75 C. What is the temperature of the water leaving the tube?

*Solution*

As the tube is noncircular, we need the hydraulic diameter. For a rectangular tube with sides of $a$ and $b$, the hydraulic diameter is

$$D_h = \frac{2ab}{a+b} = \frac{2(0.03)(0.06)}{0.03 + 0.06} = 0.04 \text{ m}$$

For water at 20 C,  $\rho = 998 \text{ kg/m}^3$

$$c_p = 4182 \text{ J/kg C}$$

$$k = 0.600 \text{ W/m C}$$

$$\mu = 1.002 \times 10^{-3} \text{ kg/ms}$$

$$\text{Pr} = 7.01$$

We need the flow velocity to get the Reynolds number. The mass flow rate is $\dot{m} = \rho V A$, where $A$ is the flow area. The flow area is $A = (0.03)(0.06) = 0.0018 \text{ m}^2$, and $V = \frac{\dot{m}}{\rho A}$. Therefore,

$$V = \frac{\dot{m}}{\rho A} = \frac{5}{(998)(0.0018)} = 2.78 \text{ m/s}$$

The Reynolds number is

$$\text{Re} = \frac{\rho V D_h}{\mu} = \frac{(998)(2.78)(0.04)}{1.002 \times 10^{-3}} = 1.108 \times 10^5$$

The flow is turbulent. Using Eq. (6.135) with $n = 0.4$ as the fluid is being heated, we have

$$\text{Nu}_D = \frac{hD}{k} = 0.023\text{Re}_D^{0.8}\text{Pr}^n = 0.023(1.108 \times 10^5)^{0.8}(7.01)^{0.4} = 544.1$$

The convective coefficient is

$$h = \frac{\text{Nu}_D k}{D} = \frac{(544.1)(0.600)}{0.04} = 8162 \text{ W/m}^2 \text{ C}$$

To get the exit temperature, we can use Eq. (6.131):

$$(T_s - T_m)_{x_2} = (T_s - T_m)_{x_1} e^{-\frac{hP}{\dot{m}c_p}(x_2 - x_1)}$$

Putting in the values, we have

$$(75 - T_m)_{x_2} = (75 - 20)_{x_1} e^{-\frac{(8162)[(2)(0.03+0.06)]}{(5)(4182)}(15)} = 19.2 \text{ C}$$

The exit temperature is $T_m$ at $x_2$, which is $75 - 19.2 = 55.8$ C.

The fluid properties should be taken at the fluid mean bulk temperature. We took properties at 20 C, but with this result, we have a mean bulk temperature of $(20 + 55.8)/2 = 37.9$ C. We should do the problem again with water properties at a higher temperature.

We did the problem again with properties at 40 C. We got a convective coefficient of 9970 W/m² C and an exit temperature of 59.8 C. With this exit temperature, the mean bulk temperature is $(20 + 59.8)/2 = 39.9$ C. Properties were taken at 40 C so no further iteration is needed. **In short, the Dittus-Boelter equation predicts that the exit temperature of the water is about 60 C.**

Let us look at this problem a little more. We used the Dittus-Boelter equation, which is the simplest of the three equations given for turbulent flow. However, the temperature difference between the tube surface and the fluid is considerable. The tube surface is at 75 C and the mean bulk water temperature is 40 C. And, some properties of water vary significantly with temperature, e.g., the Prandtl number and the viscosity. It might be interesting to see the result from the Sieder-Tate equation, which has a viscosity term to adjust for the difference between the surface and fluid temperatures.

We did the problem using the Sieder-Tate equation, Eq. (6.136), and got $h = 11,460$ W/m² C, which is 15% higher than the 9970 W/m² C value from Dittus-Boelter when we used properties at 40 C. With the Seider-Tate equation, the water exit temperature was 62.5 C.

We also did the problem using the Gnielinski equation, Eq. (6.137). This equation has the friction factor $f$. We assumed that the tube had a fairly smooth surface equivalent to drawn tubing and solved for the friction factor using the Colebrook

Equation, Eq. (6.138). Using the Excel's *Goal Seek* we found that the friction factor was 0.0164. Using this in the Gnielinski equation, we got a convective coefficient of 13,200 W/m² C and a water exit temperature of 65.0 C.

In short, the Dittus Boelter equation gave a water exit temperature of about 60 C while the Sieder-Tate and Gnielinski equations gave the exit temperature as about 64 C. Because of the temperature-dependent nature of the properties of water, we will make an engineering judgment that the Sieder-Tate and Gnielinski equations are more accurate than the Dittus-Boelter equation for this problem. **We conclude that the exit temperature of the water is about 64 C.**

### 6.4.7 Annular flow

We have discussed the equivalent diameter to use in correlation equations for flows through noncircular tubes. Flow through an annulus is a bit different as there may be two surfaces with heat transfer instead of one. Fig. 6.19 shows an annular flow area created by two concentric tubes. The area has an inside diameter $D_1$, which is the outside diameter of the inner tube, and an outside diameter $D_2$, which is the inside diameter of the outer tube.

In a double-pipe heat exchanger, one fluid flows through the inner tube and the other flows through the annular area. The fluids have different temperatures, and heat is transferred from one fluid to the other through the tube wall of the inner tube. The hydraulic diameter for the annular flow area is the difference in the diameters, i.e., $D_2 - D_1$.

$$D_h = \frac{4A}{P} = \frac{4\left[\frac{\pi}{4}\left(D_2^2 - D_1^2\right)\right]}{\pi D_1 + \pi D_2} = D_2 - D_1 \tag{6.146}$$

In most cases, one surface transfers heat and the other is perfectly insulated, i.e., adiabatic.

#### 6.4.7.1 Fully developed laminar flow

Table 6.5 [19] shows Nusselt numbers for fully developed laminar flow through an annulus when one surface is constant temperature and the other is adiabatic. $Nu_i$ is the Nusselt number on the inner surface and $Nu_o$ is that on the outer surface.

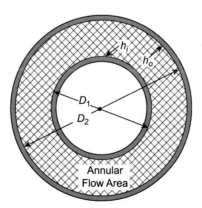

**FIGURE 6.19**

Annular flow area.

**Table 6.5 Nusselt numbers for fully developed laminar flow in an annulus.**

| $D_1/D_2$ | $Nu_i$ | $Nu_o$ |
|-----------|--------|--------|
| 0.00 | – | 3.66 |
| 0.05 | 17.46 | 4.06 |
| 0.10 | 11.56 | 4.11 |
| 0.25 | 7.37 | 4.23 |
| 0.50 | 5.74 | 4.43 |
| 1.00 | 4.86 | 4.86 |

The convective coefficients $h_i$ and $h_o$ at the inner and outer surfaces of the annulus are

$$h_i = \frac{Nu_i k}{D_h} \quad \text{and} \quad h_o = \frac{Nu_o k}{D_h} \tag{6.147}$$

### 6.4.7.2 Fully developed turbulent flow

For fully developed turbulent flow in an annulus, the convective coefficients at the two surfaces of the annulus are of the same magnitude. The above equations for turbulent flow in circular tubes may be applied with the hydraulic diameter of Eq. (6.146) used as the tube's diameter.

We conclude this chapter with an example that includes both internal and external convection.

## Example 6.12
### Heating of a warehouse

*Problem*

It is winter and workers in an unheated warehouse are cold and angry. They are freezing, and the people in the adjacent office are warm and comfy. There is a duct going through the warehouse which carries hot air to the office for heating. The warehouse workers have asked the management to provide diffusers in the duct to supply some hot air to the warehouse, but their request has been denied. The workers have decided to take matters into their own hands. They have decided to use fans to blow the warehouse air over the hot air duct and get some heating from the forced convection. The details are as follows:

The duct is 10 inches diameter and the section in the warehouse is 50 feet long. The air enters the warehouse section at a rate of 1100 cfm and a temperature of 130 F. The air in the warehouse space is at 45 F. Fans will blow the air in crossflow over the duct at a velocity of 8 m/s.

**(a)** What is the temperature of the air in the duct as it leaves the warehouse?

**(b)** How much heating will be added to the warehouse air (Btu/hr)?

**(c)** What are some practical problems with this proposed heating of the warehouse?

*Solution*

We will assume that the duct wall is thin and a constant temperature surface. To get air property values, we need to make some assumptions on temperatures. We will check these assumptions after the calculations and iterate as necessary. We will assume that the air temperature drops by 20 F from the entrance to the exit. The air enters at 130 F (54.44 C) and leaves at 110 F (43.33 C). The average temperature of the air in the duct is then (54.44 + 43.33)/2 = 48.89 C. We will also assume that the temperature of the duct wall is $T_s = 85$ F = 29.44 C.

We now proceed to get the convective coefficients $h_i$ and $h_o$ for the inside and outside surfaces of the duct.

*Internal flow.* The air enters at 1100 cfm and 130 F (54.44 C). For air at 54.44 C, $\rho = 1.075$ kg/m$^3$.
Mass flow rate

$$\dot{m} = \rho \dot{Q} = 1.075 \text{ kg/m}^3 \times 1100 \text{ ft}^3/\text{min} \times (1 \text{ min}/60 \text{ s}) \times (1 \text{ m}/3.2808 \text{ ft})^3 = 0.5581 \text{ kg/s}$$

$$D = 10 \text{ inch} \times (1 \text{ m}/39.37 \text{ inch}) = 0.254 \text{ m}$$

The average duct air temperature is 48.89 C. For air at this temperature,

$$\rho = 1.096 \text{ kg/m}^3, \quad c_p = 1007 \text{ J/kg C}, \quad k = 0.0272 \text{ W/m C}, \quad v = 17.87 \times 10^{-6} \text{ m}^2/\text{s}, \quad \text{Pr} = 0.722$$

$$\dot{m} = \rho A V = 1.096\left[(\pi/4)(0.254)^2\right] \quad V = 0.5581 \text{ and } V = 10.05 \text{ m/s}$$

$$\text{Re}_D = \frac{VD}{v} = \frac{(10.05)(0.254)}{17.87 \times 10^{-6}} = 1.428 \times 10^5 \text{(turbulent)}$$

Eq. (6.135):

$$h_i = 0.023\text{Re}_D^{0.8}\text{Pr}^{0.3}(k/D) = (0.023)(1.428 \times 10^5)^{0.8}(0.722)^{0.3}(0.0272/0.254) = 29.70 \text{ W}/\text{m}^2\text{C}$$

*External flow.*

$$D = 0.254 \text{ m} \quad \text{and} \quad = 8 \text{ m/s}$$

Eq. (6.95) with Table 6.2

$$h_o = C\text{Re}_D^m\text{Pr}^{0.37}\left(\frac{\text{Pr}}{\text{Pr}_s}\right)^{1/4}(k/D)$$

$\text{Pr}_s$ is at the duct wall temperature of 29.44 C. Air at 29.44 C, $\text{Pr}_s = 0.728$.
All other air properties are at the warehouse air temperature of 7.22 C. The properties are

$$\rho = 1.26 \text{ kg/m}^3, \quad c_p = 1006 \text{ J/kg C}, \quad k = 0.0242 \text{ W/m C}, \quad v = 14 \times 10^{-6} \text{ m}^2/\text{s}, \quad \text{Pr} = 0.735$$

$$\text{Re}_D = \frac{VD}{v} = \frac{(8)(0.254)}{14 \times 10^{-6}} = 1.451 \times 10^5$$

For this Reynolds number, from Table 6.2, the values for Eq. (6.95) are $C = 0.26$, $m = 0.6$.
Eq. (6.95):

$$h_o = C\text{Re}_D^m\text{Pr}^{0.37}\left(\frac{\text{Pr}}{\text{Pr}_s}\right)^{1/4}(k/D)$$

$$= (0.26)(1.451 \times 10^5)^{0.6}(0.735)^{0.37}(0.735/0.728)^{1/4}(0.0242/0.254) = 27.70 \text{ W/m}^2\text{ C}$$

Back to the internal flow: We use Eq. (6.131) to get the exit temperature of the duct air.

$$(T_s - T_m)_{x_2} = (T_s - T_m)_{x_1} e^{-\frac{hP}{\dot{m}c_p}(x_2 - x_1)}$$

Putting in values, we have $(29.44 - T_m)_{x_2} = (29.44 - 54.44)e^{-\frac{(29.70)[\pi(0.254)]}{(0.5581)(1007)}(50/3.2808)}$
The exit temperature of the air is $(T_m)_{x_2} = 42.59 \text{ C} = 108.7 \text{ F}$.
The heating provided to the warehouse is

$$q = \dot{m}c_p(\Delta T)_{\text{duct air}} = (0.5581)(1007)(54.44 - 42.59) = 6660 \text{ W} = 22720 \text{ Btu/h}$$

Let us now check the assumed duct wall temperature by doing an energy balance:

$$h_i A_i(T_{\text{duct air}} - T_s) = h_o A_o(T_s - T_{\text{room air}})$$

As the tube is thin, $A_i = A_o$. Putting values into the equation, we have

$$29.70\left(\frac{54.44 + 42.59}{2} - T_s\right) = 27.70(T_s - 7.22)$$

Solving for $T_s$, we get that the duct wall temperature is 28.6 C = 83.5 F.
*Summarizing.* We assumed the duct wall temperature was 85 F and we calculated it as 83.5 F.
We assumed the exit temperature of the duct air was 110 F and we calculated it as 108.7 F.
Our assumptions were close to the calculated values. We conclude that iteration is not necessary as it would have negligible impact on the results.

*Results*

**(a)** Duct air exits at 109 F.

**(b)** Heating added to the warehouse is about 22,700 Btu/hr.

**(c)** Practical problems include the following:

    The large number of fans needed. (This is not a problem if the facility is a fan factory.)

    Mounting of the fans is a major installation task.

    The noise due to the fans will be considerable.

    There will be considerable increased electrical load for the fans.

The obtained warehouse heating is not very large considering the major installation requirements and costs. We calculated that the use of the fans added 22,700 Btu/hr to the warehouse air. Actually, the impact of the fans is even less than this. Without the fans, there would be natural convection from the hot duct. Example 7.4 estimates this natural convection as 5930 Btu/hr. Hence, the fans will actually only add $22,700 - 5930 = 16,770$ Btu/hr to the air. With such a low value, it is probably better to continue lobbying the management for diffusers in the hot air duct. Or, perhaps install gas-fired infrared heaters near the ceiling or unit heaters near the work stations in the warehouse.

## 6.5 Chapter summary and final remarks

In this chapter, we first discussed the dimensionless parameters relevant to forced and natural convection: the Reynolds, Prandtl, and Nusselt numbers for forced convection and the Grashof, Prandtl, and Nusselt numbers for natural convection. We then proceeded to discuss forced convection for both external and internal flows. For external flows, we looked at the laminar and turbulent boundary layers on a flat plate and outlined the analytical solution of the flow and thermal equations for the laminar boundary layer. Besides the flat plate, we covered flows over cylinders and spheres. We also discussed flow through the tube banks of heat exchangers. For internal flows, we looked at flow through tubes and ducts. Correlation equations were presented for the different types of flows and different geometries. These equations, based on experimental studies, provide estimates for the convective coefficient $h$. Many of the equations have ranges of applicability for the Reynolds and Prandtl numbers. If a problem has parameters outside of the applicable ranges, then one should use the equation that most closely fits the problem.

Of course, the best possible way to determine the $h$-value for an object is to perform an experiment on the actual object. This is usually not possible from an economic and/or practical basis, and correlation equations have to be used. One should keep in mind that the correlations are from carefully controlled lab experiments having conditions often quite different from those encountered in practice. Therefore, the $h$-value from a correlation equation may differ significantly from the actual $h$ of an object, perhaps as much as 25% or more. (An aside to students who like to present answers with an extreme number of digits: Considering the uncertainty in the $h$-value, is this practice really appropriate?)

We continue now to Chapter 7, which discusses natural convection.

## 6.6 Problems

Notes:

- If needed information is not given in the problem statement, then use the Appendix for material properties and other information. If the Appendix does not have sufficient information, then use the Internet or other reference sources. Some good reference sources are mentioned at the end of Chapter 1. If you use the Internet, double-check the validity of the information by using more than one source.

- Your solutions should include a sketch of the problem.
- Caution: Make sure that you use absolute temperatures (Kelvin or Rankine) if the problem involves radiation.
- In all problems, unless otherwise stated, the fluid pressure is atmospheric. Gas properties in the Appendix are at atmospheric pressure. If a problem has a gas at other than atmospheric pressure, the density and kinematic viscosity should be modified accordingly through use of the ideal gas law.

**6-1** Fluid flows over a flat plate at 5 m/s. If the fluid is at 20 C, how far from the leading edge does transition from laminar flow to turbulent flow take place if
  **(a)** the fluid is air?
  **(b)** the fluid is water?
  **(c)** the fluid is engine oil?

**6-2** Air at 40 C flows over a flat plate at 10 m/s. What is the boundary layer thickness 0.25 m from the leading edge?

**6-3** Do Problem 6-2 if the air has a pressure of 2 atm.

**6-4** Air at 20 C flows at 5 m/s over a square flat plate that is 5 cm by 5 cm.
  **(a)** What is the drag force on the plate?
  **(b)** Determine the velocity components $u$ and $v$ at a location 3 cm from the leading edge. Determine these components at two vertical distances from the plate: 1/3 and 2/3 of the distance from the plate surface to the edge of the boundary layer.

**6-5** Water at 20 C flows over a rectangular heated flat plate that is at 80 C. The water velocity is 1.5 m/s.
  **(a)** What is the thickness of the hydrodynamic boundary layer 4 cm downstream of the leading edge?
  **(b)** What is the thickness of the thermal boundary layer 4 cm downstream of the leading edge?
  **(c)** If the plate is 5 cm wide and 20 cm long in the direction of flow, what is the rate of heat transfer from the plate to the water?

**6-6** Engine oil at 40 C flows at a speed of 2 m/s over a flat plate that is 20 cm wide and 10 cm in the direction of flow. Five electric strip heaters, each 2 cm wide and 20 cm long, are placed side-by-side to create the flat plate. One side of the heaters is perfectly insulated and the other side heats the oil. What is the electric power (W) needed for each heater to produce a plate that has a uniform temperature of 200 C?

**6-7** Air at 50 C and 2 atm pressure flows at a velocity of 4 m/s over a flat rectangular plate that is maintained at 100 C. The plate is 3 m long in the direction of flow.
  **(a)** What is the average convective coefficient for the plate?
  **(b)** What is the rate of convective heat transfer from the plate to the air per meter width of the plate?
  **(c)** If the plate's surface has an emissivity of 0.6, what is the rate of radiative heat transfer to the 50 C surroundings per meter width of the plate?
  **(d)** Give a statement regarding the relative values of the Part (b) and (c) results.

**6-8** Air at 20 C flows at a velocity of 0.5 m/s over a rectangular plate that is 15 cm wide and 25 cm in the direction of flow. The first 10 cm of the plate from the leading edge is perfectly insulated and the last 15 cm of the plate is maintained at a temperature of 150 C. What is the rate of heat transfer from the plate to the air?

**6-9** Air at 30 C flows over a horizontal flat plate at a speed of 2 m/s. The plate is square, 10 cm on a side. A thin electronic chip of size 1 cm by 1 cm is mounted on the plate about 3 cm downstream of the leading edge. The backside of the chip is perfectly insulated and the front surface convects to the air. If the chip produces 35 mW of power, what is the steady-state temperature of the chip?

**6-10** Nitrogen at 20 C flows at a speed of 3 m/s over a flat plate that is 10 cm wide and 40 cm long in the direction of flow. The plate's temperature is maintained at 80 C.

**(a)** What is the drag force of the nitrogen on the plate?

**(b)** What is the rate of heat transfer from the plate to the nitrogen?

**(c)** Do Parts (a) and (b) assuming that there is a trip wire at the leading edge, which makes the flow turbulent over the entire plate.

**6-11** Water at 20 C flows across a rectangular plate that is 10 cm wide and 20 cm in the direction of flow. The speed of the water is 1 m/s. The plate has an electric heater that produces a constant heat flux over the entire surface of the plate. If the plate temperature cannot exceed 70 C at any location, what is the maximum allowed power input to the heater?

**6-12** Air at 25 C flows at a velocity of 40 m/s over a square plate that is 35 cm by 35 cm. The plate is at 80 C.

**(a)** What is the drag force of the air on the plate?

**(b)** What is the convective heat transfer from the plate to the air?

**(c)** For the result of Part (b), how much heat transfer is from the laminar region and how much from the turbulent region?

**6-13** Water at 15 C flows at a velocity of 2 m/s across an unheated flat plate that is 5 cm wide and 10 cm in the direction of flow. A thin electric strip heater that is 4 mm wide and 5 cm long is placed across the plate 5 cm from the leading edge. The strip is heated to a temperature of 40 C. What is the power to the heater? Assume that all the power goes into the water.

**6-14** Air at 20 C flows at a velocity of 25 m/s over a thin horizontal circular plate of 0.5 m diameter. The plate is at 100 C. What is the rate of heat transfer from the plate to the air?

**6-15** An electrical transmission line has an outside diameter of 2 cm. The current in the line is 250 A and the line has a resistance of $3 \times 10^{-4}$ Ω per meter length. The line is outdoors and the wind is blowing over it at 20 miles per hour. If the air is at 15 C, what is the surface temperature of the line?

**6-16** Steam at 1 atm and 100 C is flowing across a 2 cm diameter cylinder at a speed of 5 m/s. The surface of the cylinder is at 200 C. What is the rate of heat transfer per meter length of the cylinder to the steam?

**6-17** Air at 20 C flows at a speed of 5 m/s across a 4 mm diameter wire. The surface of the wire is at 200 C. What is the heat transfer to the air per meter length of wire?

**6-18** Air at 15 C flows across an electrical cylindrical heater at a velocity of 3 m/s. The heater has a diameter of 0.5 cm and a length of 4 cm. The emissivity of the heater's surface is 0.9, and the surroundings are at 15 C. If the temperature of the heater's surface cannot exceed 400 C, what is the maximum allowed power input to the heater?

**6-19** An oil tank is 4 feet in diameter and 10 feet long. The tank is filled with hot oil. After the filling, the outer surface of the tank is at 140 C. To cool the oil, air at 20 C is blown over the tank's cylindrical surface in the axial direction at a speed of 3 m/s. What is the rate of heat transfer to the air at the beginning of the oil cooling? Only include the heat transfer from the tank's cylindrical surface. Do not include the tank ends.

**6-20** Air at 20 C flows across a square cylinder. The sides of the cylinder are 3 cm wide and are at 180 C. The free-stream velocity of the air is 4 m/s.

**(a)** What is the rate of heat transfer per meter length to the air if the air stream hits a flat surface of the cylinder? See Fig. P6.20A.

**(b)** What is the rate of heat transfer per meter length to the air if the air stream hits a rotated square cylinder? See Fig. P6.20B.

**(A)**

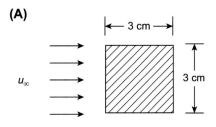

**(B)**

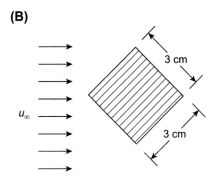

**FIGURE P6.20**

(A) Square cylinder. (B) Rotated square cylinder.

**6-21** Air at 20 C flows over the rectangular cylinder shown in Fig. P 6.21. The cylinder has a length of 3 m and its surface is at 250 C. Iaf the speed of the air is 5 m/s, what is the rate of heat transfer to the air?

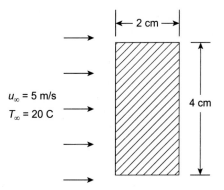

**FIGURE P6.21**

**6-22** Air at 20 C flows over a thin flat plate that is positioned normal to the air stream. The plate is square, 5 cm on a side, and its temperature is 275 C. If the air velocity is 10 m/s, what is the rate of heat transfer to the air?

**6-23** Air at 20 C flows at a velocity of 8 m/s over a 2 cm diameter aluminum sphere. The sphere's surface is at 60 C. What is the rate of heat transfer to the air?

**6-24** Water at 300 K flows at 4 m/s over a 4 cm diameter sphere whose surface temperature is 400 K. What is the rate of heat flow to the water?

**6-25** An electronic device is cooled by blowing 30 C air over it. The device can be modeled as a sphere of 0.5 cm diameter. The power dissipated by the device is 5 W. What is the needed velocity of the air if the temperature of the outer surface of the device cannot exceed 200 C? Assume that conductive and radiative heat transfers are negligible.

**6-26** It is desired to measure the temperature of hot air flowing through a duct. The air is flowing at 1.7 m/s. A thermistor is used as the thermometer. It is a 2 mm diameter sphere and has a surface emissivity of 0.85. The thermistor is inserted in the air stream, and it indicates an air temperature of 43 C when the duct walls are at 37 C. What is the actual temperature of the air? Assume that conduction through the leads of the thermistor is negligible.

**6-27** Nitrogen at 100 C enters an inline tube bank with a free-stream velocity of 5 m/s. The tube bank has 48 tubes: eight rows in the direction of flow and six tubes per row. The tubes have an outside diameter of 2 cm and their centers are 3 cm apart. The outer surfaces of the tubes are at 200 C. What is the rate of heat transfer from the tube bank to the nitrogen per meter length of the tube bank?

**6-28** Air flows through a tube bank. It enters with a temperature of 20 C and a free-stream velocity of 4 m/s. The tube bank is of staggered arrangement. The tubes are 1.25 cm OD and have a length of 1.5 m. The bank has $S_L = 4$ cm and $S_T = 3$ cm. There are 95 tubes with 10 rows in the flow direction. The outer surfaces of the tubes are at 120 C.
**(a)** What is the rate of heat transfer to the air?
**(b)** At what temperature does the air leave the tube bank?

**6-29** Water at 20 C enters a tube bank with a free-stream velocity of 1.5 m/s. Each tube has an OD of 1 cm and a length of 1 m. The tube bank is of the inline arrangement and the tube centers are 2.5 cm apart in both the flow direction and the direction perpendicular to the flow. The surfaces of the tubes are at 150 C. It is desired to have the water leave the tube bank at 50 C. If there are five tubes in each row normal to the flow direction, how many rows of tubes are needed to achieve the desired water heating?

**6-30** Fluid is flowing through a tube of pentagonal cross section. The sides of the flow area are of length "a." What is the hydraulic diameter $D_h$?

**6-31** Fluid is flowing through a tube of hexagonal cross section. The sides of the flow area are of length "a." What is the hydraulic diameter $D_h$?

**6-32** Fluid is flowing through a semicircular open channel of diameter "d." The water has a maximum depth of d/4. What is the hydraulic diameter $D_h$?

**6-33** Water flows through a 2.5 cm diameter tube at a flow rate of 0.005 kg/s. The tube is 5 m long. The water enters at 20 C and the surface of the tube is maintained at a uniform temperature of 100 C.
**(a)** What is the average convective coefficient?
**(b)** At what temperature does the water leave the tube?

**6-34** Water at a mean bulk temperature of 40 C flows at a velocity of 5 m/s through a heated 3/4 inch Type M copper tube. Determine the entrance length needed for the flow to become fully developed (a) hydrodynamically and (b) thermally.

**6-35** Engine oil at a mean bulk temperature of 40 C flows at a mass flow rate of 1.4 kg/s through a heated 3/4 inch Type M copper tube. Determine the entrance length needed for the flow to become fully developed (a) hydrodynamically and (b) thermally.

**6-36** Engine oil at 40 C enters a 1 cm diameter tube, which is maintained at a uniform temperature of 130 C. The tube is 50 m long and the oil leaves the tube at a temperature of 60 C. What is the mass flow rate of the oil?

**6-37** Air at a pressure of 1000 kPa enters a rectangular duct that has a cross section of 12 cm by 7 cm. The duct is 5 m long. The mass flow rate of the air is 0.35 kg/s. The average temperature of the air in the duct is 250 C, and the duct wall is maintained at an average temperature of 200 C. What is the decrease in air temperature as it flows through the duct?

**6-38** Air flows through a tube which has a square cross section of 1 cm by 1 cm. The tube is 20 cm long. The air enters the tube at 40 C and has a flow rate of $1.2 \times 10^{-4}$ kg/s. The surface of the tube is maintained at 150 C. What is the exit temperature of the air?

**6-39** Engine oil flows through a 2 cm diameter tube at a velocity of 0.5 m/s. The oil enters the tube at 60 C. The tube is 4 m long and its surface is maintained at 90 C.
   **(a)** What is the rate of heat transfer to the oil?
   **(b)** What is the exit temperature of the oil?

**6-40** Hot water flows through a 1/2 inch nominal Sch 40 steel pipe at a rate of 3 gpm. The pipe is 10 feet long. The water enters the pipe at 120 F and the inside surface of the pipe is at a uniform temperature of 43 F.
   **(a)** What is the temperature of the water as it leaves the pipe?
   **(b)** If the flow rate of the water doubles to 6 gpm, what is the exit temperature of the water?

**6-41** Water flows through a 3/4 inch nominal Sch 80 steel pipe at a rate of 4 gpm. The pipe is 5 feet long. The water enters at 20 C and leaves at 25 C. If the pipe surface has a uniform temperature, what is that temperature to achieve the stated heating of the water? Use Eq. (6.135) for the convective coefficient.

**6-42** Do Problem 6-41, but use Eq. (6.137) for the convective coefficient. Also use Eq. (6.138) for the friction factor.

**6-43** Air flows at a velocity of 6 m/s through a 20-m long annulus. The annulus has an inside diameter of 30 cm and an outside diameter of 45 cm. The air enters the annulus at 40 C and is heated by the inner surface of the annulus which is maintained at 150 C by condensing steam. The outer surface of the annulus is adiabatic. What is the convective coefficient for the inner surface of the annulus?

**6-44** Ethylene glycol flows at a velocity of 8 m/s through an annulus formed by concentric circular tubes. The inner diameter of the annulus is 3.5 cm and the outer diameter is 5 cm. The glycol enters at 20 C and leaves at 45 C. The inner surface of the annulus, which heats the glycol, is maintained at 90 C. The outer surface is adiabatic. How long must the annulus be to achieve the desired heating of the glycol?

**6-45** Air at 25 C flows across a flat plate that has a surface temperature of 125 C. The air velocity is 5 m/s. The plate is square, 0.5 m by 0.5 m. The drag force due to the air flow is measured and found to be 0.005 N. Using the Reynolds-Colburn analogy, estimate the convective coefficient $h$ for the plate's surface.

**6-46** Air at 20 C and a speed of 6 m/s flows across a flat plate whose surface temperature is 100 C. The plate is 1 m long in the direction of air flow and 0.5 m wide. Thermal measurements found that the average convective coefficient $h$ for the plate's surface is 8.2 $W/m^2$ C. Using the Reynolds-Colburn analogy, estimate the drag force on the plate due to the air flow.

**6-47** In Section 6.3.3, data were given for factor $F$ versus number of rows $R$ in the tube bank. The data were given for 4, 5, 7, 10, 13, and 16 rows. A fourth degree polynomial was found to fit the data very well. Using the data in the section, determine a third degree polynomial fit for the data. Compare the values of $F$ obtained from this third degree polynomial for rows 3 through 16 with those obtained from the fourth degree polynomial.

**6-48** Example 6.7 showed how the local convective coefficient varied for flow across a flat plate. We wish to investigate how the local drag coefficient $C_{fx}$ varies with downstream distance $x$ from the leading edge. Nitrogen at 50 C flows across a flat plate at a velocity of 12 m/s. The plate is at a uniform temperature of 100 C and is 5 m long in the direction of nitrogen flow. Plot $C_{fx}$ versus $x$ from the leading edge ($x = 0$) to the downstream end of the plate ($x = L = 5$ m).

**6-49** A duct is 8 m long and has a rectangular cross section of 10 cm by 20 cm. Air flows through the duct, entering at 15 C and leaving at 80 C. The duct has a uniform surface temperature of 110 C.

**(a)** What is the mass flow rate of the air (kg/s)?

**(b)** What is the rate of heat transfer from the duct to the air?

**6-50** Engine oil at 25 C enters a 2.5 m long, 3 mm diameter tube at a velocity of 2.1 m/s. The temperature of the tube wall is uniform at 50 C. What is the exit temperature of the oil?

**6-51** Ethylene glycol flows at a velocity of 7.5 m/s through a 1.3 cm diameter tube that is 20 m long. The glycol enters at 70 C and leaves at 60 C. What is the needed uniform temperature of the tube wall to effect this cooling?

**6-52** Engine oil flows at a rate of 0.05 kg/s through a section of a 2.5 cm diameter tube. The oil enters the section at a temperature of 50 C. The tube wall is kept at 20 C, and it can be assumed that the flow through the section is fully developed. The section of the tube is 5 m long.

**(a)** What is the exit temperature of the oil?

**(b)** What is the heat transfer rate from the oil to the tube wall?

**6-53** A thin-walled pipe is 15 m long and has a diameter of 2.5 cm. Water is flowing through the pipe at a rate of 0.75 kg/s. The pipe wall imparts a uniform heat flux of $7 \times 10^4$ $W/m^2$ to the water. The water enters the pipe at 20 C. Assume that the flow is fully developed.

**(a)** What is the temperature difference between the local wall temperature and the local mean water temperature.

**(b)** What is the exit temperature of the water from the pipe?

**6-54** Water flows through a 1/2″ Sch 40 steel pipe at a rate of 2 gpm. The water enters at 70 F and leaves at 90 F. A cylindrical electrical heater jacket around the pipe keeps the inner surface of the pipe at 120 F.

**(a)** How long does the pipe have to be to supply the specified water heating?

**(b)** If the heater jacket has insulation at its outer surface so that 90% of its power output goes into the water and only 10% is lost to the room air, what is the needed power input to the heater? Assume the heater is 100% efficient.

**6-55** Air flows through a duct that has a square cross section of 8 cm by 8 cm. The air enters a section of duct that is far enough downstream from the inlet so that the flow is fully developed. At the entrance to the section, the air is at 30 C and its velocity is 0.25 m/s. The section has a uniform wall heat flux to the air of 50 W/m$^2$.

**(a)** If the air exits the section at 100 C, how long is the section?

**(b)** What is the temperature of the duct surface at the section's exit?

**6-56** Water is flowing at a rate of 0.04 kg/s through a thin-walled copper tube of 2 cm diameter and 5 m length. The water enters the tube at 90 C. Air at 20 C blows across the tube at a velocity of 7 m/s.

**(a)** What is the temperature of the water as it exits the tube?

**(b)** What is the heat transfer rate from the water to the air?

# References

[1] H. Blasius, Grenzschichten in Flussigkeiten mit kleiner Reibung, Z. Angew. Math. Phys. 56 (1908) 1−37. English translation in NACA TM 1256.

[2] E. Pohlhausen, Der Warmeaustausch zwischen festen Korpern und Flussigkeiten mit kleiner Reibung und kleiner Warmeleitung, Z. Angew. Math. Mech. 1 (1921) 115−121.

[3] S.W. Churchill, H. Ozoe, Correlations for laminar forced convection in flow over an isothermal flat plate and in developing and fully developed flow in an isothermal tube, ASME J. Heat Transf. 95 (1973) 416−419.

[4] S.W. Churchill, A comprehensive correlation equation for forced convection from flat plates, AIChE J. 22 (2) (1976) 264−268.

[5] W.M. Kays, M.E. Crawford, B. Weigand, Convective Heat and Mass Transfer, fourth ed., McGraw-Hill, 2005.

[6] A. Zukauskas, Heat transfer from tubes in crossflow, Adv. Heat Tran. vol. 8 (1972) 93−160.

[7] S.W. Churchill, M. Bernstein, A correlating equation for forced convection from gases and liquids to a circular cylinder in crossflow, J. Heat Transf. 99 (1977) 300−306.

[8] E.M. Sparrow, J.P. Abraham, J.C.K. Tong, Archival correlations for average heat transfer coefficients for non-circular and circular cylinders and for spheres in cross- flow, Int. J. Heat Mass Transf. 47 (2004) 5285−5296.

[9] G.C. Vliet, G. Leppert, Forced convection heat transfer from an isothermal sphere to water, ASME J. Heat Transf. 83 (1961) 163−175.

[10] J.B. Will, N.P. Kruyt, C.H. Venner, An experimental study of forced convective heat transfer from smooth, solid spheres, Int. J. Heat Mass Transf. 109 (2017) 1059−1067.

[11] S. Whitaker, Forced convection heat transfer correlations for flow in pipes, past flat plates, single cylinders, single spheres, and for flow in packed beds and tube bundles, AIChE J. 18 (2) (1972) 361−371.

[12] A. Zukauskas, Heat transfer from tubes in crossflow, in: S. Kakac, Aung (Eds.), Handbook of Single Phase Convective Heat Transfer, Wiley Interscience, 1987.

[13] H. Hausen, Darstellung des Warmeuberganges in Rohren Durch Vergallgemeinerte Potenzbeziehungen, Z. Ver. Deut. Ing. 4 (1943) 91−98.

[14] E.N. Sieder, G.E. Tate, Heat transfer and pressure drop of liquids in tubes, Ind. Eng. Chem. 28 (1943) 1429−1435.

[15] F.W. Dittus, L.M.K. Boelter, Heat transfer in automobile radiators of the tubular type, Univ. Calif. Berkeley Publ. Eng. 2 (1930) 443−461.

[16] V. Gnielinski, New equations for heat and mass transfer in turbulent pipe and channel flow, Int. Chem. Eng. 16 (1976) 359−368.

[17] L.F. Moody, Friction factors for pipe flow, Trans. ASME 66 (1944) 671−684.

[18] C.F. Colebrook, Turbulent flow in pipes, with particular reference to the transition between the smooth and rough pipe laws, J. Inst. Civ. Eng. London 11 (1939) 133−156.

[19] R.E. Lundberg, W.C. Reynolds, W.M. Kays, Heat Transfer with Laminar Flow in Concentric Annuli with Constant and Variable Wall Temperature and Heat Flux, 1963. NASA TN D-1972.

# Natural (free) convection

## Chapter outline

## 7.1 Introduction

This chapter continues our discussion of convection, the mode of heat transfer between a surface and a fluid caused by their differences in temperature. Chapter 6 discussed forced convection, a category of convection in which the fluid has significant velocity, which is usually produced by fans, blowers, or pumps. This chapter discusses natural or free convection in which the fluid has much gentler motion caused by density gradients in the fluid. Natural convection is an important topic. It has many applications, such as radiators for heating a room, refrigeration coils, multiglazed windows, transmission lines, electric transformers, immersion heaters, and cooling of electronic devices. As flow velocities are small, in many cases natural convection may be insignificant. However, in other situations, it may be the only significant mechanism of heat transfer.

Our main objective, both in Chapter 6 and this chapter, is to determine the convective coefficient, $h$, for a variety of heat transfer situations and geometries. Much experimentation has been performed in the field of convective heat transfer. The results of these investigations have been correlated by researchers using dimensionless numbers. We will first discuss the dimensionless numbers associated with natural convection. Following that, we will present correlations for a variety of geometries: vertical plates, horizontal plates, inclined plates, cylinders, and spheres. We also will discuss natural convection in enclosed spaces such as rectangular spaces relevant to double-glazed windows and solar collectors, annular spaces between concentric cylinders, and spaces between concentric spheres.

## 7.2 Basic considerations

As discussed in Chapter 6, there are three major dimensionless numbers associated with forced convection—the Nusselt, Reynolds, and Prandtl numbers. The convective coefficient $h$ is found in the Nusselt number, Nu.

$$\mathrm{Nu} = \frac{hL}{k} \tag{7.1}$$

and the flow velocity is in the Reynolds number, Re.

$$\mathrm{Re} = \frac{\rho VL}{\mu} = \frac{VL}{v} \tag{7.2}$$

Another relevant dimensionless number is the Prandtl number, Pr, which contains fluid properties.

$$\mathrm{Pr} = \frac{c_p \mu}{k} \tag{7.3}$$

In these numbers, $h$ = convective coefficient.
$L$ = a characteristic length for the problem.
$k$ = thermal conductivity of the fluid.
$V$ = fluid velocity.
$\rho$ = density of the fluid.
$\mu$ = absolute viscosity of the fluid.
$v$ = kinematic viscosity of the fluid.
$c_p$ = specific heat at constant pressure of the fluid.
For forced convection, correlation of experimental data gives the Nusselt number as a function of the Reynolds and Prandtl numbers. That is,

$$\mathrm{Nu} = f(\mathrm{Re}, \mathrm{Pr}) \tag{7.4}$$

After the Nusselt number has been determined, the convective coefficient can be obtained from it.

$$h = \frac{k}{L} \text{Nu} \tag{7.5}$$

For natural convection, fluid flow is caused by density gradients in the fluid. The flow velocity is significantly less than for forced convection. Therefore, the Reynolds number is no longer relevant. It is replaced with the Grashof number, $\text{Gr}_L$, which contains the acceleration of gravity, $g$, and the volumetric coefficient of thermal expansion, $\beta$.

$$\text{Gr} = \frac{g\,\beta\,(T_s - T_\infty)\,L^3}{v^2} \tag{7.6}$$

$T_s$ and $T_\infty$ are the temperatures of the surface and fluid respectively. The volumetric coefficient of expansion is defined as $\beta = \frac{1}{V}\left(\frac{\partial V}{\partial T}\right)_p$. For ideal gases, $\beta = \frac{1}{T}$, where $T$ is the absolute temperature of the gas. For nonideal gases and liquids, $\beta$ is obtained from property tables.

An Important Note: The Grashof number is positive. It contains the term $(T_s - T_\infty)$ This probably should have been written $|T_s - T_\infty|$. However, carrying around the absolute value function is unwieldy. So, just remember that Gr is positive and when the fluid temperature is higher than the surface temperature, make the term $(T_\infty - T_s)$ rather than $(T_s - T_\infty)$

For natural convection, experimental data are correlated using the Grashof, Prandtl, and Nusselt numbers:

$$\text{Nu} = \frac{hL}{k} = \text{f}(\text{Gr}, \text{Pr}) \tag{7.7}$$

Once the Nusselt number for a problem is found, the convective coefficient may be obtained from Eq. (7.5).

Another dimensionless number found in natural convection correlations is the Rayleigh Number, Ra. This is the product of the Grashof and Prandtl numbers.

$$\text{Ra} = \text{Gr Pr} = \frac{g\beta(T_s - T_\infty)L^3}{v^2}\text{Pr} = \frac{g\beta(T_s - T_\infty)L^3}{v\alpha} \tag{7.8}$$

where $\alpha = $ thermal diffusivity of the fluid $= \frac{k}{\rho c_p}$.

Like forced convection, hydrodynamic and thermal boundary layers also form on surfaces associated with natural convection. Figs. 7.1 and 7.2 show the hydrodynamic boundary layer for a vertical surface. Fig. 7.1 is for a surface that is hotter than the fluid, and Fig. 7.2 is for a surface that is colder than the fluid. In both cases, the boundary layer starts at the leading edge $x = 0$. The boundary layer is first laminar, and then, if the plate is long enough, the boundary layer turns turbulent. For natural convection on a vertical surface, the transition from laminar to turbulent occurs at critical distance $x_c$ where the local Rayleigh number is about $10^9$.

$$\text{Ra}_{x_c}(\text{vertical surface}) = \frac{g\,\beta\,(T_s - T_\infty)\,x_c^3}{v\,\alpha} \approx 10^9 \tag{7.9}$$

Both figures show the velocity distribution in the laminar boundary layer. The velocity component in the $x$ direction is $u$. It is seen that $u = 0$ at both the surface and the edge of the boundary layer.

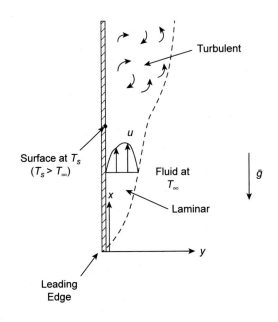

**FIGURE 7.1**

Hydrodynamic boundary layer on a vertical heated surface.

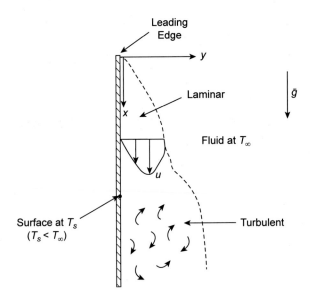

**FIGURE 7.2**

Hydrodynamic boundary layer on a vertical cooled surface.

Many experimental studies have been performed to determine the convective heat transfer for different geometries. Some correlation equations from these studies are given below. We include correlations for flat plates, cylinders, spheres, and enclosed spaces.

## 7.3 Natural convection for flat plates

The following sections include correlations for vertical, horizontal, and inclined flat plates.

### 7.3.1 Vertical plate

For a vertical plate, the characteristic length, $L$, is the height of the plate.

#### 7.3.1.1 Constant temperature surface

For a constant temperature surface, Churchill and Chu [1] recommend the following correlation equation if there is only a laminar region on the plate, i.e., if $Ra_L < 10^9$.

$$Nu = \frac{hL}{k} = 0.68 + \frac{0.670\,Ra_L^{1/4}}{\left[1 + (0.492/Pr)^{9/16}\right]^{4/9}} \tag{7.10}$$

For a constant temperature surface, Churchill and Chu [1] recommend the following correlation equation if there are both laminar and turbulent regions on the plate, i.e., if $Ra_L > 10^9$.

$$Nu = \frac{hL}{k} = \left\{0.825 + \frac{0.387\,Ra_L^{1/6}}{\left[1 + (0.492/Pr)^{9/16}\right]^{8/27}}\right\}^2 \tag{7.11}$$

Simpler equations in common use are [2–4].

$$Nu = \frac{hL}{k} = 0.59\,Ra_L^{1/4} \quad \text{for } 10^4 < Ra_L < 10^9 \tag{7.12}$$

$$\text{and } Nu = \frac{hL}{k} = 0.10\,Ra_L^{1/3} \quad \text{for } 10^9 < Ra_L < 10^{13} \tag{7.13}$$

For Eqs. (7.10) through (7.13), fluid properties should be evaluated at the film temperature $T_f = (T_s + T_\infty)/2$.

---

**Example 7.1**

**Vertical plate with constant surface temperature**

*Problem*
A vertical plate is in 25 C water. It is 10 cm wide and 20 cm high. A surface of the plate has a constant temperature of 75 C. What is the rate of heat transfer from the surface?

*Solution*
The characteristic length $L$ is the height of the plate, 20 cm. From Eq. (7.8), the Rayleigh number based on $L$ is

$$Ra_L = Gr_L\,Pr = \frac{g\beta(T_s - T_\infty)L^3}{v^2}Pr$$

Fluid properties are taken at the film temperature, $T_f = (T_s + T_\infty)/2 = (75 + 25)/2 = 50\text{ C}$
For water at 50 C, $k = 0.644$ W/m C; $v = 5.536 \times 10^{-7}$ m²/s; $Pr = 3.55$; $\beta = 0.451 \times 10^{-3}$/K
Also, the acceleration of gravity $g = 9.807$ m/s².

The Rayleigh number is therefore

$$Ra_L = \frac{(9.807)(0.451 \times 10^{-3})(75 - 25)(0.2)^3}{(5.536 \times 10^{-7})^2}(3.55) = 2.049 \times 10^{10}$$

Eq. (7.11) is an appropriate equation for convective coefficient $h$.

$$h = \left(\frac{k}{L}\right)\left\{0.825 + \frac{0.387\, Ra_L^{1/6}}{\left[1 + (0.492/Pr)^{9/16}\right]^{8/27}}\right\}^2$$

Putting values in this equation, we have

$$h = \left(\frac{0.644}{0.2}\right)\left\{0.825 + \frac{0.387(2.049 \times 10^{10})^{1/6}}{\left[1 + (0.492/3.55)^{9/16}\right]^{8/27}}\right\}^2 = 1216\ \text{W/m}^2\ \text{C}$$

The rate of heat transfer from the surface is

$$q = hA(T_s - T_\infty) = (1216)[(0.1)(0.2)](75 - 25) = \mathbf{1216\ W}$$

It is interesting (and somewhat upsetting) to note that the correlation of Eq. (7.13) gives a result of 881 W, which is 28% different from the 1216 W value we calculated using Eq. (7.11). This gives emphasis to the fact that results from existing correlations are only approximate. Accurate results can often only be obtained through experimentation on the actual surfaces in question.

## Example 7.2
### Vertical plate with constant heat flux

*Problem*
A thin vertical plate is 1.2 m high and 0.5 m wide. The left side of the plate is insulated and the right side absorbs a constant heat flux of 200 W/m². There is convection from the right side of the plate to the adjacent 20 C air. Assume that radiation from the plate is negligible.
(a) What is the temperature at the midheight of the plate, i.e., at $x = L/2 = 0.6$ m ?
(b) What is the temperature at the top of the plate, i.e., at $x = L = 1.2$ m ?

*Solution*
(a) To start the solution, we have to make an educated guess of the midheight surface temperature. Convective coefficient values for natural convection in air are small, say on the order of magnitude of 5 W/m² C. Using Eq. (7.16), we have $T_s = T_\infty + \frac{q_s''}{h_x} = 20 + \frac{200}{5} = 60$ C. The film temperature is then
$T_f = (T_s + T_\infty)/2 = (60 + 20)/2 = 40$ C.
Air properties at 40 C are $k = 0.027$ W/m C;   $v = 1.703 \times 10^{-5}$ m²/s;   Pr $= 0.710$
We will use Eq. (7.17) for the convective coefficient at $x = L/2 = 0.6$ m.

$$h_x = h_{L/2} = \left(\frac{k}{L/2}\right)\left\{0.68 + \frac{0.670\, Ra_{L/2}^{1/4}}{\left[1 + (0.437/Pr)^{9/16}\right]^{4/9}}\right\} \tag{7.17}$$

$$Ra_x = Ra_{L/2} = Gr_{L/2}Pr = \frac{g\beta(T_s - T_\infty)(L/2)^3}{v^2}Pr \tag{7.20}$$

where $g = 9.807$ m/s²; $\beta = 1/T_f = 1/(40 + 273.15) = 0.003193$ / K; and $L = 1.2$ m.
Putting values into Eq. (7.20), we have

$$Ra_{L/2} = \frac{(9.807)(0.003193)(60 - 20)(0.6)^3}{(1.703 \times 10^{-5})^2}(0.710) = 6.623 \times 10^8$$

Putting values into Eq. (7.17), we have

$$h_{L/2} = \left(\frac{0.027}{0.6}\right)\left\{0.68 + \frac{0.670(6.623 \times 10^8)^{1/4}}{\left[1 + (0.437/0.710)^{9/16}\right]^{4/9}}\right\} = 3.80\ \text{W/m}^2\ \text{C}$$

Using Eq. (7.16), we calculate the surface temperature at $x = L/2 = 0.6$ m as

$$T_s(x) = T_s(L/2) = T_\infty + \frac{q_s'}{h_{L/2}} = 20 + \frac{200}{3.80} = 72.6 \text{ C}$$

This answer is quite different from our initial guess of 60 C, so another iteration is appropriate. Let us do the calculations again with $T_s = 70$ C. The film temperature is then $T_f = (T_s + T_\infty)/2 = (70+20)/2 = 45$ C. The properties of air at 45 C are $k = 0.0274$ W/m C; $v = 1.750 \times 10^{-5}$ m$^2$/s; Pr $= 0.710$; $\beta = 0.003143$ /K
We do the calculations as before, and get Ra$_{L/2} = 7.718 \times 10^8$ and $h_{L/2} = 4.00$ W/m$^2$ C. The surface temperature at the midheight is

$$T_s(L/2) = T_\infty + \frac{q_s'}{h_{L/2}} = 20 + \frac{200}{4.00} = 70.0 \text{ C}$$

Our starting surface temperature was 70 C and we calculated 70.0 C. No further iteration is necessary. **The surface temperature at the midheight of the plate is 70.0 C.**
**(b)** To get the temperature at the top of the plate, we need $h_x$ at $x = L$.
From Eq. (7.17),

$$h_L = \left(\frac{k}{L}\right) \left\{ 0.68 + \frac{0.670 \text{ Ra}_L^{1/4}}{\left[1 + (0.437/\text{Pr})^{9/16}\right]^{4/9}} \right\} \qquad (7.21)$$

$$\text{Ra}_L = \text{Gr}_L\text{Pr} = \frac{g\beta(T_s - T_\infty)(L)^3}{v^2}\text{Pr} \qquad (7.22)$$

The film temperature is $T_f = (T_s + T_\infty)/2 = (70+20)/2 = 45$ C. The air properties at 45 C are given below: $k = 0.0274$ W/m C; $v = 1.750 \times 10^{-5}$ m$^2$/s; Pr $= 0.710$.
Putting values into Eq. (7.22), we have

$$\text{Ra}_L = \frac{(9.807)(1/(45+273.15))(70-20)(1.2)^3}{(1.750 \times 10^{-5})^2}(0.710) = 6.174 \times 10^9$$

Putting values into Eq. (7.21), we have

$$h_L = \left(\frac{0.0274}{1.2}\right) \left\{ 0.68 + \frac{0.670(6.174 \times 10^9)^{1/4}}{\left[1 + (0.437/0.710)^{9/16}\right]^{4/9}} \right\} = 3.35 \text{ W/m}^2 \text{ C}$$

Finally, $T_s(L) = T_\infty + \dfrac{q_s'}{h_L} = 20 + \dfrac{200}{3.35} = 79.7$ C

The surface temperature at the top of the plate is 79.7 C.
**Note:** We used the Churchill and Chu correlation equation, Eq. (7.17), for this solution. Alternatively, we could have used the Fujii and Fujii equation, Eq. (7.18). If the latter equation had been used, the results for the midheight and top surface temperatures would have been 77.5 and 86.0 C, respectively. The Churchill and Chu correlation gives temperatures about 7 C lower than the Fujii and Fujii correlation. Without experimentation on the actual plate, it is not possible to make a judgment of which correlation is more accurate.

### 7.3.1.2 Constant heat flux surface

For a constant heat flux surface, we are interested in determining the temperature distribution of the plate's surface, that is, we want to find $T_s(x)$. For convective heat transfer, we have the basic equation $q = hA(T_s - T_\infty)$. With a constant heat flux $q_s'$, this equation becomes

$$q_s' = \frac{q}{A} = h(T_s - T_\infty) \qquad (7.14)$$

Written for location $x$, Eq. (7.14) is

$$q_s' = h_x[T_s(x) - T_\infty] \qquad (7.15)$$

Rearranging Eq. (7.15), we get an expression for the surface temperature distribution:

$$T_s(x) = T_\infty + \frac{q_s'}{h_x} \qquad (7.16)$$

To determine the surface temperature distribution from Eq. (7.16), we need an equation for local convective coefficient $h_x$. For the laminar region, $Ra_x < 10^9$, Churchill and Chu [1] recommend

$$h_x = \left(\frac{k}{x}\right) \left\{ 0.68 + \frac{0.670\, Ra_x^{1/4}}{\left[1 + (0.437/Pr)^{9/16}\right]^{4/9}} \right\} \tag{7.17}$$

Fujii and Fujii [5] recommend

$$h_x = \left(\frac{k}{x}\right) \left(\frac{Pr^2}{4 + 9Pr^{1/2} + 10Pr} Gr_x^*\right)^{1/5} \tag{7.18}$$

$Gr_x^*$ in Eq. (7.18) is the modified Grashof number, defined as

$$Gr_x^* = \left(\frac{g\, \beta q_s'}{k\, \upsilon^2}\right) x^4 \tag{7.19}$$

Determination of $T_s(x)$ requires an iterative process as the fluid properties are taken at the film temperature, and the surface temperature is unknown at the beginning of the solution. Sparrow and Gregg [6] recommend that the film temperature be defined using the surface temperature at the midheight of the plate, i.e., at $x = L/2$. That is, $T_f = [T_s(L/2) + T_\infty]/2$.

To start the iterative solution, we guess the surface temperature at $x = L/2$. We then calculate $T_s(L/2)$ using Eq. (7.16) and either Eq. (7.17) or Eq. (7.18). We compare our calculated value with the guessed value and, if they are different, we iterate again. We continue the iterations until the guessed value is deemed close enough to the calculated value. Example 7.2 illustrates this procedure.

### 7.3.2 Horizontal plate

Consider Fig. 7.3, which shows a horizontal plate immersed in a fluid. The fluid temperature is $T_\infty$ and both surfaces of the plate are at temperature $T_s$, which is greater than $T_\infty$.

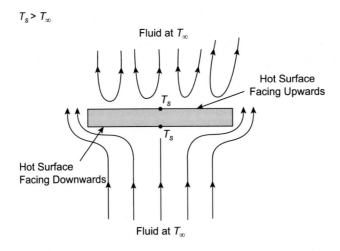

**FIGURE 7.3**

Natural convection from horizontal hot surfaces.

The upper surface of the plate shown in Fig. 7.3 is called a "hot surface facing upward," and the lower surface is called a "hot surface facing downward." Fluid motion at the two surfaces is quite different. At the upper surface, the heating of the fluid causes a decrease in the density of the fluid adjacent to the plate. This results in the rising of a heated plume through the colder fluid above it. If the temperature difference between the plate surface and the fluid above the plate is great enough, pockets of heated fluid will be created. These pockets move through the colder fluid above them, causing considerable movement and fluid mixing. At the lower surface, the heating of the fluid also causes a decrease in density of the adjacent fluid. The fluid wants to rise, but the plate is in the way. Hence, there is less motion at the lower surface than at the upper surface. This difference in fluid motion results in a difference in the convective coefficients for the two surfaces. The rate of heat transfer at the upper surface is greater than that at the lower surface.

For cold surfaces (i.e., for $T_s < T_\infty$), the heat transfer of a cold surface facing downward is the same as that of a hot surface facing upward. And, the heat transfer of a cold surface facing upward is the same as that of a hot surface facing downward.

For a horizontal surface, the characteristic length is

$$L = \frac{A}{P} \tag{7.23}$$

where $A$ is the area of the surface, and $P$ is the perimeter of the surface.

For example, for a square area of side "$a$," $L = \frac{a^2}{4a} = \frac{1}{4} a$. For a rectangular area of sides "$a$" and "$b$," $L = \frac{a\,b}{2\,(a+b)}$. For a circular area of diameter "$D$," $L = \frac{\pi D^2/4}{\pi D} = \frac{1}{4} D$.

### 7.3.2.1 Constant temperature surface

Much experimentation has been performed for natural convection from horizontal plates [2,7,8]. Correlation equations are:

For a hot surface facing upward or a cold surface facing downward,

$$\text{Nu} = \frac{hL}{k} = 0.54\ \text{Ra}_L^{1/4} \qquad \text{for}\ \ 2 \times 10^4 < \text{Ra}_L < 8 \times 10^6 \tag{7.24}$$

$$\text{Nu} = \frac{hL}{k} = 0.15\ \text{Ra}_L^{1/3} \qquad \text{for}\ \ 8 \times 10^6 < \text{Ra}_L < 2 \times 10^9 \tag{7.25}$$

For a hot surface facing downward or a cold surface facing upward,

$$\text{Nu} = \frac{h\,L}{k} = 0.27\ \text{Ra}_L^{1/4} \qquad \text{for}\ \ 10^5 < \text{Ra}_L < 10^{10} \tag{7.26}$$

Fluid properties are taken at the film temperature $T_f = (T_s + T_\infty)/2$.

### 7.3.2.2 Constant heat flux surface

Fujii and Imura [7] recommend the following correlations:

For a hot surface facing upward,

$$\text{Nu} = \frac{hL}{k} = 0.16\ \text{Ra}_L^{1/3} \qquad \text{for}\ \ \text{Ra}_L < 2 \times 10^8 \tag{7.27}$$

$$\text{Nu} = \frac{hL}{k} = 0.13\text{Ra}_L^{1/3} \quad \text{for} \quad 5 \times 10^8 < \text{Ra}_L < 10^{11} \tag{7.28}$$

For a hot surface facing downward,

$$\text{Nu} = \frac{h\,L}{k} = 0.58\ \text{Ra}_L^{1/5} \quad \text{for} \quad 10^6 < \text{Ra}_L < 10^{11} \tag{7.29}$$

In Eqs. (7.27)–(7.29), fluid properties, with the exception of $\beta$, are taken at the temperature $T_s - 0.25\ (T_s - T_\infty)$. $\beta$ is evaluated at temperature $T_\infty + 0.25\ (T_s - T_\infty)$.

## Example 7.3
### Heat flow from a heating duct

*Problem*
A horizontal heating duct is 16 inches wide and 8 inches high. The outer surface of the duct is at 110 F and the room air is at 70 F. What is the rate of convective heat transfer to the room air per foot length of duct?

*Solution*
We can either work the problem using English units or SI units. Let us do the latter.
Duct width = 16 in x (2.54 cm/in) = 40.64 cm.
Duct height = 8 inch x (2.54 cm/in) = 20.32 cm

$$T_s = 110\ F \ = \ 5/9\ (110 - 32) = 43.3\ \text{C}$$

$$T_\infty = 70\ F \ = \ 5/9\ (70\ -\ 32) = 21.1\ \text{C}$$

Fluid properties are taken at the film temperature: $T_f\ =\ (T_s + T_\infty)/2 = (43.3 + 21.1)/2 = 32.2$ C.
At this temperature, for air, $k\ = 0.0264\ \text{W}/\text{m C}; \quad v\ = 1.62 \times 10^{-5}\ \text{m}^2/\text{s}; \quad \text{Pr} =\ 0.711$
Also, $g = 9.807\ \text{m/s}^2$ and $\beta = 1/T_f\ = 1/(32.2 + 273.15)\ = 0.003275\ /\ \text{K}$
**For the upper and lower surfaces,.**

$$L = \frac{A}{P} = \frac{(0.4064)\ (length)}{[2\ (0.4064) + 2(length)]} = \frac{0.4064}{2\left(\dfrac{0.4064}{length} + 1\right)} \approx 0.2032\ \text{m}$$

We have reasonably assumed that the duct's length is much greater than its width.

$$\text{Ra}_L = \text{Gr}_L\text{Pr} = \frac{g\beta(T_s - T_\infty)(L)^3}{v^2}\text{Pr} = \frac{(9.807)(0.003275)(43.3 - 21.1)(0.2032)^3}{(1.62 \times 10^{-5})^2}(0.711)$$

$$\text{Ra}_L = 1.621 \times 10^7$$

The upper surface of the duct is a hot surface facing upward. Using Eq. (7.25), we have

$$\text{Nu} = \frac{hL}{k} = 0.15\ \text{Ra}_L^{1/3} = 0.15(1.621 \times 10^7)^{1/3} = 37.96$$

$$h = \left(\frac{k}{L}\right)\text{Nu} = \left(\frac{0.0264}{0.2032}\right)(37.96) = 4.93\ \text{W}\ /\ \text{m}^2\ \text{C}$$

The heat flow per meter length is

$$q\ /\ length = h(width)(T_s - T_\infty) = 4.93(0.4064)(43.3 - 21.1) = 44.48\ \text{W/m length}$$

The lower surface of the duct is a hot surface facing downward. Using Eq. (7.26), we have

$$\text{Nu} = \frac{hL}{k} = 0.27\ \text{Ra}_L^{1/4} = 0.27(1.621 \times 10^7)^{1/4} = 17.13$$

$$h = \left(\frac{k}{L}\right)\text{Nu} = \left(\frac{0.0264}{0.2032}\right)(17.13) = 2.23\ \text{W/m}^2\ \text{C}$$

The heat flow per meter length is

$$q / \text{length} = h(\text{width})(T_s - T_\infty) = 2.23(0.4064)(43.3 - 21.1) = 20.12 \text{ W/m length}$$

For the sides of the duct,

$$L = \text{height of duct} = 0.2032 \text{ m}$$

This characteristic length happens to be the same as that for the upper and lower surfaces, so we can use the result of the $\text{Ra}_L$ calculation above. $\text{Ra}_L = 1.621 \times 10^7$

The sides are vertical and $\text{Ra}_L < 10^9$. Therefore we use Eq. (7.10) for Nu and $h$.

$$\text{Nu} = \frac{hL}{k} = 0.68 + \frac{0.670 \, \text{Ra}_L^{1/4}}{\left[1 + (0.492/\text{Pr})^{9/16}\right]^{4/9}} = 0.68 + \frac{0.670(1.621 \times 10^7)^{1/4}}{\left[1 + (0.492/0.711)^{9/16}\right]^{4/9}} = 33.32$$

$$h = \left(\frac{k}{L}\right)\text{Nu} = \left(\frac{0.0264}{0.2032}\right)(33.32) = 4.33 \text{ W/m}^2 \text{ C}$$

The heat flow for each side per meter length is

$$q / \text{length} = h(\text{height})(T_s - T_\infty) = 4.33(0.2032)(43.3 - 21.1) = 19.53 \text{ W/m } \textit{length}$$

Total heat flow from the duct to the air is $44.48 + 20.12 + 2 (19.53) = 103.7$ W/m length.

The problem was posed in English units so we give the result as

$$103.7 \text{ W/m} \times \frac{1 \text{ m}}{3.2808 \text{ ft}} \times \frac{3.412 \text{ Btu / h}}{\text{W}} = \textbf{107.8 Btu/h  per foot length}$$

### 7.3.3 Inclined plate

Fig. 7.4 shows two orientations of an inclined flat plate. In (A), the hot surface is facing downwards and the cold surface upwards. For this case, the angle $\theta$ from the vertical is positive. In (B), the hot surface is upwards and the cold surface is downwards. The angle $\theta$ is negative. The length of the plate is $L$.

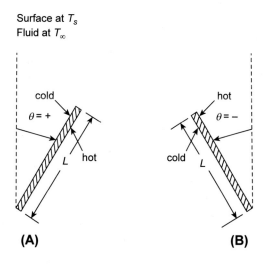

**FIGURE 7.4**

Inclined plate.

Fujii and Imura [7] recommend the following correlations:
For the hot surface facing downward or cold surface facing upward,

$$\text{Nu} = \frac{hL}{k} = 0.56 \ (\text{Ra}_L \cos \theta)^{1/4} \tag{7.30}$$

for $\quad 10^5 < \text{Ra}_L \cos \theta < 10^{11} \quad$ and $\quad 0 < \theta < 88°$

For the hot surface facing upward or cold surface facing downward,

$$\text{Nu} = \frac{hL}{k} = 0.145 \left[ (\text{Gr}_L \text{Pr})^{1/3} - (\text{Gr}_c \ \text{Pr})^{1/3} \right] + 0.56 (\text{Gr}_L \text{Pr} \cos \theta)^{1/4} \tag{7.31}$$

for $\quad 10^5 < \text{Gr}_L \text{Pr} \cos \theta < 10^{11} \quad$ and $\quad -75° < \theta < -15°$

If $\text{Gr}_L < \text{Gr}_c$, then the first term in Eq. (7.31) is deleted.

In Eq. (7.31), the subscript "c" relates to where Nu starts to separate from the characteristic of the laminar region. The values of $\text{Gr}_c$ Pr depend on the angle of inclination $\theta$. For angles of $-15, -30, -45, -60,$ and $-75°$, the $\text{Gr}_c$ Pr values are, respectively, about $5 \times 10^9$, $2 \times 10^9$, $10^9$, $10^8$, and $10^6$.

In Eqs. (7.30) and (7.31), fluid properties, with the exception of $\beta$, are taken at the temperature $T_s - 0.25 \ (T_s - T_\infty)$. $\beta$ is evaluated at temperature $T_\infty + 0.25 \ (T_s - T_\infty)$.

## 7.4 Natural convection for cylinders

The following sections include correlations for horizontal and vertical cylinders. These geometries have wide applications such as steam heating pipes for buildings, refrigeration lines, pin fins, and immersion heaters.

### 7.4.1 Horizontal cylinder

Fig. 7.5 shows natural convection from a horizontal cylinder of diameter $D$. The surface temperature of the cylinder is $T_s$, and the adjacent fluid is at $T_\infty$. The gravity vector is downwards.

Much experimentation has been performed for natural convection from cylinders. We provide two correlations from Churchill and Chu [9] for the average Nusselt Numbers at the surface. The correlations give a good fit to the experimental data over a wide range of Rayleigh Numbers. We also

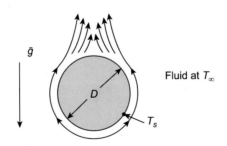

**FIGURE 7.5**

Natural convection from a horizontal cylinder.

include a simpler correlation equation from Morgan [10]. The characteristic length for the correlations is $D$, the diameter of the cylinder.

Eqs. (7.32) and (7.33) are from Churchill and Chu.

$$\text{Nu} = \frac{h\,D}{k} = 0.36 + \frac{0.518\ \text{Ra}_D^{1/4}}{\left[1 + (0.559/\text{Pr})^{9/16}\right]^{4/9}} \quad \text{for} \quad 10^{-6} < \text{Ra}_D < 10^9 \quad (7.32)$$

$$\text{Nu} = \frac{h\,D}{k} = \left\{ 0.60 + \frac{0.387\ \text{Ra}_D^{1/6}}{\left[1 + (0.559/\text{Pr})^{9/16}\right]^{8/27}} \right\}^2 \quad \text{for} \quad 10^9 < \text{Ra}_D < 10^{12} \quad (7.33)$$

These correlations may be used for both constant surface temperature and constant heat flux boundary conditions.

The correlation from Morgan, for constant surface temperature cylinders, is

$$\text{Nu} = \frac{hD}{k} = C\ \text{Ra}_D^n \quad (7.34)$$

where the $C$ and $n$ values for various $\text{Ra}_D$ ranges are

| $\text{Ra}_D$ | $C$ | $n$ |
|---|---|---|
| $10^{-10} - 10^{-2}$ | 0.675 | 0.058 |
| $10^{-2} - 10^2$ | 1.02 | 0.148 |
| $10^2 - 10^4$ | 0.850 | 0.188 |
| $10^4 - 10^7$ | 0.480 | 0.250 |
| $10^7 - 10^{12}$ | 0.125 | 0.333 |

For all three correlations, fluid properties should be taken at the film temperature $T_f = (T_s + T_\infty)/2$.

The following example deals with heating of a warehouse by natural convection from a hot horizontal duct. The example is very similar to Example 6.12. However, in that example there was forced convection at the outer surface of the duct. Here we have natural convection.

## Example 7.4
### Heat for a warehouse

*Problem*

There is a horizontal duct going through a warehouse carrying hot air to the adjacent office. The warehouse is unheated, and we would like to estimate the amount of heat provided from the duct to the warehouse air by natural convection. The details are as follows:

The duct is 10 inches diameter and the section in the warehouse is 50 feet long. The air enters the warehouse section at a rate of 1100 cfm and a temperature of 130 F. The air in the warehouse space is at 45 F.

(a) How much heat will be added to the air in the warehouse by natural convection from the outer surface of the duct (Btu/hr)?

(b) What is the temperature of the air in the duct as it leaves the warehouse?

*Solution*

We will assume that the duct wall is thin and a constant temperature surface. To get air property values, we need to make some assumptions on temperatures. We will check these assumptions after the calculations and iterate as necessary. We will assume that the duct air temperature drops by 10 F from the entrance to the exit. The air then enters at 130 F (54.44 C) and leaves at 120 F (48.89 C). The average temperature of the air in the duct is $(54.44 + 48.89)/2 = 51.67$ C. We will also assume that the temperature of the duct wall is $T_s = 110$ F $= 43.33$ C.

We now proceed to get the convective coefficients $h_i$ and $h_o$ for the inside and outside surfaces of the duct.

**Internal Flow.** The air enters at 1100 cfm and 130 F (54.44 C). For air at 54.44 C, $\rho = 1.075$ kg/m$^3$. Mass flow rate = $\dot{m} = \rho\dot{Q} = 1.075$ kg/m$^3$ $\times$ 1100 ft$^3$/min $\times$ (1 min/60 s) $\times$ (1 m/3.2808 ft)$^3 = 0.5581$ kg/s. $D = 10$ inch $\times$ (1m/39.37 inch) $= 0.254$ m.

The average duct air temperature is 51.67 C. For air at this temperature,

$$\rho = 1.086 \text{ kg/m}^3, \quad c_p = 1007 \text{ J/kg C}, \quad k = 0.0274 \text{ W/m C}, \quad v = 18.15 \times 10^{-6} \text{ m}^2/\text{s}, \quad \text{Pr} = 0.722$$

$$\dot{m} = \rho A V = 1.086\left[(\pi/4)(0.254)^2\right]V = 0.5581 \text{ and } V = 10.14 \text{ m/s}$$

$$\text{Re}_D = \frac{VD}{v} = \frac{(10.14)(0.254)}{18.15 \times 10^{-6}} = 1.419 \times 10^5 \quad \text{(turbulent)}$$

Eq. (6.135):

$$h_i = 0.023\text{Re}_D^{0.8}\text{Pr}^{0.3}(k/D)$$

$$= (0.023)(1.419 \times 10^5)^{0.8}(0.722)^{0.3}(0.0274/0.254) = 29.77 \text{ W/m}^2\text{C}$$

Let us use Eq. (6.131) to get the exit temperature of the duct air:

$$(T_s - T_m)_{x_2} = (T_s - T_m)_{x_1} \ e^{-\frac{h}{\dot{m}}\frac{P}{c_p}(x_2 - x_1)} \tag{6.131}$$

Putting in values, we have $(43.33 - T_m)_{x_2} = (43.33 - 54.44) \ e^{-\frac{(29.77)[\pi(0.254)]}{(0.5581)(1007)}(50/3.2808)}$

The exit temperature of the air is $(T_m)_{x_2} = 49.16$ C $= 120.5$ F.

**Natural Convection at Outer Surface of Duct.** $D = 0.254$ m and surface temperature $= T_s = 110$ F $= 43.33$ C. The warehouse air is at 45 F $= 7.22$ C.

The film temperature is $T_f = \frac{T_s + T_\infty}{2} = \frac{43.33 + 7.22}{2} = 25.28$ C.

For air at $T_f = 25.28$ C,

$$\rho = 1.184 \text{ kg/m}^3, \quad c_p = 1007 \text{ J/kg C}, \quad k = 0.0255 \text{ W/m C}, \quad v = 15.58 \times 10^{-6} \text{ m}^2/\text{s}, \quad \text{Pr} = 0.729$$

We will use Eq. (7.32) for natural convection from a horizontal cylinder:

$$h_o = \left[0.36 + \frac{0.518 \ \text{Ra}_D^{1/4}}{\left[1 + (0.559/\text{Pr})^{9/16}\right]^{4/9}}\right]\left(\frac{k}{D}\right) \quad \text{for} \quad 10^{-6} < \text{Ra}_D < 10^9 \tag{7.32}$$

$$\text{Ra}_D = \text{Gr}_D\text{Pr} = \frac{g\beta(T_s - T_\infty)(D)^3}{v^2}\text{Pr} = \frac{(9.807)(1/(25.28 + 273.15))(43.33 - 7.22)(0.254)^3}{(15.58 \times 10^{-6})^2}(0.729)$$

$\text{Ra}_D = 5.840 \times 10^7$. This value falls within the appropriate range for Eq. (7.32).

Putting values into Eq. (7.32), we get $h_o = 3.49$ W / m$^2$ C.

Let us now check the assumed duct wall temperature by doing an energy balance. The convection at the inner surface of the duct equals the convection at the outer surface. That is,

$$h_i \ A_i \ (T_{duct \ air} - T_s) = h_o \ A_o \ (T_s - T_{room \ air})$$

As the tube is thin, $A_i = A_o$. Putting values into the equation, we have

$$29.77\left(\frac{54.44 + 49.16}{2} - T_s\right) = 3.49(T_s - 7.22)$$

Solving for $T_s$, we get that the duct wall temperature is 47.12 C = 116.8 F.

**Summarizing:.** We assumed the duct wall temperature was 110 F and we calculated it as 116.8 F.

We assumed the exit temperature of the duct air was 120 F and we calculated it as 120.5 F.

Although the exit temperature of the duct air was close to our assumption, there was a significant different between our assumption of the duct wall temperature and our calculated value. We should therefore iterate with another assumed duct wall temperature, say 116 F, and keep the assumed exit temperature of the duct air at 120 F. Looking at the result of this iteration, we would judge whether or not a further iteration was necessary.

Instead of doing these hand calculations again, let us use Excel *Solver* to reach a solution, as follows:

We can write three equations for the heat flow $q$. One is for the convection at the inner surface, one is for the convection at the outer surface, and the final one is for the heat loss from the duct air. The equations are

$$q = h_i A_i (T_{duct\ air} - T_s) \tag{a}$$

$$q = h_o A_o (T_s - T_\infty) \tag{b}$$

$$q = \dot{m}\, c_p (\Delta T)_{duct\ air} \tag{c}$$

We can combine these equations to get two equations for the two unknowns: the duct wall temperature $T_s$ and the exit temperature of the duct air $T_{air\ out}$.

From Eqs. (a) and (c), we have

$$h_i A_i (T_{duct\ air} - T_s) = \dot{m}\, c_p (\Delta T)_{duct\ air} \tag{d}$$

From Eqs. (b) and (c), we have

$$h_o A_o (T_s - T_\infty) = \dot{m}\, c_p (\Delta T)_{duct\ air} \tag{e}$$

For solution by Excel *Solver*, we move everything to the left of the equal sign in Eqs. (d) and (e). Doing this and putting in the values for parameters, we get

$$f_1 = h_i A_i (T_{duct\ air} - T_s) \ - \ \dot{m}\, c_p (\Delta T)_{duct\ air} = 0$$

$$f_1 = 29.77 [\pi (0.254)(50/3.2808)] \left( \frac{54.44 + T_{air\ out}}{2} - T_s \right) - 0.5581 (1007)(54.44 - T_{air\ out}) = 0 \tag{f}$$

$$f_2 \quad = \quad h_o A_o (T_s - T_\infty) - \ \dot{m}\, c_p (\Delta T)_{duct\ air} = 0$$

$$f_2 = 3.49 \ [\pi (0.254)(50/3.2808)] \ (T_s - 7.22) \ - \ 0.5581 \ (1007) \ (54.44 - T_{air\ out}) = 0 \tag{g}$$

Excel *Solver* varies the values of $T_{air\ out}$ and $T_s$ until $f_1^2 + f_2^2 = 0$. This assures that $f_1$ and $f_2$ are both zero, and Eqs. (f) and (g) have been successfully solved for $T_{air\ out}$ and $T_s$.

The *Solver* results were $T_s = 48.10$ C and $T_{air\ out} = 51.35$ C.

The heat flow can be obtained from Eqs. (a), (b), or (c). Using Eq. (c), we have

$$q = \dot{m}\, c_p (\Delta T)_{duct\ air} = (0.5581) \ (1007) \ (54.44 - 51.35) \quad = \quad 1737 \ \text{W} \quad = \quad 5930 \ \text{Btu / hr}$$

**Results:**

**(a)** The rate of heat added to the warehouse air is **5930 Btu/hr**.

**(b)** The air exits the duct at **51.35 C = 124.4 F**. (Note: In the Excel program, we used the original values of the convective coefficients $h_i$ and $h_o$. We assumed this approach was appropriate since the properties of air vary minimally with temperature in the range of interest. We checked the validity of this approach by developing a more-complex Excel program which calculated the coefficients directly from Eqs. (6.135) and (7.32). The results of that program confirmed that use of the original convective coefficients gave excellent results.)

## 7.4.2 Vertical cylinder

Studies of natural convection from vertical cylinders include those of Sparrow and Gregg [17], Minkowycz and Sparrow [18], and Cebeci [19]. Let us consider a vertical cylinder of height $L$ and diameter $D$. The top and bottom surfaces can be treated as horizontal plates using the equations in Section 7.3.2. If the curvature of the cylinder is small enough, the outer surface can be considered as essentially a flat plate, and the correlation equations for a flat plate of height $L$ may be used. The criterion for modeling a vertical cylinder as a flat plate is [17].

$$D \geq \frac{35\,L}{\mathrm{Gr}_L^{1/4}} \tag{7.35}$$

For slender cylinders not meeting this criterion, Cebeci [19] provides information on the deviation of local and average Nusselt numbers of a side of a slender cylinder from those of a flat plate.

LeFevre and Ede [20], using an integral method, obtained the following equation for the laminar region ($\mathrm{Gr}_L < 10^9$), which accounts for wall curvature of the cylinder:

$$\mathrm{Nu} = \frac{h\,L}{k} = \frac{4}{3}\left[\frac{7\,\mathrm{Ra}_L \mathrm{Pr}}{5\,(20+21\mathrm{Pr})}\right]^{1/4} + \frac{4}{35}\frac{(272+315\ \mathrm{Pr})}{(64+63\ \mathrm{Pr})}\left(\frac{L}{D}\right) \tag{7.36}$$

## 7.5 Natural convection for spheres

Fig. 7.5 is also applicable for a sphere. Considerable experimentation has been performed for natural convection from spheres. Some correlation equations are as follows:

Yuge [11] performed experiments with spheres in air and obtained the equation

$$\mathrm{Nu} = \frac{hD}{k} = 2 + 0.392\ \mathrm{Gr}_D^{1/4} \qquad \text{for}\quad 1\ <\ \mathrm{Gr}_D\ <\ 10^5 \tag{7.37}$$

Eq. (7.37) contains the Grashof Number. Most correlations are based on the Rayleigh Number. The Prandtl Number for air is about 0.7. Using this, Eq. (7.37) can be changed to an equation containing the Rayleigh Number. The resulting Eq. (7.38) is more general than Eq. (7.37) and may be used for fluids other than air.

$$\mathrm{Nu} = \frac{hD}{k} = 2 + 0.43\ \mathrm{Ra}_D^{1/4} \tag{7.38}$$

Amato and Tien [12] performed experiments with spheres in water and obtained the equation

$$\mathrm{Nu} = \frac{hD}{k} = 2 + 0.50\mathrm{Ra}_D^{1/4} \qquad \text{for}\quad 3 \times 10^5\ <\ \mathrm{Ra}_D\ <\ 8 \times 10^8 \tag{7.39}$$

Churchill [13] recommends the equation

$$\mathrm{Nu} = \frac{h\,D}{k} = 2 + \frac{0.589\ \mathrm{Ra}_D^{1/4}}{\left[1\ +\ (0.469/\mathrm{Pr})^{9/16}\right]^{4/9}} \qquad \text{for}\quad \mathrm{Ra}_D < 10^{11}\ \text{and}\ \mathrm{Pr} \geq 0.7 \tag{7.40}$$

Sparrow and Stretton [14] did experiments with cubes in air and water and obtained a correlation equation that can also be applied to spheres. In the equation, Eq. (7.41), the characteristic length for spheres is $L^{**} = \pi D$.

$$\text{Nu} = \frac{h\,L^{**}}{k} = 5.748 + 0.752 \left\{ \frac{\text{Ra}_{L^{**}}}{\left[1 + (0.492/\text{Pr})^{9/16}\right]^{16/9}} \right\}^{0.252} \tag{7.41}$$

$$\text{for} \quad 200 < \text{Ra}_{L^{**}} < 1.5 \times 10^9$$

For Eqs. (7.37)–(7.41), fluid properties should be taken at the film temperature $T_f = (T_s + T_\infty)/2$.

---

## Example 7.5
### Cooling of a spherical container
*Problem*

Fig. 7.6 shows a thin-walled spherical container filled with a mixture of ice and water. There is 50% ice and 50% water by volume. The container has a diameter of 25 cm, and it is in a room where the air is at 20 C. How long will it take for all of the ice to melt?

*Solution*

Heat transfer will primarily be by natural convection from the room air to the container. There will be some radiative heat transfer to the container, but that should be insignificant as the temperatures of the problem are low. (We will check this after we determine the natural convection heat transfer.)

We first have to determine the mass of ice that needs to melt. The volume of the container is $V = (4/3)\pi r^3 = (4/3)\pi(0.125)^3 = 0.00818 \text{ m}^3$. Therefore the initial volume of the ice is $(0.00818)(0.5) = 0.00409 \text{ m}^3$. The density of ice is 916.7 kg/m³, so the initial mass of ice in the container is

$$m_{ice} = \rho_{ice} V_{ice} = (916.7 \text{ kg}/\text{m}^3)(0.00409 \text{ m}^3) = 3.75 \text{ kg}$$

The heat of fusion of ice is 333.6 kJ/kg, so the amount of heat needed to melt all of the ice in the container is (333.6 kJ/kg) × (3.75 kg) = 1251 kJ.

The temperature of the ice/water mixture remains at 0 C while the ice is melting. As the container is thin-walled, we will assume that the outer surface of the container is at 0 C. The air properties are taken at the film temperature: $T_f = (T_s + T_\infty)/2 = (0+20)/2 = 10$ C.

$$\text{For 10 C air,} \quad k = 0.0248 \text{ W/m C}; \quad \nu = 1.43 \times 10^{-5} \text{ m}^2/\text{s}; \quad \text{Pr} = 0.716$$

The Rayleigh number based on the diameter of the sphere is

$$\text{Ra}_D = \frac{g\beta(T_s - T_\infty)(D)^3}{\nu^2}\text{Pr} = \frac{(9.807)[1/(10 + 273.15)](20 - 0)(0.25)^3}{(1.43 \times 10^{-5})^2}(0.716) = 3.790 \times 10^7$$

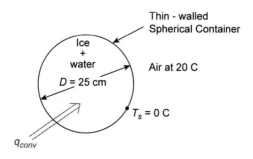

**FIGURE 7.6**

Melting of ice in a spherical container.

We will calculate convective coefficient $h$ using Eqs. (7.38), (7.40), and (7.41) and compare the results. Using Eq. (7.38), we have

$$\text{Nu} = \frac{hD}{k} = 2 + 0.43\text{Ra}_D^{1/4} = 2 + 0.43(3.790 \times 10^7)^{1/4} = 35.74$$

$$h = \frac{\text{Nu } k}{D} = \frac{(35.74)(0.0248)}{0.25} = 3.55 \text{ W} / \text{ m}^2 \text{ C}$$

Using Eq. (7.40), we have

$$\text{Nu} = \frac{hD}{k} = 2 + \frac{0.589\text{Ra}_D^{1/4}}{\left[1 + (0.469/\text{Pr})^{9/16}\right]^{4/9}} = 2 + \frac{0.589(3.790 \times 10^7)^{1/4}}{\left[1 + (0.469/0.716)^{9/16}\right]^{4/9}} = 37.69$$

$$h = \frac{\text{Nu } k}{D} = \frac{(37.69)(0.0248)}{0.25} = 3.74 \text{ W} / \text{ m}^2 \text{ C}$$

For Eq. (7.41), the characteristic length is $L^{**} = \pi D = \pi(0.25) = 0.785$ m. The Rayleigh number is

$$\text{Ra}_{L^{**}} = \frac{g\beta(T_s - T_\infty)(L^{**})^3}{v^2}\text{Pr} = \frac{(9.807)[1/(10 + 273.15)](20 - 0)(0.785)^3}{(1.43 \times 10^{-5})^2}(0.716) = 1.173 \times 10^9$$

Using Eq. (7.41), we have

$$\text{Nu} = \frac{h L^{**}}{k} = 5.748 + 0.752\left\{\frac{\text{Ra}_{L^{**}}}{\left[1 + (0.492/\text{Pr})^{9/16}\right]^{16/9}}\right\}^{0.252}$$

$$= 5.748 + 0.752\left\{\frac{1.173\text{x}10^9}{\left[1 + (0.492/0.716)^{9/16}\right]^{16/9}}\right\}^{0.252} = 117.0$$

$$h = \frac{\text{Nu} k}{L^{**}} = \frac{(117.0)(0.0248)}{0.785} = 3.70 \text{ W/m}^2 \text{ C}$$

So, for the three correlations, we have $h = 3.55$, 3.74, and 3.70 W/m² C. The values differ by a maximum of 5.4%. Let us use the average of these results for our continuing calculations.

So, $h = (3.55 + 3.74 + 3.70)/3 = 3.66$ W/m² C.

The heat flow into the container by natural convection is $q = hA(T_\infty - T_s)$, where $A = $ surface area of the container $= 4\pi R^2 = 4\pi(0.125)^2 = 0.1963$ m².

Therefore, $q = (3.66)(0.1963)(20-0) = 14.37$ W $= 14.37$ J/s.

From above, we need 1251 kJ to melt all of the ice. At a rate of 14.37 J/s, this will take

$$t = \frac{1251 \times 10^3 \text{ J}}{14.37 \text{ J} / \text{ s}} = 8.71 \times 10^4 \text{ s} \times \frac{1 \text{ h}}{3600 \text{ s}} = 24.2 \text{ hours}$$

Note that we have only considered the natural convection heat transfer. Radiation may also be significant if the outer surface of the container is not highly polished. We can calculate the radiative heat transfer from $q = \varepsilon\sigma A(T_{surr}^4 - T_s^4)$. If the outer surface of the container is highly polished, $\varepsilon$ may be as low as 0.03. The radiative heat transfer will then be

$$q = (0.03)(5.67 \times 10^{-8})(0.1963)\left[(293.15)^4 - (273.15)^4\right] = 0.61 \text{ W}$$

and the radiative heat transfer is insignificant compared with the convective heat transfer of 14.37 W. However, if the outer surface of the container is not highly polished, say $\varepsilon = 0.5$, then the radiative heat transfer will be 10.1 W. The radiative heat transfer is then the same order of magnitude as the natural convection, and it should be included in the total rate of heat flow into the container.

## 7.6 Natural convection for other objects

In this section we discuss two papers that deal with natural convection from a variety of objects.

Sparrow and Stretton [14] performed experiments on cubes in various orientations and obtained a correlation equation that can be used for bodies of unity aspect ratio such as cubes, spheres, and short cylinders. We used this equation, Eq. (7.41), for spheres

$$\text{Nu} = \frac{hL^{**}}{k} = 5.748 + 0.752 \left\{ \frac{\text{Ra}_{L^{**}}}{\left[ 1 + (0.492/\text{Pr})^{9/16} \right]^{16/9}} \right\}^{0.252} \tag{7.41}$$

$$\text{for} \quad 200 < \text{Ra}_{L^{**}} < 1.5 \times 10^9$$

Sparrow and Stretton believe that this equation can be used for situations having $\text{Ra}_{L^{**}}$ moderately beyond the above given range; for bodies having aspect ratios deviating slightly from unity; and for all Prandtl numbers, except, perhaps for liquid metals. They caution that the correlation should not be used for bodies with aspect ratios deviating significantly from unity.

The characteristic length $L^{**}$ for Eq. (7.41) is defined by the equations:

$$L^{**} = \frac{A}{D^{**}} \quad \text{and} \quad \frac{\pi (D^{**})^2}{4} = A_{\text{horiz}} \tag{7.42}$$

where $A$ is the surface area of the body and $A_{\text{horiz}}$ is the projection of the body on a horizontal plane located either above or below the body. $D^{**}$ is an equivalent diameter based on $A_{\text{horiz}}$.

Using the definitions of Eq. (7.42), the characteristic length $L^{**}$ for a sphere is $\pi D$. For a cube with its normal orientation, i.e., top and bottom on horizontal planes and sides vertical, $L^{**} = 3 S \sqrt{\pi}$, where $S$ is the length of a side of the cube. For a vertical short cylinder of length $S$ and diameter $S$, $L^{**} = (3/2) \pi S$. For a horizontal short cylinder of length $S$ and diameter S, $L^{**} = (3/4) \pi^{3/2} S$.

Lienhard [15] considered the predicted and experimental data from vertical plates, cylinders, and spheres. He proposed the following correlation for natural convection from any submerged isothermal body for fluids with Prandtl numbers on the order of unity or greater.

$$\text{Nu} = \frac{h L}{k} = 0.52 \, \text{Ra}_L^{1/4} \quad \text{for } 10^4 < \text{Ra}_L < 10^9 \tag{7.43}$$

The characteristic length $L$ in Eq. (7.43) is the distance that a fluid particle travels in the boundary layer as it goes around the object. Some examples of $L$ are shown in Fig. 7.7.

Let us apply Eq. (7.43) to the above Example 7.3 that dealt with heating from a horizontal duct. The duct is 16″ wide and 8″ high, which give a characteristic length of 16″ + 8″ = 24″ = 0.6096 m. Using this length, $\text{Ra}_L = 4.376 \times 10^8$, Nu = 75.2, and $h = 3.26 \, \text{W/m}^2$ C. This $h$ value gives a heat transfer rate of 88.2 W/m length of the duct, which is 15% lower than the 103.7 W/m result of Example 7.3 above.

We also can apply Eq. (7.43) to Example 7.5, which dealt with ice melting in a spherical container. The container has a diameter of 25 cm, which gives a characteristic length of $L = \pi D/2 = \pi(0.25)/2 = 0.3927$ m. Using this length, we get $\text{Ra}_L = 1.469 \times 10^8$, Nu = 57.3, and $h = 3.62 \, \text{W/m}^2$ C. This result for $h$ is comparable with the values we obtained in Example 7.5 from the other correlations for spheres.

Sparrow and Ansari [16] performed experiments on a vertical cylinder with height equal to diameter. They found that Eq. (7.43) underpredicted the observed heat transfer by 8%–30%,

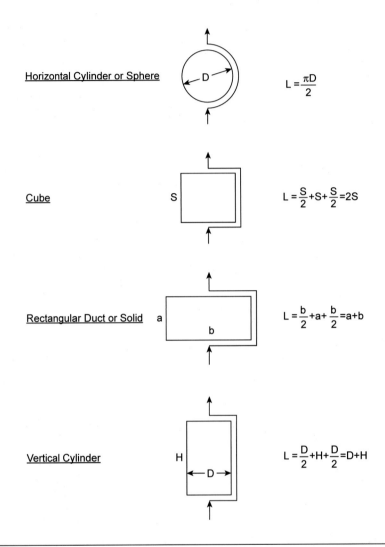

Horizontal Cylinder or Sphere $\qquad L = \dfrac{\pi D}{2}$

Cube $\qquad L = \dfrac{S}{2} + S + \dfrac{S}{2} = 2S$

Rectangular Duct or Solid $\qquad L = \dfrac{b}{2} + a + \dfrac{b}{2} = a + b$

Vertical Cylinder $\qquad L = \dfrac{D}{2} + H + \dfrac{D}{2} = D + H$

**FIGURE 7.7**

Characteristic lengths for Eq. (7.43).

depending on the Rayleigh number. Sparrow and Stretton [14] performed experiments on a cube in air and water. They found that Eq. (7.43) underpredicted the experimental data by 4%—23% for the cube in air and 8%—15% for the cube in water.

## 7.7 Natural convection for enclosed spaces

Double-glazed windows, solar collectors, and building walls incorporate enclosed rectangular spaces. Thermos bottles, solar collectors, and storage containers for cryogenic liquids often incorporate concentric cylinders or spheres. The space between the outer walls of the enclosures is filled with a liquid or gas or is evacuated to even further decrease the heat transfer. This section considers natural convection in such enclosures. Much experimental and theoretical work has been done on natural

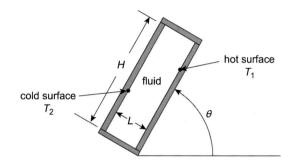

**FIGURE 7.8**

Rectangular enclosure.

convection in enclosed spaces. Catton [25] provided a comprehensive review of the research as of 1978, and many other papers have been written since then, especially due to the increasing interest in solar collectors. Correlation equations are presented below for some of the common geometries: rectangular enclosures, annular double-walled cylindrical enclosures, and annular double-walled spherical enclosures. Many more correlations can be found in research literature.

### 7.7.1 Enclosed rectangular space

Fig. 7.8 shows a rectangular enclosure. The enclosed space between isothermal walls has a height $H$ and a width $L$ between the walls. The space is tilted with an angle $\theta$ from the horizontal. The top and bottom surfaces of the enclosure are adiabatic. The hot surface is at temperature $T_1$, and the cold surface is at temperature $T_2$. To clarify the tilt angle: $\theta = 90°$ is a vertical space. $\theta = 0°$ is a horizontal space with cold surface up and hot surface down. $\theta = 180°$ is a horizontal space with hot surface up and cold surface down. For $0° < \theta < 90°$, the cold surface is up. For $90° < \theta < 180°$, the hot surface is up.

**Caution:** In the following correlation equations for rectangular enclosures, the characteristic length to use in calculating the Nusselt, Grashof, and Rayleigh numbers is the thickness of the gap, $L$, between the two sides of the enclosure. Do not get confused. In this section, we use $L$ as the gap thickness and $H$ as the height of the sides. In previous sections, $L$ was the height of surfaces. Just make sure to use the gap thickness, and not the height, in the calculation of the dimensionless numbers.

The enclosure has two plates enclosing the gaseous or liquid space plus the ends. Let us compare the basic conduction and convection equations. They are

$$q_{cond} = \frac{kA}{L}(T_1 - T_2) \quad \text{and} \quad q_{conv} = hA(T_1 - T_2) \tag{7.44}$$

We also have the definition of the Nusselt number:

$$\text{Nu} = \frac{hL}{k} \tag{7.45}$$

Putting Eqs. (7.44) and (7.45) together, we have

$$q_{conv} = (k \; \text{Nu})\frac{A}{L}(T_1 - T_2) \tag{7.46}$$

From Eq. (7.46), when Nu = 1, we have pure conduction across the space. When Nu $\neq$ 1, we have convection and an "effective" conductivity.

$$k_{eff} = k \ \text{Nu} \tag{7.47}$$

Fluid properties should be evaluated at the mean temperature of the plates, i.e., $(T_1 + T_2)/2$.

### 7.7.1.1 Horizontal rectangular enclosure

If the hot plate is at the top, there are no convective currents in the enclosure, and the heat flow through the fluid is solely by conduction. If the hot plate is at the bottom, convective currents occur when the Rayleigh number $\text{Ra}_L > 1708$. For lower values of $\text{Ra}_L$, the buoyant forces do not overwhelm the viscous forces in the fluid.

McAdams [2] recommends the following equations for air with the hot plate on the bottom:

$$\text{Nu} = \frac{hL}{k} = 0.21\text{Ra}_L^{1/4} \quad \text{for} \quad 10^4 < \text{Ra}_L < 3.2 \times 10^5 \tag{7.48}$$

$$\text{Nu} = \frac{hL}{k} = 0.075\text{Ra}_L^{1/3} \quad \text{for} \quad 3.2 \times 10^5 < \text{Ra}_L < 10^7 \tag{7.49}$$

Hollands, Unny, Raithby, and Konicek [21] proposed the following equation for air in a horizontal enclosure with a large aspect ratio ($H/L > 12$) with the hot plate on the bottom:

$$\text{Nu} = \frac{hL}{k} = 1 + 1.44\left[1 - \frac{1708}{\text{Ra}_L}\right]^* + \left[\left(\frac{\text{Ra}_L}{5830}\right)^{1/3} - 1\right]^* \quad \text{for} \quad \text{Ra}_L < 10^8 \tag{7.50}$$

**Important Note:** For the terms marked $[\ ]^*$ in Eq. (7.50): If the value inside the bracket is negative, the term should be set to zero.

Globe and Dropkin [22] proposed the following equation for liquids ($0.02 < \text{Pr} < 8750$) in a horizontal enclosure with the hot plate on the bottom:

$$\text{Nu} = \frac{hL}{k} = 0.069\text{Ra}_L^{1/3} \ \text{Pr}^{0.074} \quad \text{for} \quad 1.5 \times 10^5 < \text{Ra}_L < 6.8 \times 10^8 \tag{7.51}$$

### 7.7.1.2 Vertical rectangular enclosure

McAdams [2] recommends the following equations for a vertical enclosed air space:

$$\text{Nu} = \frac{hL}{k} = 0.20\left(\frac{H}{L}\right)^{-1/9}\text{Ra}_L^{1/4} \quad \text{for} \quad 2 \times 10^4 < \text{Ra}_L < 2.1 \times 10^5 \tag{7.52}$$

$$\text{Nu} = \frac{hL}{k} = 0.071\left(\frac{H}{L}\right)^{-1/9}\text{Ra}_L^{1/3} \quad \text{for} \quad 2.1 \times 10^5 < \text{Ra}_L < 1.1 \times 10^7 \tag{7.53}$$

ElSherbiny, Raithby, and Hollands [23] recommend the following equations for a vertical enclosed air space for different aspect ratios ($H/L$).

$$\text{Aspect Ratio} \ (H \ / \ L) = 5 \quad \text{and} \quad \text{Ra}_L < 10^8$$

$$\text{Nu} = \frac{hL}{k} = \left\{\left[1 + \left(\frac{0.193\ \text{Ra}_L^{1/4}}{1 + (1800/\text{Ra}_L)^{1.289}}\right)^3\right]^{1/3}, \quad 0.0605\ \text{Ra}_L^{1/3}\right\}_{\text{max}} \tag{7.54}$$

There are two expressions inside the parentheses. Use the larger value.

$$\text{Aspect Ratio } (H/L) = 10 \quad \text{and} \quad \text{Ra}_L < 9.7 \times 10^6$$

$$\text{Nu} \quad = \quad \frac{h\,L}{k} \quad = \left\{ \left[ 1 + (0.125 \text{ Ra}_L^{0.28})^9 \right]^{1/9}, \quad 0.061 \text{ Ra}_L^{1/3} \right\}_{\text{max}} \tag{7.55}$$

There are two expressions inside the parentheses. Use the larger value.

$$\text{Aspect Ratio } (H/L) = 20 \quad \text{and} \quad \text{Ra}_L < 2 \times 10^6$$

$$\text{Nu} \quad = \quad \frac{h\,L}{k} \quad = \left[ 1 + \left( 0.064 \text{ Ra}_L^{1/3} \right)^{6.5} \right]^{1/6.5} \tag{7.56}$$

$$\text{Aspect Ratio } (H/L) = 40 \quad \text{and} \quad \text{Ra}_L < 2 \times 10^5$$

$$\text{Nu} \quad = \quad \frac{h\,L}{k} \quad = \left[ 1 + \left( 0.0303 \text{ Ra}_L^{0.402} \right)^{11} \right]^{1/11} \tag{7.57}$$

$$\text{Aspect Ratio } (H/L) = 80 \quad \text{and} \quad \text{Ra}_L < 3 \times 10^4$$

$$\text{Nu} \quad = \quad \frac{h\,L}{k} \quad = \left[ 1 + \left( 0.0227 \text{ Ra}_L^{0.438} \right)^{18} \right]^{1/18} \tag{7.58}$$

$$\text{Aspect Ratio } (H/L) = 110 \quad \text{and} \quad \text{Ra}_L < 1.2 \times 10^4$$

$$\text{Nu} \quad = \quad \frac{h\,L}{k} \quad = \left[ 1 + \left( 0.0607 \text{ Ra}_L^{1/3} \right)^{18} \right]^{1/18} \tag{7.59}$$

The above correlations are for vertical spaces with air. The following two correlations for constant heat flux on the hot surface are recommended by MacGregor and Emery [24] for fluids with higher Prandtl numbers than air, e.g., liquids.

$$\text{Nu} = \frac{h\,L}{k} = \quad 0.42 \text{ Ra}_L^{1/4} \text{Pr}^{0.012} \left( \frac{H}{L} \right)^{-0.3} \tag{7.60}$$

Eq. (7.60) is for $10 < H/L < 40 \qquad 1 < \text{Pr} < 2 \times 10^4 \qquad 10^4 < \text{Ra}_L < 10^7$

$$\text{Nu} = \frac{h\,L}{k} = \quad 0.46 \text{ Ra}_L^{1/3} \tag{7.61}$$

Eq. (7.61) is for $1 < H/L < 40 \qquad 1 < \text{Pr} < 20 \qquad 10^6 < \text{Ra}_L < 10^9$

### 7.7.1.3 Inclined rectangular enclosure
Based on their experiments, Hollands, Unny, Raithby, and Konicek [21] proposed the following equation for air in an inclined rectangular enclosure of large aspect ratio ($H/L > 12$) with the hot plate

on the bottom. It is valid for angles $\theta \le 60°$ and should give good results for angles as large as 75 degrees.

$$Nu = \frac{hL}{k} = 1 + 1.44 \left[ 1 - \frac{1708}{Ra_L \cos \theta} \right]^* \left( 1 - \frac{1708(\sin 1.8\,\theta)^{1.6}}{Ra_L \cos \theta} \right)$$
$$+ \left[ \left( \frac{Ra_L \cos \theta}{5830} \right)^{1/3} - 1 \right]^* \tag{7.62}$$

for $Ra_L < 10^5$

**Important Note:** For the terms marked $[\ ]^*$ in Eq. (7.62): If the value inside the bracket is negative, the term should be set to zero.

## Example 7.6
### Solar collector

*Problem*
The flat plate solar collector shown in Fig. 7.9 consists of a glass cover and an absorber plate. The collector is 2 m high and 1.5 m wide and is tilted at an angle of 60 degrees from the horizontal. The air space between the cover and the absorber plate is 3 cm thick. The absorber plate temperature is 75 C and the cover temperature is 35 C. What is the heat transfer rate by natural convection from the absorber plate?

*Solution*
We will use Eq. (7.62) for the solution.

$$Nu = \frac{hL}{k} = 1 + 1.44 \left[ 1 - \frac{1708}{Ra_L \cos \theta} \right]^* \left( 1 - \frac{1708(\sin 1.8\,\theta)^{1.6}}{Ra_L \cos \theta} \right) + \left[ \left( \frac{Ra_L \cos \theta}{5830} \right)^{1/3} - 1 \right]^* \tag{7.62}$$

The average temperature of the cover and absorber plate is $(35 + 75)/2 = 55$ C.
Air at 55 C has the following properties: $k = 0.028$ W/mC; $v = 1.84 \times 10^{-5}$ m²/s; Pr $= 0.708$.
The Rayleigh number based on the air gap $L$ is

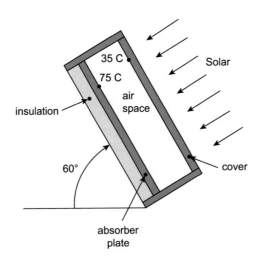

35 C
75 C
air space
insulation
Solar
60°
cover
absorber plate

**FIGURE 7.9**

Solar collector for Example 7.6.

$$\text{Ra}_L = \frac{g\beta(T_s - T_\infty)(L)^3}{v^2}\text{Pr} = \frac{(9.807)[1/(55 + 273.15)](75 - 35)(0.03)^3}{(1.84 \times 10^{-5})^2}(0.708) = 6.750 \times 10^4$$

Looking at Eq. (7.62), we need values of

$$\text{Ra}_L \cos 60° = (6.750 \times 10^4)(0.5) = 3.375 \times 10^4$$
$$\sin(1.8\theta) = \sin[(1.8)(60)] = \sin 108° = 0.951$$

Putting the values into Eq. (7.62), we have

$$\text{Nu} = \frac{hL}{k} = 1 + 1.44\left[1 - \frac{1708}{3.375 \times 10^4}\right]^*\left(1 - \frac{1708(0.951)^{1.6}}{3.375 \times 10^4}\right) + \left[\left(\frac{3.375 \times 10^4}{5830}\right)^{1/3} - 1\right]^* = 3.0989$$

$$h = \frac{\text{Nu } k}{L} = \frac{(3.0989)(0.028)}{0.03} = 2.892 \text{ W/m}^2 \text{ C}$$

The natural convection heat transfer is

$$q_{conv} = h \ A \ (T_1 - T_2) = (2.892) \ [(2)(1.5)] \ (75 - 35) = \textbf{347 W}$$

## 7.7.2 Annular space between concentric cylinders

Concentric cylinders are shown in Fig. 7.10. The inner cylinder has an outer radius $r_1$, and the outer cylinder has an inner radius $r_2$. The surfaces of the cylinders are at temperatures $T_1$ and $T_2$, respectively. The annular space between the cylinders is filled with a gas or a liquid. The cylinders are horizontal and have a length $L$.

If the space between the cylinders contained a solid, we showed in an earlier chapter that the conduction through a cylindrical shell is

$$q = \frac{2\pi kL}{\ln \ (r_2/r_1)}(T_1 - T_2) \tag{7.63}$$

With gas or liquid in the annular space, we have motion and convection rather than conduction. In Eq. (7.63), the thermal conductivity $k$ of the motionless fluid is replaced by an effective conductivity $k_{eff}$ for the fluid in motion.

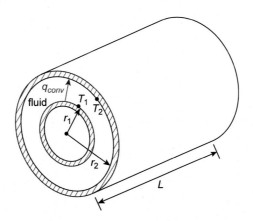

**FIGURE 7.10**

Concentric cylinders.

Raithby and Hollands [26] recommend the following equation for the effective conductivity for use with Eq. (7.63):

$$\frac{k_{eff}}{k} = 0.386 \left( \frac{Pr}{0.861 + Pr} \right)^{1/4} Ra_{cyl}^{1/4} \tag{7.64}$$

where : The length to use in $Ra_{cyl}$ is $L_{cyl} = \dfrac{2 \; [\ln \; (r_2/r_1)]^{4/3}}{\left( r_1^{-3/5} + r_2^{-3/5} \right)^{5/3}}$ \hfill (7.65)

Eq. (7.64) should give good results for $0.7 \leq Pr \leq 6000$ and $Ra_{cyl} \leq 10^7$. As $k_{eff}$ has to be greater than $k$: If $k_{eff}/k$ calculated by Eq. (7.64) is less than unity, $k_{eff}$ should be made equal to $k$.

Fluid properties are taken at the average temperature of the surfaces, i.e., $(T_1 + T_2)/2$.

### 7.7.3 Space between concentric spheres

Concentric spheres are shown in Fig. 7.11. The inner sphere has an outer radius $r_1$ and the outer sphere has an inner radius $r_2$. The surfaces of the spheres are at temperatures $T_1$ and $T_2$, respectively. The annular space between the spheres is filled with a gas or a liquid.

If the space between the spheres had been solid, we showed in Chapter 3 that the conduction through a spherical shell is

$$q = \frac{4\pi k}{(1/r_1) - (1/r_2)} \; (T_1 - T_2) \tag{7.66}$$

With gas or liquid in the annular space, we have motion and convection rather than conduction. In Eq. (7.66), the thermal conductivity $k$ of the motionless fluid is replaced by an effective conductivity $k_{eff}$ for the fluid in motion.

Raithby and Hollands [26] recommend the following equation for the effective conductivity for use with Eq. (7.66):

$$\frac{k_{eff}}{k} = 0.74 \left( \frac{Pr}{0.861 + Pr} \right)^{1/4} Ra_{sph}^{1/4} \tag{7.67}$$

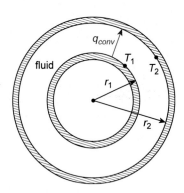

**FIGURE 7.11**

Concentric spheres.

$$\text{where : The length to use in Ra}_{sph} \text{ is } L_{sph} = \frac{[(1/r_1) - (1/r_2)]^{4/3}}{2^{1/3} \left( r_1^{-7/5} + r_2^{-7/5} \right)^{5/3}} \qquad (7.68)$$

Eq. (7.67) should give good results for $0.7 \leq \text{Pr} \leq 4000$ and $\text{Ra}_{sph} \leq 10^4$. As $k_{eff}$ has to be greater than $k$: If $k_{eff}/k$ calculated by Eq. (7.67) is less than unity, $k_{eff}$ should be made equal to $k$. Fluid properties are taken at the average temperature of the surfaces, i.e., $(T_1 + T_2)/2$.

## Example 7.7
### Double-walled spherical container

*Problem*

A liquid at 50 C is stored in a double-walled spherical container. The inner sphere is thin-walled and has a diameter of 1.5 m. The outer sphere is at 20 C. The space between the spheres is 5 cm thick and is filled with air at 0.5 atm. What is the rate of natural convection heat flow out of the liquid?

*Solution*

$$r_1 = D_1/2 = 1.5/2 = 0.75 \text{ m} \quad \text{and} \quad r_2 = r_1 + \text{thickness of air gap} = 0.75 + 0.05 = 0.80 \text{ m}$$

The properties of the air in the gap are taken at the average temperature $T_{avg} = (50+20)/2 = 35$ C and 0.5 atm pressure. Air properties are given in the tables for 1 atm pressure. The thermal conductivity and absolute viscosity are not significantly affected by pressure. However, the kinematic viscosity is affected as its definition includes density. The air may be considered to be an ideal gas. The ideal gas law is $P = \rho RT$. This shows that, with a constant temperature, the pressure is directly proportional to the density. In our case, the pressure has been halved. Therefore, the density at 0.5 atm is half of the density at 1 atm.

Regarding the kinematic viscosity $v$ : $v$ at 0.5 atm $= \dfrac{\mu}{\rho \text{ at 0.5 atm}} = \dfrac{\mu}{(0.5)(\rho \text{ at 1 atm})}$

Putting in values for $\mu$ and $\rho$ for 1 atm and 35 C, we have

$$v \text{ at 0.5 atm} = \frac{\mu}{\rho \text{ at 0.5 atm}} = \frac{\mu}{(0.5)(\rho \text{ at 1 atm})} = \frac{1.89 \times 10^{-5}}{(0.5)(1.150)} = 3.287 \times 10^{-5} \text{ m}^2/\text{s}$$

The other properties of air at 35 C are $k = 0.0267$ W/m C and Pr $= 0.711$.
We will use Eqs. (7.66), (7.67), and (7.68).

From Eq. (7.68), $L_{sph} = \dfrac{[(1/0.75)-(1/0.80)]^{4/3}}{2^{1/3}[(0.75)^{-7/5}+(0.80)^{-7/5}]^{5/3}} = 0.005006$

Using this value to get $\text{Ra}_{sph}$,

$$\text{Ra}_{sph} = \frac{g\beta(T_s - T_\infty)(L_{sph})^3}{v^2}\text{Pr} = \frac{(9.807)[1/(35+273.15)](50-20)(0.005006)^3}{(3.287 \times 10^{-5})^2}(0.711) = 78.82$$

From Eq. (7.67),

$$\frac{k_{eff}}{k} = 0.74 \left(\frac{\text{Pr}}{0.861 + \text{Pr}}\right)^{1/4} \text{Ra}_{sph}^{1/4} = 0.74 \left(\frac{0.711}{0.861 + 0.711}\right)^{1/4}(78.82)^{1/4} = 1.808$$

Finally, from Eq. (7.66),

$$q_{conv} = \frac{4\pi k_{eff}}{(1/r_1) - (1/r_2)}(T_1 - T_2) = \frac{4\pi[(1.808)(0.0267)]}{(1/0.75) - (1/0.80)}(50 - 20) = \mathbf{218 \text{ W}}$$

## 7.8 Natural convection between vertical fins

The addition of fins to a surface enhances the heat transfer for the surface. Fins were discussed in Chapter 3. However, that chapter only discussed single fins convecting to an open adjacent fluid. This section discusses fins that are closely spaced. In this situation, the flow field in the fluid adjacent to a fin impacts the flow field of the fin next to it. Such interaction affects the convective coefficient for the fins and hence affects the heat transfer.

In particular, we will be looking at vertical straight rectangular fins on a vertical surface. A picture of such fins were shown in Fig. 3.27; specifically, the picture of the fins on the power transformer.

Fig. 7.12 shows three vertical fins attached to a vertical surface. Only three fins are shown, but there can indeed be many fins on the surface. The fins have thickness $t$, and the space between the fins is $S$. The fins are $L$ high by $b$ wide. The surface to which the fins are attached is $L$ high and $W$ wide.

Elenbaas [27] considered the heat transfer in vertical channels both analytically and experimentally. Bar-Cohen and Rohsenow [28] later provided the following Nusselt Number correlation equation for isothermal vertical plates:

$$\mathrm{Nu}_s = \left( \frac{576}{[\mathrm{Ra}_s(S/L)]^2} + \frac{2.873}{[\mathrm{Ra}_s(S/L)]^{1/2}} \right)^{-1/2} \tag{7.69}$$

where

$$\mathrm{Nu}_s = \frac{h}{k} \frac{S}{} \tag{7.70}$$

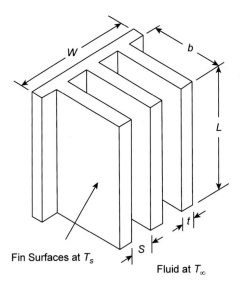

Fin Surfaces at $T_s$

Fluid at $T_\infty$

**FIGURE 7.12**

Vertical rectangular fins on a surface.

$$Ra_s = \frac{g \beta (T_s - T_\infty) S^3}{v^2} \; Pr \tag{7.71}$$

The material properties in Eqs. (7.69)–(7.71) are those of the fluid. They should be taken at the film temperature $T_f = \frac{T_s + T_\infty}{2}$.

After obtaining $h$ from Eq. (7.70), the convective heat transfer for an array of $N$ fins on a surface is

$$q = h \, [(2N)(b)(L)] \; (T_s - T_\infty) \tag{7.72}$$

(Note: Eq. (7.72) ascribes the same convective coefficient to all surfaces of the fins in the array. Actually, the two outer surfaces of the fin array would have a different $h$ value than the surfaces between fins. Assuming that the two outer surfaces have the same $h$ as the other surfaces should, in most cases, have insignificant impact on the result.)

Bar-Cohen and Rohsenow also determined the optimum spacing, $S_{opt}$, between fins. This is the spacing that maximizes the heat transfer from an array of fins. The optimum spacing is

$$S_{opt} = 2.714 \left(\frac{Ra_s}{S^3 L}\right)^{-1/4} \tag{7.73}$$

We close this section with two examples for fins.

## Example 7.8
### Fins on a tank wall

*Problem*

Fins have been added to the vertical wall of a small tank to assist in cooling the liquid inside the tank. Eleven vertical fins have been added to a square section of the tank wall that is 10 cm by 10 cm. The fins are 0.8 mm thick, 10 cm high, and 2 cm wide. They are equally spaced on the total width of the wall section. The fins are at a temperature of 130 C, and the adjacent air is at 30 C. What is the rate of heat removal by the fins?

*Solution*

$$T_f = \frac{T_s + T_\infty}{2} = \frac{130 + 30}{2} = 80 \; C$$

For air at 80 C, $k = 0.0295 \; W/m \; C$, $v = 21.01 \times 10^{-6} \; m^2/s$, $Pr = 0.715$, $\beta = \frac{1}{T_f} = \frac{1}{80 + 273.15} = 0.00283/K$

There are 11 fins equally spaced over the width of the wall section that is 10 cm wide. The spacing between fins is

$$S = \frac{W - N \, t}{N - 1} = \frac{0.10 - 11 \, (0.0008)}{10} = 0.00912 \; m$$

From Eq. (7.71),

$$Ra_s = \frac{g\beta(T_s - T_\infty)S^3}{v^2} Pr = \frac{(9.807)(0.00283)(130 - 30)(0.00912)^3}{(21.01 \times 10^{-6})^2}(0.715) = 3412$$

From Eq. (7.69),

$$Nu_s = \left(\frac{576}{[Ra_s(S/L)]^2} + \frac{2.873}{[Ra_s(S/L)]^{1/2}}\right)^{-1/2}$$

For this problem, $Ra_s = 3412$, $S = 0.00912 \; m$, and $L = 0.10 \; m$. Putting these values into Eq. (7.69), we get $Nu_s = 2.434$.

From Eq. (7.70), $h = \frac{k \; Nu_s}{S} = \frac{(0.0295)(2.434)}{0.00912} = 7.88 \; W/m^2 \; C$.

Finally, we get the heat transfer rate from Eq. (7.72):

$$q = h[(2N)(b)(L)](T_s - T_\infty) = (7.88)(2)(11)(0.02)(0.10)(130 - 30) = 34.7 \text{ W}$$

The rate of heat removal by the 11 fins is 34.7 W.

The second fin example deals with designing a fin array for removing heat at a specified rate from a plate.

## Example 7.9
### Fins for cooling of electronic components

*Problem*

Heat-producing electronic components are mounted on one side of a vertical plate. Vertical fins are to be added on the other side of the plate to increase the convective heat transfer from the plate to the adjacent air. The fins need to remove heat at a rate of 25 W from the plate. The plate is 8 cm wide and 12 cm high. The fins have a thickness of 1 mm and are 12 cm high and 4 cm wide. They are equally spaced. The plate and the fins are at 155 C and the adjacent air is at 25 C.

(a) How many fins are needed to remove the specified heat, and what is the spacing S between them?

(b) If the fins had not been added to the plate, what would have been the natural convection heat transfer rate from the surface?

*Solution*

$$T_f = \frac{T_s + T_\infty}{2} = \frac{155 + 25}{2} = 90 \text{ C}$$

For air at 90 C,

$$k = 0.0302 \text{ W/m C}, \ v = 22.02 \times 10^{-6} \text{ m}^2/\text{s}, \ \text{Pr} = 0.713, \ \beta = \frac{1}{T_f} = \frac{1}{90 + 273.15} = 0.00275/\text{K}$$

The heat transfer rate due to the fins is given by Eq. (7.72):

$$q = h \ [(2N)(b)(L)] \ (T_s - T_\infty) \tag{7.72}$$

For this problem, $q = 25$ W, $b = 0.04$ m, $L = 0.12$ m, and $(T_s - T_\infty) = 155 - 25 = 130$ C.
Putting the values into Eq. (7.72), we get the requirement

$$h \ N = 20.03 \tag{7.74}$$

The convective coefficient $h$ is a function of the spacing $S$ between fins. A spacing that results in a higher convective coefficient will require a lesser number of fins to achieve the desired heat transfer $q$. Let us continue with our analysis:
From Eq. (7.71),

$$\text{Ra}_s = \frac{g\beta(T_s - T_\infty)S^3}{v^2}\text{Pr} = \frac{(9.807)(0.00275)(155 - 25)S^3}{(22.02 \times 10^{-6})^2}(0.713)$$

$$\text{and Ra}_s = 5.162 \times 10^9 \ S^3 \tag{7.75}$$

Combining Eqs. (7.69) and (7.70), we have

$$h = \left(\frac{576}{[\text{Ra}_s(S/L)]^2} + \frac{2.873}{[\text{Ra}_s(S/L)]^{1/2}}\right)^{-1/2} \left(\frac{k}{S}\right) \tag{7.76}$$

For this problem, $k = 0.0302$ W/m C and $L = 0.12$ m. Using these values and Eqs. (7.74) and (7.75), Eq. (7.76) can be changed to

$$20.03 = \frac{0.0302N}{S} \left(\frac{576}{(4.302 \times 10^{10} \ S^4)^2} + \frac{2.873}{(4.302 \times 10^{10} \ S^4)^{1/2}}\right)^{-1/2} \tag{7.77}$$

For a desired number of fins $N$, Eq. (7.77) may be solved for the required spacing $S$ between the fins. Of course, we have to make sure that the overall width of the fin array is less than or equal to the width $W$ of the plate. That is, we need to verify that

$$(N-1) S \ + \ N t \ \leq \ W \tag{7.78}$$

Summarizing: There are multiple answers to Part (a) of this problem. We first decide how many fins we want. Then we solve Eq. (7.77) for the required fin spacing that will give us the specified heat removal rate of 25 W. For example, let us say that we want 15 fins on the wall. Then N = 15 and Eq. (7.77) becomes

$$20.03 = \frac{0.0302(15)}{S} \left( \frac{576}{(4.302 \times 10^{10} \ S^4)^2} + \frac{2.873}{(4.302 \times 10^{10} \ S^4)^{1/2}} \right)^{-1/2} \tag{7.79}$$

Eq. (7.79) can be solved for spacing $S$. One way to do this is by using Excel *Goal Seek*. If we wish to use *Goal Seek*, we rearrange Eq. (7.79) to the form

$$20.03 - \frac{0.0302(15)}{S} \left( \frac{576}{(4.302 \times 10^{10} \ S^4)^2} + \frac{2.873}{(4.302 \times 10^{10} \ S^4)^{1/2}} \right)^{-1/2} = 0 \tag{7.80}$$

We tell *Goal Seek* to make the left side of Eq. (7.80) zero by changing the value of $S$. When we did this, we got the result $S = 0.002924$ m = 2.924 mm. Using Eq. (7.78), we find that the total width of the fin array is 5.3 cm and the fin array will indeed fit on the 8-cm-wide plate.

We have seen that an array of 15 fins will remove the specified 25 W from the plate. But there are several possible solutions to Part (a) of this problem. Indeed, we could have any number of fins from 3 to 25 to remove the specified heat. We would just input the number $N$ of the fins into Eq. (7.77) and solve the equation for the needed fin spacing $S$. We did this for several numbers of fins and got the results: For 3 fins, the needed spacing is 6.009 mm; for 5 fins, the spacing is 4.398 mm; for 10 fins, the spacing is 3.367 mm; and for 20 fins, the spacing is 2.652 mm. It turns out that 25 fins is the maximum number of fins that can be used. A greater number of fins give a fin array width that, per Eq. (7.78), is greater than the width of the plate.

**(b)** We can use Eq. (7.10) for solution of Part (b):

$$Nu = \frac{h \ L}{k} = 0.68 + \frac{0.670 \ Ra_L^{1/4}}{\left[ 1 + (0.492/Pr)^{9/16} \right]^{4/9}} \tag{7.10}$$

For this problem, $L = 0.12$ m. Calculating $Ra_L$, we have

$$Ra_L = \frac{g\beta(T_s - T_\infty)L^3 Pr}{v^2} = \frac{(9.807)(0.00275)(155 - 25)(0.12)^3(0.713)}{(22.02 \times 10^{-6})^2} = 8.909 \times 10^6$$

From Eq. (7.10), we have

$$h = \left[ 0.68 + \frac{0.670 \ Ra_L^{1/4}}{\left[ 1 + (0.492/Pr)^{9/16} \right]^{4/9}} \right] \left( \frac{k}{L} \right) = \left[ 0.68 + \frac{(0.670) \ (8.909 \times 10^6)^{1/4}}{\left[ 1 + (0.492/0.713)^{9/16} \right]^{4/9}} \right] \left( \frac{0.0302}{0.12} \right)$$

$$h = 7.245 \ \text{W/m}^2 \ \text{C}$$

The heat flow from the unfinned plate is

$$q = h \ A \ (T_s - T_\infty) = (7.245)[(0.08) \ (0.12)] \ (155 - 25) = 9.04 \ \text{W}$$

Without fins, the heat transfer by natural convection from the plate would have been 9.04 W.

# 7.9 Chapter summary and final remarks

In this chapter, we discussed natural (free) convection for a variety of surfaces and enclosures. Much experimentation and analysis has been done on this topic, and we have provided several equations for predicting the convective coefficients. One must keep in mind, however, that the correlation equations come from carefully-controlled laboratory experiments. Laboratory conditions, in most cases, are not duplicated in practice. Hence, the results from the correlation equations may differ by as much as 25% or more from results actually obtained in practice. The best situation, of course, would be to obtain convective coefficients from experimentation on the actual objects of interest. This, however, is usually not possible due to cost, schedule, and other constraints. Although the results from the correlation equations are "rough," they do indeed provide valuable information for conceptual and design purposes.

We did not include a discussion of combined free and forced convection. In some situations, fluid flow is such that convection is not entirely natural (free) or not entirely forced. Flow velocities may be in excess of those caused by the density gradients of natural convection but may not be large enough to completely overwhelm the effects of the density gradients. In these cases of *mixed* convection, the Nusselt Numbers for sole natural convection and sole forced convection have to be combined in some fashion to accurately model the situation.

In general, combined free and forced convection should be considered if $\frac{Gr_L}{Re_L^2} \approx 1$.

If $\frac{Gr_L}{Re_L^2} \ll 1$, free convection is negligible and the Nusselt number is a function of the Reynolds and Prandtl numbers. That is $Nu_L = f\ (Re_L, Pr)$

If $\frac{Gr_L}{Re_L^2} \gg 1$, forced convection is negligible and the Nusselt number is a function of the Grashof and Prandtl numbers. That is, $Nu_L = f\ (Gr_L, Pr)$

Combined free and forced convection is discussed in Refs. [29–35].

We now move on to the next chapter that deals with heat exchangers. The chapter is very application-oriented and should be of much interest to many of you.

## 7.10 Problems

Notes:

- If needed information is not given in the problem statement, then use the Appendix for material properties and other information. If the Appendix does not have sufficient information, then use the Internet or other reference sources. Some good reference sources are mentioned at the end of Chapter 1. If you use the Internet, double-check the validity of the information by using more than one source.
- Your solutions should include a sketch of the problem.
- Caution: Make sure that you use absolute temperatures (Kelvin or Rankine) if the problem involves radiation.
- In all problems, unless otherwise stated, the fluid pressure is atmospheric. Gas properties in the Appendix are at atmospheric pressure. If a problem has a gas at other than atmospheric pressure, affected properties such as density and kinematic viscosity should be modified accordingly through use of the ideal gas law.

**7-1** A thin vertical flat plate is 50 cm high and 20 cm wide. Its two surfaces are at 200 C and the plate convects to the surrounding 30 C air. What is the heat transfer from the two sides of the plate to the air?

**7-2** A thin, square electrical heater is vertical and is 20 cm by 20 cm. One side of the heater is perfectly insulated and the other side has convection to the room air. The air is at 25 C and the heater dissipates 40 W. The active surface of the heater can be considered as a constant flux surface.

**(a)** What is the convective coefficient at 10, 15, and 20 cm from the bottom of the plate?

**(b)** What is the average convective coefficient for the plate?

**7-3** A vertical square plate is 20 cm by 20 cm. Its two surfaces are at 80 C. The plate is in a tank of 30 C water. What is the rate of heat transfer from the plate to the water?

**7-4** A square flat plate is 1.5 m by 1.5 m. Its two surfaces are maintained at 120 C. The plate convects from its two surfaces to the room air which is at 20 C.

(a) What is the rate of heat transfer to the air if the plate is vertical?

(b) What is the rate of heat transfer to the air if the plate is inclined 45 degrees with the horizontal?

**7-5** A thin flat plate is 10 cm by 20 cm. It is in a tank of 20 C water. The 10 cm edge is horizontal and the plate is inclined at an angle of 30 degrees with the horizontal. The plate's two surfaces are at 90 C. What is the rate of convective heat transfer from the plate to the water?

**7-6** A vertical cylinder of 5 cm diameter and 20 cm length has a surface temperature of 150 C. It is in room air at 20 C.

(a) In determining the heat transfer from the cylinder, can the surface of the cylinder be treated as a vertical flat plate?

(b) What is the rate of heat transfer from the cylinder to the air?

**7-7** A circular plate, 20 cm in diameter, is suspended horizontally in 20 C air. If the top and bottom of the plate are both at 100 C, what is the rate of heat transfer to the air?

**7-8** A square flat plate is 15 cm by 15 cm. It is oriented horizontally in 20 C air. The top and bottom of the plate are at 80 C. What is the rate of heat transfer from the plate to the air?

**7-9** A thin flat plate has the shape of an equilateral triangle with sides of length 15 cm. The plate is suspended horizontally in a tank of 40 C water. The top of the plate is at 80 C and the bottom is at 60 C. What is the rate of heat transfer from the plate to the water?

**7-10** A circular flat plate heater has a diameter of 25 cm. It is suspended vertically in 30 C water. If the surface of the heater cannot exceed 95 C, what is the maximum allowed power to the heater?

**7-11** A square plate is 1 m by 1 m. It is inclined 30 degrees with the horizontal. The downward facing surface of the plate is perfectly insulated and the upward facing surface receives solar radiation. At equilibrium, the upper surface convects to the surrounding 20 C air with a constant flux of 700 W/m². What is the equilibrium mean temperature of the upper surface?

**7-12** A horizontal plate is 30 cm by 50 cm. Its bottom surface is perfectly insulated and its top surface is a constant flux surface convecting to the adjacent 25 C air. The mean temperature of the top surface of the plate is 55 C. What is the rate of convective heat transfer from the plate to the air?

**7-13** A metal street sign is about 0.5 m by 0.5 m. Solar radiation at a rate of 250 W/m² hits the sign and heats it. If the sign has an absorptivity of 0.6 for the solar radiation and an emissivity of 0.7, what is the surface temperature of the sign? Assume that the air is quiet and the air and surroundings are at 30 C?

**7-14** A horizontal cylinder has a diameter of 1.5 cm and is in water at 25 C. The surface of the cylinder is at 90 C. What is the heat transfer from the cylinder to the water per meter length of cylinder?

**7-15** Steam is flowing through a horizontal 2″ Sch 40 steel pipe. The outer surface of the pipe has a temperature of 225 F. The pipe is used to help heat a warehouse.

(a) If the pipe is 70 feet long, estimate the rate of convective heat transfer to the 65 F room air.

**(b)** If the pipe's outer surface has an emissivity of 0.5, what is the rate of radiative heat transfer to the room? Assume the room walls are at the same temperature as the room air, that is, 65 F.

**7-16** Oil at 60 C flows through a horizontal thin-walled copper tube of 1.9 cm outside diameter. The tube is in 25 C water. What is the rate of heat transfer to the water per meter length of tube?

**7-17** A natural gas hot water heater is in the basement of a house. The heater is 18$''$ diameter and 55$''$ high. The air in the basement is at 65 F. If the top and side surfaces of the heater are at 95 F, estimate the heat loss from the heater. Only include heat transfer from the top and side of the heater. The heat transfer from the bottom of the heater is negligible.

**7-18** A cylindrical immersion heater has a rating of 700 W. It has a diameter of 2.5 cm and a length of 40 cm and is oriented horizontally in a large tank of water. The water is at 20 C.
**(a)** At steady state, what is the surface temperature of the heater?
**(b)** If the power to the heater is accidentally turned on when the heater is in 20 C air, what would be the steady-state temperature of the heater's surface?

**7-19** An exhaust duct from a furnace is uninsulated and provides heating to a factory. The duct is horizontal and of 20 cm diameter. The gases inside the duct are at 250 C and the room air is at 15 C. What is the rate of convective heat transfer to the air per meter length of duct?

**7-20** An incandescent light bulb is very inefficient. Only about 5% of the input power produces light. The remainder is absorbed by the glass bulb, which then dissipates the heat by convection to the surrounding air and by radiation to the surroundings. A 75 W light bulb is modeled as a 10 cm diameter sphere. The emissivity of the glass is about 0.9, and both the air and the walls of the room are at 25 C. Estimate the temperature of the glass bulb.

**7-21** A 5 cm diameter sphere has a surface temperature of 50 C. It is in 20 C water. What is the rate of heat transfer from the sphere to the water?

**7-22** An electric heater is inside a hollow aluminum sphere. The sphere's outside diameter is 3 cm, and the sphere is in air at 20 C. What electric power input to the heater is needed for the sphere's outer surface to be at 80 C?

**7-23** Do Problem 7-22 if the sphere is in 20 C water rather than air.

**7-24** An incandescent light bulb can be modeled as a sphere of 10 cm diameter. Operating in 20 C air, the outer surface of the light bulb is at 150 C. Assuming that the emissivity of the bulb's surface is 0.8 and that the walls of the room are at 20 C, what is the rate of heat transfer to the room air and surroundings? Include both convective and radiative modes.

**7-25** An electric heater is inside an aluminum cube that is 8 cm by 8 cm by 8 cm. The cube is on a table and it may be assumed that heat transfer from the cube's bottom surface is negligible. Power to the heater causes all five exposed surfaces of the cube to be at 150 C. The room air is at 15 C. What is the rate of heat transfer from the cube to the air?

**7-26** Thin electronic components that generate heat are sandwiched between two aluminum plates. The plates are rectangular, 20 cm by 30 cm. The assembly is positioned horizontally. As the plates are of high conductivity, we will assume that both plates and the components are at the same temperature. The components will fail if their temperature reaches 80 C.
**(a)** If the assembly is in 20 C air, what is the maximum allowable power to the components?
**(b)** If the assembly is in 20 C water, what is the maximum allowable power to the components?

**7-27** A horizontal heating duct is 30 cm wide and 15 cm high. The outer surface of the duct is at 65 C and the temperature of the room air is 22 C. What is the rate of convective heat transfer to the room air per meter length of duct?

**7-28** A cube of copper, 5 cm by 5 cm by 5 cm, cools in a tank of 20 C water. What is the rate of heat transfer from the cube to the water when the cube's surface temperature is 80 C?

**7-29** In the winter, a homeowner adds storm windows to his house. Consider a vertical window that is 0.9 m wide and 1.5 m high, which has a storm window on it. The window/storm window combination forms an enclosed cavity having a 5 cm air gap between the windows. The inside of the storm window is at −5 C, and the outside of the house window is at 15 C. What is the rate of convective heat transfer through the air gap?

**7-30** Instead of using a storm window, the homeowner of Problem 7-29 decides to replace the house window with a double-glazed window. The space between the glass panes is 1.25 cm and it is filled with argon. For the same temperature conditions as Problem 7-29, determine the rate of convective heat transfer through the argon space if (a) the argon is at 1 atm pressure and (b) the argon is at 0.5 atm pressure.

**7-31** A solar collector consists of a glass cover plate and an absorber plate separated by a 4 cm thick air space. The assembly is a sealed square box of sides 2 m by 2 m. The collector is inclined at an angle of 50 degrees from the horizontal. If the temperature of the cover plate is 35 C and the absorber plate is at 65 C, what is the rate of convective heat transfer across the air space?

**7-32** A rectangular sealed assembly has two parallel square plates (60 cm by 60 cm) separated by an air space that is 4 cm thick. The hotter plate is at 70 C and the cooler plate is at 0 C. Determine the convective heat transfer between the plates if the assembly is

**(a)** vertical.

**(b)** horizontal with the cooler plate on the bottom.

**(c)** horizontal with the hotter plate on the bottom.

**7-33** A vertical wall section of a house consists of two plywood layers separated by a 9 cm thick air gap. The wall section is 2.5 m high and 5 m wide. The temperature of the two surfaces in contact with the air gap are 20 C and −5 C, respectively.

**(a)** What is the rate of convective heat transfer across the air gap?

**(b)** Let us say that a horizontal baffle is placed in the wall section halfway up the wall. The baffle changes the air gap from a single air gap 2.5 m high and 5 m wide to two gaps that are each 1.25 m high and 5 m wide. What is the new heat transfer rate if temperatures remain the same?

**7-34** A solar collector has a glass cover plate and an absorber plate separated by an air space. The collector is square, 1 m by 1 m, and is tilted at an angle of 20 degrees from the horizontal. The cover plate is at 30 C and the absorber plate is at 60 C. What is the rate of convective heat transfer across the air space if

**(a)** the air space is 2 cm thick?

**(b)** the air space is 4 cm thick?

**7-35** An annulus is formed by two concentric horizontal cylinders. The outer diameter of the inner cylinder is 70 mm and the inner diameter of the outer cylinder is 100 mm. The space between the cylinders is filled with water. The inner and outer surfaces of the annulus are at 20 and 80 C, respectively. What is the rate of convective heat transfer per meter length of the cylinders?

**7-36** A solar collector consists of a glass outer tube and a concentric inner absorber tube. The collector is horizontal, with the inner tube having an outer diameter of 10 cm and the outer tube having an inner diameter of 14 cm. The solar collector is 2 m long. The glass tube has a temperature of 25 C and the absorber tube has a temperature of 75 C. If air fills the annular space, what is the rate of convective heat transfer through the air?

**7-37** A container for liquid oxygen consists of two concentric spheres. The inner sphere, which holds the $LO_2$, is thin-walled and has a diameter of 1 m. The outer sphere has an inner diameter of 1.05 m. The space between the spheres is filled with air. With the container filled with $LO_2$, the inner sphere has a temperature of 90.2 K and the inner surface of the outer sphere has a temperature of 287 K. The latent heat of vaporization of oxygen is $2.14 \times 10^5$ J/kg. What is the rate (kg/s) at which gaseous oxygen leaves the container through a small vent on the top of the container?

**7-38** The space between two thin concentric spheres contains air at 5 atm pressure. The spheres have diameters of 20 and 25 cm and respective temperatures of 40 and 100 C. What is the rate of convective heat transfer through the air space?

**7-39** A house has a gable roof. Each of the two roof sections is 20 feet by 30 feet. The 20 foot edge is parallel to the ground and the 30 foot edge is inclined 25° from the horizontal. The outside air is quiet and at 10 F. The outer surface of the roof is at 40 F. What is the rate of convective heat loss from the roof?

**7-40** A wire of 0.3 mm diameter is horizontal in a tank of water which is at 25 C. A current passing through the wire heats the outer surface of the wire to 90 C. What is the rate of heat convection to the water per meter length of wire?

**7-41** Water at 40 C flows over a 10 cm diameter sphere at a velocity of 0.1 m/s. The sphere's surface is at 80 C. Should both free and forced convection be considered or is one insignificant compared with the other?

**7-42** Air at 20 C flows across the top surface of a horizontal heated plate at a free stream velocity of 0.5 m/s. The plate is 10 cm wide and 20 cm in the direction of flow. The temperature of the top of the plate is 120 C. Should both free and forced convection be considered or is one insignificant compared with the other?

**7-43** A vertically-mounted aluminum heat sink is 7.5 cm high and 15 cm wide. The vertical fins are equally spaced across the 15 cm width of the heat sink. There are 20 vertical fins forming 19 vertical channels for air flow. Each fin is 0.5 mm thick, 7.5 cm high, and 2.5 cm wide. The fins are at 80 C and the adjacent air is at 20 C. What is the heat transfer from the fins to the air (W)?

**7-44** A vertically-mounted aluminum heat sink is 9 cm high and 12 cm wide. There are equally spaced vertical fins across the entire 12 cm width of the heat sink, and the fins are spaced at their optimal spacing for maximum heat transfer. Each fin is 0.5 mm thick, 9 cm high, and 1.5 cm wide. The fins are at 80 C and the adjacent air is at 20 C.
**(a)** How many fins are there?
**(b)** What is the heat transfer from the fins to the air?

**7-45** It is desired to remove 55 W of heat from a tank wall through the use of vertical fins. There are 17 fins and they are equally spaced at a spacing of 6 mm. The fins are 0.7 mm thick and 3 cm wide. The fins are at 80 C and the adjacent air is at 10 C. What is the needed vertical height of the fins for the required 55 W heat removal?

**7-46** A student is performing an experiment in a thermal lab. He is measuring the natural convective heat transfer from a horizontal section of duct that has an equilateral triangular cross section. The measured heat transfer will be compared with the predictions from existing correlation equations. The duct section is 2 m long, and each of the three sides of the duct is 25 cm wide. There is an electric heater inside the duct, which keeps the duct wall at 90 C. The room air is at 25 C.

  **(a)** If the duct is positioned with one of its flat surfaces facing up, what is the predicted rate of natural convection to the room air?

  **(b)** If the duct is positioned with one of its flat surfaces facing down, what is the predicted rate of natural convection to the room air?

  **(c)** Compare the results of Part (a) and Part (b) with the predicted rate of natural convection from a horizontal cylindrical duct section having the same surface area as the triangular duct section.

**7-47** Water is flowing at a rate of 0.04 kg/s through a horizontal, thin-walled copper tube of 2 cm diameter and 5 m length. The water enters the tube at 90 C. The room air is at 20 C, and the tube transfers heat to the room air by natural convection.

  **(a)** What is the temperature of the water as it exits the tube?

  **(b)** What is the heat transfer rate from the water to the room air?

# References

[1] S.W. Churchill, H.H.S. Chu, Correlation equations for laminar and turbulent free convection from a vertical plate, Int. J. Heat Mass Transf. 18 (1975) 1323−1328.

[2] W.H. McAdams, Heat Transmission, 3rd ed., McGraw-Hill, 1954.

[3] C.Y. Warner, V.S. Arpaci, An experimental investigation of turbulent natural convection in air at low pressures for a vertical heated flat plate, Int. J. Heat Mass Transf. 11 (1968) 397.

[4] F.J. Bayley, An analysis of turbulent free convection heat transfer, Proc. Inst. Mech. Eng. 169 (1955) 361.

[5] T. Fujii, M. Fujii, The dependence of local Nusselt number on Prandtl number in the case of free convection along a vertical surface with uniform heat flux, Int. J. Heat Mass Transf. 19 (1976) 121−122.

[6] E.M. Sparrow, J.L. Gregg, Laminar free convection from a vertical plate with uniform surface heat flux, Trans. ASME 78 (1956) 435−440.

[7] T. Fujii, H. Imura, Natural convection heat transfer from a plate with arbitrary inclination, Intl. J. Mass Heat Transf. 15 (1972) 755−767.

[8] J.R. Lloyd, W.R. Moran, Natural convection adjacent to horizontal surface of various planforms, J. Heat Transf. 96 (1974) 443−447.

[9] S.W. Churchill, H.H.S. Chu, Correlating equations for laminar and turbulent free convection from a horizontal cylinder, Int. J. Heat Mass Transf. 18 (1975) 1049−1053.

[10] V.T. Morgan, The overall convective heat transfer from smooth circular cylinders, Adv. Heat Transf. 11 (1975) 199−264.

[11] T. Yuge, Experiments in heat transfer from spheres including combined natural and forced convection, J. Heat Transf. 82 (1960) 214−220.

[12] W.S. Amato, C. Tien, Free convection heat transfer from isothermal spheres in water, Int. J. Heat Mass Transf. 15 (1972) 327−339.

[13] S.W. Churchill, Free convection around immersed bodies, in: G.F. Hewitt (Ed.), Heat Exchanger Design Handbook, Hemisphere Publishing, Washington, D. C., 1983, pp. 2.5.7−24.

[14] E.M. Sparrow, A.J. Stretton, Natural convection from variously oriented cubes and from other bodies of unity aspect ratio, Int. J. Heat Mass Transf. 28 (4) (1985) 741−752.

[15] J.H. Lienhard, On the commonality of equations for natural convection from immersed bodies, Int. J. Heat Mass Transf. 16 (1973) 2121−2123.

[16] E.M. Sparrow, M.A. Ansari, A refutation of King's rule for multi-dimensional external natural convection, Int. J. Heat Mass Transf. 26 (1983) 1357−1364.

[17] E.M. Sparrow, J.L. Gregg, Laminar free convection from the outer surface of a vertical cylinder, Trans. ASME 78 (1956) 1823−1829.

[18] W.J. Minkowycz, E.M. Sparrow, Local nonsimilar solutions for NAtural convection on a vertical cylinder, J. Heat Transf. 96 (1974) 178−183.

[19] T. Cebeci, Laminar free convective heat transfer from the outer surface of a vertical slender circular cylinder, in: Proc. Fifth Intl. Heat Transfer Conference, Tokyo, Paper NC 1.4, September 1974, pp. 15−19.

[20] E.J. LeFevre, A.J. Ede, Laminar free convection from the outer surface of a vertical circular cylinder, in: Proc. Ninth Intl. Congress Applied Mechanics, Brussels, Vol. 4, 1956, pp. 175−183.

[21] K.G.T. Hollands, T.E. Unny, G.D. Raithby, L. Konicek, Free convective heat transfer across inclined air layers, J. Heat Transf. 98 (1976) 189−193.

[22] S. Globe, D. Dropkin, Natural-convection heat transfer in liquids confined by two horizontal plates and heated from below, J. Heat Transf. 81 (1959) 24−28.

[23] S.M. ElSherbiny, G.D. Raithby, K.G.T. Hollands, Heat transfer by natural convection across vertical and inclined air layers, J. Heat Transf. 104 (1982) 96−102.

[24] R.K. MacGregor, A.P. Emery, Free convection through vertical plane layers: moderate and high Prandtl number fluids, J. Heat Transf. 91 (1969) 391−403.

[25] I. Catton, Natural convection in enclosures, in: Proc. Sixth Intl. Heat Transfer Conference, Toronto, Canada, vol. 6, 1978, pp. 13−31.

[26] G.D. Raithby, K.G.T. Hollands, A general method of obtaining approximate solutions to laminar and turbulent free convection problems, Adv. Heat Transf. 11 (1975) 265−315.

[27] W. Elenbaas, Heat dissipation of parallel plates by free convection, Physica 9 (1942) 1−28.

[28] A. Bar-Cohen, W.M. Rohsenow, Thermally optimum spacing of vertical natural convection cooled parallel plates, J. Heat Transf. 106 (1984) 116−123.

[29] B. Metais, E.R.G. Eckert, Forced, mixed, and free convection regimes, J. Heat Transf. 86 (1964) 295.

[30] C.K. Brown, W.H. Gauvin, Combined free and forced convection I, II, Can. J. Chem. Eng. 43 (6) (1965) 306−313.

[31] C.A. Depew, J.L. Franklin, C.H. Ito, Combined free and forced convection in horizontal, uniformly heated tubes, ASME Paper 75-HT-19 (August 1975).

[32] S.W. Churchill, Combined free and forced convection around immersed bodies, in: Heat Exchanger Design Handbook, Section 2.5.9, Hemisphere Publishing, New York, 1983.

[33] S.W. Churchill, Combined free and forced convection in channels, in: Heat Exchanger Design Handbook, Section 2.5.10, Hemisphere Publishing, New York, 1983.

[34] D.G. Osborne, F.P. Incropera, Experimental study of mixed convection heat transfer for transitional and turbulent flow between horizontal, parallel plates, Int. J. Heat Mass Transf. 28 (1985) 1337.

[35] J.R. Lloyd, E.M. Sparrow, Combined forced and free convection flow on vertical surfaces, Int. J. Heat Mass Transf. 13 (1970) 434−438.

## Chapter outline

## 8.1 Introduction

Heat exchangers are devices used to transfer thermal energy between two fluids. As the fluids flow through an exchanger, one fluid gains heat and the other loses heat. If the process does not include evaporation or condensation, both fluids will experience a change in temperature—one fluid increasing in temperature and the other decreasing in temperature. Heat exchangers have a myriad of applications. In power plants, they serve as steam condensers, feedwater heaters, steam generators, and preheaters. Heat exchangers see extensive use in chemical process applications. In addition, they are extensively used in the HVAC industry, where they are boilers, condensers, evaporators, heaters, air conditioners, radiators, and heating/cooling coils.

## 8.2 Types of heat exchangers

There are three main types of heat exchangers: direct contact, regenerator, and recuperator.

In a direct contact exchanger, the two inlet fluid streams combine and mix together. Examples of direct contact exchangers are cooling towers, open feedwater heaters in power plants, and chemical processes where fluids can be mixed together. There is no wall separating the fluids to corrode and foul

*Heat Transfer Principles and Applications.* **https://doi.org/10.1016/B978-0-12-802296-2.00008-1**

and degrade the heat exchanger performance. In addition, the pressure drop of the exchanger is less than that of exchangers incorporating tubes. Finally, the direct contact exchanger can be considerably more economical than other types of exchangers. One major disadvantage, however, is that the fluids mix together. In many applications, contamination of a fluid by the other fluid is highly undesirable. For example, one fluid may be radioactive; the other nonradioactive. Or, one may be poisonous; the other nonpoisonous. In these examples, direct contact exchangers are definitely unsuitable.

In a regenerator, heat is transferred to a storage medium, the "core" or "matrix" of the exchanger, by the hotter fluid, and then the cooler fluid gains heat from the storage medium. It is a periodic process, with the two fluids alternating in contact with the core. Regenerators are of the fixed-matrix or rotary types. Regenerators have a high surface area per volume, and the high surface area makes the exchanger particularly attractive for gas-to-gas applications. A disadvantage is that there can be minor mixing of the fluid streams.

In recuperators, the two fluids are separated by a wall. The fluids transfer heat by convection to and from the wall and conduction through the wall. The major advantage of a recuperator is that the two fluids are indeed separated, with no cross-contamination if the wall maintains its integrity. Disadvantages include the higher capital cost of recuperators and the increased pressure drop, which results in higher pumping costs.

Our discussion in this chapter is limited to recuperators. In particular, we will discuss double-pipe, shell-and-tube, and crossflow heat exchangers.

The simplest type of recuperator is the double-pipe heat exchanger shown in Fig. 8.1A. It consists of two concentric tubes. One fluid flows in the inner tube and the other fluid flows in the annular area between the inner and outer tubes. If the fluids flow in the same direction, as they do in Fig. 8.1A, then the flow is said to be parallel flow. If the fluids flow in opposite directions, the flow is counterflow. Each fluid makes a single pass through the exchanger; they only go through the exchanger once.

In a shell-and-tube heat exchanger, one fluid flows through the tubes, and the other fluid flows around the tubes, enclosed by the "shell" of the exchanger. Fig. 8.1B shows an exchanger with one shell pass and two tube passes. Exchangers can also have more than two tube passes and more than one shell pass. For example, Fig. 8.1C shows a two-shell pass, four-tube-pass exchanger. Baffles are placed in shell-and-tube heat exchangers to channel the flow of the shell fluid across the tubes. The baffles enhance contact of the shell fluid with the tubes. They minimize bypassing of tubes by the shell fluid.

In crossflow heat exchangers, one fluid flows through the tubes and the other fluid flows crosswise to the tubes. The crossflow increases the heat transfer relative to flows that are longitudinal along the tubes, such as in the double-pipe exchanger. Fig. 8.1D shows two crossflow heat exchangers. In one exchanger, a fluid flows through the tubes and the other fluid flows unrestrained over the tubes. It is said that the tube fluid is "unmixed" as it is confined to the tubes. The other fluid is "mixed" as it is somewhat unconfined and can mix with itself. In the second exchanger, there are fins on the outside of the tubes. One fluid flows through the tubes and is "unmixed." The other fluid flows through the channels formed by the fins. This fluid is also said to be "unmixed" as it is confined to the channels, which hinder its mixing.

Figs. 8.2 and 8.3 are photographs of heat exchangers. Fig. 8.2A shows shell-and-tube heat exchangers used in a chemical plant to cool hot gases with water. Fig. 8.2B shows a tube bundle for a shell-and-tube heat exchanger. In the foreground is the tube sheet to which the tubes are attached. Metal plates, or "baffles," are seen further back in the photo. These baffles direct the shell-side fluid around the outside of the tubes, assuring that the fluid contacts the tubes as much as possible.

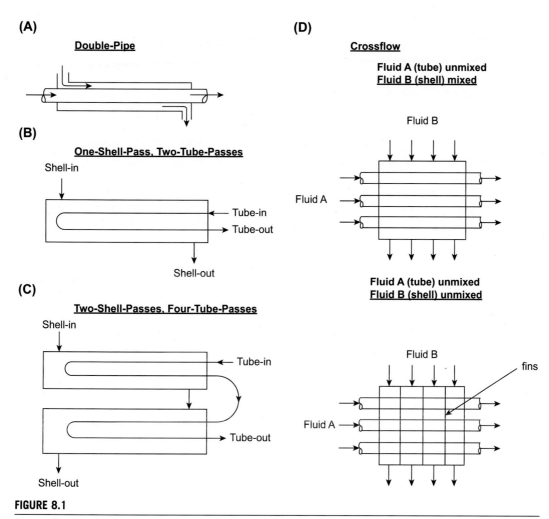

**FIGURE 8.1**

Types of heat exchangers.

Fig. 8.2C shows U-tubes for a one-shell-pass, two-tube-pass heat exchanger. In the left of the photo is a shell-and-tube heat exchanger. Finally, Fig. 8.2D is a crossflow heat exchanger. One fluid flows through the copper tubes and the other fluid flows over the tubes. The aluminum fins on the tubes inhibit mixing of the outer fluid. Hence, it is said that both fluids are "unmixed."

Fig. 8.3A and B show a small heat exchanger manufactured by Exergy LLC of Garden City, NY. The exchanger is 13.5 inches long, 3 inches diameter, and of nickel brazed 316L stainless steel. This model has 253 tubes, and its heat transfer area is $0.60$ $m^2$. Uses of the exchanger include chemical processing, water for injection for pharmaceutical plants, and semiconductor manufacturing. Many thanks to Exergy for providing the photos.

**FIGURE 8.2**

(A) Shell-and-tube heat exchangers. (B) Heat exchanger tube sheet and tubes. (C) U-tube bundle.
(D) Crossflow heat exchanger (both fluids unmixed).

Finally, Fig. 8.4 shows a badly corroded tube sheet and tubes. The effect of corrosion on heat transfer is discussed in Section 8.4 below. Corroded heat exchangers can often be repaired and returned to service through tube replacement. Several companies in the United States and overseas provide such service.

## 8.2.1 Temperature distribution in double-pipe heat exchangers

Fig. 8.5 shows a double-pipe heat exchanger, which consists of two concentric pipes (or tubes). Fluid A flows through the inner tube, and Fluid B flows in the annular area between the tubes, which is called the "shell." The outer surface of the outer tube is well-insulated so that heat transfer between the heat exchanger and the environment is insignificant compared with the heat transfer between the fluids. Fig. 8.5 shows both fluids flowing in the same direction, left to right. This is a double-pipe *parallel flow* exchanger. If the flow direction of one of the fluids was reversed, the exchanger would be a double-pipe *counterflow* exchanger.

**(A)**                     **(B)**

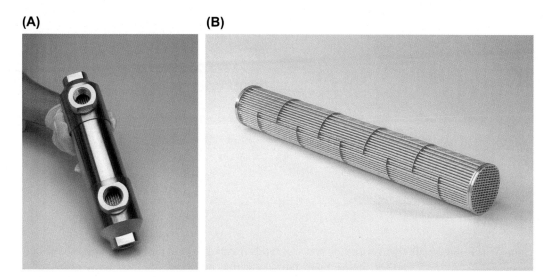

**FIGURE 8.3**

(A) Small heat exchanger (outside). (B) Small heat exchanger (inside).

**FIGURE 8.4**

Corroded tube sheet and tubes.

    Let us say that the temperatures of the fluids, $T_A$ and $T_B$, are different, with $T_A$ being greater than $T_B$. Then there will be heat transfer across the tube wall from Fluid A to Fluid B. As discussed Chapter 3, the rate of heat transfer $q$ is

$$q = \frac{\Delta T_{\text{overall}}}{\sum R} = \frac{T_A - T_B}{\sum R} \tag{8.1}$$

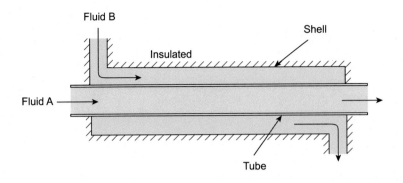

**FIGURE 8.5**

Double-pipe heat exchanger.

where $\sum R$ is the sum of the resistances between fluids $A$ and $B$. These resistances are the convective resistance at the inner surface of the tube, the conductive resistance of the tube's wall, and the convective resistance at the outer surface of the tube.

The rate of heat transfer between the fluids can also be expressed in terms of the overall heat transfer coefficient $U$. That is,

$$q = UA(\Delta T)_{avg} \tag{8.2}$$

where $(\Delta T)_{avg}$ is the "average" temperature difference between the fluids and $A$ is the heat transfer area between the fluids.

A double-pipe heat exchanger does not have a single, unique area $A$. It has the area of the inner surface of the tube and the area of the outer surface of the tube, which are different. The inner surface has area $A_i = 2\pi r_i L$, where $r_i$ is the radius of the inner surface and $L$ is the length of the tube. Similarly, the outer surface of the tube has an area $A_o = 2\pi r_o L$, where $r_o$ is the radius of the outer surface and $L$ is the tube's length. As the areas are different, the $U$ values must be referenced to a specific area: $U_i$ is referenced to the inner surface area and $U_o$ is referenced to the outer surface area. The rate of heat transfer in Eq. (8.2) is therefore

$$q = U_i A_i (\Delta T)_{avg} = U_o A_o (\Delta T)_{avg} \tag{8.3}$$

From Eq. (8.3), it is seen that $U_i A_i = U_o A_o$. As $A_i = 2\pi r_i L$ and $A_o = 2\pi r_o L$, the coefficients are related by

$$U_o = \left(\frac{r_i}{r_o}\right) U_i \tag{8.4}$$

Let us continue our discussion of the double-pipe heat exchanger. Fig. 8.6 shows two double-pipe exchangers. The one on the left is parallel flow; the one on the right is counterflow. Under each figure is a diagram showing how the fluid temperatures change as they travel through the exchanger. The hotter fluid enters the exchanger at $T_{hi}$ and leaves at $T_{ho}$. The colder fluid enters at $T_{ci}$ and leaves at $T_{co}$. We will be calling the left end of the exchanger "Location 1" and the right end of the exchanger "Location 2." Some interesting items: For parallel flow exchanger, the colder fluid cannot exit the

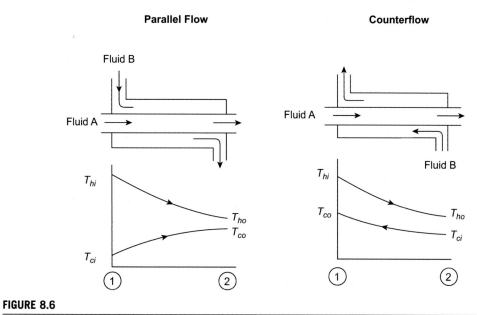

**FIGURE 8.6**

Temperature distribution in double-pipe exchangers.

exchanger at a higher temperature than the hotter fluid exits. However, for the counterflow exchanger, the exit temperature of the colder fluid *can* be higher than the exit temperature of the hotter fluid. Also, for the counterflow exchanger, the temperature curves are not generally parallel. That is, the temperature difference between the hot and cold fluids usually varies with location in the exchanger. If, however, one fluid is evaporating at constant temperature and the other fluid is condensing at constant temperature, then the temperature difference will be constant throughout the exchanger. Such is also the case if the products of the mass flow rate and the specific heat for the two fluids are equal.

Let us now determine the average temperature difference between the fluids, which is needed in Eqs. (8.2) and (8.3). Consider the parallel flow exchanger in Fig. 8.6. We will take a vertical slice of the exchanger between locations $x$ and $x + dx$. Fig. 8.7 shows control volumes for Fluids A and B on both sides of the tube wall. The area of the wall between $x$ and $x + dx$ is $dA$. We will assume that the temperature of Fluid A is greater than the temperature of Fluid B. As the fluids flow through their respective control volumes, Fluid A gives heat to Fluid B and Fluid A's enthalpy decreases. Fluid B experiences an increase in enthalpy due to the heat received from Fluid A. The various energy flows are shown in Fig. 8.7.

For Fluid A, the energy balance is

$$\dot{m}_A c_A T_A = U dA (T_A - T_B) + \dot{m}_A c_A (T_A + dT_A) \tag{8.5}$$

For Fluid B, the energy balance is

$$\dot{m}_B c_B T_B + U dA (T_A - T_B) = \dot{m}_B c_B (T_B + dT_B) \tag{8.6}$$

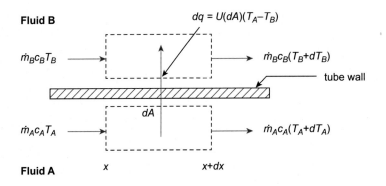

**FIGURE 8.7**

Energy flows for differential control volumes.

In these two equations, $\dot{m}$ is the mass flow rate, $c$ is the specific heat, $T$ is the fluid temperature, $U$ is the overall heat transfer coefficient of the exchanger, and $dA$ is the differential wall area between the fluids. Subscripts indicate the particular fluid: the hot fluid (Fluid A) and the cold fluid (Fluid B).

Rearranging Eq. (8.5), we get

$$dT_A = -\frac{UdA(T_A - T_B)}{\dot{m}_A c_A} \tag{8.7}$$

Rearranging Eq. (8.6), we get

$$dT_B = \frac{UdA(T_A - T_B)}{\dot{m}_B c_B} \tag{8.8}$$

Subtracting Eq. (8.8) from Eq. (8.7), we have

$$d(T_A - T_B) = - UdA(T_A - T_B)\left(\frac{1}{\dot{m}_A c_A} + \frac{1}{\dot{m}_B c_B}\right) \tag{8.9}$$

Dividing both sides of Eq. (8.9) by $(T_A - T_B)$, we arrive at

$$\frac{d(T_A - T_B)}{T_A - T_B} = - U\left(\frac{1}{\dot{m}_A c_A} + \frac{1}{\dot{m}_B c_B}\right) dA \tag{8.10}$$

Both sides of Eq. (8.10) can now be integrated over the entire length of the heat exchanger, from the left end (Location 1) to the right end (Location 2). Assuming that $U$, the mass flow rates, and the specific heats are constants, we have

$$\ln\left(\frac{(T_A - T_B)_2}{(T_A - T_B)_1}\right) = - UA\left(\frac{1}{\dot{m}_A c_A} + \frac{1}{\dot{m}_B c_B}\right) \tag{8.11}$$

As the differential heat flow $dq$ for the infinitesimal control volumes shown in Fig. 8.7 is equal to $UdA(T_A - T_B)$, Eq. (8.7) may be written as

$$dq = - \dot{m}_A c_A dT_A \tag{8.12}$$

and Eq. (8.8) may be written as

$$dq = \dot{m}_B c_B dT_B \tag{8.13}$$

Integrating Eqs. (8.12) and (8.13) over the entire length of the heat exchanger; i.e., from the left end (Location 1) to the right end (Location 2), we get expressions for the total rate of heat flow $q$ for the exchanger:

$$q = \dot{m}_A c_A \left[ (T_A)_1 - (T_A)_2 \right] \tag{8.14}$$

$$\text{and } q = \dot{m}_B c_B \left[ (T_B)_2 - (T_B)_1 \right] \tag{8.15}$$

By using Eqs. (8.14) and (8.15) in Eq. (8.11), Eq. (8.11) then becomes a third expression for the heat flow $q$ of the exchanger:

$$q = UA \frac{\left[ (T_A - T_B)_1 - (T_A - T_B)_2 \right]}{\ln \left( \frac{(T_A - T_B)_1}{(T_A - T_B)_2} \right)} \tag{8.16}$$

Comparing this equation to Eq. (8.2), it is seen that the temperature factor is the average temperature difference between the two fluids in the heat exchanger. As the factor is the average (or mean) temperature difference and contains a log function, the temperature term is called the Log Mean Temperature Difference (LMTD). Summarizing, we have

$$q = UA(LMTD) \tag{8.17}$$

where the LMTD is

$$LMTD = \frac{\left[ (T_h - T_c)_1 - (T_h - T_c)_2 \right]}{\ln \left( \frac{(T_h - T_c)_1}{(T_h - T_c)_2} \right)} \tag{8.18}$$

In Eq. (8.18), we changed the subscripts on the temperatures from $A$ and $B$ to the more commonly used subscripts $h$ and $c$, respectively. We can do this as Fluid A is the hot fluid and Fluid B is the cold fluid.

Calculation of the *LMTD* looks difficult, but it really is not. The term $(T_A - T_B)_1$ is the temperature difference of the fluids at the left end of the exchanger and the term $(T_A - T_B)_2$ is the temperature difference of the fluids at the right end. Eq. (8.18) says that LMTD is the temperature difference of the fluids at the left end of the exchanger minus the temperature difference of the fluids at the right end of the exchanger divided by the natural log of the ratio of the two temperature differences. Indeed, Eq. (8.18) can be written even more simply as

$$LMTD = \frac{(\Delta T)_1 - (\Delta T)_2}{\ln \left( \frac{(\Delta T)_1}{(\Delta T)_2} \right)} \tag{8.19}$$

We derived the above equations using a parallel flow double-pipe heat exchanger. If we had done the derivation for a counterflow double-pipe exchanger, and the cold fluid went right to left, the only change in the equations would be Eq. (8.15), which would be replaced with

$$q = \dot{m}_B c_B \left[ (T_B)_1 - (T_B)_2 \right] \tag{8.20}$$

Just keep in mind that the temperature differences in the equations (and $q$) are positive.

Caution: For a counterflow exchanger, $\Delta T_1$ can be equal to $\Delta T_2$. If this is the case, Eq. (8.19) calculates the *LMTD* as 0/0. For this special case, the temperature difference between the fluids is the same throughout the exchanger, and $\Delta T_1 = \Delta T_2$ should be used as the value for $(\Delta T)_{avg}$ instead of the calculated *LMTD*.

Example 8.1 shows how easy it is to calculate the LMTD for a double-pipe heat exchanger. It also shows that the LMTD for a parallel flow exchanger is different from the LMTD for a counterflow heat exchanger having the same entrance and exit fluid temperatures.

## Example 8.1
### LMTD for parallel and counterflow double-pipe exchangers
*Problem*

In a double-pipe heat exchanger, the hotter fluid enters at 150 C and leaves at 80 C and the colder fluid enters at 20 C and leaves at 60 C. Determine the LMTD if the heat exchanger is
(a) parallel flow and (b) counterflow.

*Solution*

**Parallel flow.** The first step in the solution is to draw a sketch of the temperature distribution of the exchanger and put the temperatures and flow directions on the sketch. Looking at Fig. 8.6, we have the parallel heat exchanger in the figure below.

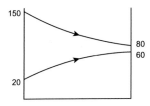

From Eq. (8.19), we have,

$$LMTD = \frac{(\Delta T)_1 - (\Delta T)_2}{\ln\left(\dfrac{(\Delta T)_1}{(\Delta T)_2}\right)} = \frac{(150 - 20) - (80 - 60)}{\ln\left(\dfrac{150 - 20}{80 - 60}\right)} = 58.8 \text{ C}$$

**Counterflow.** Again using Fig. 8.6, the sketch for the counterflow exchanger is given below.

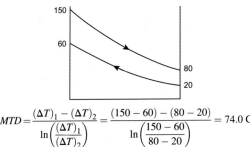

$$LMTD = \frac{(\Delta T)_1 - (\Delta T)_2}{\ln\left(\dfrac{(\Delta T)_1}{(\Delta T)_2}\right)} = \frac{(150 - 60) - (80 - 20)}{\ln\left(\dfrac{150 - 60}{80 - 20}\right)} = 74.0 \text{ C}$$

It is seen that the *LMTD*'s are different for the two different types of double-pipe heat exchangers even though the entering and leaving fluid temperatures are the same.

Another interesting observation is that the *LMTD* for the counterflow exchanger is greater than the *LMTD* for the parallel flow exchanger. This is a general characteristic and the reason why counterflow exchangers are usually preferred over parallel exchangers. For the same *U, A,* and fluid entering and leaving temperatures, a counterflow exchanger will have a greater heat transfer rate $q$ than a parallel flow exchanger.

## 8.3 The overall heat transfer coefficient

Let us now look at the overall heat transfer coefficient in some detail. Fig. 8.8 shows a cross-sectional view of the inner tube of a double-pipe heat exchanger. The convective coefficient is $h_i$ at the inner surface of the tube and $h_o$ at the outer surface. The tube has inner and outer radii $r_i$ and $r_o$ and a length $L$ perpendicular to the plane of the figure. Fluid A in inside the tube; Fluid B is in the shell; and Fluid A has a higher temperature than Fluid B.

From Eqs. (8.1) through (8.3), we have

$$q = \frac{\Delta T}{\sum R} = U_i A_i \Delta T = U_o A_o \Delta T \tag{8.21}$$

Resistances were discussed in Chapter 3. At the inner surface of the tube, the convective resistance is $\frac{1}{h_i A_i}$. At the outer surface of the tube, the convective resistance is $\frac{1}{h_o A_o}$.

And, the resistance of the tube wall is $\frac{\ln(r_o/r_i)}{2\pi k L}$, where $k$ is the thermal conductivity of the tube material and $L$ is the length of the tube. The circumferential areas in Eq. (8.21) are $A_i = 2\pi r_i L$ and $A_o = 2\pi r_o L$. Inserting these resistances into Eq. (8.21), we have

$$U_i = \frac{1}{\dfrac{1}{h_i} + \dfrac{r_i \ln(r_o/r_i)}{k} + \dfrac{r_i}{h_o r_o}} \tag{8.22}$$

$$\text{and } U_o = \frac{1}{\dfrac{r_o}{h_i r_i} + \dfrac{r_o \ln(r_o/r_i)}{k} + \dfrac{1}{h_o}} \tag{8.23}$$

In many cases, one term in the denominators of these equations predominates. The term associated with the resistance of the tube wall is usually much less than the other two terms. And, convective

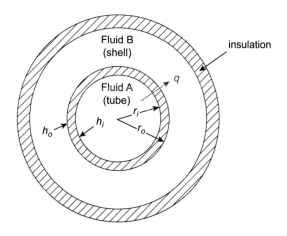

**FIGURE 8.8**

Convective coefficients for a double-pipe heat exchanger.

**Table 8.1 Typical ranges of the overall heat transfer coefficient $U$.**

| Type of exchanger | $U$ (W/m$^2$ C) |
|---|---|
| Water-to-water | 850–2500 |
| Water-to-oil | 100–400 |
| Gas-to-gas | 10–50 |
| Gas-to-water | 10–250 |
| Steam-to-fuel oil | 100–400 |
| Oil-to-oil | 50–400 |
| Feedwater heater | 1000–8500 |
| Steam condenser | 1000–6000 |
| Finned tube (water/air) | 30–60 |
| Finned tube (steam/air) | 30–300 |

coefficients for liquids are usually quite higher than those for gases. Therefore, if one fluid is a liquid and the other is a gas, the term related to the liquid is often negligible.

Typical ranges of the overall heat transfer coefficient are given in Table 8.1.

The overall heat transfer coefficients in Eqs. (8.22) and (8.23) are for new, clean heat exchangers. With extended operation, the coefficients may appreciably decrease due to degradation of the heat transfer surfaces through corrosion, deposits, and possibly biological fouling such as algae. Both the inner and outer surfaces of the tube may be affected. Additional resistance terms may be added to the denominators of Eqs. (8.22) and (8.23) to account for the change in the overall heat transfer coefficient. The revised equations are

$$U_i = \frac{1}{\dfrac{1}{h_i} + R_{fi} + \dfrac{r_i \ln(r_o/r_i)}{k} + \dfrac{r_i}{h_o r_o} + R_{fo}(r_i/r_o)} \tag{8.24}$$

$$\text{and } U_o = \frac{1}{\dfrac{r_o}{h_i r_i} + R_{fi}(r_o/r_i) + \dfrac{r_o \ln(r_o/r_i)}{k} + \dfrac{1}{h_o} + R_{fo}} \tag{8.25}$$

where $R_{fi}$ is the fouling factor (or resistance) on the inner surface of the tube and $R_{fo}$ is the fouling factor (or resistance) on the outer surface of the tube.

By comparing Eq. (8.22) to Eq. (8.24) and Eq. (8.23) to Eq. (8.25), we arrive at the following relations between the overall heat transfer coefficients for a new (clean) exchanger and one with fouling:

$$\frac{1}{U_i(\text{fouled})} - \frac{1}{U_i(\text{clean})} = R_{fi} + R_{fo}\left(\frac{r_i}{r_o}\right) \tag{8.26a}$$

$$\frac{1}{U_o(\text{fouled})} - \frac{1}{U_o(\text{clean})} = R_{fo} + R_{fi}\left(\frac{r_o}{r_i}\right) \tag{8.26b}$$

Typical fouling factors are given in Table 8.2. More extensive information is given in Refs. [1,2].

| Table 8.2 Typical fouling factors $R_f$. | |
|---|---|
| **Fluid** | $R_f$ **(m² C/W)** |
| Distilled water and boiler feedwater | 0.0001 (below 50 C) |
| Distilled water and boiler feedwater | 0.0002 (above 50 C) |
| Water (river) | 0.0004 |
| Fuel Oil | 0.0009 |
| Oil (hydraulic, lubricating, and transformer) | 0.0002 |
| Ethylene glycol | 0.00035 |
| Steam | 0.0001 |
| Air | 0.0004 |
| Engine exhaust gases | 0.0018 |
| Refrigerants (liquid) | 0.0002 |
| Refrigerants (vapor) | 0.0004 |

## Example 8.2

### Overall heat transfer coefficient and fouling factor

*Problem*

A refrigerant is flowing through the tube of a double-pipe counterflow heat exchanger. The tube is 1-inch Type L copper. Air flows in the annular region of the exchanger. The convective coefficient on the inner surface of the tube is 140 W/m² C and the coefficient on the outer surface of the tube is 40 W/m² C.

**(a)** What is the overall heat transfer coefficient based on the area of the outer surface of the tube?

**(b)** After a year of service, the inner surface of the tube has a fouling factor of 0.0002 m² C/W and the outer surface has a fouling factor of 0.00015 m² C/W. By what percentage has the overall heat transfer coefficient decreased from that calculated for the clean exchanger in Part (a)?

*Solution*

From the Internet, the copper tubing has an outside diameter of 1.125 inch and a wall thickness of 0.05 inch. Also, copper has a thermal conductivity of 400 W/m C. Therefore,

$$r_o = (1.125 \,/\, 2) \text{ in} \times \frac{1 \text{ m}}{39.37 \text{ in}} = 0.01429 \text{ m}$$

$$r_i = r_o - 0.05 = (1.125 \,/\, 2) - 0.05 = 0.5125 \text{ in} \times \frac{1 \text{ m}}{39.37 \text{ in}} = 0.01302 \text{ m}$$

**(a)** We will use Eq. (8.23) for the overall heat transfer coefficient.

$$U_o = \frac{1}{\dfrac{r_o}{h_i r_i} + \dfrac{r_o \ln(r_o/r_i)}{k} + \dfrac{1}{h_o}} \tag{8.23}$$

From the problem statement, $h_i = 140$ W/m² C and $h_o = 40$ W/m² C. Putting values into Eq. (8.23), we have

$$U_o(\text{new}) = \frac{1}{\dfrac{0.01429}{(140)(0.01302)} + \dfrac{0.01429 \ln(0.01429/0.01302)}{400} + \dfrac{1}{40}} = \frac{1}{0.03284} = 30.45 \text{ W/m}^2 \text{ C}$$

The new exchanger has an overall heat transfer coefficient of 30.45 W/m$^2$ C based on the outer surface area of the tube.

(b) Eq. (8.25) includes terms for the fouling factors.

$$U_o = \cfrac{1}{\cfrac{r_o}{h_i r_i} + R_{fi}(r_o/r_i) + \cfrac{r_o \ln(r_o/r_i)}{k} + \cfrac{1}{h_o} + R_{fo}} \tag{8.25}$$

Comparing Eq. (8.25) to Eq. (8.23), it is seen that Eq. (8.25) has two additional resistance terms in the denominator. They are

$$R_{fi}(r_o \, / \, r_i) = 0.0002(0.01429 \, / \, 0.01302) = 0.00022 \quad \text{and} \quad R_{fo} = 0.00015.$$

Adding these values to the denominator in the calculation for $U_o$(new) above, we have, for the year-old exchanger

$$U_o(\text{year} - \text{old}) = \frac{1}{0.03284 + 0.00022 + 0.00015} = 30.11 \text{ W/m}^2 \text{ C}$$

The % decrease in $U_o$ is $\frac{(30.45-30.11)}{30.45} \times 100 = 1.12\%$.

It is seen that the effect of the fouling is very, very minimal.

# 8.4 Analysis methods

The two most common methods to solve heat exchanger problems are the LMTD method and the effectiveness—number of transfer unit ($\varepsilon$-NTU) method. Although both methods can be used for a given problem, each method has its particular strong points. The LMTD method is especially useful in the design and sizing of heat exchangers. If inlet and outlet temperatures and the overall heat transfer coefficient are given, or are easily determined, then the LMTD can be readily calculated and the LMTD method can be used to determine the heat transfer rate and the required surface area for the exchanger. If, on the other hand, only the inlet temperatures and flow rates are known, then the LMTD cannot be easily determined. Use of the LMTD method will entail tedious iterations to ultimately determine the LMTD and achieve a successful problem solution. The $\varepsilon$-NTU method is greatly superior in these performance-type problems where the outlet temperatures are to be determined. We will now discuss the two methods and give examples to illustrate their use.

## 8.4.1 Log mean temperature difference method

### 8.4.1.1 Double-pipe heat exchangers

The outer surfaces of heat exchangers are well-insulated and the heat transfer to the surrounding environment is very small. It is assumed that this heat transfer is negligible and that the total amount of heat leaving the hot fluid is gained by the cold fluid. There are three equations for $q$, which can be used to solve problems by the LMTD method:

For the heat leaving the hot fluid, we have

$$q = \dot{m}_h c_{ph}(\Delta T)_h = \dot{m}_h c_{ph}(T_{hi} - T_{ho}) \tag{8.27}$$

For the heat gained by the cold fluid, we have

$$q = \dot{m}_c c_{pc}(\Delta T)_c = \dot{m}_c c_{pc}(T_{co} - T_{ci}) \tag{8.28}$$

and the equation with the overall heat transfer coefficient $U$ and the *LMTD* is

$$q = UA(LMTD) \tag{8.29}$$

The temperature subscripts in Eqs. (8.27) and (8.28) are

$hi$ = hot fluid entering (going in)

$ho$ = hot fluid leaving (going out)

$ci$ = cold fluid entering (going in)

$co$ = cold fluid leaving (going out)

Example 8.3 shows the application of Eqs. (8.27)–(8.29) in determining the required heat transfer area of an exchanger and the flow rate for one of the fluids.

## Example 8.3

### LMTD method for parallel and counterflow double-pipe exchangers

*Problem*

A double-pipe heat exchanger heats water from 20 to 65 C. The heating is done by hot gases ($c_p$ = 900 J/kg C) that enter the exchanger at 250 C and leave at 75 C. The flow rate of the water is 2 kg/s. The overall heat transfer coefficient of the heat exchanger is 280 W/m$^2$ C.

**(a)** What is the surface area required if the heat exchanger is parallel flow?

**(b)** What is the surface area required if the heat exchanger is counterflow?

**(c)** What is the flow rate of the hot gases?

*Solution*

**(a)** We first sketch the temperature diagram for a double-pipe parallel heat exchanger, putting the temperatures and flow directions on the sketch.

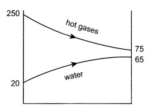

Givens in this problem include $U$ = 280 W/m$^2$ C and $\dot{m}_c$ = 2 kg/s.

Using Eq. (8.19),

$$LMTD = \frac{(\Delta T)_1 - (\Delta T)_2}{\ln\left(\frac{(\Delta T)_1}{(\Delta T)_2}\right)} = \frac{(250 - 20) - (75 - 65)}{\ln\left(\frac{(250 - 20)}{(75 - 65)}\right)} = 70.16 \text{ C}$$

$$q = \dot{m}_c c_{pc}(\Delta T)_c = 2(4180)(65 - 20) = 3.762 \times 10^5 \text{ W}$$

$$q = UA(LMTD) = 280A(70.16) = 3.762 \times 10^5$$

Solving this for $A$, we have $A$ = **19.15 m$^2$**.

**(b)** Similarly, we sketch the temperature diagram for a double-pipe counterflow heat exchanger.

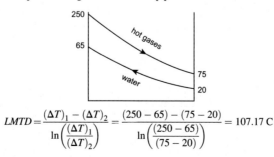

$$LMTD = \frac{(\Delta T)_1 - (\Delta T)_2}{\ln\left(\frac{(\Delta T)_1}{(\Delta T)_2}\right)} = \frac{(250 - 65) - (75 - 20)}{\ln\left(\frac{(250 - 65)}{(75 - 20)}\right)} = 107.17 \text{ C}$$

From Part (a), $q = 3.762 \times 10^5$ W

$q = UA\,(LMTD) = (280)\,A\,(107.17) = 3.762 \times 10^5$

Solving for $A$, we have $A = 12.54\ \text{m}^2$.

It is seen that, for the same heat transfer, the double-pipe counterflow heat exchanger requires less heat transfer area than a parallel exchanger.

**(c)** $q = \dot{m}_h c_{ph}(\Delta T)_h = \dot{m}_h(900)(250 - 75) = 3.762 \times 10^5$

$$\dot{m}_h = 2.39\ \text{kg/s}$$

The hot gases flow at a rate of **2.39 kg/s.**

### 8.4.1.2 Non−double-pipe heat exchangers

Not all heat exchangers are of the double-pipe type. For example, three non−double-pipe exchangers were shown in Fig. 8.1B−D. If a heat exchanger is not a double pipe, Eqs. (8.27) and (8.28) are still applicable. However, Eq. (8.29) has to be modified with the addition of a correction factor, $F$. The equation becomes

$$q = UAF(LMTD)_{dpcf} \tag{8.30}$$

Take special note of the subscript on the *LMTD* term. It stands for "double-pipe counterflow." The correction factor $F$ is a correction to the *LMTD* which we would have if the heat exchanger were of the double-pipe counterflow type rather than the non−double-pipe type. This is further illustrated in the examples presented below.

The value of $F$ is provided for some common types of heat exchangers in Figs. 8.9−8.11 below. In these graphs, $F$ is a function of parameters $P$ and $R$, which are

$$P = \frac{t_o - t_i}{T_i - t_i} \tag{8.31}$$

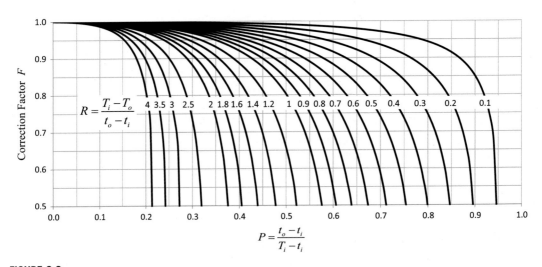

**FIGURE 8.9**

Log mean temperature difference (LMTD) correction factor *F* for one-shell-pass, multiple of two-tube-passes heat exchanger.

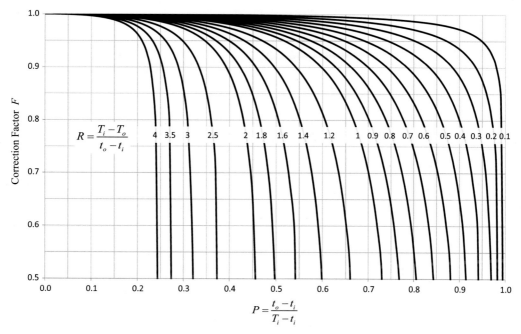

**FIGURE 8.10**

Log mean temperature difference (LMTD) correction factor $F$ for two-shell-passes, multiple of four-tube-passes heat exchanger.

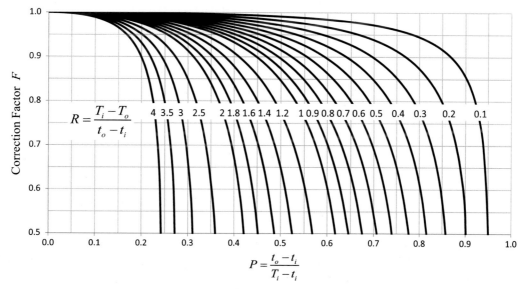

**FIGURE 8.11**

Log mean temperature difference (LMTD) correction factor $F$ for crossflow heat exchanger: one fluid mixed and one fluid unmixed.

$$R = \frac{T_i - T_o}{t_o - t_i} \tag{8.32}$$

In Eqs. (8.31) and (8.32), the $t$ temperatures are for the tube-side fluid and the $T$ temperatures are for the shell-side fluid. However, it makes no difference which fluid is assigned to the tube or to the shell. If the problem is not specific, you have the freedom to assign either of the fluids to the tube or the shell. The subscripts "$i$" and "$o$" refer to the fluid's inlet and outlet.

Functions for $F$ for some common types of exchangers are given in Table 8.3.

Graphs and functions for $F$ were given in a pioneering 1940 paper by Bowman, Mueller, and Nagle [3]. More recent papers on the LMTD method and the correction factor are in Refs. [4–7]. Compilations of graphs and functions for $F$ are in Refs. [8,9]. Please note that a graph and function have not been included above for crossflow exchangers with both fluids unmixed. Such exchangers are discussed in Refs. [3,5]. Problems involving a crossflow exchanger with both fluids unmixed may be treated using the $\varepsilon$-NTU method of Section 8.4.2.

**Table 8.3 Correction factor $F$ for log mean temperature difference (LMTD) method.**

| | |
|---|---|
| Parallel flow double-pipe | $F = 1$ |
| Counterflow double-pipe | $F = 1$ |
| One or both fluids condensing or evaporating | $F = 1$ |
| | In the below relations, $P = \frac{t_o - t_i}{T_i - t_i}$ $R = \frac{T_i - T_o}{t_o - t_i}$ ($t$ = tube, $T$ = shell) |
| One-shell-pass and multiple of two-tube-passes | $F = \frac{\sqrt{1+R^2}}{1-R} \dfrac{\ln[(1-PR)/(1-P)]}{\ln\left[\dfrac{2 - P\left(1 + R - \sqrt{1 + R^2}\right)}{2 - P\left(1 + R + \sqrt{1 + R^2}\right)}\right]}$ |
| | $F = \frac{P}{1-P} \dfrac{\sqrt{2}}{\ln\left[\dfrac{2/P - 2 + \sqrt{2}}{2/P - 2 - \sqrt{2}}\right]} \quad \text{for} \quad R = 1$ |
| Two-shell-passes and multiple of four-tube-passes | $F = \frac{\sqrt{1+R^2}}{2(1-R)} \dfrac{\ln[(1-PR)/(1-P)]}{\ln\left[\dfrac{2 - P\left(1 + R - \sqrt{1 + R^2}\right) + 2\sqrt{(1-P)(1-PR)}}{2 - P\left(1 + R + \sqrt{1 + R^2}\right) + 2\sqrt{(1-P)(1-PR)}}\right]}$ |
| | $F = \frac{\sqrt{2}}{2} \dfrac{P/(1-P)}{\ln\left[\dfrac{2 - P(2 - \sqrt{2}) + 2\sqrt{(1-P)^2}}{2 - P(2 + \sqrt{2}) + 2\sqrt{(1-P)^2}}\right]} \quad \text{for} \quad R = 1$ |
| Crossflow with one fluid mixed one fluid unmixed | $F = -\frac{1}{1-R} \dfrac{\ln[(1-PR)/(1-P)]}{\ln[1+(1/R)\ln(1-PR)]}$ |
| | $F = -\dfrac{P/(1-P)}{\ln[1+\ln(1-P)]} \quad \text{for} \quad R = 1$ |

There are a couple of special cases: First, if the heat exchanger is a double-pipe exchanger, then $F = 1$ and the *LMTD* in Eq. (8.30) is the one appropriate for the type of double-pipe exchanger, i.e., parallel flow or counterflow. Also, if one or both of the fluids is condensing or evaporating, then $F = 1$ and there is no need to use the $F$ graphs or $F$ equations for the specific type of heat exchanger.

---

### Example 8.4
#### LMTD method with correction factor $F$
*Problem*

A shell-and-tube heat exchanger with one shell pass and two tube passes has water in both the shell and the tubes. The hot water enters at 90 C and is cooled to 60 C. The cold water enters at 10 C and is heated to 50 C. The rating of the heat exchanger is 80 kW and the overall heat transfer coefficient is 950 W/m² C.

**(a)** What is the needed surface area of the exchanger?

**(b)** What are the flow rates of the two water streams?

*Solution*

This is a non–double-pipe exchanger so we need correction factor $F$. Graphs for $F$ are based on a double-pipe counterflow exchanger with the same entering and leaving temperatures as those for the non–double-pipe exchanger of our problem. Therefore, we first sketch the temperature diagram for a double-pipe counterflow exchanger and put temperatures and flow directions on the diagram.

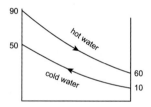

**(a)**  $q = UAF(LMTD)_{dpcf}$

From the diagram, $LMTD_{dpcf} = \dfrac{(\Delta T)_1 - (\Delta T)_2}{\ln\left(\dfrac{(\Delta T)_1}{(\Delta T)_2}\right)} = \dfrac{(90-50)-(60-10)}{\ln\left(\dfrac{(90-50)}{(60-10)}\right)} = 44.81$ C

Factor $F$ for this exchanger is in Fig. 8.9. Let us put the hot water in the tubes and the cold water in the shell. (Choice is arbitrary as it was not specified in the problem statement.)

For Fig. 8.9, $P = \frac{t_o - t_i}{T_i - t_i} = \frac{60-90}{10-90} = 0.38$   and   $R = \frac{T_i - T_o}{t_o - t_i} = \frac{10-50}{60-90} = 1.33$.

For these $P$ and $R$ values, $F = 0.89$.

$$q = UAF(LMTD)_{dpcf} = 80,000 = (950)(A)(0.89)(44.81)$$

Solving for A, we get $A = 2.11$ m².

**(b)** For the cold water, $q = \dot{m}_c c_{pc}(\Delta T)_c = 80,000 = \dot{m}_c(4180)(50 - 10)$.

Solving for $\dot{m}_c$, we get $\dot{m}_c = 0.478$ kg/s.

For the hot water, $q = \dot{m}_h c_{ph}(\Delta T)_h = 80,000 = \dot{m}_h(4180)(90 - 60)$.

Solving for $\dot{m}_h$, we get $\dot{m}_h = 0.638$ kg/s.

The cold water flows at 0.478 kg/s and the hot water flows at 0.638 kg/s.

## Example 8.5
### LMTD method with condensing fluid
*Problem*

It is desired to heat water using condensing steam in a crossflow, unfinned heat exchanger. The water is in the tubes and the steam flows across the tubes. The steam is at 130 C. The water flow rate is 3 kg/s and the water is heated from 20 to 90 C. The overall heat transfer coefficient is 3500 W/m² C. What is the required surface area of the exchanger?

*Solution*

This is a non–double-pipe exchanger so the equation for the heat flow rate is $q = UAF(LMTD)_{dpcf}$. Normally, we would have to use a graph or equation to find $F$ for the crossflow heat exchanger. However, we have a fluid that is condensing. This is a special case, and $F = 1$. The temperature diagram is given in the figure below.

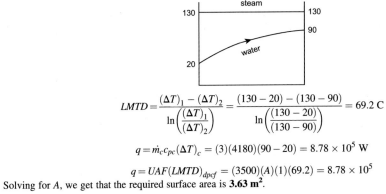

$$LMTD = \frac{(\Delta T)_1 - (\Delta T)_2}{\ln\left(\frac{(\Delta T)_1}{(\Delta T)_2}\right)} = \frac{(130 - 20) - (130 - 90)}{\ln\left(\frac{(130 - 20)}{(130 - 90)}\right)} = 69.2 \text{ C}$$

$$q = \dot{m}_c c_{pc}(\Delta T)_c = (3)(4180)(90 - 20) = 8.78 \times 10^5 \text{ W}$$

$$q = UAF(LMTD)_{dpcf} = (3500)(A)(1)(69.2) = 8.78 \times 10^5$$

Solving for $A$, we get that the required surface area is **3.63 m²**.

## Example 8.6
### LMTD method with iteration
*Problem*

A double-pipe counterflow heat exchanger uses exhaust gases ($c_p = 1000$ J/kg C) to heat 2 kg/s of water from 15 to 40 C. The gases enter the exchanger at 250 C. The overall heat transfer coefficient of the exchanger is 350 W/m² C and the heat transfer area of the exchanger is 5 m². What is the flow rate of the exhaust gases?

*Solution*

The first step is to draw the temperature diagram for the exchanger.

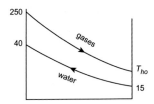

We have three equations for the heat transfer rate

$$q = \dot{m}_c c_{pc}(\Delta T)_c \tag{1}$$

$$q = \dot{m}_h c_{ph}(\Delta T)_h \tag{2}$$

$$q = UA(LMTD) \tag{3}$$

We can get the heat transfer rate from Eq. (1):

$$q = \dot{m}_c c_{pc}(\Delta T)_c = 2(4180)(40 - 15) = 2.09 \times 10^5 \text{ W}$$

Using this in Eq. (2), we have

$$2.09 \times 10^5 = \dot{m}_h(1000)(250 - T_{ho}) \tag{4}$$

The LMTD is $LMTD = \dfrac{(\Delta T)_1 - (\Delta T)_2}{\ln\left(\dfrac{(\Delta T)_1}{(\Delta T)_2}\right)} = \dfrac{(250 - 40) - (T_{ho} - 15)}{\ln\left(\dfrac{250 - 40}{T_{ho} - 15}\right)}$

Therefore, Eq. (3) becomes

$$2.09 \times 10^5 = (350)(5)\ \dfrac{(250 - 40) - (T_{ho} - 15)}{\ln\left(\dfrac{250 - 40}{T_{ho} - 15}\right)} \tag{5}$$

Simplifying Eq. (5), we get

$$119.4 = \dfrac{225 - T_{ho}}{\ln\left(\dfrac{210}{T_{ho} - 15}\right)} \tag{6}$$

Iteration is needed to solve Eq. (6) for the exit temperature of the exhaust gases. Once that is known, Eq. (4) can be used to obtain the desired flow rate of the gases. For the iteration, we could solve Eq. (6) by trial and error: Pick a value for $T_{ho}$, evaluate the right side of Eq. (6) and see if it is equal to 119.4. If not, pick another value of $T_{ho}$ and repeat the process until the right side of Eq. (6) achieves 119.4. This can be very tedious. A better way is to use Excel's *Goal Seek* to do the iterations and get an answer.

On the Excel spreadsheet, put the initial guessed value for $T_{ho}$ in one cell, say cell A1. (Note: $T_{ho}$ has to be less than 250. We used a guess of 40. The value is not critical.) Put the right side of Eq. (6) in cell A2. That is, cell A2 will have **=(225-A1)/LN(210/(A1-15))**.

Call up *Goal Seek* and tell it to make cell A2 equal to 119.4 by changing cell A1.

From *Goal Seek*, we got that $T_{ho} = 74.6$ C. Using this in Eq. (4), we find that the flow rate of the exhaust gases is **1.192 kg/s**.

---

## Example 8.7

### LMTD method when outlet temperatures are not given
*Problem*

Hot water is available at 75 C in a factory. It is desired to use the water in a double-pipe counterflow heat exchanger to heat cold water. The cold water enters the exchanger at 15 C and flows at a rate of 2 kg/s through the exchanger. The hot water flows at a rate of 1.5 kg/s. The overall heat transfer coefficient of the exchanger is 500 W/m² C and the surface area of the exchanger is 8 m².

**(a)** What is the rating of the exchanger (kW)?

**(b)** What are the exit temperatures of the two water streams?

*Solution*

The first step is to draw the temperature diagram for the exchanger.

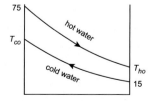

We have three equations for the heat transfer rate:

$$q = \dot{m}_c c_{pc} (\Delta T)_c \tag{1}$$

$$q = \dot{m}_h c_{ph} (\Delta T)_h \tag{2}$$

$$q = UA(LMTD) \tag{3}$$

Putting the known values into these equations, we have

$$q = 2(4180)(T_{co} - 15) \tag{4}$$

$$q = 1.5(4180)(75 - T_{ho}) \tag{5}$$

$$q = (500)(8) \frac{[(75 - T_{co}) - (T_{ho} - 15)]}{\ln\left(\dfrac{75 - T_{co}}{T_{ho} - 15}\right)} \tag{6}$$

We thus have three equations, Eqs. (4)–(6), to solve for the three unknowns $q$, $T_{co}$, and $T_{ho}$. Iteration is needed. We could solve the equations by trial and error. That is, guess $T_{co}$, use Eq. (4) to get $q$, put the $q$ value into Eq. (5) to get $T_{ho}$, use the $T_{co}$ and $T_{ho}$ in Eq. (6) to calculate $q$, and see if that $q$ value equals the $q$ value of Eqs. (4) and (5). If not, pick another $T_{co}$ and repeat the process until the $q$'s match. This is a potentially very tedious process. Fortunately, we can use software to do the iteration. One way is to use Excel's *Solver* add-in program. To do this, we rearrange the equations to create functions. Let us call the functions $f_1, f_2$, and $f_3$.

From Eq. (4), we have $f_1 = q - 2(4180)(T_{co} - 15) = 0$.

From Eq. (5), we have $f_2 = q - 1.5(4180)(75 - T_{ho}) = 0$.

From Eq. (6), we have $f_3 = q - (500)(8) \frac{[(75 - T_{co}) - (T_{ho} - 15)]}{\ln\left(\dfrac{75 - T_{co}}{T_{ho} - 15}\right)} = 0$.

If all three functions are zero, we have the solution to the three equations. And, making the quantity $f_1^2 + f_2^2 + f_3^2$ equal to zero will assure that all three functions are indeed zero. In an Excel spreadsheet, we use three cells for the guesses for $T_{co}$, $T_{ho}$, and $q$. Let us say we use cells A1, A2, and A3. (Reasonable guesses could be 40, 40, and 1e5.) In three other cells, say cells F1, F2, and F3, we put the three functions written with cell addresses. For example, cell F1 will have = **A3-2∗4180∗(A1-15)**. Finally, there is a test cell, say cell F4, with = **F12 + F22 + F32** in it. We call up *Solver* and tell it to make cell F4 zero by changing cells A1 through A3. If there is no convergence to a solution, tell *Solver* to make cell F4 a minimum. The criterion that F4 is zero is often too stringent.

From *Solver* we got the results $q = 1.54 \times 10^5$ W, $T_{co} = 33.4$ C, and $T_{ho} = 50.5$ C.

(a) The exchanger's rating is **154 kW**.

(b) The cold water exits at **33.4 C** and the hot water exits at **50.5 C**.

## 8.4.2 Effectiveness—number of transfer unit method

In the last two examples, iteration was needed for problem solution. The LMTD method, although excellent for heat exchanger sizing problems, requires iteration for performance-type problems in which both the inlet and outlet temperatures are not given. One can only imagine the difficulty in solving iteration problems in the 1950s and earlier without the assistance of modern-day calculators and computers.

Kays and London [10] developed the Effectiveness-Number of Transfer Units ($\varepsilon$-NTU) method that eliminates the need for iteration for most performance-type problems. The method is as follows:

The "heat capacity rate" is $C_c$ for the cold fluid and $C_h$ for the hot fluid. It is defined as the product of the mass flow rate and the specific heat.

$$\text{For the cold fluid, } C_c = \dot{m}_c c_{pc} \tag{8.33}$$

$$\text{For the hot fluid, } C_h = \dot{m}_h c_{ph} \tag{8.34}$$

For the two fluids in a heat exchanger, one has the higher heat capacity rate, $C_{max}$, and the other has the lower rate, $C_{min}$. The fluid with the higher heat capacity rate is called the "maximum fluid" and the fluid with the lower heat capacity rate is called the "minimum fluid".

The "effectiveness" of a heat exchanger is defined as

$$\varepsilon = \frac{\text{actual heat transfer}}{\text{maximum heat transfer}} = \frac{q}{q_{max}} \tag{8.35}$$

The actual heat transfer is given by Eqs. (8.27) and (8.28).

$$q = \dot{m}_c c_{pc}(T_{co} - T_{ci}) = \dot{m}_h c_{ph}(T_{hi} - T_{ho}) \tag{8.36}$$

Using the definition of the heat capacity rate, Eq. (8.36) can be written as

$$q = C_c(T_{co} - T_{ci}) = C_h(T_{hi} - T_{ho}) \tag{8.37}$$

The maximum heat transfer is that where the minimum fluid undergoes the maximum possible temperature change in the exchanger. The maximum temperature in an exchanger is that of the incoming hot fluid and the minimum temperature in an exchanger is that of the incoming cold fluid. Therefore, the maximum possible temperature change in an exchanger is $T_{hi} - T_{ci}$, and

$$q_{max} = C_{min}(T_{hi} - T_{ci}) \tag{8.38}$$

From Eqs. (8.35) and (8.38), we have

$$\varepsilon = \frac{q}{q_{max}} = \frac{q}{C_{min}(T_{hi} - T_{ci})} \tag{8.39}$$

Looking at Eq. (8.39), we see that if the effectiveness $\varepsilon$, the heat capacity rate for the minimum fluid, and the incoming temperatures are known, then the actual heat transfer of the heat exchanger can be determined. The outlet temperatures of the fluids do not need to be known.

Using Eqs. (8.37) and (8.39), the relation for the effectiveness can be written solely in terms of the fluid temperatures.

If the cold fluid is the minimum fluid, then

$$\varepsilon = \frac{T_{co} - T_{ci}}{T_{hi} - T_{ci}} \tag{8.40}$$

If the hot fluid is the minimum fluid, then

$$\varepsilon = \frac{T_{hi} - T_{ho}}{T_{hi} - T_{ci}} \tag{8.41}$$

Another important dimensionless parameter for the $\varepsilon$-NTU method is the "number of transfer units," (NTU). This is defined as

$$NTU = \frac{UA}{C_{min}} \tag{8.42}$$

Looking at Eq. (8.42), we see that the NTU is directly proportional to the heat transfer area $A$. The larger the NTU, the larger the heat exchanger area $A$.

The effectiveness $\varepsilon$ for a given type of heat exchanger is a function of the ratio of heat capacity rates $C_{ratio} = \frac{C_{min}}{C_{max}}$ and the number of transfer units NTU. Effectiveness graphs for various types of heat exchangers are in Figs. 8.12 through 8.17A and B. Effectiveness equations are in Table 8.4.

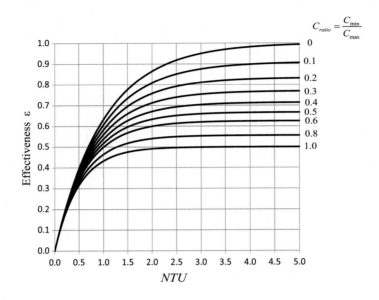

**FIGURE 8.12**

Effectiveness $\varepsilon$ for parallel flow.

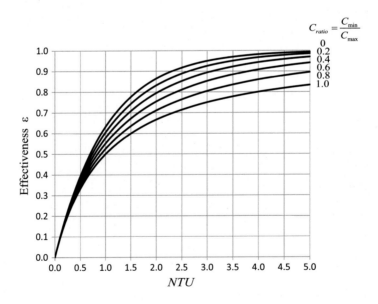

**FIGURE 8.13**

Effectiveness $\varepsilon$ for counterflow heat exchanger.

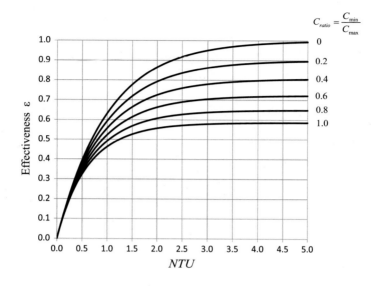

**FIGURE 8.14**

Effectiveness $\varepsilon$ for one-shell-pass, multiple of two-tube-passes heat exchanger.

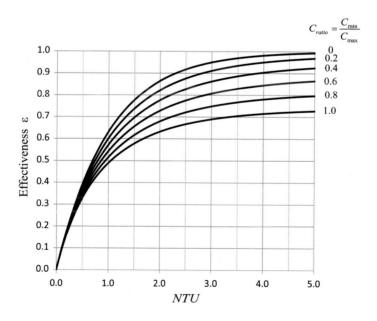

**FIGURE 8.15**

Effectiveness $\varepsilon$ for two-shell-passes, multiple of four-tube-passes heat exchanger.

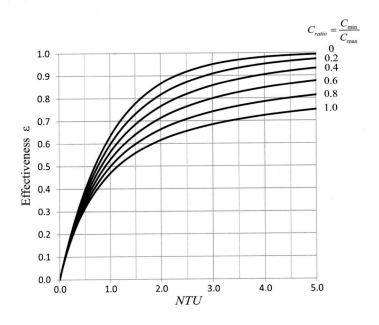

**FIGURE 8.16**

Effectiveness $\varepsilon$ for crossflow heat exchanger: both fluids unmixed.

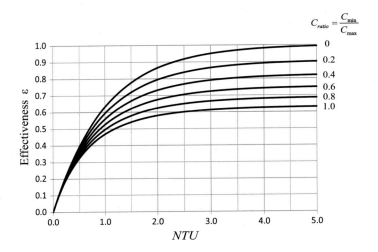

**FIGURE 8.17A**

Effectiveness $\varepsilon$ for crossflow heat exchanger: one fluid mixed and one fluid unmixed ($C_{max}$ fluid mixed, $C_{min}$ fluid unmixed).

$$C_{ratio} = \frac{C_{min}}{C_{max}}$$

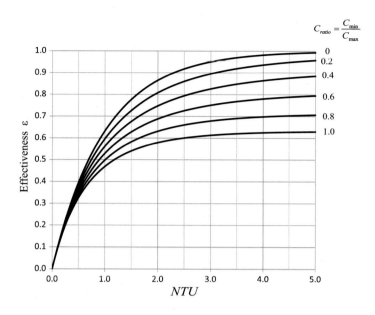

**FIGURE 8.17B**

Effectiveness $\varepsilon$ for crossflow heat exchanger: one fluid mixed and one fluid unmixed ($C_{min}$ fluid mixed, $C_{max}$ fluid unmixed).

We now present a problem that would have required iteration if it had been solved by the LMTD method. The $\varepsilon$-NTU method does not require iteration.

## Example 8.8

**Effectiveness—number of transfer unit method when outlet temperatures are not given**

*Problem*

This is the same problem as Example 8.7. In that example, we showed that use of the LMTD method required iteration for a solution. In this example, we show how much easier it is to use the $\varepsilon$-NTU method when outlet temperatures are not given. Iteration is not needed for the solution.

Please refer to the problem statement and temperature diagram for Example 8.7.

*Solution*

Givens in this problem are $\dot{m}_c = 2 \text{ kg}/\text{s}$  $\dot{m}_h = 1.5 \text{ kg}/\text{s}$  $c_{pc} = c_{ph} = 4180 \text{ J}/\text{kg C}$.

$$T_{hi} = 75 \text{ C} \quad T_{ci} = 15 \text{ C} \quad U = 500 \text{ W}/\text{m}^2 \text{ C} \quad A = 8 \text{ m}^2$$

To find: $q$,  $T_{co}$, and  $T_{ho}$.

We first determine the capacity rates:

$$C_h = \dot{m}_h c_{ph} = 1.5(4180) = 6270 \text{ W}/\text{C} \quad C_c = \dot{m}_c c_{pc} = 2(4180) = 8360 \text{ W}/\text{C}$$

It is seen that $C_{min} = C_h = 6270$ and $C_{mas} = C_c = 8360$.

Therefore, $C_{min}/C_{max} = 6270/8360 = 0.75$.

$$NTU = \frac{UA}{C_{min}} = \frac{(500)(8)}{6270} = 0.64$$

With $C_{min}/C_{max} = 0.75$ and $NTU = 0.64$, we use Fig. 8.13 and find that the effectiveness $\varepsilon \approx 0.40$.

**Table 8.4 Effectiveness $\varepsilon$ for effectiveness–number of transfer unit ($\varepsilon$-NTU) method.**

$$NTU = \frac{UA}{C_{min}} \qquad C_{ratio} = \frac{C_{min}}{C_{max}}$$

Double-pipe parallel flow

$$\varepsilon = \frac{1-\exp[-NTU(1+C_{ratio})]}{1+C_{ratio}}$$

Double-pipe counterflow

$$\varepsilon = \frac{1-\exp[-NTU(1-C_{ratio})]}{1-C_{ratio}\exp[-NTU(1-C_{ratio})]}$$

$$\varepsilon = \frac{NTU}{NTU+1} \quad \text{for} \quad C_{ratio}=1$$

One-shell-pass and multiple of two-tube-passes

$$\varepsilon = \frac{2}{1+C_{ratio}+\sqrt{1+C_{ratio}^2}\,(1+e^{-\Gamma})/(1-e^{-\Gamma})}$$

$$\text{where } \Gamma = NTU\sqrt{1+C_{ratio}^2}$$

Two-shell-passes and multiple of four-tube-passes

$$\varepsilon = \frac{[(1-\varepsilon_1 C_{ratio})/(1-\varepsilon_1)]^2-1}{[(1-\varepsilon_1 C_{ratio})/(1-\varepsilon_1)]^2-C_{ratio}}$$

$$\varepsilon = \frac{2\varepsilon_1}{1+\varepsilon_1} \quad \text{for} \quad C_{ratio}=1$$

Where: $\varepsilon_1$ is The effectiveness of each of the two shell passes. That is, $\varepsilon_1$ is the effectiveness given immediately above for a one-shell-pass exchanger. The $NTU$ for the two-shell-pass exchanger is twice the $NTU$ for the one-shell-pass exchanger.

Crossflow with both fluids unmixed

$$\varepsilon = 1 - \exp\left\{\frac{NTU^{0.22}}{C_{ratio}}\left[\exp\left(-C_{ratio}NTU^{0.78}\right)-1\right]\right\}$$

Crossflow with One fluid mixed
One fluid unmixed

For $C_{max}$ mixed and $C_{min}$ unmixed $\varepsilon = \frac{1}{C_{ratio}}\left(1-\exp\{-C_{ratio}[1-\exp(-NTU)]\}\right)$

For $C_{min}$ mixed and $C_{max}$ unmixed $\varepsilon = 1-\exp\left\{-\frac{1}{C_{ratio}}[1-\exp(-C_{ratio}NTU)]\right\}$

One or both fluids condensing or evaporating ($C_{ratio}=0$)

$$\varepsilon = 1 - \exp(-NTU)$$

The effectiveness relation is $\varepsilon = \frac{C_c(T_{co}-T_{ci})}{C_{min}(T_{hi}-T_{ci})} = \frac{(8360)(T_{co}-15)}{(6270)(75-15)} = 0.02222(T_{co}-15)$ and $\varepsilon = 0.4 = 0.02222(T_{co}-15)$.
Solving for $T_{co}$, we get $T_{co} = 33.0$ C.
For the cold fluid, $q = C_c(T_{co}-T_{ci}) = (8360)(33.0-15) = 1.505 \times 10^5$ W.
For the hot fluid, $q = C_h(T_{hi}-T_{ho}) = (6270)(75-T_{ho}) = 1.505 \times 10^5$ W.
Solving for $T_{ho}$, we get $T_{ho} = 51.0$ C.
In summary, the heat transfer rate of the exchanger is **1.505 × 10⁵ W**.
The cold water leaves the exchanger at **33.0 C**.
The hot water leaves the exchanger at **51.0 C**.
These answers are almost identical to those obtained via the LMTD solution in Example 8.7. The differences are due to the minor inaccuracy of reading the graphs. It is seen that no iteration was needed for the ε-NTU solution.

We close this section with three more examples using the ε-NTU method.

## Example 8.9
### Effectiveness—number of transfer unit method for determining surface area
*Problem*
A crossflow heat exchanger uses exhaust gases ($c_p$ = 1020 J/kg C) to heat water. The water flows through the tubes and the gases flow over the outside of the tubes, which are finned. The water flows at 1.5 kg/s. It enters the exchanger at 20 C and leaves at 70 C. The exhaust gases enter at 190 C and leave at 85 C. The overall heat transfer coefficient is 250 W/m² C. What is the needed surface area?

*Solution*

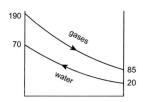

We will first find the flow rate of the exhaust gases, $\dot{m}_h$.

$$C_c = \dot{m}_c c_{pc} = (1.5)(4180) = 6270 \text{ W/C}$$

$$q = C_c(\Delta T)_c = C_h(\Delta T)_h$$

$$q = (6270)(70-20) = \dot{m}_h(1020)(190-85)$$

Solving for $\dot{m}_h$, we get $\dot{m}_h = 2.93$ kg/s and $C_h = \dot{m}_h c_{ph} = (2.93)(1020) = 2990$.

$$C_{min} = C_h = 2990(\text{exhaust gases}) \quad C_{max} = C_c = 6270(\text{water})$$

$$C_{min}/C_{max} = 2990/6270 = 0.48$$

As the minimum fluid is the exhaust gases, the equation for the effectiveness is

$$\varepsilon = \frac{C_c(T_{co}-T_{ci})}{C_{min}(T_{hi}-T_{ci})} = \frac{(6270)(70-20)}{(2990)(190-20)} = 0.62$$

With $C_{min}/C_{max} = 0.48$ and $\varepsilon = 0.62$, we go into Fig. 8.16 and find that $NTU = 1.3$.
$NTU = \frac{UA}{C_{min}} = 1.3 = \frac{(250)A}{2990}$ Solving for $A$, we get $A = 15.5$ m².
The needed surface area of the exchanger is **15.5 m²**.
**Note:** As both the inlet and exit temperatures of the fluids were given, the LMTD method can also be readily used to solve this problem. The amount of work needed for both solutions is about the same.

## Example 8.10

### Effectiveness—number of transfer unit method for a steam condenser

*Problem*

A shell-and-tube heat exchanger uses condensing steam to heat water. Both the steam and the water make a single pass through the exchanger. The water flows at a rate of 4 kg/s and is heated from 25 to 70 C. The steam is condensing at 120 C. The overall heat transfer coefficient of the exchanger is 1100 W/m² C.

**(a)** What is the needed area of the exchanger?

**(b)** If the flow rate of the water is reduced to 2 kg/s and everything else remains the same, what is the outlet temperature of the water?

*Solution*

**(a)** As always, we first draw the temperature diagram for the exchanger.

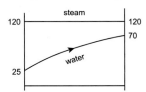

For a condenser, $C_{min}/C_{max} = 0$ and the noncondensing fluid is the minimum fluid.

$$C_{min} = C_c = \dot{m}_c c_{pc} = (4)(4180) = 16720$$

To get the surface area, we need the number of transfer units $NTU$ as $NTU = \frac{UA}{C_{min}}$.

The effectiveness $\varepsilon$ is $\varepsilon = \frac{q}{q_{max}} = \frac{C_c(T_{co}-T_{ci})}{C_{min}(T_{hi}-T_{ci})}$.

And, for a condenser, the relationship between $\varepsilon$ and $NTU$ is $\varepsilon = 1 - e^{-NTU}$.

Using these three relationships, we have

$$\varepsilon = \frac{(16720)(70 - 25)}{(16720)(120 - 25)} = 0.4737 = 1 - e^{-NTU}$$

$$e^{-NTU} = 1 - 0.4737 = 0.5263$$

Taking the natural log of both sides of the equation, we have

$$\ln(e^{-NTU}) = -NTU = \ln(0.5263) = -0.642 \text{ and } NTU = 0.642$$

$$NTU = \frac{UA}{C_{min}} = 0.642 = \frac{(1100)A}{16720}$$

Solving for $A$, we get $A = \textbf{9.76 m}^2$.

**(b)** With the reduction in flow rate, $C_{min}$ becomes

$$C_{min} = C_c = \dot{m}_c c_{pc} = (2)(4180) = 8360$$

The number of transfer units is $NTU = \frac{UA}{C_{min}} = \frac{(1100)(9.76)}{8360} = 1.284$.

The effectiveness is $\varepsilon = 1 - e^{-NTU} = 1 - e^{-1.284} = 0.723$.

$$\varepsilon = \frac{q}{q_{max}} = \frac{C_c(T_{co} - T_{ci})}{C_{min}(T_{hi} - T_{ci})} = 0.723 = \frac{(8360)(T_{co} - 25)}{(8360)(120 - 25)}$$

Solving for $T_{co}$, we get. $T_{co} = 93.7$ C.

The outlet temperature of the water is **93.7 C.**

(Note: If we had solved this problem using the LMTD method, we would have needed iterations due to the log term in the *LMTD*.)

The final ε-NTU problem is a more complicated one. We mentioned that the ε-NTU method was especially useful if the LMTD solution for a problem required iteration. Well, here is an ε-NTU solution that also requires iteration.

## Example 8.11

### Effectiveness—number of transfer unit method involving iteration

*Problem*

Air is being heated by hot oil in a crossflow heat exchanger. The air enters at 20 C and 1 atm pressure and is heated to 40 C. The flow rate of the air is 1.3 m³/s. The oil ($c_p$ = 2100 J/kg C) enters the exchanger at 120 C. The oil flows in the tubes of the exchanger and the air flows across the unfinned tubes. The overall heat transfer coefficient is 75 W/m²C and the heat transfer area is 12 m². What is the heat transfer rate of the exchanger and the exit temperature of the oil?

*Solution*

We first draw the temperature diagram.

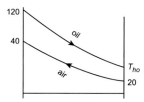

We then have to change the volumetric flow rate of the air to a mass flow rate. Using the ideal gas law, we have

$$\rho = \frac{P}{RT} = \frac{1.01325 \times 10^5}{(287)(293.15)} = 1.204 \text{ kg/m}^3$$

$$\dot{m}_c = \rho\dot{V} = \left(1.204 \text{ kg/m}^3\right)\left(1.3 \text{ m}^3/\text{s}\right) = 1.565 \text{ kg/s}$$

The hot fluid is the oil and the cold fluid is the air.

$$\text{(air) } C_c = \dot{m}_c c_{pc} = (1.565)(1007) = 1576$$

$$\text{(oil) } C_h = \dot{m}_h c_{ph} = \dot{m}_h(2100)$$

The heat transfer rate is $q = C_c(\Delta T)_c = (1576)(40 - 20) = \mathbf{3.152 \times 10^4}$ **W**.

As we do not have $\dot{m}_h$, we do not know which fluid is the minimum fluid. We have the value of the capacity rate for the air, so let us assume that the air is the minimum fluid.

Then, $C_{min} = C_c = 1576$.

The number of transfer units is $NTU = \frac{UA}{C_{min}} = \frac{(75)(12)}{1576} = 0.571$.

The effectiveness is $\varepsilon = \frac{q}{q_{max}} = \frac{q}{C_{min}(T_{hi} - T_{ci})} = \frac{3.152 \times 10^4}{(1576)(120 - 20)} = 0.2$.

With these values, we go into Fig. 8.17B to get $C_{min}/C_{max}$. We find that there is no solution for our $NTU$ and $\varepsilon$ values. Hence, the oil, not the air, must be the minimum fluid.

With the oil as the minimum fluid, we have

$$NTU = \frac{UA}{C_{min}} = \frac{UA}{\dot{m}_h c_{ph}} = \frac{(75)(12)}{\dot{m}_h(2100)} = \frac{0.4286}{\dot{m}_h} \tag{1}$$

$$\varepsilon = \frac{q}{q_{max}} = \frac{q}{C_{min}(T_{hi} - T_{ci})} = \frac{3.152 \times 10^4}{\dot{m}_h(2100)(120 - 20)} = \frac{0.1501}{\dot{m}_h} \tag{2}$$

To determine the flow rate of the oil, $\dot{m}_h$, we need an iterative process. We make an initial guess for $\dot{m}_h$ and calculate $NTU$ and $\varepsilon$ from Eqs. (1) and (2). With these values of $NTU$ and $\varepsilon$, we can go into Fig. 8.17A and determine the corresponding value of $C_{ratio} = C_{min}/C_{max}$. We can get $\dot{m}_h$ from the value of this ratio as $C_{max} = C_c = 1576$ and $C_{min} = \dot{m}_h(2100)$. So,

$$C_{ratio} = \frac{C_{min}}{C_{max}} = \frac{\dot{m}_h(2100)}{1576} = 1.332\dot{m}_h \tag{3}$$

We compare the calculated value for $\dot{m}_h$ with the initial guessed value. If they agree, we have reached a solution. If not, we make another guess for $\dot{m}_h$ and iterate again and again until the calculated value agrees with the guessed value. (Note: We are using Fig. 8.17A as the minimum fluid, the oil, is the unmixed fluid.)

Alternatively, we can use the equations in Table 8.4 to reach the solution. For a crossflow exchanger with one fluid unmixed and the other mixed, and with the unmixed fluid being the minimum fluid, the appropriate equation is

$$\varepsilon = (1 / C_{ratio})\{1 - \exp[-C_{ratio}(1 - e^{-NTU})]\} \tag{4}$$

The variables in this equation are all functions of $\dot{m}_h$ as given in Eqs. (1)–(3). Therefore, Eq. (4) is a function of only one variable, $\dot{m}_h$. Eq. (4) is rearranged to

$$\varepsilon - (1 / C_{ratio})\{1 - \exp[-C_{ratio}(1 - e^{-NTU})]\} = 0 \tag{5}$$

Eq. (5) can be solved by Matlab's *fzero* function, Excel's *Goal Seek*, or Excel's *Solver*. The software iterates with changing values of $\dot{m}_h$ until the left side of Eq. (5) is zero.

It turns out that Eq. (5) has multiple solutions. We want the solution that has $\varepsilon \leq 1$ and $C_{ratio} \leq 1$. If Excel is used, it turns out that *Solver* is better than *Goal Seek* as *Solver* allows constraints (for example, $\varepsilon \leq 1$ and $C_{ratio} \leq 1$). *Goal Seek* does not have the capability of constraints. Depending on the initial guess of $\dot{m}_h$, the solution by *Goal Seek* may give a solution where $\varepsilon$ and/or $C_{ratio}$ are greater than unity. Yes, it is a solution to Eq. (5), but it is not the one we are looking for. Using *Solver*, we got the result that $\dot{m}_h = 0.186$ kg/s. Other parameters for the solution were

$C_{ratio} = C_{min}/C_{max} = 0.248$, $\varepsilon = 0.807$, and $NTU = 2.3$.
For the oil, $q = \dot{m}_h c_{ph}(\Delta T)_h$.

$$3.152 \times 10^4 = (0.186)(2100)(120 - T_{ho})$$

Solving for $T_{ho}$, we get $T_{ho} = 39.3$ C.
The exit temperature of the oil is **39.3 C**.

## 8.5 Chapter summary and final remarks

In this chapter, we discussed the thermal aspects of some common types of heat exchangers. We discussed the determination of the overall heat transfer coefficient for an exchanger and the impact of the fouling of heat exchanger surfaces on the coefficient. We then outlined two methods of thermal analysis: the LMTD method and the ε-NTU method. Although either method could be used for thermal analysis, we found that the LMTD method was especially useful for design and sizing of the heat exchangers, while the ε-NTU method was especially useful for determining heat exchanger performance.

There are several factors other than thermal, which must be considered in the selection of a heat exchanger for a particular purpose. Consideration must be given to the cost of the exchanger—both the capital cost and the operating costs. For example, there will be costs for cleaning of the exchanger to maintain the overall heat transfer coefficient, and electrical costs for pumping the fluids through the exchanger. An important factor is scheduling. Is the exchanger off-the-shelf or must it be custom-designed and custom-manufactured? Does its delivery schedule impact the overall schedule for the project? There are structural aspects such as the support for the exchanger and the design of the baffles to prevent or minimize vibration problems while still maintaining the desired fluid flow. The exchanger will also have physical constraints. It must fit in its allocated space in a plant. Indeed, there are many factors to be considered in the selection of a heat exchanger. Thermal design is a major factor, but there are other factors of significance.

## 8.6 Problems

Notes:

- If needed information is not given in the problem statement, then use the Appendix for material properties and other information. If the Appendix does not have sufficient information, then use

the Internet or other reference sources. Some good reference sources are mentioned at the end of Chapter 1. If you use the Internet, double-check the validity of the information by using more than one source.
- Your solutions should include a sketch of the problem.
- In all problems, unless otherwise stated, the fluid pressure is atmospheric. Gas properties in the Appendix are at atmospheric pressure. If a problem has a gas at other than atmospheric pressure, affected properties such as density and kinematic viscosity should be modified accordingly through use of the ideal gas law.

**8-1** For a double-pipe heat exchanger: Hot water at an average temperature of 70 C flows through the tube and colder water at an average temperature of 30 C flows through the shell. The tube is a 1" Sch 10 carbon steel pipe. The convective coefficient is 5800 W/m² C at the inner surface of the pipe and 2000 W/m² C at the outer surface. What is the overall heat transfer coefficient of the exchanger?

**8-2** After a few years of operation, the exchanger of Problem 8-1 has a fouling resistance of 0.0002 m² C/W on its outer surface. What is the new overall heat transfer coefficient of the exchanger?

**8-3** For a double-pipe heat exchanger: The tube is 1" Type M copper. The convective coefficient is 650 W/m² C on the inside surface and 200 W/m² C on the outside surface. What is the overall heat transfer coefficient of the exchanger?

**8-4** After a period of use, the tube of Problem 8-3 has a fouling resistance of 0.002 m² C/W on its inside surface. What is the new overall heat transfer coefficient of the exchanger?

**8-5** A double-pipe heat exchanger heats water from 15 to 65 C. The heating is done by hot gases ($c_p = 900$ J/kg C) that enter the exchanger at 200 C. The flow rate of the water is 2 kg/s and the flow rate of the hot gases is 5 kg/s. The overall heat transfer coefficient is 310 W/m² C.
**(a)** What is the exit temperature of the hot gases?
**(b)** What is the surface area required if the heat exchanger is parallel flow?
**(c)** What is the surface area required if the heat exchanger is counterflow?

**8-6** A double-pipe counterflow heat exchanger uses hot water to heat cold water. The cold water enters at 20 C and leaves at 50 C. The hot water enters at 80 C and leaves at 40 C. The mass flow rate of the cold water is 1.5 kg/s. If the overall heat transfer coefficient is 250 W/m² C,
**(a)** What is the required surface area of the exchanger?
**(b)** What is the flow rate of the hot water (kg/s)?
**(c)** What is the rate of heat transfer to the cold water?

**8-7** A double-pipe parallel flow heat exchanger uses hot oil ($c_p = 2400$ J/kg C) to heat water. The water enters at 15 C and leaves at 80 C. The water has a flow rate of 1.5 kg/s. The oil enters at 150 C and has a flow rate of 3 kg/s. The overall heat transfer coefficient is 350 W/m² C.
**(a)** What is the exit temperature of the oil?
**(b)** What is the needed surface area of the heat exchanger?

**8-8** A double-pipe counterflow exchanger uses hot oil ($c_p = 2000$ J/kg C) to heat water from 30 to 80 C. The mass flow rate of the water is 0.8 kg/s. The oil flow rate is 0.9 kg/s and the oil enters the heat exchanger at 180 C. The overall heat transfer coefficient is 400 W/m² C.

(a) What is the exit temperature of the oil?

(b) What is the required surface area of the exchanger?

**8-9** Water at a flow rate of 2.5 kg/s flows through the tube of a double-pipe counterflow heat exchanger. The water is heated from 20 to 70 C by condensing saturated steam at 100 C in the shell side of the exchanger. The overall heat transfer coefficient of the exchanger is 2500 W/m$^2$ C.

(a) What surface area is required to achieve the desired heating of the water?

(b) What is the rate of condensation of the steam (kg/s)?

**8-10** Do Problem 8-9 if the exchanger were a single-shell-pass, two-tube-pass exchanger with the water in the tube and the steam in the shell.

**8-11** It is desired to heat water at a flow rate of 0.1 kg/s from 30 to 80 C. The heating will be done by hot oil ($c_p$ = 2000 J/kg C) entering the exchanger at 190 C. The flow rate of the oil is 0.15 kg/s. The overall heat transfer coefficient is 400 W/m$^2$ C. What heat exchanger area is needed?

**8-12** Hot water at a flow rate of 0.2 kg/s is needed in a factory. The water is heated from 15 to 45 C in a double-pipe counterflow heat exchanger. Heating is accomplished by hot air entering the exchanger at 250 C and flowing at a rate of 0.2 kg/s. Two heat exchangers are available in the factory: Exchanger 1 has U = 450 W/m$^2$ C and A = 0.42 m$^2$. Exchanger 2 has U = 225 W/m$^2$ C and A = 0.55 m$^2$. Which exchanger should be used?

**8-13** A double-pipe counterflow heat exchanger heats water from 15 to 45 C. The heating is accomplished by oil ($c_p$ = 2100 J/kg C) entering the exchanger at 90 C and leaving at 60 C. The exchanger's rating is 70 kW and its overall heat transfer coefficient is 360 W/m$^2$ C.

(a) What is the surface area of the exchanger?

(b) What are the flow rates of the oil and the water?

**8-14** Exhaust gases ($c_p$ = 1040 J/kg C) are used to preheat water in a single-pass, crossflow heat exchanger with one fluid mixed. The water flows through the tubes and the exhaust gases flow over the tubes. The water flows at 2 kg/s and is heated from 20 to 75 C. The gases are cooled from 250 to 90 C. The overall heat transfer coefficient is 220 W/m$^2$ C. Using the LMTD method, determine the surface area of the exchanger.

**8-15** Do Problem 8-14 using the ε-NTU method.

**8-16** Condensing steam at 150 C heats oil ($c_p$ = 2000 J/kg C) flowing at 2 kg/s from 15 to 80 C. The exchanger has two shell passes and four tube passes. The overall heat transfer coefficient is 2500 W/m$^2$ C. What is the needed surface area of the exchanger?

**8-17** A double-pipe counterflow heat exchanger uses water to cool hot air. The water enters at a rate of 0.15 kg/s and a temperature of 25 C. The air enters at 125 C and is cooled to 75 C. The flow rate of the air is 1 kg/s. The overall heat transfer coefficient is 275 W/m$^2$ C.

(a) What is the surface area needed?

(b) What is the temperature of the water leaving the exchanger?

**8-18** A crossflow heat exchanger with both fluids unmixed is used to heat water with engine oil. The water enters at 25 C and leaves at 70 C. The oil ($c_p$ = 2300 J/kg C) enters the exchanger at 140 C. Both the water and the oil have a flow rate of 1.5 kg/s. The heat exchanger has a surface area of 50 m$^2$. What is the overall heat transfer coefficient for the exchanger?

**8-19** A crossflow heat exchanger with both fluids unmixed is used to heat pressurized water with exhaust gases. The hot gases ($c_p$ = 1040 J/kg C) enter the exchanger at 350 C with a flow rate of 1 kg/s. The water enters at 40 C and leaves at 145 C. The water flow rate is 0.3 kg/s. The surface area of the exchanger is 4 m$^2$. What is the overall heat transfer coefficient?

**8-20** A double-pipe counterflow heat exchanger has a cold fluid entering at 25 C and a hot fluid entering at 140 C. For the cold fluid, $C_c = 12,000$ W/C, and for the hot fluid, $C_h = 9000$ W/C. The surface area of the exchanger is 20 m$^2$ and the overall heat transfer coefficient is 500 W/m$^2$C.

**(a)** What is the heat transfer rate for the exchanger?

**(b)** What are the exit temperatures for the two fluids?

**8-21** For a double-pipe heat exchanger, air at an average temperature of 200 C flows through the tube and water at an average temperature of 60 C flows through the shell side. The tube is thin-walled and has a diameter of 5 cm. It is of stainless steel. The convective coefficient is 75 W/m$^2$ C on the inside surface of the tube and 3500 W/m$^2$ C on the outside surface.

**(a)** What is the overall heat transfer coefficient of the exchanger?

**(b)** After a long period of use, it is found that the overall heat transfer coefficient has decreased 20% from the value you calculated in Part (a). Assuming all the fouling is on the outside of the tube, what is the fouling factor?

**8-22** Water enters a heat exchanger at 40 C and has a flow rate of 4 kg/s. The water is heated by condensing steam at 1 atm pressure. The heat exchanger has an effectiveness of 80%, and its overall heat transfer coefficient is 950 W/m$^2$ C.

**(a)** What is the area of the exchanger?

**(b)** What is the exit temperature of the water?

**8-23** A crossflow exchanger heats air from 20 to 50 C using water that enters the exchanger at 90 C and leaves at 65 C. Both fluids are unmixed. The flow rate of the water is 3 kg/s and the overall heat transfer coefficient is 75 W/m$^2$ C. What is the surface area of the exchanger?

**8-24** A shell-and-tube heat exchanger has one shell pass and two tube passes. Oil ($c_p = 2100$ J/kg C) heats water. The oil enters the exchanger at 280 C and has a flow rate of 6 kg/s. The water enters the exchanger at 20 C and leaves at 85 C. The water has a flow rate of 8 kg/s. The overall heat transfer coefficient of the exchanger is 625 W/m$^2$ C. What is the required surface area of the exchanger?

**8-25** A double-pipe counterflow heat exchanger heats 2.5 kg/s of water from 20 to 55 C. Heating is accomplished by oil ($c_p = 2100$ J/kg C) that enters the exchanger at 200 C and exits at 150 C. The overall heat transfer coefficient of the exchanger is 145 W/m$^2$ C.

**(a)** What is the flow rate of the oil?

**(b)** If the water flow rate is reduced to 1 kg/s, what flow rate of oil is needed to keep the exit temperature of the water at 55 C? Assume that the overall heat transfer coefficient stays at 145 W/m$^2$ C.

**8-26** A crossflow heat exchanger heats water from 25 to 50 C with oil ($c_p = 2100$ J/kg C) entering at 130 C. The oil flows in the tubes of the exchanger and the water flows over the outside of the tubes, which are unfinned. The flow rate of the oil is 1.5 kg/s and the water flow rate is 0.5 kg/s. The overall heat transfer coefficient of the exchanger is 225 W/m$^2$ C. What is the required surface area to achieve the desired heating of the water?

**8-27** A crossflow heat exchanger heats water from 25 to 50 C with air entering at 200 C. The water flows in the tubes of the exchanger and the air flows over the outside of the tubes, which are unfinned. The flow rate of the water is 0.4 kg/s and the air flow rate is 0.4 kg/s. The overall heat transfer coefficient of the exchanger is 80 W/m$^2$ C. What is the required surface area to achieve the desired heating of the water?

**8-28** A crossflow heat exchanger heats air using hot water. The water is in the tubes and the air flows across the outside of the tubes. There are no fins on the tubes. The surface area of the exchanger is 4.5 m$^2$ and the overall heat transfer coefficient is 75 W/m$^2$ C. The water enters the exchanger at 95 C and the air enters at 20 C. Both the water and air flow rates are 0.25 kg/s.

**(a)** At what temperatures do the fluids leave the exchanger?

**(b)** What is the heat transfer rate for the exchanger?

(Do this problem by the ε-NTU method.)

**8-29** Do Problem 8-28 using the LMTD method.

**8-30** A shell-and-tube counterflow heat exchanger with both fluids having a single pass cools water with ethylene glycol. The glycol has a flow rate of 1.2 kg/s and enters the exchanger at 15 C. The flow rate of the water is 0.8 kg/s and the water is cooled from 85 to 40 C. The overall heat transfer coefficient is 900 W/m$^2$ C.

**(a)** What is the required surface area of the exchanger to achieve the stated cooling of the water?

**(b)** If the tubes are 3/4 inch Type L copper, and the exchanger must fit into a 30-foot space in the factory, how many tubes are needed?

**8-31** A one-shell-pass, two-tube-pass heat exchanger cools water with oil ($c_p = 2000$ J/kg C). The oil has a flow rate of 1.2 kg/s and enters the exchanger at 20 C. The flow rate of the water is 0.8 kg/s, and the water enters the exchanger at 85 C. The overall heat transfer coefficient is 200 W/m$^2$ C and the heat transfer area of the exchanger is 25 m$^2$.

**(a)** What are the exit temperatures of the water and the oil?

**(b)** What is the heat transfer rate for the exchanger?

**8-32** A shell-and-tube single-shell pass, two-tube pass heat exchanger is designed to heat 0.25 kg/s of water from 25 to 60 C. Heating is accomplished by hot oil ($c_p = 2200$ J/kg C) flowing through the exchanger at 0.2 kg/s. The oil enters the exchanger at 160 C. The design $U$ of the exchanger is 320 W/m$^2$ C. After considerable use, it is found that the water can only be heated to 50 C. What is the degraded value of $U$ for the exchanger?

**8-33** A crossflow heat exchanger uses air to cool water. The air is mixed; the water is unmixed. The air enters the exchanger at 30 C, leaves at 65 C, and has a flow rate of 3 kg/s. The water has a flow rate of 2 kg/s and enters the exchanger at 80 C. What is the $UA$ value for the exchanger?

**8-34** Water is used to heat oil. The water enters the heat exchanger at 90 C and leaves at 50 C. The oil is heated from 20 to 70 C. What is the effectiveness of the heat exchanger?

**8-35** Hot water is used to heat cold water. The hot water enters at 80 C and leaves at 50 C. The cold water enters at 10 C and is heated to 50 C. What is the effectiveness of the heat exchanger?

# References

[1] Standards of the Tubular Exchanger Manufacturers Association, seventh ed., Tubular Exchanger Manufacturers Association, New York, 1988.

[2] E.F.C. Somerscales, J.G. Knudsen, Fouling of Heat Transfer Equipment, Hemisphere Publishing, 1981.

[3] R.A. Bowman, A.C. Mueller, W.M. Nagle, Mean temperature difference in design, Trans. ASME 62 (1940) 283–294.

[4] K. Gardner, J. Taborek, Mean temperature difference: a reappraisal, AIChE J. 23 (1977) 777–786.

[5] A.S. Tucker, The LMTD correction factor for single-pass crossflow heat exchangers with both fluids unmixed, J. Heat Transf. 118 (1996) 488–490.

[6] A. Fakheri, Log mean temperature correction factor: an alternative representation, Proc. ASME IMECE (2002) 111–115. New Orleans, 2002.

[7] A. Fakheri, A general expression for the determination of the log mean temperature correction factor for shell and tube heat exchangers, J. Heat Transf. 125 (2003) 527–530.

[8] L.C. Thomas, Heat Transfer — Professional Version, second ed., Capstone Publishing, 1999.

[9] R.K. Shah, D.P. Sekulić, Fundamentals of Heat Exchanger Design, Wiley, 2003.

[10] W. Kays, A.L. London, Compact Heat Exchangers, McGraw-Hill, 1955.

# Radiation heat transfer

## Chapter outline

## 9.1 Introduction

The three modes of heat transfer are conduction, convection, and radiation. We discussed details of conduction and convection in previous chapters and briefly discussed radiative heat transfer in Chapter 1. We presented the following equation for radiation from a surface at temperature $T_s$ to a large surrounding enclosure at $T_{surr}$:

$$q_{rad} = \varepsilon \sigma A \left( T_s^4 - T_{surr}^4 \right) \tag{9.1}$$

where

$\varepsilon$ is the emissivity of the surface

$A$ is the area of the surface

$\sigma$ is the Stefan $-$ Boltzmann constant $= 5.67 \times 10^{-8}$ W/m$^2$ K$^4 = 0.1714 \times 10^{-8}$ Btu/h ft$^2$ R$^4$.

Temperatures $T_s$ and $T_{surr}$ must be in absolute temperature units (Kelvin or Rankine).

In this chapter, we discuss radiative heat transfer in considerably more detail. Topics covered include blackbody radiation, radiation properties, radiation shape factors, exchange of radiant energy between surfaces, and radiation shields.

## 9.2 Blackbody emission

The surfaces of all objects emit thermal radiation by virtue of their temperature being above absolute zero. This radiation is emitted as electromagnetic radiation in the wavelength range of about 0.1$-$100 $\mu$m (1 $\mu$m $= 10^{-6}$ m $=$ 1 micron). Some other areas of the electromagnetic spectrum are visible light (about 0.4$-$0.8 $\mu$m), X-rays (about $10^{-11}$ m to $2 \times 10^{-8}$ m), microwaves (about 1 mm$-$10 m), and radio waves (about 10 m$-$30 km). Unlike conduction and convection, which require a medium for their transport, radiation is transported unhindered through a vacuum. Air, oxygen, and nitrogen are also essentially transparent to radiation. Some other gases, however, (e.g., water vapor, carbon dioxide, and some hydrocarbon gases), have appreciable absorption of radiation in specific wavelength regions. We will not be considering radiation transport through these latter types of gases.

The wavelength $\lambda$ of the radiation is related to the frequency $v$ of the radiation by the speed of light $c_o$:

$$c_o = \lambda v \tag{9.2}$$

where $c_o = 2.9979 \times 10^8$ m/s in a vacuum.

A "blackbody" has a surface that emits radiation at the maximum possible rate. The emission is over all wavelengths and is given by **Planck's law**, which is

$$E_{b\lambda}(\lambda, T) = \frac{2\pi hc_o^2 \lambda^{-5}}{\exp\left(\dfrac{hc_o}{\lambda kT}\right) - 1} \tag{9.3}$$

where

$$h = \text{Planck's constant} = 6.62607 \times 10^{-34} \text{ J} \cdot \text{s}$$

$$k = \text{Boltzmann's constant} = 1.38065 \times 10^{-23} \text{ J/K}$$

$$E_{b\lambda} = \text{spectral emissive power of a blackbody}$$

The spectral emissive power is the rate of thermal emission per unit surface area per unit wavelength. If the wavelength of the radiation is in $\mu$m, then the units of the spectral emissive power are W/(m$^2$ $\mu$m).

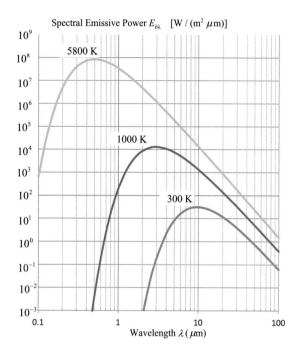

**FIGURE 9.1**

Blackbody spectral emissive power.

Putting in the values for the various parameters, Eq. (9.3) can be expressed as

$$E_{b\lambda}(\lambda, T) = \frac{C_1 \lambda^{-5}}{\exp(C_2/\lambda T) - 1} \tag{9.4}$$

where $C_1 = 3.7417 \times 10^8 (\text{W/m}^2) \cdot (\mu\text{m})^4$ and $C_2 = 1.4388 \times 10^4 \ \mu\text{m} \cdot \text{K}$.

The blackbody spectral emissive power is shown in Fig. 9.1 for blackbodies having temperatures of 300 K, 1000 K, and 5800 K. The 300 K curve is for bodies at typical room temperature and the 5800 K curve approximates the emission from the sun.

The spectral emissive power of a blackbody given in Eq. (9.4) is a function of both the wavelength of the radiation and the absolute temperature of the blackbody. The blackbody emits over all wavelengths. If we integrate spectral emissive power $E_{b\lambda}$ over all wavelengths, we get the total emissive power $E_b$ which is a function of only the temperature $T$ of the blackbody.

$$E_b(T) = \int_0^\infty E_{b\lambda}(\lambda, T) d\lambda = \sigma T^4 \tag{9.5}$$

where $\sigma = 5.670 \times 10^{-8} \ \text{W/m}^2 \ \text{K}^4 = \text{Stefan} - \text{Boltzmann constant}$

Eq. (9.5) is called the **Stefan−Boltzmann law**.

We have used the adjectives "spectral" and "total". "Spectral" is for a parameter that depends on the wavelength. A parameter with "total" is independent of wavelength.

Looking at Fig. 9.1, it is seen that the maximum (or peak) emission from a blackbody occurs at lower wavelengths for higher temperatures. For 5800 K, the peak emission is at about 0.5 μm. For 1000 K, it is at about 3 μm. And, for 300K, it is at about 9 μm. This is **Wien's displacement law**, which is

$$\lambda_{max} T = 2898 \ \mu m \cdot K \tag{9.6}$$

where $\lambda_{max}$ is the wavelength of peak emission.

Using Eq. (9.6), we get the following wavelengths of peak emission that we previously estimated by looking at Fig. 9.1:

$$\text{For 5800 K,} \quad \lambda_{max} = 2898/5800 = 0.500 \ \mu m$$

$$\text{For 1000 K,} \quad \lambda_{max} = 2898/1000 = 2.90 \ \mu m$$

$$\text{For 300 K,} \quad \lambda_{max} = 2898/300 = 9.66 \ \mu m$$

This shifting of wavelengths with temperature causes the "greenhouse effect" We have all been in hot, uncomfortable cars and rooms on hot, sunny days. This is primarily due to the transmissivity property of window glass. Glass has high transmission at the low wavelengths of solar radiation (5800 K blackbody) and very low transmission at the higher wavelengths of the 300 K radiation inside the car or room. The solar radiation easily goes through the glass into the inside air space, but it cannot leave the space through the essentially opaque glass. The temperature of the space increases to a very uncomfortable level if air conditioning is not provided.

We have seen that blackbody emission depends on wavelength and temperature. Sometimes we are interested in finding the emissive power from a blackbody for a given wavelength band. This can be determined by integrating the spectral emissive power over the wavelength range. For example, if we want the total emissive power for a blackbody at temperature $T$ for the wavelength band $\lambda_1$ to $\lambda_2$, we have

$$E_b(\lambda_1 \rightarrow \lambda_2, T) = \int_{\lambda_1}^{\lambda_2} E_{b\lambda}(\lambda, T) d\lambda \tag{9.7}$$

If we want the fraction of the radiation emitted by a blackbody that is in wavelength band $\lambda_1$ to $\lambda_2$, we have the total emissive power for the wavelength band divided by the total emissive power for all wavelengths. That is,

$$F_{\lambda_1 \rightarrow \lambda_2} = \frac{\int_{\lambda_1}^{\lambda_2} E_{b\lambda}(\lambda, T) d\lambda}{\int_0^\infty E_{b\lambda}(\lambda, T) d\lambda} \tag{9.8}$$

where $F_{\lambda_1 \rightarrow \lambda_2}$ is the fraction of blackbody emission that is in wavelength range $\lambda_1$ to $\lambda_2$. From the Stefan–Boltzmann Law, the denominator is equal to $\sigma T^4$, so Eq. (9.8) may be written as

$$F_{\lambda_1 \rightarrow \lambda_2} = \frac{\int_{\lambda_1}^{\lambda_2} E_{b\lambda}(\lambda, T) d\lambda}{\sigma T^4} \tag{9.9}$$

Using Eq. (9.4), Eq. (9.9) becomes

$$F_{\lambda_1 \rightarrow \lambda_2} = \frac{\int_{\lambda_1}^{\lambda_2} \dfrac{C_1 \lambda^{-5}}{\exp(C_2/\lambda T) - 1} d\lambda}{\sigma T^4} \tag{9.10}$$

The integral in Eq. (9.10) may be broken into two integrals, each starting at $\lambda = 0$. That is,

$$F_{\lambda_1 \to \lambda_2} = F_{0 \to \lambda_2} - F_{0 \to \lambda_1} = \frac{\int_0^{\lambda_2} \frac{C_1 \lambda^{-5}}{\exp(C_2/\lambda T) - 1} d\lambda}{\sigma T^4} - \frac{\int_0^{\lambda_1} \frac{C_1 \lambda^{-5}}{\exp(C_2/\lambda T) - 1} d\lambda}{\sigma T^4} \quad (9.11)$$

The integrals in Eqs. (9.10) and (9.11) are over wavelengths $\lambda$ for a given blackbody temperature $T$. The variable of integration can be changed from $\lambda$ to $\lambda T$, which modifies Eq. (9.11) to

$$F_{\lambda_1 T \to \lambda_2 T} = F_{0 \to \lambda_2 T} - F_{0 \to \lambda_1 T} = \int_0^{\lambda_2 T} \frac{C_1/\sigma}{(\lambda T)^5 [\exp(C_2/\lambda T) - 1]} d(\lambda T)$$

$$- \int_0^{\lambda_1 T} \frac{C_1/\sigma}{(\lambda T)^5 [\exp(C_2/\lambda T) - 1]} d(\lambda T) \quad (9.12)$$

Table 9.1 gives the blackbody radiation function $F_{0 \to \lambda T} = \int_0^{\lambda T} \frac{C_1/\sigma}{(\lambda T)^5 [\exp(C_2/\lambda T) - 1]} d(\lambda T)$ for different values of $\lambda T$. The use of Table 9.1 in determining the fraction of blackbody radiation in the wavelength band of $\lambda_1$ to $\lambda_2$ is illustrated in Example 9.1.

**Table 9.1 Blackbody radiation function.**

| $\lambda T$ (μm·K) | $F_{0\text{-}\lambda T}$ | $\lambda T$ (μm·K) | $F_{0\text{-}\lambda T}$ | $\lambda T$ (μm·K) | $F_{0\text{-}\lambda T}$ |
|---|---|---|---|---|---|
| 1000 | 0.0003 | 5000 | 0.6338 | 10,500 | 0.9238 |
| 1200 | 0.0021 | 5200 | 0.6580 | 11,000 | 0.9320 |
| 1400 | 0.0078 | 5400 | 0.6804 | 11,500 | 0.9390 |
| 1600 | 0.0197 | 5600 | 0.7011 | 12,000 | 0.9452 |
| 1800 | 0.0393 | 5800 | 0.7202 | 13,000 | 0.9552 |
| 2000 | 0.0667 | 6000 | 0.7379 | 14,000 | 0.9630 |
| 2200 | 0.1009 | 6200 | 0.7542 | 15,000 | 0.9691 |
| 2400 | 0.1403 | 6400 | 0.7692 | 16,000 | 0.9739 |
| 2600 | 0.1831 | 6600 | 0.7833 | 18,000 | 0.9809 |
| 2800 | 0.2279 | 6800 | 0.7972 | 20,000 | 0.9857 |
| 3000 | 0.2733 | 7000 | 0.8082 | 25,000 | 0.9923 |
| 3200 | 0.3181 | 7200 | 0.8193 | 30,000 | 0.9954 |
| 3400 | 0.3618 | 7400 | 0.8296 | 40,000 | 0.9981 |
| 3600 | 0.4036 | 7600 | 0.8392 | 50,000 | 0.9998 |
| 3800 | 0.4434 | 7800 | 0.8481 | 75,000 | 0.9998 |
| 4000 | 0.4809 | 8000 | 0.8563 | 100,000 | 1.0000 |
| 4200 | 0.5161 | 8500 | 0.8747 | | |
| 4400 | 0.5488 | 9000 | 0.8901 | | |
| 4600 | 0.5793 | 9500 | 0.9031 | | |
| 4800 | 0.6076 | 10,000 | 0.9143 | | |

## Example 9.1
### Blackbody emission in a wavelength band

*Problem*
What fraction of the emission from a blackbody at 5800 K lies in the visible range of the electromagnetic spectrum, i.e., in the range from 0.4 to 0.8 μm?

*Solution*
There are different ways to solve this problem. We will first solve it using Table 9.1.
The fraction in the band from $\lambda_1 = 0.4$ μm to $\lambda_2 = 0.8$ μm is $F_{0 \to \lambda_2 T} - F_{0 \to \lambda_1 T}$

$$\lambda_1 T = (0.4)(5800) = 2320$$
$$\lambda_2 T = (0.8)(5800) = 4640$$

From Table 9.1, $F_{0 \to \lambda_2 T} = 0.5850$ and $F_{0 \to \lambda_1 T} = 0.1245$

$$\text{Fraction} = F_{0 \to \lambda_2 T} - F_{0 \to \lambda_1 T} = 0.5850 - 0.1245 = 0.4605$$

46.1% of the radiation from a blackbody at 5800 K lies in the wavelength band from 0.4 to 0.8 μm.
This problem can also be solved without using Table 9.1. For example, the *quad* function of Matlab can be used to determine the integral in Eq. (9.10) directly. For this problem, we have

$$F_{\lambda_1 \to \lambda_2} = \frac{\int_{\lambda_1}^{\lambda_2} \frac{C_1 \lambda^{-5}}{\exp(C_2/\lambda T) - 1} d\lambda}{\sigma T^4} \qquad (9.10)$$

Putting in the values for the limits and constants for this problem, we have

$$F_{0.4 \to 0.8 \text{ μm}} = \frac{\int_{0.4}^{0.8} \frac{3.7417 \times 10^8 \lambda^{-5}}{\exp(1.4388 \times 10^4 / \lambda(5800)) - 1} d\lambda}{(5.67 \times 10^{-8})(5800)^4} \qquad (9.13)$$

Eq. (9.13) simplifies to

$$F_{0.4 \to 0.8 \text{ μm}} = 5.8314 \int_{0.4}^{0.8} \frac{\lambda^{-5}}{\exp(2.4807/\lambda) - 1} d\lambda \qquad (9.14)$$

The integral in Eq. (9.14) can be determined using the following Matlab statement interactively:

$$\gg S = \text{quad}\left('x.\hat{\ }(-5)./(\exp(2.4807./x) - 1)', 0.4, 0.8\right)$$

(Note that the periods for element-by-element operations are needed in the Matlab statement.)
The result is $S = 0.07907$, which, using Eq. (9.14), gives the same result as we got using Table 9.1.

$$F_{0.4 \to 0.8 \text{ μm}} = (5.8314)(0.07907) = 0.461$$

Excel could also have been used to determine the integral in Eq. (9.14). Indeed, the spreadsheet could have been programmed to determine the integral using, for example, Simpson's rule. Such programming, however, is a bit involved, and this is a study of heat transfer, not computer programming. Therefore, we will determine the integral in an alternative way, as follows:

Open an Excel spreadsheet. Our wavelength limits are 0.4–0.8. Using an increment of 0.02, we put the wavelength values in Column A, cells A1 through A21. In Column B, we put the values of the integrand in Eq. (9.14) corresponding to the wavelength values of Column A. The integrand values are in cells B1 through B21. We highlight Columns A and B and do Insert of a scatter chart. Clicking on the plot area, using the Layout tab, we Insert a trendline of a polynomial of order 3. We click the appropriate boxes to display the trendline equation and the R-squared value on the chart. Doing this, we got the trendline equation $y = 3.5663x^3 - 7.3632x^2 + 4.7057x - 0.7328$ and an R-squared value of 0.9991, which showed that the third order polynomial was an excellent fit for the integrand. We chose the polynomial form as a polynomial is easy to integrate by hand. Integrating the integrand function, we get

$$\int_{0.4}^{0.8} y \, dx = \frac{3.5663}{4}(0.8^4 - 0.4^4) - \frac{7.3632}{3}(0.8^3 - 0.4^3) + \frac{4.7057}{2}(0.8^2 - 0.4^2) - 0.7328(0.8 - 0.4) = 0.0791$$

From Eq. (9.14), $F_{0.4 \to 0.8 \text{ μm}} = (5.8314)(0.0791) = 0.461$.
In conclusion, the same result was obtained using Table 9.1, Matlab, and Excel.

## 9.3 Radiation properties

In this section, we define the following radiation properties: emissivity (or "emittance"), reflectivity (or "reflectance"), absorptivity (or "absorptance"), and transmissivity (or "transmittance").

The surface of an ideal blackbody absorbs all radiation that hits it and emits the maximum possible radiant energy. A real surface emits less than a black surface at the same temperature. The total emissivity $\varepsilon$ of a real surface is the ratio of the emissive power $E$ of the surface to the emissive power $E_b$ of a black surface at the same temperature. The emissivity of a black surface is 1. The emissivity of real surfaces is less than 1.

$$\text{Total Emissivity} = \varepsilon(T) = \frac{E(T)}{E_b(T)} = \frac{\int_0^\infty E_\lambda(\lambda, T)d\lambda}{\int_0^\infty E_{b\lambda}(\lambda, T)d\lambda} \tag{9.15}$$

This is the "total" emissivity of the surface as the emissive powers have been integrated over all wavelengths. The spectral emissivity of the surface is the ratio of the spectral emissive power of the real surface to the spectral emissive power of a black surface for the same wavelength $\lambda$ and same temperature $T$. That is,

$$\text{Spectral Emissivity} = \varepsilon_\lambda(\lambda, T) = \frac{E_\lambda(\lambda, T)}{E_{b\lambda}(\lambda, T)} \tag{9.16}$$

Fig. 9.2 shows the spectral emissive powers for black, gray, and real surfaces at 2000 K. The gray surface in the figure has an emissivity of 0.6. For a real surface at a given temperature, the emissivity varies with wavelength. For a gray surface, which is an idealization like the black surface, the emissivity does not vary with wavelength. The spectral emissive power of a gray surface at all wavelengths is a constant fraction of the emissive power of the black surface at the corresponding wavelength. That is,

$$\text{For a gray surface: } E_\lambda(\lambda, T) = \varepsilon\, E_{b\lambda}(\lambda, T) \tag{9.17}$$

where emissivity $\varepsilon$ is a constant.

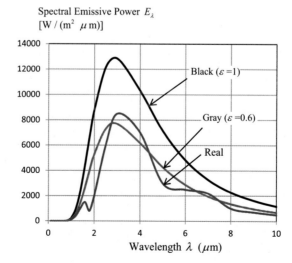

**FIGURE 9.2**

Spectral emissive power for black, gray, and real surfaces.

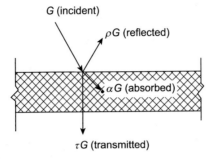

G (incident)

ρG (reflected)

αG (absorbed)

τG (transmitted)

**FIGURE 9.3**

Radiation incident on an object.

Fig. 9.3 shows radiation hitting an object. The incident radiation flux is $G$ (W/m$^2$). Some of the incident radiation is reflected from the surface, some is absorbed by the body, and some might be transmitted through the object.

The reflectivity $\rho$ is the fraction of incident radiation that is reflected. The absorptivity $\alpha$ is the fraction of incident radiation that is absorbed. And, the transmissivity $\tau$ is the fraction of incident radiation that is transmitted through the object. Therefore, if the incident flux is $G$, the amount $\rho G$ is reflected, $\alpha G$ is absorbed, and $\tau G$ is transmitted. The energy balance is

$$G = \rho G + \alpha G + \tau G \qquad (9.18)$$

Factoring out $G$, we have

$$1 = \rho + \alpha + \tau \qquad (9.19)$$

The radiation properties in Eq. (9.19) are "total" properties, i.e., properties for all wavelengths. Eq. (9.19) could also be written for specific wavelengths. If $G_\lambda$ is the incident radiation flux at wavelength $\lambda$, the corresponding equations for Eqs. (9.18) and (9.19) are

$$G_\lambda = \rho_\lambda G_\lambda + \alpha_\lambda G_\lambda + \tau_\lambda G_\lambda \qquad (9.20)$$

$$\text{and } 1 = \rho_\lambda + \alpha_\lambda + \tau_\lambda \qquad (9.21)$$

If no radiation is transmitted through the object, the object is said to be "opaque" and transmissivities $\tau$ and $\tau_\lambda$ are zero.

We now discuss **Kirchoff's law,** which relates the emissivity and absorptivity of a surface. Let us put a small object inside a black enclosure that is at temperature $T$. When thermal equilibrium is reached, both the object and the enclosure are at temperature $T$. For this equilibrium, the object's emission at a given wavelength must equal the radiation the object absorbs at that wavelength. That is,

$$E_\lambda = \alpha_\lambda G_{b\lambda} \qquad (9.22)$$

where $G_{b\lambda}$ is the incident radiation on the object from the walls of the black enclosure.

From Eq. (9.16), the spectral emissive power of the object is $E_\lambda = \varepsilon_\lambda E_{b\lambda}$ so Eq. (9.22) becomes

$$\varepsilon_\lambda E_{b\lambda} = \alpha_\lambda G_{b\lambda} \qquad (9.23)$$

If the object inside the enclosure is a blackbody, then $E_\lambda = E_{b\lambda}$ and $\alpha_\lambda = 1$ as a black surface absorbs all incident radiation. Eq. (9.22) becomes

$$E_{b\lambda} = G_{b\lambda} \qquad (9.24)$$

Putting Eq. (9.24) into Eq. (9.23), we arrive at

$$\varepsilon_\lambda(\lambda, T) = \alpha_\lambda(\lambda, T) \tag{9.25}$$

The spectral emissivity of a surface at a given wavelength and temperature equals the spectral absorptivity of the surface at the same wavelength and temperature.

For a gray surface, defined above, radiation properties are independent of wavelength. For such a surface, Eq. (9.25) becomes

$$\varepsilon(T) = \alpha(T) \tag{9.26}$$

Eq. (9.26) is also valid when the incident and emitted radiation have the same spectral distribution.

---

## Example 9.2
### Equilibrium temperature of a plate

*Problem*

Sunlight is incident on one side of a thin metal plate at the rate of 800 W/m$^2$. The other side of the plate is perfectly insulated. The plate has an absorptivity of 0.85 for solar wavelengths and 0.1 for long wavelengths. The air adjacent to the plate is at 15 C, and the surroundings are also at 15 C. The convective coefficient at the plate's surface is 20 W/m$^2$ C. What is the equilibrium temperature of the plate?

*Solution*

Given: $\alpha_{solar} = 0.85$, $\alpha_{long} = 0.1$, and $G = 800$ W / m$^2$.

As radiation is involved, we need to use absolute temperature units.

Therefore, $T_\infty = T_{surr} = 15$ C $= 15 + 273.15 = 288.15$ K.

For equilibrium, an energy balance on the plate gives: Energy Flux In = Energy Flux Out

$$\text{Energy Flux In} = \text{radiant energy absorbed by the plate} = \alpha_{solar} G$$

$$\text{Energy Flux Out} = \text{convection to surrounding air} + \text{radiation to surroundings}$$

$$\text{Energy Flux Out} = (T - T_\infty) + \varepsilon\sigma\left(T^4 - T_{surr}^4\right)$$

From the energy balance, we have

$$\alpha_{solar} G = h(T - T_\infty) + \varepsilon\sigma\left(T^4 - T_{surr}^4\right) \tag{9.27}$$

The emission from the plate surface is at long wavelengths, so we have $\varepsilon = \varepsilon_{long}$, and by Kirchoff's Law, Eq. (9.25), $\varepsilon_{long} = \alpha_{long}$, so Eq. (9.27) becomes

$$\alpha_{solar} G = h(T - T_\infty) + \alpha_{long}\sigma\left(T^4 - T_{surr}^4\right) \tag{9.28}$$

Putting values into Eq. (9.28) and rearranging the equation, we get

$$680 - 20(T - 288.15) - (0.1)\left(5.67 \times 10^{-8}\right)\left(T^4 - 288.15^4\right) = 0 \tag{9.29}$$

Eq. (9.29) can be solved for the plate temperature $T$ by trial and error, Matlab's *fzero* function, or Excel's *Goal Seek*. The plate temperature is found to be 321.1 K = **47.9 C**.

So far, we have discussed "spectral" radiation properties and "total" radiation properties. Spectral properties are wavelength-dependent. Total properties are wavelength-independent and are obtained by integrating the spectral property over a wavelength range. Total emissivity was defined in Eq. (9.15). Total absorptivity is defined as

$$\text{Total Absorptivity} = \alpha = \frac{\int_0^\infty \alpha_\lambda G_\lambda d\lambda}{\int_0^\infty G_\lambda d\lambda} \tag{9.30}$$

where $\alpha_\lambda$ is the spectral absorptivity.

$G_\lambda$ is the spectral distribution of the incident radiation (W/m$^2$ $\mu$m).

Total reflectivity and transmissivity are defined similarly.

## Example 9.3
### Total absorptivity of a plate

*Problem*

Radiation is incident on a plate's surface with the following intensity and spectral distribution:

$$
\begin{aligned}
G_\lambda &= 300 \text{ W/m}^2\mu\text{m} && \text{for } 0 < \lambda < 2.5 \ \mu\text{m} \\
&\ \ 1200 \text{ W/m}^2\mu\text{m} && \text{for } 2.5 \ \mu\text{m} < \lambda < 5 \ \mu\text{m} \\
&\ \ 500 \text{ W/m}^2\mu\text{m} && \text{for } 5 \ \mu\text{m} < \lambda < 10 \ \mu\text{m} \\
&\ \ 0 && \text{for } \lambda > 10 \ \mu\text{m}
\end{aligned}
$$

The spectral absorptivity of the surface is

$$
\begin{aligned}
\alpha_\lambda &= 0 && \text{for } 0 < \lambda < 1 \ \mu\text{m} \\
&\ \ 0.73 && \text{for } 1 \ \mu\text{m} < \lambda < 4 \ \mu\text{m} \\
&\ \ 0.24 && \text{for } \lambda > 4 \ \mu\text{m}
\end{aligned}
$$

**(a)** What is the total absorptivity of the surface?

**(b)** What is the rate of radiant energy absorption by the surface for the given incident radiation (W/m$^2$)?

*Solution*

From Eq. (9.30),

$$
\alpha = \frac{\int_0^\infty \alpha_\lambda G_\lambda d\lambda}{\int_0^\infty G_\lambda d\lambda} = \frac{A}{B}
$$

where

$$
A = \int_0^1 (0)(300)d\lambda + \int_1^{2.5} (0.73)(300)d\lambda + \int_{2.5}^4 (0.73)(1200)d\lambda + \cdots
$$

$$
+ \int_4^5 (0.24)(1200)d\lambda + \int_5^{10} (0.24)(500)d\lambda + \int_{10}^\infty (0)(0.24)d\lambda
$$

$$
= 0 + (0.73)(300)(2.5 - 1) + (0.73)(1200)(4 - 2.5) + (0.24)(1200)(5 - 4) + \cdots
$$

$$
+ (0.24)(500)(10 - 5) + 0 = 2530.5
$$

$$
B = \int_0^{2.5} (300)d\lambda + \int_{2.5}^5 (1200)d\lambda + \int_5^{10} (500)d\lambda
$$

$$
= (300)(2.5 - 0) + (1200)(5 - 2.5) + (500)(10 - 5) = 6250
$$

**(a)** $\alpha = \frac{A}{B} = \frac{2530.5}{6250} = \mathbf{0.405}$

**(b)** $\frac{q}{A} = \alpha G = (0.405)(6250) = \mathbf{2530 \ W/m^2}$

## Example 9.4
### Radiant transmission through a window

*Problem*

A kiln has a 6 cm diameter window in one of its walls. The kiln environment is at 1200 C and can be assumed to be black. The window has the following spectral transmissivity to radiation from a black environment at 1200 C:

$$\tau_\lambda = 0.6 \qquad \text{for } 0 < \lambda < 2 \text{ μm}$$
$$0.75 \qquad \text{for } 2 \text{ μm} < \lambda < 4 \text{ μm}$$
$$0 \qquad \text{for } \lambda > 4 \text{ μm}$$

**(a)** What is the total transmissivity of the glass window for black radiation at 1200 C?

**(b)** What is the rate of radiant energy transmitted through the window to the outside of the kiln?

*Solution*

The total transmissivity for incident black radiation is

$$\tau = \frac{\int_0^\infty \tau_\lambda E_{b\lambda} d\lambda}{\int_0^\infty E_{b\lambda} d\lambda} = \tau = \frac{(0.6) \int_0^{2\ \mu m} E_{b\lambda} d\lambda + (0.75) \int_{2\ \mu m}^{4\ \mu m} E_{b\lambda} d\lambda}{\sigma T^4}$$

$$T = 1200\ C + 273.15 = 1473.15\ K$$
$$\lambda_1 T = (2)(1473.15) = 2946\ \mu m\ K$$
$$\lambda_2 T = (4)(1473.15) = 5893\ \mu m\ K$$

From Table 9.1, $F_{0 \to \lambda_1 T} = 0.2610$ and $F_{0 \to \lambda_2 T} = 0.7284$

**(a)** $\tau = (0.6)(0.2610) + (0.75)(0.7284 - 0.2610) = \mathbf{0.507}$

**(b)** $q = \tau A \sigma T^4 = (0.507)\left[(\pi/4)(0.06)^2\right](5.67 \times 10^{-8})(1473.15)^4 = \mathbf{383\ W}$

Radiation properties can also be directional, varying with the direction of the radiation to and from a surface. For example, we can have "normal" emission or reflection in a direction normal, or perpendicular, to a surface. There are also "hemispherical" radiation properties, with the properties integrated over the hemisphere from the surface, that is, in all directions from or to the surface. In this text, we will not consider directional characteristics of radiation. We will assume that radiation is hemispherical. For example, when we use the symbol $\varepsilon_\lambda$, we mean the *hemispherical spectral emissivity* and when we use $\varepsilon$, we mean the *hemispherical total emissivity*. Typical hemispherical emissivities for various surfaces are given in Appendix F.

# 9.4 Radiation shape factors

Radiative heat transfer between surfaces will be discussed in the next section. However, before doing so, we must discuss the Radiation Shape Factor, also called the Angle Factor, View Factor, or Configuration Factor.

Radiation leaves surfaces by both emission and the reflection of incident radiation. Reflection from a surface can be specular, like reflection from a mirror, or diffuse with reflection over all directions. Or, it can be partially specular and partially diffuse. We will assume that both emission and reflection from a surface is diffuse, with radiation leaving a surface over all the angles of a hemisphere.

In determining the transfer of radiation between surfaces, one major factor is the portion of the radiation leaving a surface that is intercepted by the other surface. This is a geometric factor that depends on the relative orientations of the two surfaces and their sizes. The **radiation shape factor** $F_{i-j}$ is defined as the fraction of radiant energy leaving Surface $i$ that hits Surface $j$ directly.

Sometimes determination of shape factors is quite easy. For example, if two infinite parallel plates (Plates 1 and 2) are exchanging radiation, then $F_{1-2} = 1$ and $F_{2-1} = 1$ as all the radiation from one plate will hit the other plate. As another example: Say we have two infinitely long concentric cylinders and we have radiation between the outer surface of the inner cylinder (Surface 1) and the inner surface of the outer cylinder (Surface 2). All of the radiation from Surface 1 will hit Surface 2 and $F_{1-2} = 1$.

However, not all of the radiation from Surface 2 will hit Surface 1 as some of the radiation from Surface 2 will hit Surface 2. Therefore, $F_{2-1} \neq 1$. The actual value of $F_{2-1}$ depends on the diameters of the cylinders.

More complex shape factors are obtained through double integration over the two involved surfaces. Figs. 9.4–9.7 give radiation shape factors for parallel rectangles, perpendicular rectangles with a common edge, parallel coaxial disks, and concentric cylinders.

Table 9.2 gives the relations that were used to generate Figs. 9.4–9.7.

There is another item that needs to be discussed before we consider radiative transfer between surfaces, as follows:

Let us say we have two surfaces, $i$ and $j$, which are exchanging radiation. There is a relationship between the areas of the surfaces and the respective shape factors of the surfaces. This relationship, called the **reciprocity theorem**, is

$$A_i F_{i-j} = A_j F_{j-i} \tag{9.31}$$

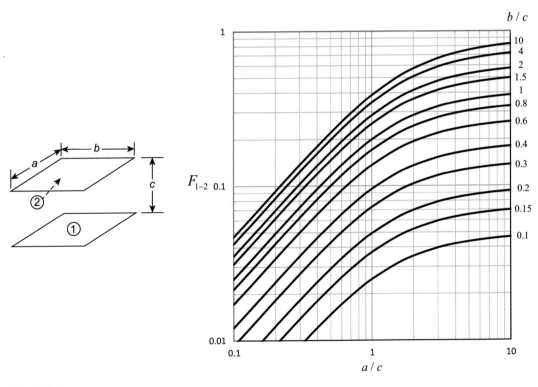

**FIGURE 9.4**

Radiation shape factor for parallel rectangles.

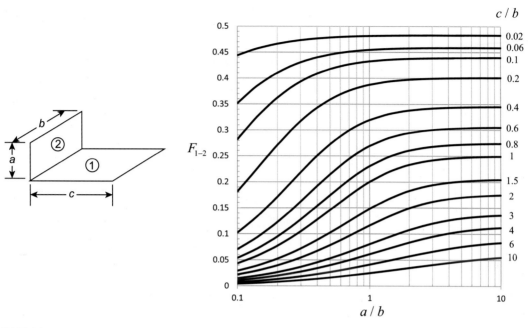

**FIGURE 9.5**

Radiation shape factor for perpendicular rectangles with a common edge.

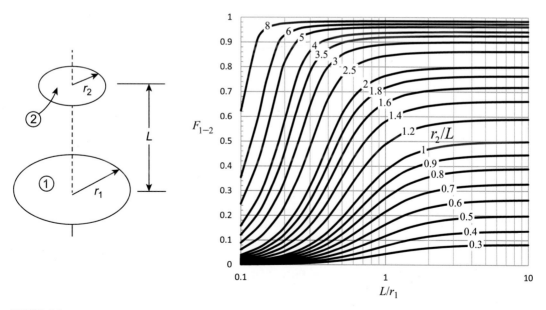

**FIGURE 9.6**

Radiation shape factor for two parallel coaxial disks.

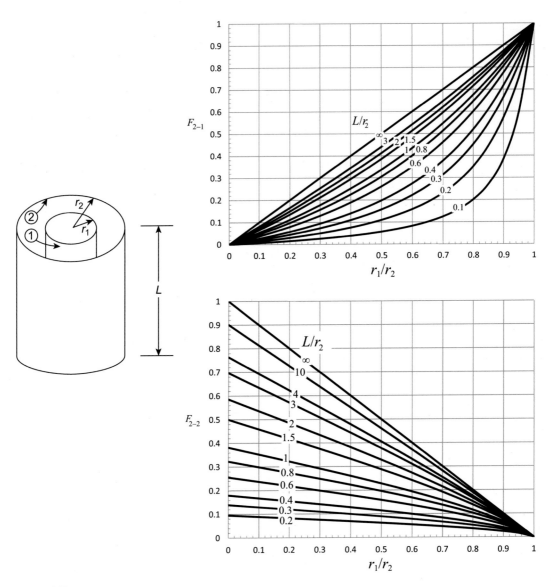

**FIGURE 9.7**

Radiation shape factors for concentric cylinders.

Figs. 9.4–9.7 are very useful in common problems involving radiative transfer. The following example shows the usefulness of the shape factor figures for determining shape factors for other geometries. In particular, we will be using Fig. 9.5 to determine the shape factor for perpendicular, unconnected rectangles.

---

**Table 9.2  Radiation shape factor relations for Figs. 9.4–9.7.**

**Fig. 9.4—Parallel rectangles**

$x = a/c \quad y = b/c$

$$F_{1-2} = \frac{2}{\pi xy} \left\{ \begin{array}{l} \ln\left[\dfrac{(1+x^2)(1+y^2)}{1+x^2+y^2}\right]^{1/2} + x(1+y^2)^{1/2}\tan^{-1}\left(\dfrac{x}{(1+y^2)^{1/2}}\right) \\[2ex] + y(1+x^2)^{1/2}\tan^{-1}\left(\dfrac{y}{(1+x^2)^{1/2}}\right) - x\tan^{-1}(x) - y\tan^{-1}(y) \end{array} \right\}$$

**Fig. 9.5—Perpendicular rectangles with common edge**

$x = a/b \quad y = c/b$

$$F_{1-2} = \frac{1}{\pi y}\left( \begin{array}{l} y\tan^{-1}(1/y) + x\tan^{-1}(1/x) - (x^2+y^2)^{1/2}\tan^{-1}\left(\dfrac{1}{(x^2+y^2)^{1/2}}\right) \\[2ex] + \dfrac{1}{4}\ln\left\{\dfrac{(1+y^2)(1+x^2)}{1+y^2+x^2}\left[\dfrac{y^2(1+y^2+x^2)}{(1+y^2)(y^2+x^2)}\right]^{y^2}\left[\dfrac{x^2(1+x^2+y^2)}{(1+x^2)(x^2+y^2)}\right]^{x^2}\right\} \end{array} \right)$$

**Fig. 9.6—Parallel coaxial disks**

$x = L/r_1 \quad y = r_2/L \quad S = 1 + x^2(1+y^2)$

$$F_{1-2} = \frac{1}{2}\left\{ S - \left[S^2 - 4(xy)^2\right]^{1/2}\right\}$$

**Fig. 9.7—Concentric cylinders**

$x = r_1/r_2 \quad y = L/r_2$

$$F_{2-1} = x - \frac{x}{\pi}\left\{ \cos^{-1}\left(\frac{x^2+y^2-1}{1+y^2-x^2}\right) - \frac{x}{2y}\left\{ \begin{array}{l} \left[\left(\frac{1}{x^2}+\frac{y^2}{x^2}-1\right)^2 + 4\left(\frac{1}{x^2}+\frac{y^2}{x^2}-1\right) - \frac{4}{x^2}+4\right]^{1/2} \\[2ex] \cos^{-1}\left[\frac{x(x^2+y^2-1)}{1+y^2-x^2}\right] + \left(\frac{y^2}{x^2}-\frac{1}{x^2}+1\right)\sin^{-1}(x) \\[2ex] -\pi\left(\frac{1}{x^2}+\frac{y^2}{x^2}-1\right)/2 \end{array} \right\} \right\}$$

$$F_{2-2} = 1 - x + \frac{2x}{\pi}\tan^{-1}\left[2\left(\frac{1-x^2}{x^2}\right)^{1/2}\frac{x}{y}\right] - \frac{y}{2\pi}\left\{ \begin{array}{l} \left(\frac{4+y^2}{x^2}\right)^{1/2}\frac{x}{y}\sin^{-1}\left[\frac{4\left(\frac{1-x^2}{x^2}\right)+y^2\left(\frac{1-2x^2}{x^2}\right)}{\left(\frac{y}{x}\right)^2+4\left(\frac{1-x^2}{x^2}\right)}\right] \\[2ex] -\sin^{-1}(1-2x^2) + \frac{\pi}{2}\left[\frac{x}{y}\left(\frac{4+y^2}{x^2}\right)^{1/2} - 1\right] \end{array} \right\}$$

## Example 9.5

**Radiation shape factor for perpendicular, unconnected rectangles**

*Problem*
Determine the shape factor $F_{1-4}$ for the Surfaces 1 and 4 in the following sketch:

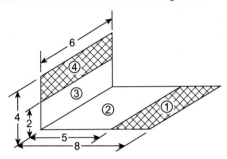

*Solution*
Fig. 9.5 is for perpendicular rectangles with a common edge, so it cannot be applied directly to our problem as Surfaces 1 and 4 do not have a common edge. However, the figure can be used along with shape factor algebra to determine the desired shape factor.

From the definition of shape factor, $F_{1-4}$ is the fraction of radiation from Surface 1 that hits Surface 4 directly. The fraction of radiation from Surface 1 that hits the vertical surface of Surfaces 3 and 4 is $F_{1-(3+4)}$, which is equal to the fraction of radiation from Surface 1 that hits Surface 3 plus the fraction of radiation from Surface 1 that hits Surface 4. That is,

$$F_{1-(3+4)} = F_{1-3} + F_{1-4} \tag{a}$$

Rearranging, we get $F_{1-4} = F_{1-(3+4)} - F_{1-3}$ (b)

To determine $F_{1-4}$, we need to get the shape factors on the right side of Eq. (b). These shape factors are not for surfaces sharing a common edge. Therefore, Fig. 9.5 cannot be used directly to get them. However, through shape factor algebra, we can convert Eq. (b) into an equation that has only the unknown factor $F_{1-4}$ and shape factors that can be obtained from Fig. 9.5.

Note that Eqs. (a) and (b) do not contain the areas of the surfaces. Such is not the case if a surface comprised of more than one surface radiates to a single surface. For example, let us consider the radiation from the combined surface of Surfaces 3 and 4 to Surface 1. That is, we want $F_{(3+4)-1}$.

Multiplying Eq. (a) by the area $A_1$ of Surface 1, we get

$$A_1 F_{1-(3+4)} = A_1 F_{1-3} + A_1 F_{1-4} \tag{c}$$

From reciprocity,

$$A_1 F_{1-(3+4)} = A_{(3+4)} F_{(3+4)-1} \tag{d}$$

$$A_1 F_{1-3} = A_3 F_{3-1}$$

$$A_1 F_{1-4} = A_4 F_{4-1}$$

Substituting Eq. (d) into Eq. (c), we have

$$A_{(3+4)} F_{(3+4)-1} = A_3 F_{3-1} + A_4 F_{4-1} \tag{e}$$

That is, the radiation from Surfaces 3 and 4 to Surface 1 equals the radiation from Surface 3 to Surface 1 plus the radiation from Surface 4 to Surface 1.

Following Eq. (e), we have, for the radiation from Surfaces 3 and 4 to Surfaces 1 and 2,

$$A_{(3+4)} F_{(3+4)-(1+2)} = A_3 F_{3-(1+2)} + A_4 F_{4-(1+2)} \tag{f}$$

Similarly, for the combined horizontal Surfaces 1 and 2 radiating to surface 4, we have

$$A_{(1+2)}F_{(1+2)-4} = A_1F_{1-4} + A_2F_{2-4} \qquad (g)$$

Also,

$$F_{2-(3+4)} = F_{2-3} + F_{2-4} \qquad (h)$$

Substituting $F_{2-4}$ from Eq. (h) into Eq. (g), we have

$$A_{(1+2)}F_{(1+2)-4} = A_1F_{1-4} + A_2\left(F_{2-(3+4)} - F_{2-3}\right) \qquad (i)$$

Rearranging Eq. (f), we have

$$A_4F_{4-(1+2)} = A_{(3+4)}F_{(3+4)-(1+2)} - A_3F_{3-(1+2)} \qquad (j)$$

Because of reciprocity, the left sides of Eqs. (i) and (j) are equal. Hence, we can equate the right sides of the equations, which gives

$$A_1F_{1-4} + A_2\left(F_{2-(3+4)} - F_{2-3}\right) = A_{(3+4)}F_{(3+4)-(1+2)} - A_3F_{3-(1+2)} \qquad (k)$$

Rearranging Eq. (k) for the desired $F_{1-4}$, we get

$$F_{1-4} = \frac{1}{A_1}\left[A_{(3+4)}F_{(3+4)-(1+2)} - A_3F_{3-(1+2)} - A_2\left(F_{2-(3+4)} - F_{2-3}\right)\right] \qquad (l)$$

All of the shape factors in Eq. (l) are for two surfaces having a common edge. Hence, Fig. 9.5 can be used for their evaluation.

In summary: Through the use of shape factor algebra, we have been able to obtain an expression for the desired $F_{1-4}$ in terms of shape factors that can be evaluated using Fig. 9.5.

Looking at the figure for this problem, $A_1 = 18$; $A_2 = 30$; $A_3 = 12$; and $A_{(3+4)} = 24$.

For shape factor $F_{(3+4)-(1+2)}$ using Fig. 9.5, $a = 8$; $b = 6$; $c = 4$. $c/b = 0.6667$; $a/b = 1.3333$; and $F_{(3+4)-(1+2)} = 0.27$ from the figure. The other shape factors, from Fig. 9.5, are $F_{3-(1+2)} = 0.35$; $F_{2-(3+4)} = 0.19$; and $F_{2-3} = 0.13$. Putting these values into Eq. (l), we get

$$F_{1-4} = \frac{1}{18}[24(0.27) - 12(0.35) - 30(0.19 - 0.13)] = \mathbf{0.027}$$

# 9.5 Radiative heat transfer between surfaces

Let us now discuss the radiation coming into a surface and the radiation leaving a surface. The incoming radiation is irradiation $G$ and the outgoing radiation is radiosity $J$. The surface is named "Surface $i$."

$G_i$ = irradiation = rate at which radiation is incident on surface $i$ per unit surface area

The radiation leaving a surface consists of two parts: the portion of the incident radiation that is reflected from the surface and the radiation emitted by the surface. That is,

$$J_i = \rho_i G_i + \varepsilon_i E_{bi} \qquad (9.32)$$

where $J_i$ = radiosity of surface $i$.
   $G_i$ = irradiation of surface $i$.
   $\rho_i$ = reflectivity of surface $i$.
   $\varepsilon_i$ = emissivity of surface $i$.
   $E_{bi}$ = blackbody emissive power of surface $i = \sigma\, T_i^4$.
   The net rate of heat loss from surface $i$ is

$$q_i = A_i(J_i - G_i) \tag{9.33}$$

where $A_i$ is the area of surface $i$.

Rearranging Eq. (9.32), we get

$$G_i = \frac{J_i - \varepsilon_i E_{bi}}{\rho_i} \tag{9.34}$$

Substituting $G_i$ from Eq. (9.34) into Eq. (9.33), we get

$$q_i = A_i\left(J_i - \frac{J_i - \varepsilon_i E_{bi}}{\rho_i}\right) \tag{9.35}$$

As we have a gray surface, $\rho_i = 1 - \varepsilon_i$, and Eq. (9.35) becomes

$$q_i = A_i\left(J_i - \frac{J_i - \varepsilon_i E_{bi}}{1 - \varepsilon_i}\right) \tag{9.36}$$

Rearranging Eq. (9.36), we finally arrive at

$$q_i = \frac{A_i \varepsilon_i}{1 - \varepsilon_i}(E_{bi} - J_i) \tag{9.37}$$

Let us look again at Eq. (9.33):

$$q_i = A_i(J_i - G_i) \tag{9.33}$$

We would like to get $G_i$ in terms of $J_i$. Let us consider the enclosure of N surfaces shown in Fig. 9.8. The N surfaces are exchanging radiant energy with each other. (It should be mentioned that these surfaces do not need to be solid surfaces. Some could be open or closed windows through which radiation passes. If such is the case, these windows would be assigned equivalent blackbody temperatures corresponding to the rate of radiation going through them.)

Let us look at the radiant interactions for one of the surfaces of the enclosure, i.e., the interactions for surface $i$.

In Section 9.4, we discussed radiation shape factors. Shape factor $F_{i-j}$ is the fraction of radiant energy leaving surface $i$ that hits surface $j$. We assume that radiation from a surface is diffuse and the radiation leaves a surface over the entire hemisphere. Fig. 9.9 shows three different types of surfaces: flat, convex, and concave. If a surface is flat or convex, none of the radiation leaving it will return to hit the surface. However, if a surface is concave, some of the radiation leaving the surface will hit the same surface. That is, for flat or convex surfaces, $F_{i-i} = 0$. For concave surfaces, $F_{i-i} \neq 0$.

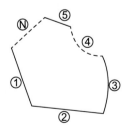

**FIGURE 9.8**

An enclosure of N surfaces.

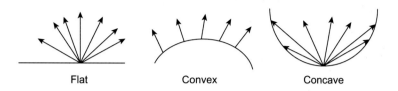

Flat       Convex       Concave

**FIGURE 9.9**

Radiation from flat, convex, and concave surfaces.

There is a very useful relation between the shape factors of the surfaces of an enclosure, as follows:
Radiation leaving a surface of an enclosure will hit all of the surfaces of the enclosure that have a nonzero shape factor with the surface sending out the radiation. That is, if the enclosure has N surfaces, and Surface $i$ is sending out the radiation, the relation is

$$F_{i-1} + F_{i-2} + F_{i-3} + \cdots + F_{i-N} = 1$$

Using the summation symbol, this relation is $\sum_{j=1}^{N} F_{i-j} = 1$.

Let us now look at the radiation coming into and leaving one of the surfaces of the enclosure. From previous definitions, $G_i$ is the rate at which radiant energy is incident on Surface $i$ per unit area of surface, and $J_i$ is the rate at which radiant energy leaves Surface $i$ per unit area of surface. Surface $i$ has a surface area of $A_i$. Therefore, $G_i A_i$ is the rate at which radiation is incident on Surface $i$.

Consider the enclosure of N surfaces. Surface $i$ gets radiation from the other surfaces and, if concave, can get radiation from itself. From the definition of radiosity, the rate of radiation leaving Surface 1 is $J_1 A_1$, and, from the definition of shape factor, the rate at which the radiation from Surface 1 hits surface $i$ is $J_1 A_1 F_{1-i}$ . Therefore, looking at radiation coming in from all N surfaces of the enclosure to surface $i$, we have

$$G_i A_i = J_1 A_1 F_{1-i} + J_2 A_2 F_{2-i} + J_3 A_3 F_{3-i} + \cdots + J_N A_N F_{N-i} \tag{9.38}$$

Using the summation symbol, Eq. (9.38) is

$$G_i A_i = \sum_{j=1}^{N} J_j A_j F_{j-i} \tag{9.39}$$

Substituting this into Eq. (9.33), we get

$$q_i = A_i J_i - \sum_{j=1}^{N} J_j A_j F_{j-i} \tag{9.40}$$

We discussed earlier the **Reciprocity Theorem** for two Surfaces $i$ and $j$, namely,

$$A_i F_{i-j} = A_j F_{j-i} \tag{9.41}$$

Using this in the summation term of Eq. (9.40), the equation becomes

$$q_i = A_i J_i - \sum_{j=1}^{N} J_j A_i F_{i-j} \qquad (9.42)$$

Factoring out $A_i$, Eq. (9.42) becomes

$$q_i = A_i \left( J_i - \sum_{j=1}^{N} J_j F_{i-j} \right) \qquad (9.43)$$

From above, we also had Eq. (9.37)

$$q_i = \frac{A_i \varepsilon_i}{1 - \varepsilon_i} (E_{bi} - J_i) \qquad (9.37)$$

where $\overset{\cdot}{E}_{bi}$ = total emissive power of a black surface = $\sigma T_i^4$.

Eqs. (9.43) and (9.37) may be written for all N surfaces of the enclosure. We have N equations from Eq. (9.43) and N equations from Eq. (9.37). These 2N simultaneous equations may be solved for the N $q_i$, s and the N $J_i$, s.

We could decrease the number of simultaneous equations from 2N to N by doing the following: each surface has two equations for $q$: Eqs. (9.37) and (9.43). If we equate the right sides of these equations, we get N equations for the $J_i$. After we solve the N equations and get the $J_i$, we can then use either Eq. (9.43) or Eq. (9.37) to get the heat flows $q_i$.

## 9.5.1 Radiation heat transfer for a two-surface enclosure

If an enclosure has only two surfaces exchanging radiation, Eqs. (9.43) and (9.37) are

### 9.5.1.1 For surface 1

$$q_1 = A_1 \left( J_1 - \sum_{j=1}^{2} J_j F_{1-j} \right) = A_1 (J_1 - J_1 F_{1-1} - J_2 F_{1-2}) \qquad (9.44)$$

$$q_1 = \frac{A_1 \varepsilon_1}{1 - \varepsilon_1} (E_{b1} - J_1) \qquad (9.45)$$

### 9.5.1.2 For surface 2

$$q_2 = A_2 \left( J_2 - \sum_{j=1}^{2} J_j F_{2-j} \right) = A_2 (J_2 - J_1 F_{2-1} - J_2 F_{2-2}) \qquad (9.46)$$

$$q_2 = \frac{A_2 \varepsilon_2}{1 - \varepsilon_2} (E_{b2} - J_2) \qquad (9.47)$$

Eqs. (9.44) through (9.47) may be solved simultaneously for $q_1, q_2, J_1,$ and $J_2$. The results for the heat flows are

$$q_1 = \frac{\sigma\left(T_1^4 - T_2^4\right)}{\dfrac{1-\varepsilon_1}{\varepsilon_1 A_1} + \dfrac{1}{A_1 F_{1-2}} + \dfrac{1-\varepsilon_2}{\varepsilon_2 A_2}} \quad \text{and} \quad q_2 = -q_1 \tag{9.48}$$

Note that Eq. (9.48) only contains one shape factor. The other shape factors have been eliminated by applying the shape factor relations $F_{1-1} + F_{1-2} = 1$ and $F_{2-1} + F_{2-2} = 1$. The reciprocity theorem $A_1 F_{1-2} = A_2 F_{2-1}$ and the Stefan–Boltzmann Law $E_b = \sigma T^4$ were also used in reaching the final Eq. (9.48).

As a special case, if shape factor $F_{1-2} = 1$, then Eq. (9.48) becomes

$$q_1 = \frac{\sigma A_1 \left(T_1^4 - T_2^4\right)}{\dfrac{1}{\varepsilon_1} + \left(\dfrac{A_1}{A_2}\right)\left(\dfrac{1-\varepsilon_2}{\varepsilon_2}\right)} \tag{9.49}$$

Common applications of Eq. (9.49) are the following:

**(A)** Small object in a large enclosure

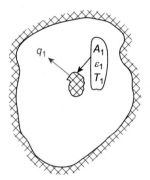

Conditions : $A_1 \ll A_2$ and $F_{1-2} = 1$

or $F_{1-2} = 1$ and $\varepsilon_2 = 1 \,(\text{black surface})$

$$q_1 = \varepsilon_1 \sigma A_1 \left(T_1^4 - T_2^4\right) \tag{9.50}$$

**(B)** Infinite parallel plates

$$A_2 \; \varepsilon_2 \; T_2$$
$$q_1 \uparrow$$
$$A_1 \; \varepsilon_1 \; T_1$$

Conditions : $A_1 = A_2 = A$ and $F_{1-2} = 1$

$$q_1 = \frac{\sigma A \left( T_1^4 - T_2^4 \right)}{\dfrac{1}{\varepsilon_1} + \dfrac{1}{\varepsilon_2} - 1} \tag{9.51}$$

**(C)** Infinitely long concentric cylinders

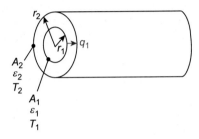

Conditions : $A_1 / A_2 = (2\pi r_1 L)/(2\pi r_2 L) = r_1/r_2$ and $F_{1-2} = 1$

$A_1 = 2\pi r_1 L$ where $L =$ length of cylinders

$$q_1 = \frac{\sigma A_1 \left( T_1^4 - T_2^4 \right)}{\dfrac{1}{\varepsilon_1} + \left( \dfrac{r_1}{r_2} \right) \left( \dfrac{1 - \varepsilon_2}{\varepsilon_2} \right)} \tag{9.52}$$

**(D)** Concentric spheres

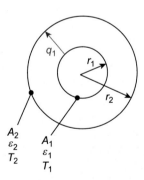

Conditions : $A_1 / A_2 = (4\pi r_1^2)/(4\pi r_2^2) = (r_1/r_2)^2$  $F_{1-2} = 1$

$A_1 = 4\pi r_1^2$

$$q_1 = \frac{\sigma A_1 \left( T_1^4 - T_2^4 \right)}{\dfrac{1}{\varepsilon_1} + \left( \dfrac{r_1}{r_2} \right)^2 \left( \dfrac{1 - \varepsilon_2}{\varepsilon_2} \right)} \tag{9.53}$$

## Example 9.6
### Radiant heat transfer for a cubical storage container

*Problem*

A thin-walled cubical container that is 0.5 m by 0.5 m by 0.5 m is filled with ice water. The container is inside another cubical container and the 3 cm gap between the containers is a vacuum. The outer surface of the inner container has an emissivity of 0.1 and a temperature of 1 C. The inner surface of the outer container has an emissivity of 0.3 and a temperature of 19 C.

**(a)** What is the rate of heat transfer to the ice water?

**(b)** If the container initially contains 1/2 ice and 1/2 water by volume, how long will it take for all the ice to be melted?

*Solution*

This is a two-surface radiation problem. Let Surface 1 be the outer surface of the inner cube and Surface 2 be the inner surface of the outer cube.

$$A_1 = (6)(0.5)^2 = 1.5 \text{ m}^2$$

$$A_2 = (6)[0.5 + 2(0.03)]^2 = 1.8816 \text{ m}^2$$

$$\varepsilon_1 = 0.1 \text{ and } T_1 = 1 \text{ C} = 274.15 \text{ K} \quad \varepsilon_2 = 0.3 \text{ and } T_2 = 19 \text{ C} = 292.15 \text{ K}$$

**(a)** All of the radiation leaving the inner cube will hit the outer cube. Therefore $F_{1-2} = 1$.

Hence, Eq. (9.49) is the appropriate equation for this two-surface problem.

$$q_1 = \frac{\sigma A_1 \left(T_1^4 - T_2^4\right)}{\dfrac{1}{\varepsilon_1} + \left(\dfrac{A_1}{A_2}\right)\left(\dfrac{1 - \varepsilon_2}{\varepsilon_2}\right)} \tag{9.49}$$

Putting values into Eq. (9.49), we have

$$q_1 = \frac{\sigma A_1 \left(T_1^4 - T_2^4\right)}{\dfrac{1}{\varepsilon_1} + \left(\dfrac{A_1}{A_2}\right)\left(\dfrac{1 - \varepsilon_2}{\varepsilon_2}\right)} = \frac{(5.67 \times 10^{-8})(1.5)(274.15^4 - 292.15^4)}{\dfrac{1}{0.1} + \left(\dfrac{1.5}{1.8816}\right)\left(\dfrac{1 - 0.3}{0.3}\right)} = -11.73 \text{ W}$$

The minus sign for $q_1$ means that the heat flow is into Surface 1, that is, into the ice water.

The rate of heat transfer to the ice water is 11.73 W.

**(b)** The volume of the container is $(0.5 \text{ m})^3 = 0.125 \text{ m}^3$. The volume of the ice is half of this or 0.0625 m$^3$. The density of ice is 916.7 kg/m$^3$, so the initial mass of ice in the container is

$$m_{Ice} = (916.7 \text{ kg/m}^3)(0.0625 \text{ m}^3) = 57.3 \text{ kg}$$

The heat of fusion of ice is 333.6 kJ/kg, so the amount of heat needed to melt all of the ice in the container is (333.6 kJ/kg) x (57.3 kg) = 19115 kJ.

From Part (a), $q = 11.73 \text{ W} = 11.73 \text{ J/s}$.

The time need to melt all the ice is

$$t = \frac{19115 \times 10^3 \text{ J}}{11.73 \text{ J/s}} = 1.630 \times 10^6 \text{ s} \times \frac{1 \text{ h}}{3600 \text{ s}} \times \frac{1 \text{ day}}{24 \text{ h}} = 18.9 \text{ days}$$

It will take about 19 days to melt all the ice.

An alternative way to get equations to solve for the radiosities and heat flows is to write nodal equations using the electrical circuit diagram. We assume that heat flows are directed into the nodes, and the sum of the heat flows into a node has to equal zero for steady state. Looking at Fig. 9.10, we have the following equations for nodes $J_1$ and $J_2$:

$$\text{Node } J_1 : \frac{E_{b1} - J_1}{\left(\dfrac{1 - \varepsilon_1}{\varepsilon_1 A_1}\right)} + \frac{J_2 - J_1}{\left(\dfrac{1}{A_1 F_{1-2}}\right)} = 0 \tag{9.54}$$

$$\text{Node } J_2 : \frac{E_{b2} - J_2}{\left(\dfrac{1 - \varepsilon_2}{\varepsilon_2 A_2}\right)} + \frac{J_1 - J_2}{\left(\dfrac{1}{A_1 F_{1-2}}\right)} = 0 \tag{9.55}$$

$$E_{b1} = \sigma T_1{}^4 \qquad J_1 \qquad J_2 \qquad E_{b2} = \sigma T_2{}^4$$

$$q_1 \longrightarrow \bullet \!\!-\!\!\bigvee\!\!\bigvee\!\!-\!\!\bullet \!\!-\!\!\bigvee\!\!\bigvee\!\!-\!\!\bullet \!\!-\!\!\bigvee\!\!\bigvee\!\!-\!\!\bullet\!\!\longleftarrow q_2$$

$$\frac{1-\varepsilon_1}{\varepsilon_1 A_1} \qquad \frac{1}{A_1 F_{1-2}} \qquad \frac{1-\varepsilon_2}{\varepsilon_2 A_2}$$

**FIGURE 9.10**

Resistive circuit for a two-surface enclosure.

Eq. (9.54) and (9.55) can be solved simultaneously for $J_1$ and $J_2$. Eq. (9.44) or Eq. (9.45) may then be used to get $q_1$. Or Eq. (9.46) or Eq. (9.47) may be used to get $q_2$. When the heat flow from one surface is known, it is easy to get the heat flow for the other surface, as, for a two-surface enclosure, $q_1 = -q_2$.

The Electric-Heat Analogy and the Resistance Concept were discussed in Chapter 3. Radiant exchange in a two-surface enclosure can be modeled by the resistive circuit shown in Fig. 9.10. The end potentials are $E_{b1} = \sigma T_1^4$ and $E_{b2} = \sigma T_2^4$. Intermediate potentials are $J_1$ and $J_2$.

## Example 9.7

### Radiant heat transfer for a semicylindrical furnace

*Problem*

A very long semicylindrical furnace has a radius of 2 m. The curved top surface, whose emissivity is 0.75, has built-in radiant heaters that provide a uniform heat flux of 8000 W/m² to the furnace space. The bottom surface is at 150 C and has an emissivity of 0.9.

**(a)** What is the temperature of the curved top surface?

**(b)** What is the rate of heat flow to the bottom surface per meter length of the furnace?

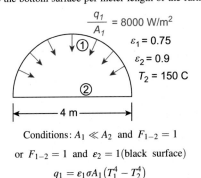

$$\frac{q_1}{A_1} = 8000 \text{ W/m}^2$$
$$\varepsilon_1 = 0.75$$
$$\varepsilon_2 = 0.9$$
$$T_2 = 150 \text{ C}$$

$$\longleftarrow 4 \text{ m} \longrightarrow$$

Conditions: $A_1 \ll A_2$ and $F_{1-2} = 1$

or $F_{1-2} = 1$ and $\varepsilon_2 = 1$ (black surface)

$$q_1 = \varepsilon_1 \sigma A_1 \left( T_1^4 - T_2^4 \right)$$

*Solution*

This is a two-surface problem as the furnace is very long. If the furnace had been short, some radiation would have escaped through the ends of the furnace, and the problem would have been a three-surface or four-surface problem.

**(a)** The applicable equation for this two-surface problem is Eq. (9.48).

$$q_1 = \frac{\sigma \left( T_1^4 - T_2^4 \right)}{\dfrac{1-\varepsilon_1}{\varepsilon_1 A_1} + \dfrac{1}{A_1 F_{1-2}} + \dfrac{1-\varepsilon_2}{\varepsilon_2 A_2}} \qquad (9.48)$$

For this problem, $\varepsilon_1 = 0.75$, $\varepsilon_2 = 0.9$, and $T_2 = 150 \text{ C} = 423.15 \text{ K}$.

$$q_1 / A_1 = 8000 \text{ W/m}^2$$

The surface areas are $A_1 = \pi r L$ and $A_2 = 2rL$.where $r$ = radius and $L$ = length of the furnace.

Hence, $A_1/A_2 = \frac{\pi r L}{2 r L} = \frac{\pi}{2} = 1.5708$.

Looking at Eq. (9.48), we need shape factor $F_{1-2}$.

For radiation from the bottom Surface 2, we have $F_{2-1} + F_{2-2} = 1$. However $F_{2-2} = 0$ as Surface 2 is flat. Therefore, $F_{2-1} = 1$. From reciprocity, $A_1 F_{1-2} = A_2 F_{2-1}$ and

$$F_{1-2} = \left(\frac{A_2}{A_1}\right) F_{2-1} = \left(\frac{1}{1.5708}\right)(1) = 0.6366$$

Dividing both sides of Eq. (9.48) by $A_1$, we have

$$\frac{q_1}{A_1} = \frac{\sigma(T_1^4 - T_2^4)}{\frac{1 - \varepsilon_1}{\varepsilon_1} + \frac{1}{F_{1-2}} + \frac{1 - \varepsilon_2}{\varepsilon_2}\left(\frac{A_1}{A_2}\right)} \tag{9.56}$$

Putting values into Eq. (9.56), we have

$$8000 = \frac{5.67 \times 10^{-8}\left[T_1^4 - (423.15)^4\right]}{\frac{1 - 0.75}{0.75} + \frac{1}{0.6366} + \frac{1 - 0.9}{0.9}(1.5708)} \tag{9.57}$$

Solving Eq. (9.57) for $T_1$, we get $T_1 = 755$ K $= 482$ C.

The top surface of the furnace is at 482 C.

**(b)** The heat flux from Surface 1 is $\frac{q_1}{A_1} = 8000$ W/m$^2$ and the area of Surface 1 is $A_1 = \pi r L = \pi(2)L = 6.2832L$.

Therefore, $\frac{q_1}{L} = (8000)(6.2832) = 50265$ W/m length.

By energy conservation, the rate of heat flow from Surface 1 is equal to the rate of heat flow to Surface 2. Hence, the bottom surface of the furnace (Surface 2) receives a heat flow of **50,265 W/m length.**

## 9.5.2 Radiation heat transfer for a three-surface enclosure

If an enclosure has three surfaces exchanging radiation, Eqs. (9.43) and (9.37) are

### 9.5.2.1 For surface 1

$$q_1 = A_1\left(J_1 - \sum_{j=1}^{3} J_j F_{1-j}\right) = A_1(J_1 - J_1 F_{1-1} - J_2 F_{1-2} - J_3 F_{1-3}) \tag{9.58}$$

$$q_1 = \frac{A_1 \varepsilon_1}{1 - \varepsilon_1}(E_{b1} - J_1) \tag{9.59}$$

### 9.5.2.2 For surface 2

$$q_2 = A_2\left(J_2 - \sum_{j=1}^{3} J_j F_{2-j}\right) = A_2(J_2 - J_1 F_{2-1} - J_2 F_{2-2} - J_3 F_{2-3}) \tag{9.60}$$

$$q_2 = \frac{A_2 \varepsilon_2}{1 - \varepsilon_2}(E_{b2} - J_2) \tag{9.61}$$

### 9.5.2.3 For surface 3

$$q_3 = A_3 \left( J_3 - \sum_{j=1}^{3} J_j F_{3-j} \right) = A_3 (J_3 - J_1 F_{3-1} - J_2 F_{3-2} - J_3 F_{3-3}) \tag{9.62}$$

$$q_3 = \frac{A_3 \varepsilon_3}{1 - \varepsilon_3} (E_{b3} - J_3) \tag{9.63}$$

Eqs. (9.58) through (9.63) may be solved simultaneously for $q_1, q_2, q_3, J_1, J_2,$ and $J_3$.

Radiant exchange in a three-surface enclosure can be modeled by the resistive circuit shown in Fig. 9.11. The end potentials are $E_{b1} = \sigma T_1^4$, $E_{b2} = \sigma T_2^4$, and $E_{b3} = \sigma T_3^4$. Intermediate potentials are $J_1$, $J_2$, and $J_3$.

An alternative way to get equations to solve for the radiosities and heat flows is to write nodal equations using the electrical circuit diagram. We assume that heat flows are directed into the nodes, and the sum of the heat flows into a node has to equal zero for steady state. Looking at Fig. 9.11, we have the following equations for nodes $J_1$, $J_2$, and $J_3$:

$$\text{Node } J_1: \frac{E_{b1} - J_1}{\left(\dfrac{1 - \varepsilon_1}{\varepsilon_1 A_1}\right)} + \frac{J_2 - J_1}{\left(\dfrac{1}{A_1 F_{1-2}}\right)} + \frac{J_3 - J_1}{\left(\dfrac{1}{A_1 F_{1-3}}\right)} = 0 \tag{9.64}$$

$$\text{Node } J_2: \frac{E_{b2} - J_2}{\left(\dfrac{1 - \varepsilon_2}{\varepsilon_2 A_2}\right)} + \frac{J_1 - J_2}{\left(\dfrac{1}{A_1 F_{1-2}}\right)} + \frac{J_3 - J_2}{\left(\dfrac{1}{A_2 F_{2-3}}\right)} = 0 \tag{9.65}$$

$$\text{Node } J_3: \frac{E_{b3} - J_3}{\left(\dfrac{1 - \varepsilon_3}{\varepsilon_3 A_3}\right)} + \frac{J_1 - J_3}{\left(\dfrac{1}{A_1 F_{1-3}}\right)} + \frac{J_2 - J_3}{\left(\dfrac{1}{A_2 F_{2-3}}\right)} = 0 \tag{9.66}$$

Eqs. (9.64) through (9.66) can be solved simultaneously for $J_1$, $J_2$, and $J_3$. Eqs. (9.58) through (9.63) may then be used to get the heat flows $q_1, q_2,$ and $q_3$.

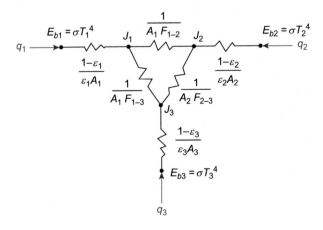

**FIGURE 9.11**

Resistive circuit for a three-surface enclosure.

## Example 9.8

### Heat transfer from a radiant heater

*Problem*

A radiant heater on the ceiling of a factory is used to heat a work table directly under it. Both the heater and table top are 2 m by 2 m. The heater is at 500 C and has an emissivity of 0.9. The table top is at 30 C and has an emissivity of 0.5. The temperature of the factory is 15 C. The distance between the heater and the table top is 3.6 m.

**(a)** What is the required power output of the radiant heater?

**(b)** What is the radiant heat flux received by the table top?

*Solution*

This is a three-surface problem. Surface 1 is the heater; Surface 2 is the table top; and Surface 3 is the factory environment. The electrical circuit diagram is

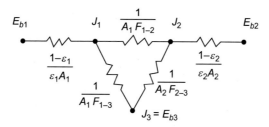

(Note: Fig. 9.11 shows that the resistance between $J_3$ and $E_{b3}$ is normally $(1-\varepsilon_3)/(\varepsilon_3 A_3)$. However, in this problem, the area of Surface 3 is very large. Hence, the resistance between $J_3$ and $E_{b3}$ is negligible. Hence, $J_3 = E_{b3}$. Alternatively, we could have considered the factory environment, i.e., Surface 3, to be black.)

$$E_{b1} = \sigma T_1^4 = (5.67 \times 10^{-8})(500 + 273.15)^4 = 20260$$

$$E_{b2} = \sigma T_2^4 = (5.67 \times 10^{-8})(30 + 273.15)^4 = 478.9$$

$$E_{b3} = \sigma T_3^4 = (5.67 \times 10^{-8})(15 + 273.15)^4 = 390.9$$

$$\frac{1 - \varepsilon_1}{\varepsilon_1 A_1} = \frac{1 - 0.9}{(0.9)(4)} = 0.02778 \qquad \frac{1 - \varepsilon_2}{\varepsilon_2 A_2} = \frac{1 - 0.5}{(0.5)(4)} = 0.25$$

We can get shape factor $F_{1-2}$ from Fig. 9.4 with a = b = 2 and c = 3.6.
From the figure, we get $F_{1-2} = 0.080$. From reciprocity, $F_{2-1} = 0.080$.
Looking at the electrical circuit, we also need $F_{1-3}$ and $F_{2-3}$.
Shape factor equations for the three-surface enclosure are

$$F_{1-1} + F_{1-2} + F_{1-3} = 1$$
$$F_{2-1} + F_{2-2} + F_{2-3} = 1$$
$$F_{3-1} + F_{3-2} + F_{3-3} = 1$$

From the first equation, $F_{1-1} = 0$ as the surface is flat; $F_{1-3} = 1 - F_{1-2} = 1 - 0.08 = 0.92$.
From the second equation, $F_{2-2} = 0$ as the surface is flat; $F_{2-3} = 1 - F_{2-1} = 1 - 0.08 = 0.92$.
Looking at the electrical circuit, we see that we need three additional resistances. They are

$$\frac{1}{A_1 F_{1-3}} = \frac{1}{A_2 F_{2-3}} = \frac{1}{(4)(0.92)} = 0.272 \text{ and } \frac{1}{A_1 F_{1-2}} = \frac{1}{(4)(0.08)} = 3.125$$

Redrawing the electrical circuit to show the resistance values, we have

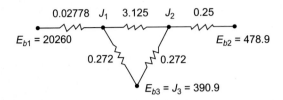

The nodal equations are

$$\text{For Node } J_1: \frac{20260 - J_1}{0.02778} + \frac{J_2 - J_1}{3.125} + \frac{390.9 - J_1}{0.272} = 0$$

$$\text{For Node } J_2: \frac{J_1 - J_2}{3.125} + \frac{478.9 - J_2}{0.25} + \frac{390.9 - J_2}{0.272} = 0$$

Solving the nodal equations simultaneously for $J_1$ and $J_2$, we get

$$J_1 = 18280 \quad \text{and} \quad J_2 = 1150$$

**(a)** $q_1 = \dfrac{E_{b1} - J_1}{0.02778} = \dfrac{20260 - 18280}{0.02778} = 71270 \text{ W}$

The required power output of the radiant heater is 71.3 kW.

**(b)** $q_2 = \dfrac{E_{b2} - J_2}{0.25} = \dfrac{478.9 - 1150}{0.25} = -2684 \text{ W}$

(Note: Negative value means that heat flow is inward to the surface.)

$$\frac{q_2}{A_2} = \frac{2684}{4} = 671 \text{ W/m}^2$$

The radiative heat flux to the table top is 671 W/m².

### 9.5.2.4 Three-surface enclosure with an insulated surface

Sometimes surfaces of an enclosure are perfectly insulated on their back side. For these surfaces, the energy incident on the surface is equal to the energy leaving it. That is, the irradiation $G$ equals the radiosity $J$. As shown by Eq. (9.33), the heat loss $q$ from the surface is zero. This adiabatic surface is also called a *reradiating* or *refractory* surface.

If surface $i$ is the adiabatic surface, then $G_i = J_i$. The definition of radiosity $J_i$ is $J_i = (1 - \varepsilon_i)G_i + \varepsilon_i E_{bi}$. As $G_i = J_i$, it follows that $J_i = E_{bi} = \sigma T_i^4$ for the adiabatic surface.

Fig. 9.12 shows the electrical circuit diagram for a three-surface enclosure with Surface 3 being the perfectly insulated surface.

**FIGURE 9.12**

Resistive circuit for a three-surface enclosure with Surface 3 adiabatic.

The nodal equations (Eqs. 9.64–9.66) still apply, with the first term of Eq. (9.66) deleted as $J_3 = E_{b3}$. Once the equations are solved and $J_3$ has been determined, the temperature of the insulated surface may be found as $J_3 = E_{b3} = \sigma T_3^4$. It should be noted that the emissivity of the insulated surface is irrelevant and does not factor into the calculations.

Looking at the electrical circuit of Fig. 9.12, one can obtain the following expression for the heat flow from Surface 1 to Surface 2:

$$q_1 = -q_2 = \frac{\sigma\left(T_1^4 - T_2^4\right)}{\dfrac{1-\varepsilon_1}{\varepsilon_1 A_1} + \dfrac{1}{A_1 F_{1-2} + \dfrac{1}{\frac{1}{A_1 F_{1-3}} + \frac{1}{A_2 F_{2-3}}}} + \dfrac{1-\varepsilon_2}{\varepsilon_2 A_2}} \tag{9.67}$$

Example 9.9 illustrates the solution of a problem involving three surfaces, with one of the surfaces being perfectly insulated.

## Example 9.9
### Radiant heat transfer for a truncated cone with an insulated surface

*Problem*
A truncated cone has a height of 4 cm. The closed bottom of the cone has a diameter of 4 cm, a temperature of 800 C, and an emissivity of 0.8. The top surface has a diameter of 3.5 cm and is open to a large room that is at 25 C. The side wall of the cone is perfectly insulated and has an emissivity of 0.2.
**(a)** What is the temperature of the side wall?
**(b)** What is the rate at which heat is transferred into the room?

*Solution*
This is a problem involving a three-surface enclosure. The bottom and side surfaces are closed surfaces, and the top surface is an open window. The side surface is a perfectly insulated or "refractory" surface.

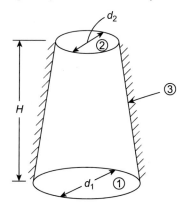

$$d_1 = 0.04 \text{ m} \qquad \varepsilon_1 = 0.8 \qquad T_1 = 800 \text{ C} = 1073.15 \text{ K}$$
$$d_2 = 0.035 \text{ m} \quad H = 0.04 \text{ m} \qquad \varepsilon_3 = 0.2$$
Shape factor $F_{1-2}$ is determined from Fig. 9.6. For this figure,

$$r_1 = 0.02 \quad r_2 = 0.0175 \qquad\qquad L = 0.04$$
$$L/r_1 = 0.04/0.02 = 2 \quad r_2/L = 0.0175/0.04 = 0.44 \text{ and } F_{1-2} = 0.14$$

$$F_{1-1} + F_{1-2} + F_{1-3} = 1$$
$$F_{1-1} = 0 \text{ since flat surface, and } F_{1-3} = 1 - F_{1-2} = 1 - 0.14 = 0.86$$

$$A_1 = \frac{\pi}{4}d_1^2 = \frac{\pi}{4}(0.04)^2 = 1.257 \times 10^{-3} \text{ m}^2$$

$$A_2 = \frac{\pi}{4}d_2^2 = \frac{\pi}{4}(0.035)^2 = 9.621 \times 10^{-4} \text{ m}^2$$

$$A_3 = \pi(r_1 + r_2)\sqrt{H^2 + (r_1 - r_2)^2} = 4.722 \times 10^{-3} \text{ m}^2$$

From reciprocity,

$$A_2 F_{2-1} = A_1 F_{1-2}$$

$$F_{2-1} = \frac{A_1}{A_2} F_{1-2} = \frac{1.257 \times 10^{-3}}{9.621 \times 10^{-4}}(0.14) = 0.183$$

$$F_{2-1} + F_{2-2} + F_{2-3} = 1$$

$F_{2-2} = 0$ as flat surface, and $F_{2-3} = 1 - F_{2-1} = 1 - 0.183 = 0.817$

The electrical circuit diagram for this problem is

$$E_{b1} = \sigma T_1^4 = (5.67 \times 10^{-8})(800 + 273.15)^4 = 75201$$

$$E_{b2} = \sigma T_2^4 = (5.67 \times 10^{-8})(25 + 273.15)^4 = 448$$

$$\frac{1 - \varepsilon_1}{\varepsilon_1 A_1} = \frac{1 - 0.8}{(0.8)(1.257 \times 10^{-3})} = 198.9$$

$$\frac{1}{A_1 F_{1-3}} = \frac{1}{(1.257 \times 10^{-3})(0.86)} = 925$$

$$\frac{1}{A_1 F_{1-2}} = \frac{1}{(1.257 \times 10^{-3})(0.14)} = 5682$$

$$\frac{1}{A_2 F_{2-3}} = \frac{1}{(9.621 \times 10^{-4})(0.817)} = 1272$$

**Nodal Equations.**

$$\text{Node } J_1: \frac{E_{b1} - J_1}{\left(\frac{1 - \varepsilon_1}{\varepsilon_1 A_1}\right)} + \frac{J_2 - J_1}{\left(\frac{1}{A_1 F_{1-2}}\right)} + \frac{J_3 - J_1}{\left(\frac{1}{A_1 F_{1-3}}\right)} = 0 \tag{a}$$

$$\text{Node } J_3: \frac{J_1 - J_3}{\left(\frac{1}{A_1 F_{1-3}}\right)} + \frac{J_2 - J_3}{\left(\frac{1}{A_2 F_{2-3}}\right)} = 0 \tag{b}$$

Putting in the known values,
For Eq. (a),

$$\frac{75201 - J_1}{198.9} + \frac{448 - J_1}{5682} + \frac{J_3 - J_1}{925} = 0 \tag{c}$$

For Eq. (b),

$$\frac{J_1 - J_3}{925} + \frac{448 - J_3}{1272} = 0 \tag{d}$$

Solving Eq. (c) and (d) simultaneously, we get $J_1 = 66863$ and $J_3 = 38901$.

**(a)**

$$J_3 = E_{b3} = \sigma T_3^4 T_3 = \left(\frac{J_3}{\sigma}\right)^{0.25} = \left(\frac{38901}{5.67 \times 10^{-8}}\right)^{0.25} = 910.1 \text{ K} = 637 \text{ C}$$

The side wall is at 637 C.

**(b)** The side wall is adiabatic. Therefore, the heat flow into the room equals the heat flow from the bottom surface of the truncated cone. This heat flow is

$$q_1 = \frac{E_{b1} - J_1}{\left(\dfrac{1 - \varepsilon_1}{\varepsilon_1 A_1}\right)} = \frac{75201 - 66863}{198.9} = 41.9 \text{ W}$$

The rate of radiant heat flow into the room is 41.9 W.

Below is another circuit diagram for this problem. We have put the resistance and potential values on the diagram. We have also included the heat flows for the various surfaces. It is seen that the side wall is indeed adiabatic, with heat in = heat out. Alternatively, Eq. (9.67), with the last term in the denominator deleted, could have been used to get this result.

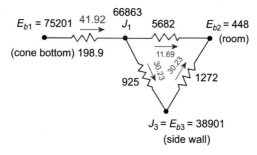

# 9.6 Radiation shields

Sometimes we want to decrease the radiant heat transfer between two surfaces. This can be done by inserting a thin sheet of material between the two surfaces. The material can be in the form of a plate, cylinder, sphere, or any other shape appropriate to the surfaces being affected. Usually, the added sheet has low emissivities on its two surfaces. The inserted sheet is called a **radiation shield.** Multiple layers of shields are used in insulations for cryogenic fluid containers. Shields are also used to improve the accuracy of temperature measurement. For example, say we are measuring the temperature of a gas flowing through a pipe. If the temperature of the pipe wall is different from that of the gas, the measured gas temperature may be significantly in error due to radiative transfer between the pipe wall and the temperature sensor. Placing a radiation shield between the sensor and the pipe wall will reduce the measurement error.

Let us look at a couple of examples that illustrate the effectiveness of radiation shields in decreasing radiant transfer between surfaces.

Fig. 9.13 shows a radiation shield between two large plane surfaces. The left surface (Surface 1) has a temperature $T_1$ and an emissivity $\varepsilon_1$. The right surface (Surface 2) has temperature $T_2$ and emissivity $\varepsilon_2$. The radiation shield has emissivity $\varepsilon_{31}$ on its side facing Surface 1 and emissivity $\varepsilon_{32}$ on its side facing Surface 2.

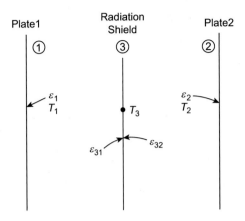

**FIGURE 9.13**

Radiation shield between two large plates.

Let us look at the impact of the radiation shield on the radiative transfer between Surfaces 1 and 2. The heat transfer from Surface 1 to Surface 2 without the shield in place is given by Eq. (9.51).

$$q_1 = \frac{\sigma A \left( T_1^4 - T_2^4 \right)}{\dfrac{1}{\varepsilon_1} + \dfrac{1}{\varepsilon_2} - 1} \tag{9.51}$$

For steady heat transfer, with the shield in place, the heat transfer from Surface 1 to the shield equals the heat transfer from the shield to Surface 2. That is,

$$q_{1\rightarrow3} = q_{3\rightarrow2} = q_{1\rightarrow2} \tag{9.68}$$

Using Eq. (9.51), we have

$$q_{1\rightarrow3} = \frac{\sigma A \left( T_1^4 - T_3^4 \right)}{\dfrac{1}{\varepsilon_1} + \dfrac{1}{\varepsilon_{31}} - 1} = q_{3\rightarrow2} = \frac{\sigma A \left( T_3^4 - T_2^4 \right)}{\dfrac{1}{\varepsilon_{32}} + \dfrac{1}{\varepsilon_2} - 1} \tag{9.69}$$

With temperatures $T_1$ and $T_2$ and all the emissivities known, Eq. (9.69) may be solved for unknown temperature $T_3$. Through algebra, it is found that $T_3$ is

$$T_3 = \left[ \frac{\left( \dfrac{1}{\varepsilon_{32}} + \dfrac{1}{\varepsilon_2} - 1 \right) T_1^4 + \left( \dfrac{1}{\varepsilon_1} + \dfrac{1}{\varepsilon_{31}} - 1 \right) T_2^4}{\dfrac{1}{\varepsilon_1} + \dfrac{1}{\varepsilon_2} + \dfrac{1}{\varepsilon_{31}} + \dfrac{1}{\varepsilon_{32}} - 2} \right]^{0.25} \tag{9.70}$$

After $T_3$ is determined, the heat flow for the system can be obtained from Eq. (9.69).

If there are multiple radiation shields, equations like Eq. (9.69) can be written between adjacent shields and the resulting equations can be solved simultaneously for the temperatures of the shields. Then one of the heat flow equations can be used to get the heat flow through the system.

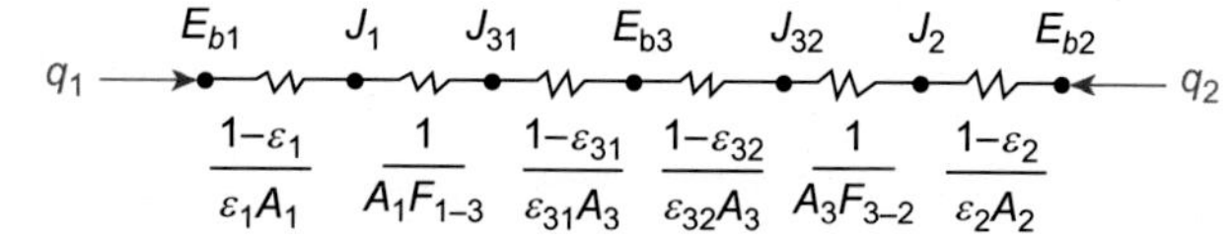

**FIGURE 9.14**

Circuit diagram for a one radiation shield system.

Another way to get the heat flow is through use of the electrical circuit for the system. The circuit has resistances in series as the heat flow is the same through all segments of the system. Fig. 9.10 showed the circuit diagram for radiative transfer between two surfaces. There are two surface resistances and one space resistance. Addition of a radiation shield between the surfaces adds two surface resistances and one space resistance to the diagram. The resulting circuit for one radiation shield is shown in Fig. 9.14. If there are more than one shield between the end surfaces, additional surface and space resistances can be added to the circuit for the additional shield(s).

We can use this circuit diagram like we previously used an electrical circuit diagram when we discussed conduction through composite materials. Using the circuit diagram of Fig. 9.14, the heat flow is the overall difference in potential (i.e., difference in emissive power $E_{b1} - E_{b2} = \sigma(T_1^4 - T_2^4)$) divided by the sum of the resistances between $E_{b1}$ and $E_{b2}$. That is,

$$q_{1\to2} = \frac{\sigma(T_1^4 - T_2^4)}{\dfrac{1-\varepsilon_1}{\varepsilon_1 A_1} + \dfrac{1}{A_1 F_{1-3}} + \dfrac{1-\varepsilon_{31}}{\varepsilon_{31} A_3} + \dfrac{1-\varepsilon_{32}}{\varepsilon_{32} A_3} + \dfrac{1}{A_3 F_{3-2}} + \dfrac{1-\varepsilon_2}{\varepsilon_2 A_2}} \tag{9.71}$$

Eq. (9.71) may also be used for concentric cylinders and concentric spheres.

For parallel plates, all of the areas are equal and all the shape factors in Eq. (9.71) are unity. Eq. (9.71) then reduces to

$$q_{1\to2} = \frac{\sigma A(T_1^4 - T_2^4)}{\left(\dfrac{1}{\varepsilon_1} + \dfrac{1}{\varepsilon_2} - 1\right) + \left(\dfrac{1}{\varepsilon_{31}} + \dfrac{1}{\varepsilon_{32}} - 1\right)} \tag{9.72}$$

Let us say we want to find the temperature $T_3$ of the shield. The heat flow is the same through all segments of the system and we can write the resistance equation for $q$ between any two points in the circuit of Fig. 9.14. Therefore, if we want the temperature of the shield, we can write an equation between the unknown potential $E_{b3}$ and one of the known end potentials ($E_{b1}$ or $E_{b2}$). For example, between $E_{b1}$ and $E_{b3}$ we have

$$q_{1\to2} = \frac{\sigma(T_1^4 - T_3^4)}{\dfrac{1-\varepsilon_1}{\varepsilon_1 A_1} + \dfrac{1}{A_1 F_{1-3}} + \dfrac{1-\varepsilon_{31}}{\varepsilon_{31} A_3}} \tag{9.73}$$

and between $E_{b3}$ and $E_{b2}$, we have

$$q_{1\to2} = \frac{\sigma(T_3^4 - T_2^4)}{\dfrac{1-\varepsilon_{32}}{\varepsilon_{32} A_3} + \dfrac{1}{A_3 F_{3-2}} + \dfrac{1-\varepsilon_2}{\varepsilon_2 A_2}} \tag{9.74}$$

Equating the right sides of Eqs. (9.73) and (9.74), we have

$$\frac{\sigma\left(T_1^4 - T_3^4\right)}{\dfrac{1-\varepsilon_1}{\varepsilon_1 A_1} + \dfrac{1}{A_1 F_{1-3}} + \dfrac{1-\varepsilon_{31}}{\varepsilon_{31} A_3}} = \frac{\sigma\left(T_3^4 - T_2^4\right)}{\dfrac{1-\varepsilon_{32}}{\varepsilon_{32} A_3} + \dfrac{1}{A_3 F_{3-2}} + \dfrac{1-\varepsilon_2}{\varepsilon_2 A_2}} \tag{9.75}$$

Eq. (9.75) may be solved for its only unknown, the shield temperature $T_3$.

Example 9.10 illustrates the reduction of radiant heat transfer between surfaces through the use of radiation shields.

## Example 9.10
### Radiant heat transfer for a parallel plate system with a radiation shield

*Problem*
Two large parallel plates are exchanging radiation. One plate has a temperature of 1000 K and an emissivity of 0.8. The other plate is at 350 K and has an emissivity of 0.4.
**(a)** What is the radiant heat flux between the plates?
**(b)** It is desired to reduce the heat transfer between the plates by inserting a thin radiation shield between them. The shield has an emissivity of 0.1 on one side and 0.05 on the other. What is the new rate of heat transfer after the shield is in place, and what is the temperature of the shield?
**(c)** An additional shield like the one of Part (b) is placed in the system to further reduce the heat transfer. What is the resulting heat flux, and what are the temperatures of the two shields?

*Solution*
**(a)** $\varepsilon_1 = 0.8$; $\varepsilon_2 = 0.4$; $T_1 = 1000$ K; $T_2 = 350$ K.
Using Eq. (9.51),

$$\frac{q_1}{A} = \frac{\sigma\left(T_1^4 - T_2^4\right)}{\dfrac{1}{\varepsilon_1} + \dfrac{1}{\varepsilon_2} - 1} = \frac{5.67 \times 10^{-8}\left(1000^4 - 350^4\right)}{\dfrac{1}{0.8} + \dfrac{1}{0.4} - 1} = 20310 \text{ W/m}^2$$

**(b)** Consider Fig. 9.13. For this problem, $\varepsilon_1 = 0.8$; $\varepsilon_2 = 0.4$; $\varepsilon_{31} = 0.1$; and $\varepsilon_{32} = 0.05$.

$$T_1 = 1000 \text{ K} \quad \text{and} \quad T_2 = 350 \text{ K}.$$

Using Eq. (9.70),

$$T_3 = \left[\frac{\left(\dfrac{1}{\varepsilon_{32}} + \dfrac{1}{\varepsilon_2} - 1\right)T_1^4 + \left(\dfrac{1}{\varepsilon_1} + \dfrac{1}{\varepsilon_{31}} - 1\right)T_2^4}{\dfrac{1}{\varepsilon_1} + \dfrac{1}{\varepsilon_2} + \dfrac{1}{\varepsilon_{31}} + \dfrac{1}{\varepsilon_{32}} - 2}\right]^{0.25} = \left[\frac{\left(\dfrac{1}{0.05} + \dfrac{1}{0.4} - 1\right)1000^4 + \left(\dfrac{1}{0.8} + \dfrac{1}{0.1} - 1\right)350^4}{\dfrac{1}{0.8} + \dfrac{1}{0.4} + \dfrac{1}{0.1} + \dfrac{1}{0.05} - 2}\right]^{0.25}$$

$$T_3 = 908.8 \text{ K}$$

Now that the shield temperature is known, one of the relations in Eq. (9.69) may be used to get the heat flux.

$$\frac{q_{1\to2}}{A} = \frac{\sigma\left(T_1^4 - T_3^4\right)}{\dfrac{1}{\varepsilon_1} + \dfrac{1}{\varepsilon_{31}} - 1} = \frac{\sigma\left(T_3^4 - T_2^4\right)}{\dfrac{1}{\varepsilon_{32}} + \dfrac{1}{\varepsilon_2} - 1}$$

Using the first relation, we have

$$\frac{q_{1\to2}}{A} = \frac{5.67 \times 10^{-8}\left(1000^4 - 908.8^4\right)}{\dfrac{1}{0.8} + \dfrac{1}{0.1} - 1} = 1760 \text{ W/m}^2$$

Alternatively, Eq. (9.72), which was obtained from the circuit diagram, could have been used to get the heat flux without having to first find the shield temperature.

$$\frac{q_{1\rightarrow 2}}{A} = \frac{\sigma\left(T_1^4 - T_2^4\right)}{\left(\dfrac{1}{\varepsilon_1} + \dfrac{1}{\varepsilon_2} - 1\right) + \left(\dfrac{1}{\varepsilon_{31}} + \dfrac{1}{\varepsilon_{32}} - 1\right)} = \frac{5.67 \times 10^{-8}\left(1000^4 - 350^4\right)}{\left(\dfrac{1}{0.8} + \dfrac{1}{0.4} - 1\right) + \left(\dfrac{1}{0.1} + \dfrac{1}{0.05} - 1\right)} = 1760 \text{ W/m}^2$$

The temperature of the shield is 908.8 K and the heat flux is 1760 W/m$^2$.

**(c)** A two-shield system is shown in Fig. 9.15 and the circuit diagram for the system is shown in Fig. 9.16. Let us look at Fig. 9.15.

For this problem, $\varepsilon_1 = 0.8$; $\varepsilon_2 = 0.4$; $\varepsilon_{31} = 0.1$; $\varepsilon_{32} = 0.05$; $\varepsilon_{41} = 0.1$; and $\varepsilon_{42} = 0.05$.

$$T_1 = 1000 \text{ K} \text{ and } T_2 = 350 \text{ K}.$$

We can use the circuit diagram of Fig. 9.16 to get equations to solve for $q_{1\rightarrow 2}$, $T_3$, and $T_4$. The heat flow is the difference in overall $E_b$ divided by the resistances between the end $E_b$s. That is,

$$q_{1\rightarrow 2} = \frac{\sigma\left(T_1^4 - T_2^4\right)}{\dfrac{1-\varepsilon_1}{\varepsilon_1 A_1} + \dfrac{1}{A_1 F_{1-3}} + \dfrac{1-\varepsilon_{31}}{\varepsilon_{31} A_3} + \dfrac{1-\varepsilon_{32}}{\varepsilon_{32} A_3} + \dfrac{1}{A_3 F_{3-4}} + \dfrac{1-\varepsilon_{41}}{\varepsilon_{41} A_4} + \dfrac{1-\varepsilon_{42}}{\varepsilon_{42} A_4} + \dfrac{1}{A_4 F_{4-2}} + \dfrac{1-\varepsilon_2}{\varepsilon_2 A_2}} \tag{9.76}$$

For this problem, $A_1 = A_2 = A_3 = A_4 = A$ and all the shape factors are unity. Eq. (9.76) then reduces to

$$q_{1\rightarrow 2} = \frac{\sigma A\left(T_1^4 - T_2^4\right)}{\dfrac{1}{\varepsilon_1} + \dfrac{1}{\varepsilon_{31}} + \dfrac{1}{\varepsilon_{32}} + \dfrac{1}{\varepsilon_{41}} + \dfrac{1}{\varepsilon_{42}} + \dfrac{1}{\varepsilon_2} - 3} \tag{9.77}$$

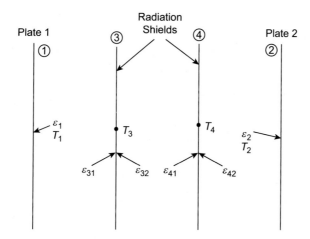

**FIGURE 9.15**

Two-shield system.

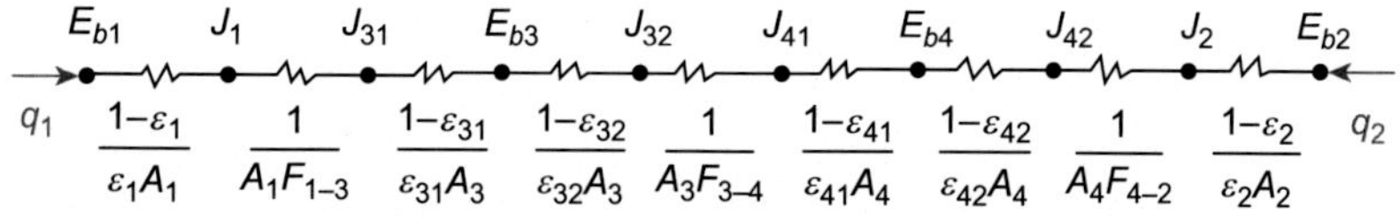

**FIGURE 9.16**

Circuit diagram for two-shield system of Fig. 9.15.

Putting in values for the parameters of Eq. (9.77), we get $\frac{q_{1\to2}}{A} = 919 \text{ W}/\text{m}^2$.

We can also use Fig. 9.16 to get the shield temperatures. For $T_3$, we can write an equation either between nodes $E_{b1}$ and $E_{b3}$ or between nodes $E_{b3}$ and $E_{b2}$. The equation between nodes $E_{b1}$ and $E_{b3}$ is

$$q_{1\to2} = \frac{\sigma(T_1^4 - T_3^4)}{\dfrac{1 - \varepsilon_1}{\varepsilon_1 A_1} + \dfrac{1}{A_1 F_{1-3}} + \dfrac{1 - \varepsilon_{31}}{\varepsilon_{31} A_3}} \qquad (9.78)$$

With equal areas and unity shape factors, Eq. (9.78) becomes

$$\frac{q_{1\to2}}{A} = \frac{\sigma(T_1^4 - T_3^4)}{\dfrac{1}{\varepsilon_1} + \dfrac{1}{\varepsilon_{31}} - 1} \qquad (9.79)$$

Using the values we have for $\frac{q_{1\to2}}{A}$, $T_1$, $\varepsilon_1$, and $\varepsilon_{31}$, we can solve Eq. (9.79) for shield temperature $T_3$. It is found that $T_3 = 955.6$ K.

For shield temperature $T_4$, we can write an equation between nodes $E_{b4}$ and $E_{b2}$ of Fig. 9.16. This equation is

$$q_{1\to2} = \frac{\sigma(T_4^4 - T_2^4)}{\dfrac{1 - \varepsilon_{42}}{\varepsilon_{42} A_4} + \dfrac{1}{A_4 F_{4-2}} + \dfrac{1 - \varepsilon_2}{\varepsilon_2 A_2}} \qquad (9.80)$$

With equal areas and unity shape factors, Eq. (9.80) becomes

$$\frac{q_{1\to2}}{A} = \frac{\sigma(T_4^4 - T_2^4)}{\dfrac{1}{\varepsilon_{42}} + \dfrac{1}{\varepsilon_2} - 1} \qquad (9.81)$$

Using the values we have for $\frac{q_{1\to2}}{A}$, $T_2$, $\varepsilon_{42}$, and $\varepsilon_2$, we can solve Eq. (9.81) for shield temperature $T_4$. It is found that $T_4 = 776.5$ K.

For the two-shield system, the heat flux is 919 W/m² and the shield temperatures are 955.6 and 776.5 K.

To summarize:

Without radiation shields, the heat flux between the two plates is 20310 W/m². With one shield, the heat flux is 1760 W/m². With two shields, the heat flux is 919 W/m².

It is seen that use of one shield greatly decreases the heat flux. Use of an additional shield further decreases the heat flux but not to as great an extent as achieved by the first shield.

We mentioned at the beginning of this section that radiation shields are effective in increasing the accuracy of temperature measurements in certain situations. Example 9.11 illustrates this.

## Example 9.11
### Radiation shield for temperature sensor

*Problem (a)*

A thermocouple is used to measure the temperature of hot air flowing through a large duct. The emissivity of the thermocouple is 0.6 and the duct walls are at a temperature of 500 K. The convective coefficient at the thermocouple's surface is 100 W/m² C. The thermocouple indicates an air temperature of 1025 K. What is the actual temperature of the air?

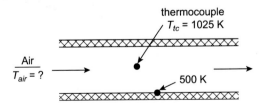

*Solution (a)*

From an energy balance on the thermocouple at steady state:
The heat gained by convection from the air equals the heat lost by radiation to the duct walls. That is,

$$hA(T_{air} - T_{tc}) = \varepsilon\sigma A(T_{tc}^4 - T_{duct}^4)$$

Putting in values, we have

$$100(T_{air} - 1025) = 0.6(5.67 \times 10^{-8})(1025^4 - 500^4)$$

Solving this we get $T_{air} = 1379$ K.

The actual air temperature is 1379 K, but the thermocouple indicates a temperature of 1025 K. This difference of 354 K is due to the duct walls being at a much lower temperature than the air There is radiative transfer from the thermocouple to the duct walls, thereby lowering the thermocouple's temperature. The relatively high emissivity of the thermocouple also contributes to the large difference between the measured and actual air temperature.

*Problem (b)*

A thin cylindrical shield is placed over the thermocouple to shield the thermocouple from the duct walls. The shield has an emissivity of 0.1 on both of its surfaces, and the convective coefficient is 50 W/m² C on both surfaces of the shield. Determine the effect of the shield on the accuracy of the thermocouple reading.

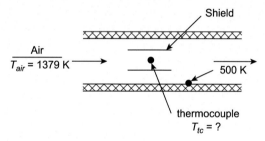

*Solution (b)*

We will realistically assume that the surface area of the thermocouple is very small compared with the inner surface area of the shield and that the outer surface area of the shield is small compared with the surface area of the duct walls.
The heat balance on the shield is

$$\varepsilon_{tc}\sigma A_{tc}(T_{tc}^4 - T_s^4) + 2h_s A_s(T_{air} - T_s) = \varepsilon_s\sigma A_s(T_s^4 - T_{duct}^4)$$

where $A_s$ is the surface area of one side of the shield.
Dividing this equation by area $A_s$, we have

$$\varepsilon_{tc}\sigma(A_{tc}/A_s)(T_{tc}^4 - T_s^4) + 2h_s(T_{air} - T_s) = \varepsilon_s\sigma(T_s^4 - T_{duct}^4)$$

As $A_{tc} \ll A_s$, the first term of this equation can be deleted, which leaves us with

$$2h_s(T_{air} - T_s) = \varepsilon_s\sigma(T_s^4 - T_{duct}^4)$$

Putting in values, we have $2(50)(1379 - T_s) = 0.1(5.67 \times 10^{-8})(T_s^4 - 500^4)$.
Solving for shield temperature $T_s$, we get $T_s = 1246$ K.
The heat balance on the thermocouple is

$$h_{tc}A_{tc}(T_{air} - T_{tc}) = \varepsilon_{tc}\sigma A_{tc}(T_{tc}^4 - T_s^4)$$

Putting in values, we have $(100)(1379 - T_{tc}) = (0.6)(5.67 \times 10^{-8})(T_{tc}^4 - 1246^4)$.
Solving for the thermocouple temperature, we get $T_{tc} = 1281$ K.
The actual air temperature is 1379 K and the indicated temperature by the thermocouple is 1281 K, giving an error of 98 K.
Use of the radiation shield reduced the thermocouple error from 354 to 98 K. This is a very significant improvement.

## 9.7 Sky radiation and solar collectors

Radiant emission from the sun has a spectral distribution similar to that of a blackbody at about 5800 K. The sun radiates about $3.8 \times 10^{23}$ kW of energy, of which about $1.7 \times 10^{14}$ kW is intercepted by the earth. At the outer edge of the earth's atmosphere, the solar irradiance, at normal incidence, is about 1360 W/m². This is called the "Solar Constant." As the radiation goes through the atmosphere, its intensity is decreased due to absorption and scattering by gases, particulates, and water vapor. On clear, sunny days, the solar insolation at the surface of the earth can be as high as about 1000 W/m². It is much lower on cloudy or smoggy days.

Let us now briefly discuss solar collectors. A solar collector will typically have a cover plate in contact with the environment and an absorber plate deeper within the device. We want as much radiation as possible to enter the collector and the least possible amount to leave it. For the cover plate, therefore, we want high transmittance for solar radiation and a low transmittance for radiation at the infrared wavelengths of the collector. Fortunately, several plastics and glasses have these characteristics, including polycarbonate, plexiglass, clear low-iron oxide glass. For float glass and tempered glass, the solar transmittance is about 0.8 and the infrared transmittance is estimated at 0.2 [1].

For the absorber plate, we want high absorptance and low emittance. This is often achieved by putting coatings on the absorber plates. One very good coating is black chrome on copper, aluminum, or stainless steel. This coating has an absorptance of about 0.87 and an emittance of about 0.08. The coating must be commercially applied. However, there are spray coatings available, which can be applied by the user, which also have high absorptance and low emittance. One might think that flat black paint would be a very good coating for an absorber plate. Such is not the case. Indeed, the paint has a high absorptance (about 0.95), but it unfortunately also has a high emittance (about 0.9) [1].

In predicting the performance of solar collectors and systems, it is necessary to have data on the amount of solar radiation reaching a particular location. Such data are available from the National Renewable Energy Laboratory in Golden, Colorado. One of its publications [2] gives the solar radiation incident on many US cities. The publication gives incident solar radiation (kWh/m²/day) for several types of solar collectors: flat-plate collectors (fixed, 1-axis tracking, 2-axis tracking) and concentrating collectors. As an example for New York City, for a fixed flat-plate collector facing south, the average solar radiation is 1.6 kWh/m² for a day in December and 6.0 kWh/m² for a day in June.

We have mentioned scattering and absorption of solar radiation by gases and particulates. Such materials also emit radiation. The spectral emission is unlike blackbody radiation, but it is often useful to treat the atmosphere as a blackbody having a temperature called the "effective sky temperature." This temperature depends on parameters such as altitude, humidity, cloud coverage, and particles in the air such as dust and pollution. The sky temperature is lower than the air temperature at the earth's surface. The sky temperature is typically between about 230 and 290 K. The lower temperatures are for cold, cloudless conditions while the higher temperatures are for warm, cloudy conditions.

Discussion of an interesting (and fun) phenomenon closes this chapter. The author has observed ice in the street in the morning when the air temperature overnight has been above freezing. This can be explained by radiative heat transfer. The phenomenon is illustrated in Example 9.12.

## Example 9.12
**Water freezing in street**

*Problem*

There are shallow puddles of water in the street during daytime. At night, the average air temperature is 5 C and the effective sky temperature is 250 K. The convective coefficient between the water and the air is 10 W/m$^2$ C and the emissivity of the water surface is about 0.95. Will the water freeze overnight?

*Solution*

For equilibrium:

Heat into the water by convection from the air = heat out of the water by radiation to the sky

$$hA(T_{air} - T_{water}) = \varepsilon\sigma A\left(T_{water}^4 - T_{sky}^4\right)$$

Putting in values, we have

$$(10)(278.15 - T_{water}) = (0.95)\left(5.67 \times 10^{-8}\right)\left(T_{water}^4 - 250^4\right)$$

Solving for the water temperature, we get $T_{water} = 270.4$ K. As $T_{water} < 273.15$ K, the water will freeze.

## 9.8 Chapter summary and final remarks

In this chapter, we discussed the fundamentals of radiation heat transfer. We first considered emission from blackbodies. We looked at the wavelength distribution of the emission as given by Planck's law. We discussed Wien's law and its importance with respect to the "greenhouse effect." We defined radiation properties (emissivity, reflectivity, absorptivity, and transmissivity), irradiation, radiosity, and the gray body. We discussed radiation shape factors that are used in determining the radiant heat transfer between surfaces. We covered radiant exchange between gray, diffuse surfaces of an enclosure, with special emphasis on two- and three-surface enclosures. We also observed the usefulness of electrical circuit diagrams in determining the equations for such radiant exchange. We discussed the use of radiation shields to decrease radiative transfer between surfaces. Finally, we briefly discussed sky radiation and solar collectors.

In short, we covered a lot of material. However, there are more-advanced topics of radiation heat transfer that were not covered. These include the angular dependence of radiation properties, radiative exchange for specular surfaces, gaseous radiation, and radiation transmission through media that absorb and scatter radiation. Information on these topics may be obtained from the references in the "Further reading" section at the end of the chapter.

## 9.9 Problems

Notes:

- If needed information is not given in the problem statement, then use the Appendix for material properties and other information. If the Appendix does not have sufficient information, then use the Internet or other reference sources. Some good reference sources are mentioned at the end of Chapter 1. If you use the Internet, double-check the validity of the information by using more than one source.
- Your solutions should include a sketch of the problem.
- Unless specifically mentioned, the problems only involve radiant heat transfer, and there is no convection heat transfer.

**9-1** What is the total emissive power of a blackbody at 1500 C?

**9-2** What percentage of the emission from a blackbody at 2000 K lies in the wavelength band of 2 to 6 μm?

**9-3** What is the rate of emission from a 85 F Gy body with $\varepsilon = 0.5$ between wavelengths 3 to 15 μm?

**9-4** Radiation from a 600 C blackbody is incident on a window. The window transmits 85% of incident radiation between the wavelengths 0.3 to 3.5 μm. What heat flux $(W/m^2)$ will be transmitted through the window?

**9-5** A surface at 1500 K has the following spectral emissivities:
For $0 < \lambda < 1.5$ μm, $\varepsilon_\lambda = 0.2$,
For $1.5 < \lambda < 4$ μm, $\varepsilon_\lambda = 0.45$,
For $4 < \lambda < \infty$ μm, $\varepsilon_\lambda = 0.65$.
**(a)** What is the total emissivity $\varepsilon$ of the surface?
**(b)** What is the emissive power of the surface?

**9-6** Obtain Wien's law (Eq. 9.6) by differentiating Eq. (9.4) with respect to wavelength $\lambda$ and setting the result to zero.

**9-7** A square black surface is at 800 C. It is 40 cm by 40 cm.
**(a)** What is the rate (W) at which the surface emits thermal radiation?
**(b)** What is the spectral emissive power of the surface at a wavelength of 8 μm?

**9-8** Three possible glass cover plates are available for a solar collector. Plate A has a transmissivity of 0.85 for wavelengths of 0.25 to 1.5 μm and zero transmission for wavelengths outside this range. Plate B has a transmissivity of 0.95 for wavelengths of 0.4 to 1 μm and zero transmission for wavelengths outside this range. Plate C has a transmissivity of 0.72 for all wavelengths. Which plate will have the greatest transmission of solar radiation into the collector? Assume the sun is a blackbody at 5800 K.

**9-9** A glass window has a transmissivity of 0.87 for wavelengths between 0.2 and 3.5 μm. It is opaque outside this wavelength range. For a blackbody source at 1500 K, what is the radiant heat flux transmitted through the window?

**9-10** A thermos bottle consists of two thin concentric cylindrical walls with a vacuum between them. The inner cylinder has a diameter of 10 cm and the outer cylinder has a diameter of 11.5 cm. The thermos is 30 cm high. The inner cylindrical container holds coffee at 70 C, and the outer cylinder is essentially at the room temperature of 20 C. Both wall surfaces facing the vacuum space have an emissivity of 0.1. What is the rate of heat transfer from the coffee?

**9-11** A thermistor temperature sensor is spherical in shape, 4 mm diameter, and is used to measure the temperature of dry air flowing in a 25 cm diameter pipe. The thermistor's surface has an emissivity of 0.9 and the emissivity of the pipe wall is 0.8. The convective coefficient between the thermistor and the air is 50 $W/m^2$ C. The pipe wall is at 250 C, and the sensor's temperature is 540 C. What is the temperature of the air?

**9-12** A flat-plate solar collector has an absorptivity of 0.91 and an emissivity of 0.12. The plate is square, 5 m by 5 m, and the convective coefficient at the plate's surface is 4 $W/m^2$ C. The solar insolation on the plate is 750 $W/m^2$, the plate's temperature is 60 C, and the air temperature is 16 C.
**(a)** What is the rate of energy absorbed by the plate?
**(b)** What is the plate's collection efficiency?

**9-13** A gray body having area $A_1 = 1.5$ m$^2$ and emissivity $\varepsilon_1 = 0.8$ is being heated in a rectangular furnace having dimensions 3 m by 3.5 m by 2.5 m. The inside surface of the furnace has an emissivity $\varepsilon_2 = 0.65$. The surface of the body has a temperature of 750 C, and the inside surface of the furnace has a temperature of 250 C. The shape factor between the body's surface and the enclosure is 1.

(a) What is the rate of radiant heat transfer from the gray body to the enclosure?

(b) What would be the % error if we had assumed that $A_1 \ll A_2$ in calculating the solution to Part (a)?

**9-14** A furnace is cylindrical in shape: 2 m long and 1.5 m diameter. There is a window at one end of the furnace. The window is on the axis of the furnace and is 25 mm diameter. A pyranometer placed at the window receives 2.3 W of radiation from the opposite end of the furnace. The temperature of the pyranometer is 60 C. What is the temperature of that opposite end of the furnace?

**9-15** A cubic oven has inside dimensions 0.45 m by 0.45 m by 0.45 m. The emissivity of all six walls of the oven is 0.7. One wall of the oven is at 300 C. The other walls of the oven are at 150 C. What is the rate of radiant heat transfer from the 300 C wall?

**9-16** A square room has floor and ceiling dimensions of 3.5 m by 3.5 m and a height of 2.5 m. The floor is at 300 K, the ceiling is at 285 K, and the walls are perfectly insulated. The emissivity of all surfaces is 0.7.

(a) What is the rate of radiant heat transfer from the floor?

(b) What is the rate of radiant heat transfer to the ceiling?

(c) What is the temperature of the walls?

**9-17** An environmental testing chamber is rectangular, with a 4 m by 5 m floor and walls 3 m high. One 5 m by 3 m side wall is at 250 C and the floor is at 80 C. The other four surfaces of the chamber are perfectly insulated. If all the surfaces of the chamber are black, what is the net radiant heat transfer between the heated wall and the floor?

**9-18** A long horizontal pipe 10 cm in diameter runs through a space with air at 20 C. The pipe surface is at 90 C and has an emissivity of 0.65. What is the radiant heat loss from the pipe per meter length?

**9-19** A horizontal pipe of 6.35 cm outer diameter and 8 m length runs through a large room where the air temperature is 20 C. The outer surface of the pipe is at 110 C. The walls of the room are at 25 C and have an emissivity of 0.3. The convective coefficient at the outer surface of the pipe is 3 W/m$^2$ C and the surface has an emissivity of 0.5. What is the total rate of heat loss from the pipe by convection and radiation?

**9-20** A long solid cylinder has a diameter of 2.5 cm. Its temperature is 500 C and its surface has an emissivity of 0.3. A long thin-walled cylinder is placed concentrically over the first cylinder. This second cylinder has a diameter of 6 cm and an emissivity of 0.15 on both of its surfaces. The two-cylinder arrangement is in a vacuum chamber whose surfaces are at 28 C.

(a) What is the rate of energy lost by the inner cylinder?

(b) What is the equilibrium temperature of the outer cylinder?

**9-21** Energy is being transferred from one spaceship to another. This is done by two square plates 2 m by 2 m; one plate on each spaceship. The ships are maneuvered so that the plates are parallel and 40 cm apart. One plate has emissivity 0.6 and temperature 800 C. The other plate has emissivity 0.85 and temperature 250 C. Assume that the space environment is equivalent to a blackbody at 0 K.

**(a)** What is the net rate of energy transferred from one spaceship to the other?

**(b)** What is the rate of energy loss from the hot plate?

**9-22** A vertical cylindrical furnace has dimensions of 80 cm diameter by 50 cm high. A heater in the bottom of the furnace produces a heat input of 3000 W to the furnace. The bottom surface of the furnace has an emissivity of 0.8. The top surface of the furnace has an emissivity of 0.55 and is maintained at 425 K. The side wall of the furnace is nonconducting and reradiating. What are the temperatures of the bottom and the side of the furnace?

**9-23** Two steel plates are welded together, perpendicular to each other, with a common edge. Plate No. 1 is 25 cm by 50 cm and has a temperature of 600 C and an emissivity of 0.55. Plate No. 2 is 40 cm by 50 cm, has an emissivity of 0.4, and is perfectly insulated on its back side. The common edge between the plates is 50 cm long. The inner surfaces of the plates face into a large room at 25 C.

**(a)** What is the rate of heat lost by Plate No. 1?

**(b)** What is the temperature of Plate No. 2?

**9-24** Two large plates are parallel and facing each other. One plate has temperature 1000 K and emissivity 0.3. The other plate has temperature 400 K and emissivity 0.6. A thin radiation shield having an emissivity of 0.04 on both sides is placed between the two plates.

**(a)** What is the heat flux (W/m²) between the plates without the shield?

**(b)** What is the heat flux with the shield in place?

**(c)** What is the temperature of the shield when it is between the plates?

**9-25** A long duct has a rectangular cross section. The height is 30 cm and the width is 50 cm. The top surface has a heat flux of 1000 W/m² and an emissivity of 0.3, and the bottom is perfectly insulated. One side wall has temperature 250 C and emissivity 0.45. The other side wall has temperature 150 C and emissivity 0.7.

**(a)** What are the temperatures of the top and bottom surfaces?

**(b)** What are the heat fluxes for the side walls?

**9-26** An environmental test chamber is cubic in shape, 7 m by 7 m by 7 m. The floor is electrically heated, with a uniform heat flux of 500 W/m² and emissivity 0.8. The ceiling is perfectly insulated. The side walls are all at 300 K and have an emissivity of 0.3.

**(a)** What are the temperatures of the floor and ceiling?

**(b)** What are the heat fluxes for the four side walls?

**9-27** A long duct has a cross section of an equilateral triangle. The sides of the triangle are 1.5 feet long. One side of the duct is perfectly insulated. Another side has emissivity 0.2 and temperature 400 F. The third side has emissivity 0.5 and temperature 800 F.

**(a)** What is the temperature of the insulated side?

**(b)** What are the heat flows per foot length for the two other sides?

**9-28** A room is rectangular in shape, with the floor and ceiling being 3 m by 5 m and the height being 2.5 m. The floor is electrically heated and has a temperature of 35 C and an emissivity of 0.75. The side walls are perfectly insulated and have an emissivity of 0.8. The ceiling has an emissivity of 0.9. It is poorly insulated and has a temperature of 15 C.

**(a)** What is the temperature of the side walls?

**(b)** What is the required power input to the floor heater to keep the floor at 35 C?

**9-29** Two very long parallel plates are facing each other. Plate 1 is at 600 K and the surface facing Plate 2 has an emissivity of 0.6. Plate 2 is at 400 K and its surface facing Plate 1 has an emissivity of 0.3. A thin radiation shield (Plate 3) is placed between the other two plates. This plate has sides A and B, which are of different emissivities. When side A faces Plate 1, Plate 3 has a temperature of 520 K. When side B faces Plate 1, Plate 3 has a temperature of 567 K.
(a) What is the heat flux between Plates 1 and 2 without the radiation shield in place?
(b) What are the emissivities of sides A and B of the radiation shield?
(c) When side A of the shield is facing Plate 1, what is the heat flux for the three-plate system?
(d) When side B of the shield is facing Plate 1, what is the heat flux for the three-plate system?

**9-30** A plane surface having temperature $T_1$ and emissivity $\varepsilon_1$ is radiating to a large enclosure at temperature $T_2$. A thin, plane radiation shield is placed between the plane surface and the enclosure's surface. The side of the shield facing the plane surface has emissivity $\varepsilon_2$ and the side of the shield facing the enclosure's surface has emissivity $\varepsilon_3$.
(a) Show that, with the shield in place, the heat flux from the plane surface to the enclosure is

$$\frac{q_{1-2}}{A} = \frac{\sigma\left(T_1^4 - T_2^4\right)}{(1/\varepsilon_1) + (1/\varepsilon_2) + (1/\varepsilon_3) - 1}$$

(b) If $\varepsilon_1 = 0.4$, $\varepsilon_2 = 0.15$, $\varepsilon_3 = 0.05$, $T_1 = 500\,C$, and $T_2 = 25\,C$, what is the heat flux from the plane surface to the enclosure and what is the temperature of the shield?

**9-31** A rectangular oven has a floor and ceiling 3.75 m by 5.25 m and walls of 4 m height. The upper half of the four walls are radiant heaters having temperature 600 C and emissivity 0.9. The floor has a temperature of 100 C and an emissivity of 0.8. The ceiling and the lower unheated halves of the walls are perfectly insulated. What is the rate of radiant heat transfer to the floor?

**9-32** Two concentric, parallel disks having diameters of 0.5 m are 10 cm apart and are in a large room that is at 20 C. One disk has a temperature of 250 C and an emissivity of 0.7. The other disk is perfectly insulated on its back side.
(a) What is the rate of radiant heat transfer from the heated disk?
(b) What is the temperature of the insulated disk?

**9-33** A long electric cylindrical heater has a diameter of 1.5 cm and produces 40 W per meter length. It is on the centerline of a concentric tube of 2.5 cm inside diameter that is at 30 C. The space between the heater and the tube is evacuated so there is only radiant heat transfer. The surface of the heater has an emissivity of 0.8 and the inner surface of the tube has an emissivity of 0.7. What is the surface temperature of the heater?

**9-34** A thin-walled spherical tank has an outer diameter of 1 m and is filled with liquid nitrogen at 77 K. The tank is enclosed within another spherical tank of inner diameter 1.04 m, and the space between the tanks is a vacuum so that radiative heat transfer is the only mode of heat transfer between the tanks. The emissivities of the outer surface of the inner tank and the inner surface of the outer tank are both 0.08. The temperature of the outer tank is 20 C.
(a) What is the rate of heat transfer to the LN$_2$?
(b) What is the rate of evaporation of the LN$_2$ if the heat of vaporization of the liquid nitrogen is 200 kJ/kg?

**9-35** A long steam pipe passes through the center of a long rectangular duct. The outer surface of the pipe is at 200 C and has an emissivity of 0.8. The duct walls are at 25 C and have an emissivity of 0.6. The OD of the pipe is 10 cm and the duct cross section is 45 cm by 60 cm. What is the radiation loss from the pipe per meter length?

**9-36** An incandescent light bulb has a tungsten filament that can be modeled as a gray body at 2800 K. What fraction of the radiant energy emitted by the filament is in the visible wavelengths of 0.4–0.8 μm?
(Note: This problem shows the inefficiency of incandescent light bulbs. Most of the energy emitted by the tungsten filament of the bulb is not in the visible region of the electromagnetic spectrum. Rather, it is in the infrared region.)

**9-37** Two concentric, parallel disks are radiating between themselves and the space between them. The disks are 0.3 m apart. Disk 1 has a diameter of 0.7 m, temperature 700 K, and emissivity 0.9. Disk 2 has a diameter of 0.5 m, temperature 450 K, and emissivity 0.6. The space between them may be considered as a black medium at 300 K.
**(a)** What is the rate of radiative heat transfer from Disk 1 to Disk 2?
**(b)** What is the rate of heat transfer to the environment between the disks?

**9-38** A thin flat plate is 1 m by 1 m. One side of the plate is exposed to solar radiation and also convection to the adjacent 300 K environment with a convective coefficient of 15 W/m² C. The other side of the plate is perfectly insulated. The side of the plate facing the sun has an absorptivity of 0.7 for long-wavelength radiation and an absorptivity of 0.93 for solar radiation. If the solar irradiation of the plate is 1000 W/m², what is the temperature of the plate?

**9-39** A shallow pan of water is placed outdoors on a night when the effective sky temperature is 230 K. The air temperature is 5 C, which is above the freezing temperature of 0 C for water. The water's surface has a convective coefficient of 30 W/m² C and an emissivity of 0.8. Will the water in the pan freeze?

**9-40** An object has an emissivity of 0.5 between wavelengths 0.1 and 3 μm and an emissivity of 0.2 between wavelengths 3 and 10 μm. If the object is at 4000 K, what is its total emissive power?

**9-41** Regular window glass has a transmissivity of 0.9 for wavelengths from 0.2 to 3 μm. Tinted glass has a transmissivity of 0.9 for wavelengths from 0.5 to 1 μm. Assume that the transmissivity outside these wavelength ranges is zero. Consider the sun to be a blackbody at 5800 K.
**(a)** What is the fraction of incident solar energy that is transmitted through each glass?
**(b)** What is the fraction of incident solar energy in the visible range of 0.4 to 0.8 μm that is transmitted through each glass?

**9-42** Hot air is flowing through a 15 cm diameter duct. A thermistor is used to measure the air temperature. The thermistor is spherical in shape. It has a diameter of 2 mm and an emissivity of 0.8. The surface of the thermistor has a convective coefficient of 100 W/m² C. The surface of the duct has a temperature of 150 C and an emissivity of 0.7.
**(a)** If the thermistor indicates a temperature of 325 C, what is actual temperature of the air?
**(b)** It is decided to place a radiation shield between the thermistor and the duct wall to decrease the error between the thermistor reading and the actual air temperature. The shield is a thin-walled cylinder of 2 cm diameter and 4 cm length. Both sides of the shield have an emissivity of 0.15 and a convective coefficient of 80 W/m² C. Assuming that the shield does not affect the convective coefficient for the thermistor, what will be the thermistor reading for the actual gas temperature found in Part (a)?

**9-43** A hemispherical furnace has a flat bottom surface of 3 m diameter. The curved top surface has a temperature of 800 C and an emissivity of 0.9. The bottom surface has a temperature of 100 C and an emissivity of 0.6. What is the rate of radiative heat transfer from the top surface to the bottom surface?

**9-44** A horizontal disk has a diameter of 8 cm. There is a horizontal ring parallel to and above the disk. The ring is concentric with the disk and 10 cm above it. The ring has an inner diameter of 12 cm and an outer diameter of 14 cm. If the disk is Surface 1 and the ring is Surface 2, what is the radiation shape factor $F_{1-2}$?

**9-45** An evacuated multilayered insulation is used for a liquid nitrogen tank. The outer layers of the insulation are aluminum with an emissivity of 0.15. They are at temperatures 80 and 300 K. There are four internal polished aluminum shields, each of which has an emissivity of 0.04 on both sides. The shields are 1.5 mm apart and the total thickness of the insulation is 7.5 mm.
(a) What is the radiant heat transfer across the insulation?
(b) What is the effective conductivity and R-value for the insulation?

**9-46** A long duct has a square cross section, 45 cm on a side. The four inside surfaces of the duct have different boundary conditions as follows: Surface 1 has a temperature of 750 K and an emissivity of 0.75. Surface 2 is adjacent to Surface 1 and has a temperature of 500 K and an emissivity of 0.6. Surface 3, which is across from Surface 1, has a heat flux of 450 W/m$^2$ and an emissivity of 0.4. Surface 4, which is across from Surface 2, is perfectly insulated.
(a) What are the heat flows per meter length of duct for Surfaces 1 and 2?
(b) What are the temperatures of Surfaces 3 and 4?

**9-47** A solar collector has a metal absorber plate in contact with the atmospheric air with water tubes directly behind the absorber plate. Solar radiation is incident on the plate at a rate of 700 W/m$^2$. The solar absorptance of the plate is 0.89 and the emittance is 0.12. The adjacent air is at 22 C and the effective sky temperature is 15 C. The convective coefficient for the plate is 8 W/m$^2$ C. The absorber plate's temperature is 65 C. What is the rate of heat transfer to the tubes behind the absorber plate per square meter of plate surface area?

**9-48** The solar collector of Problem 9-47 is not in service and no heat is being removed by the water in the tubes behind the plate. What is the temperature of the absorber plate?

**9-49** A glass window has a transmissivity of 0.91 for wavelengths of 0.25 to 4 μm. It has zero transmissivity for other wavelengths.
(a) What is the total transmissivity of the window for solar radiation? Assume blackbody radiation at 5800 K.
(b) What is the total transmissivity of the glass for room temperature radiation? Assume blackbody radiation at 300 K.
(c) If solar radiation is incident on the window at a rate of 700 W/m$^2$, what is the rate of radiation transmission through the window?

**9-50** Radiation incident on a surface has the following intensity:

$$\begin{aligned}
G_\lambda = 400 \text{ W/m}^2\mu\text{m} && \text{for } 0 < \lambda < 2 \ \mu\text{m}\\
1500 \text{ W/m}^2\mu\text{m} && \text{for } 2 \ \mu\text{m} < \lambda < 4 \ \mu\text{m}\\
600 \text{ W/m}^2\mu\text{m} && \text{for } 4 \ \mu\text{m} < \lambda < 7 \ \mu\text{m}\\
0 && \text{for } \lambda > 7 \ \mu\text{m}
\end{aligned}$$

The spectral absorptivity of the surface is

$$\alpha_\lambda = 0 \qquad \text{for } 0 < \lambda < 3 \ \mu\text{m}$$
$$0.85 \qquad \text{for } 3 \ \mu\text{m} < \lambda < 5 \ \mu\text{m}$$
$$0.12 \qquad \text{for } \lambda > 5 \ \mu\text{m}$$

**(a)** What is the total absorptivity of the surface?
**(b)** What is the rate of radiant energy absorption by the surface for the given incident radiation (W/m$^2$)?

**9-51** A kiln has a 7 cm diameter window in one of its walls. The kiln environment is at 1800 C and may be assumed to be black. The room environment is at 23 C and may be assumed to be black. The window has the following radiation properties:

$$\text{For } \lambda < 3 \ \mu\text{m}, \quad \varepsilon_\lambda = 0.1, \quad \tau_\lambda = 0.88, \quad \rho_\lambda = 0.02$$
$$\text{For } \lambda > 3 \ \mu\text{m}, \quad \varepsilon_\lambda = 0.6, \quad \tau_\lambda = 0.15, \quad \rho_\lambda = 0.25$$

**(a)** Assuming conduction and convection are negligible, what is the temperature of the window?
**(b)** What is the net radiant heat transfer through the window from the furnace to the room?

**9-52** It is desired to pick the glass for the cover of a solar collector that has the greatest transmission of solar radiation. The sun is modeled as a blackbody at 5800 K. The two options are as follows:

Glass A

$$\tau_\lambda = 0 \qquad \text{for } 0 < \lambda < 0.3 \ \mu\text{m}$$
$$0.9 \qquad \text{for } 0.3 \ \mu\text{m} < \lambda < 3 \ \mu\text{m}$$
$$0 \qquad \text{for } \lambda > 3 \ \mu\text{m}$$

Glass B

$$\tau_\lambda = 0 \qquad \text{for } 0 < \lambda < 0.5 \ \mu\text{m}$$
$$0.95 \qquad \text{for } 0.5 \ \mu\text{m} < \lambda < 5 \ \mu\text{m}$$
$$0 \qquad \text{for } \lambda > 5 \ \mu\text{m}$$

Which glass should be chosen?

**9-53** It is desired to pick the coating for the absorber plate of a solar collector that has the greatest absorption of solar radiation. The sun is modeled as a blackbody at 5800 K. The three options are as follows:

Coating A

$$\alpha_\lambda = 0.9 \qquad \text{for } 0 < \lambda < 3 \ \mu\text{m}$$
$$0.15 \qquad \text{for } 3 \ \mu\text{m} < \lambda < 8 \ \mu\text{m}$$
$$0 \qquad \text{for } \lambda > 8 \ \mu\text{m}$$

Coating B

$$\alpha_\lambda = 0.15 \qquad \text{for } 0 < \lambda < 3 \ \mu m$$
$$0.9 \qquad \text{for } 3 \ \mu m < \lambda < 8 \ \mu m$$
$$0 \qquad \text{for } \lambda > 8 \ \mu m$$

Coating C

$$\alpha_\lambda = 0.72 \ \text{ for all wavelengths}$$

Which coating should be chosen?

# References

[1] F. Kreith, A. Rabl, Solar energy, in: W.M. Rohsenow, J.P. Hartnett, E.N. Ganic (Eds.), Handbook of Heat Transfer Applications, second ed., McGraw-Hill, 1985. Chapter 7.
[2] Solar Radiation Data Manual for Flat-Plate and Concentrating Collectors, National Renewable Energy Laboratory (Golden, CO), NREL/TP-463-5607, April 1994. Available online, rredc.nrel.gov/solar/pubs/redbook/.

# Further reading

[1] R. Siegel, J.R. Howell, Thermal Radiation Heat Transfer, third ed., Hemisphere, 1993.
[2] E.M. Sparrow, R.D. Cess, Radiation Heat Transfer, Hemisphere, 1978.
[3] H.C. Hottel, in: W.C. McAdams (Ed.), Heat Transmission, third ed., McGraw-Hill, 1954. Chpt. 4.
[4] H.C. Hottel, R.B. Egbert, Radiant heat transmission from water vapor, AIChE Trans 38 (1942) 531–565.
[5] C.L. Tien, Thermal radiation properties of gases, Adv. Heat Tran. 5 (1968) 254–321.
[6] D.K. Edwards, R. Matovosian, Scaling rules for total absorptivity and emissivity of gases, J. Heat Transf. 106 (1984) 685–689.
[7] E.R.G. Eckert, Radiation relations and properties, in: W.M. Rohsenow, J.P. Hartnett (Eds.), Handbook of Heat Transfer, Section 15, McGraw-Hill, 1973.

# Multimode heat transfer

## 10.1 Introduction

Like essentially all heat transfer texts, the organization of this text is segmental. The first chapter is general, and presents an introduction to heat transfer and its three modes: conduction, convection, and radiation. Next are three chapters concentrating on conduction. There are two later chapters on convection, and then a chapter on radiation. This organization is very good from a pedagogical viewpoint. However, concentrating on a single mode at a given time is not very realistic. Most problems involve more than one mode of heat transfer, and these modes act simultaneously. For example, heat transfer through a building wall involves conduction through the wall, convection at the surfaces of the wall, and sometimes radiation from the wall surfaces, all acting together.

At the time of the early chapters on conduction, the details of convection had not yet been discussed. Therefore, values of convection coefficients had to be given in the problem statements. Examples and problems in this chapter are designed to be much more realistic, and convective coefficients are not provided. Indeed, for real engineering problems, an engineer has to determine the convective coefficients by his/her self. If quick, approximate answers are sufficient, the engineer can use estimates based on prior experience. If more accurate results are needed, then the appropriate equations can be used to determine the coefficients. At this point in the text, the reader has the capability to determine convective coefficients by his/her self. Problems can be more like "real world" problems.

The problems in this chapter involve multiple modes of heat transfer. The reader has to determine which modes of heat transfer are relevant for a given problem. He/she also has to somehow obtain any needed information that is lacking in the problem statement. In addition, a decision must be made on the method to be used for the solution of the problem.

Good engineers are resourceful and persistent. This chapter should enhance the reader's problem-solving abilities. It should also increase the reader's confidence in being able to solve heat transfer problems. Finally, the examples and problems in this chapter should provide the reader with a worthwhile review of material covered earlier in the book.

## 10.2 Procedure for solution of multimode problems

In solving the problems, the following questions must be answered:

What modes of heat transfer (conduction, convection, and/or radiation) are significant?

If convection is significant, is it natural convection or is it forced convection?

Is the problem steady-state or transient?

Does the problem statement contain all necessary information? If not, where will the missing information be obtained?

How should the problem be solved? Should it be solved analytically or numerically? Does the problem involve differential equations? Does the problem involve simultaneous equations? What software, if any, should be used in the solution?

After these questions are answered, the solution can proceed in earnest.

## 10.3 Examples

The following eight examples each incorporate more than one mode of heat transfer. The examples are based on earlier examples and problems in this book. At the time they were earlier discussed, concentration was focused on one specific mode of heat transfer. For example, Chapters 2, 3, and 4 dealt with conduction. For problems in these chapters involving convection boundary conditions, the values of the convective coefficients were givens in the problems. Now, with the experience of Chapters 6 and 7, the reader has the ability to determine convective coefficients by his/her self. There is no need for the convective coefficients to be given.

The below examples involve the solution of nonlinear simultaneous equations. We have found that Microsoft Excel's *Solver* add-in is very useful. (This add-in program comes with Excel. It just has to be enabled.) Sometimes convergence depends on the initial guesses of the variables. However, we have found *Solver* to be successful in almost all situations. We have provided details regarding the Excel spreadsheets used in the solution of the examples. We have also provided listings of the Matlab programs used for solution of some of the examples.

Property information was needed for the determination of forced and natural convective coefficients for air. In particular, we needed the temperature variation of air's thermal conductivity, kinematic viscosity, and Prandtl number. We took tabular data for these parameters and fit the data to polynomial functions. This was done through use of Excel spreadsheets. Tabular data were input to the spreadsheet, scatter charts were inserted, and polynomial trendlines and equations were obtained. We found that third-degree polynomial fits were excellent for thermal conductivity and kinematic viscosity. A fourth-degree polynomial was needed to accurately fit the data for the Prandtl number. The obtained equations were:

Thermal conductivity of air (0–1000 C):

$k(W/m\ C)$ and $T(C)$: $k = 4.629 \times 10^{-12}T^3 - 2.520 \times 10^{-8}T^2 + 7.561 \times 10^{-5}T + 0.02364$

$$(10.1)$$

$k(W/mK)$ and $T(K)$: $k = 4.629 \times 10^{-12}T^3 - 2.899 \times 10^{-8}T^2 + 9.041 \times 10^{-5}T + 0.001009$

$$(10.2)$$

Kinematic viscosity of air (0–1000 C):

$v(m^2/s)$ and $T(C)$: $v = -1.725 \times 10^{-14}T^3 + 8.853 \times 10^{-11}T^2 + 8.947 \times 10^{-8}T + 1.329 \times 10^{-5}$

$$(10.3)$$

$v(m^2/s)$ and $T(K)$: $v = -1.725 \times 10^{-14}T^3 + 1.027 \times 10^{-10}T^2 + 3.725 \times 10^{-8}T - 4.196 \times 10^{-6}$

$$(10.4)$$

Prandtl number of air (0–1000 C):

$T(C)$: $Pr = 2.916 \times 10^{-13}T^4 - 8.291 \times 10^{-10}T^3 + 8.653 \times 10^{-7}T^2 - 3.387 \times 10^{-4}T + 0.7371$

$$(10.5)$$

$T(K)$: $Pr = 2.916 \times 10^{-13}T^4 - 1.148 \times 10^{-9}T^3 + 1.675 \times 10^{-6}T^2 - 1.021 \times 10^{-3}T + 0.9126$

$$(10.6)$$

## Example 10.1
### Cylindrical heater

*Problem*

A horizontal electric cylindrical heater has an output of 50 W. The heater has a diameter of 4 cm and a length of 30 cm. The ends of the cylinder are insulated so all of the heat output leaves through the cylindrical side of the heater. The heater is in a large room whose walls are at 30 C, and the room air is at 20 C. The surface of the heater has an emissivity of 0.5.
(a) What is the temperature of the surface of the heater?
(b) What are the respective heat flows by convection and radiation from the surface of the heater?

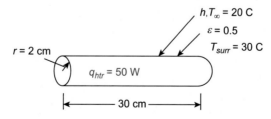

*Solution*

Preliminary: There is convection and radiation at the surface of the cylinder. The convection is natural convection. We have steady state. All necessary information is given. The heater output equals the convection and radiation from the surface.

Because radiation is involved, temperatures must be in absolute temperature units. Eqs. (10.2), (10.4), (10.6) will be used for the properties of air.

The energy conservation equation at the cylinder's surface is

Heater output $=$ Natural convection and radiation from the surface

$$q = 50 \text{ W} = hA(T_s - T_\infty) + \varepsilon\sigma A(T_s^4 - T_{surr}^4) \tag{10.7}$$

where $T_s =$ surface temperature of the cylinder

Also,

$$A = \pi DL = \pi(0.04)(0.3) = 0.0377 \text{ m}^2$$

$\varepsilon = 0.5 \quad \sigma = 5.67 \times 10^{-8} \text{ W/m}^2\text{K}^4 \quad T_\infty = 20 + 273.15 = 293.15 \text{ K} \quad T_{surr} = 30 + 273.15 = 303.15 \text{ K}$

An appropriate equation for the natural convection is Eq. (7.32):

$$\text{Nu} = \frac{hD}{k} = 0.36 + \frac{0.518\text{Ra}_D^{1/4}}{\left[1 + (0.559/\text{Pr})^{9/16}\right]^{4/9}} \quad \text{for} \quad 10^{-6} < \text{Ra}_D < 10^9 \tag{7.32}$$

where

$$\text{Ra}_D = \text{Gr}_D \text{ Pr} = \frac{g\beta(T_s - T_\infty)D^3}{v^2}\text{Pr}$$

(Note: At the end of the solution, we will check that the $\text{Ra}_D$ limits for Eq. (7.32) have been satisfied.)

The air properties are taken at the film temperature $T_{film} = \frac{T_s + T_\infty}{2}$ and

$$g = 9.807 \text{ m/s}^2 \quad \beta = \frac{1}{T_{film}}$$

We can use a Microsoft Excel spreadsheet for the solution. In particular, we can use Excel's *Goal Seek* program. Rearranging Eq. (10.7) for such a solution, we get function $f_1$:

$$f_1 = 50 - hA(T_s - T_\infty) - \varepsilon\sigma A(T_s^4 - T_{surr}^4) = 0 \tag{10.8}$$

Putting the various constants and parameters into this equation and getting convective coefficient $h$ from Eq. (7.32), we have a function that only depends on the unknown surface temperature $T_s$. If function $f_1$ equals zero, we have a solution. In the spreadsheet, we give a guess for the value of $T_s$. *Goal Seek* iterates on this value until a value that satisfies $f_1 = 0$ is hopefully obtained. We now give a detailed description of the Excel spreadsheet. Readers uninterested in the details can go directly to the end of this description for the results.

### Excel spreadsheet for Example 10.1

The general approach is to put labels in one column and their values/expressions in the adjacent column. Entries are in SI units and temperatures in Kelvin.

| | A | B | D | F |
|---|---|---|---|---|
| 1 | Ts | 400 (initial guess) | =0.518*B16^0.25 | =B6-B20-B21 |
| 2 | Tinf | 293.15 | =1+(0.559/B13)^(9/16) | |
| 3 | Tsurr | 303.15 | | |
| 4 | D | 0.04 | | |
| 5 | L | 0.3 | | |
| 6 | q htr | 50 | | |
| 7 | emiss | 0.5 | | |
| 8 | sigma | 5.67E-8 | | |
| 9 | g | 9.807 | | |
| 10 | Tfilm | =(B1+B2)/2 | | |
| 11 | k | =4.629E-12*B10^3−2.899E-8*B10^2 +9.041E-5*B10+0.001009 | | |
| 12 | nu | =−1.725E-14*B10^3+1.027E-10*B10^2 +3.725E-8*B10-4.196E-6 | | |

| 13 | Pr | =2.916E-13*B10^4-1.148E-9*B10^3+1.675E-6*B10^2-1.021E-3*B10+0.9126 |
| 14 | beta | =1/B10 |
| 15 | Gr | =B9*B14*(B1−B2)*B4^3/(B12^2) |
| 16 | Ra | =B15*B13 |
| 17 | Nu | =0.36+D1/D2^(4/9) |
| 18 | h | =B11/B4*B17 |
| 19 | A | =PI()*B4*B5 |
| 20 | qconv | =B18*B19*(B1−B2) |
| 21 | qrad | =B7*B8*B19*(B1^4−B3^4) |

Calling up *Goal Seek*, we tell it to change cell B1 until cell F1 is zero. Convergence is reached, and the value in B1 becomes 404.977 K. Cell B16 is checked, and its value is within the appropriate limits for Eq. (7.32). The values in Cells B20 and B21 are, respectively, the heat flows by convection and radiation. The value in B20 is 30.28 and the value in B21 is 19.72.

(a) The temperature of the surface of the cylinder is 405.0 K = 131.9 C.

(b) The convective heat flow at the surface is 30.28 W, and the radiative heat flow at the surface is 19.72 W. When the program is executed, the values are found in cells B20 and B21, respectively.

Note: In Chapter 3, we discussed combined convection and radiation from a surface. We defined a radiative coefficient $h_{rad}$ that could be added to the convective coefficient $h$ to obtain a combined coefficient $h_{combined} = h + h_{rad}$. This combined coefficient could then be used to get the total heat transfer from the surface. That is $q = h_{combined}A(T_s - T_\infty)$. Eq. (3.35) defined the radiative coefficient as $h_{rad} = \varepsilon\sigma(T_s^2 + T_{surr}^2)(T_s + T_{surr})$.

This technique of combining the convective and the radiative coefficients gives exact results if $T_\infty = T_{surr}$. It can be applied when $T_\infty \neq T_{surr}$, but there will be some error. In this problem, $T_\infty$ and $T_{surr}$ were different. Specifically, $T_\infty = 20$ C and $T_{surr} = 30$ C. We modified the above Excel program slightly to solve the problem using the combined-coefficient approach and got results very close to the above exact results. The surface temperature was 401.8 K, the convective heat transfer was 29.26 W, and the radiative heat transfer was 20.74 W.

*Matlab program for Example 10.1*

```
% Example 10.1
% Cylindrical Heater
clear, clc
% Kelvin temperatures since we have radiation
tsurr=303.15;
tinf=293.15;
D=0.04;
L=0.3;
A=pi*D*L;
g=9.807;
sigma=5.67e-8;
eps=0.5;
q_heater=50;
% Begin looping on ts; Each pass increases the surface temperature by 0.01 K
for ts=350:0.01:450
    tfilm=(ts+tinf)/2;
    kair=4.629e-12*tfilm^3-2.899e-8*tfilm^2+9.041e-5*tfilm+.001009;
    nuair=-1.725e-14*tfilm^3+1.027e-10*tfilm^2+3.725e-8*tfilm-4.196e-6;
    Pr=2.916e-13*tfilm^4-1.148e-9*tfilm^3+1.675e-6*tfilm^2-1.021e-3*tfilm+0.9126;
    beta=1/tfilm;
    Gr=g*beta*abs(ts-tinf)*D^3/nuair^2;
```

```
Ra=Gr*Pr;
d1=(1+(.559/Pr)(9/16))(4/9);
Nu=0.36+0.518*(Ra)0.25/d1;
h=Nu*kair/D;
f1=q_heater-h*A*(ts-tinf)-eps*sigma*A*(ts^4-tsurr^4)
% Stop program when f1 <=0. This is the solution
if f1<=0
    ts
    qcond=h*A*(ts-tinf)
    qrad=eps*sigma*A*(ts^4-tsurr^4)
    break
end
end
```

The Matlab program gave the same results as the Excel spreadsheet.

# Example 10.2
## Liquid nitrogen container

### Problem
A spherical liquid nitrogen container has a diameter of 1.5 m. The container is thin-walled and is surrounded by another thin-walled hollow sphere of diameter 1.6 m. The spheres are of stainless steel and have an emissivity of 0.2. The space between the spheres is evacuated. The container is in a room where the room air and the room enclosure are at 22 C. The temperature of the liquid nitrogen is 77 K and the heat of vaporization of nitrogen is 200 kJ/kg.

(a) What is the rate of heat transfer to the liquid nitrogen?

(b) If a two-inch-thick layer of fiberglass insulation (k = 0.036 W/m C) is added to the outside of the outer sphere, what is the new rate of heat transfer to the liquid nitrogen? Assume that the emissivity at the outer surface of the insulation is 0.85.

(c) Was the addition of the fiberglass insulation worthwhile? Consider the evaporation rate of the liquid nitrogen in your answer.

### Solution
**Part (a).**

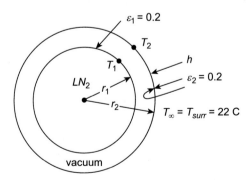

Preliminary: There is radiation between the two spheres. There is convection between the outer surface of the outer sphere and the room air. As no information is given regarding the flow of air over the sphere, we will assume that the convection is natural. There is also radiation between the outer sphere's outer surface and the room's enclosure. We have steady state. All necessary information is given. We have series heat flow. The rate of radiative heat transfer from the outer sphere to the inner sphere is equal to the rate of convection and radiation from the room air and enclosure to the outer surface of the outer sphere. As radiation is involved, temperatures must be in absolute units. Eqs. (10.2), (10.4), (10.6) will be used for the properties of air.

$$r_1 = 0.75 \text{ m} \quad r_2 = 0.8 \text{ m} \quad \varepsilon_1 = \varepsilon_2 = 0.2$$

$$A_1 = 4\pi r_1^2 = 4\pi(0.75)^2 = 7.069 \text{ m}^2 \quad A_2 = 4\pi r_2^2 = 4\pi(0.8)^2 = 8.042 \text{ m}^2$$

$$T_\infty = T_{surr} = 22 + 273.15 = 295.15 \text{ K}$$

Let $q_1$ = the radiative transfer from the inner sphere to the outer sphere

$q_2$ = the convective transfer from the outer sphere to the room air

$q_3$ = the radiative transfer from the outer sphere to the room enclosure.

$$\text{The energy conservation equation is } q_1 = q_2 + q_3 \tag{10.9}$$

Eq. (9.53) can be used for $q_1$.

$$q_1 = \frac{\sigma A_1 \left(T_1^4 - T_2^4\right)}{\dfrac{1}{\varepsilon_1} + \left(\dfrac{r_1}{r_2}\right)^2 \left(\dfrac{1 - \varepsilon_2}{\varepsilon_2}\right)} \tag{9.53}$$

As the container is thin-walled, $T_1$ = liquid nitrogen temperature = 77 K. Putting values into Eq. (9.53), we have

$$q_1 = \frac{\left(5.67 \times 10^{-8}\right)(7.069)\left(77^4 - T_2^4\right)}{\dfrac{1}{0.2} + \left(\dfrac{0.75}{0.8}\right)^2 \left(\dfrac{1 - 0.2}{0.2}\right)} = 4.7068 \times 10^{-8}\left(77^4 - T_2^4\right) \tag{10.10}$$

An appropriate equation for natural convection for spheres is Eq. (7.40):

$$\text{Nu} = \frac{hD}{k} = 2 + \frac{0.589 \text{Ra}_D^{1/4}}{\left[1 + (0.469/\text{Pr})^{9/16}\right]^{4/9}} \quad \text{for} \quad \text{Ra}_D < 10^{11} \text{ and Pr} \geq 0.7 \tag{7.40}$$

where

$$\text{Ra}_D = \text{Gr}_D \, \text{Pr} = \frac{g\beta(T_s - T_\infty)D^3}{v^2}\text{Pr}$$

(Note: At the end of the solution, we will check that the $\text{Ra}_D$ and Pr limits for Eq. (7.40) have been satisfied.) The air properties are taken at the film temperature $T_{film} = \frac{T_s + T_\infty}{2}$, and

$$g = 9.807 \text{ m/s}^2 \quad \beta = \frac{1}{T_{film}}$$

Using the $h$ from Eq. (7.40), we have

$$q_2 = hA_2(T_2 - T_\infty) \tag{10.11}$$

Lacking any other information, we will assume the room is large and that the radiation from the outer surface of the sphere to the room enclosure is modeled by Eq. (9.50):

$$q_3 = \varepsilon_2 \sigma A_2\left(T_2^4 - T_{surr}^4\right) \tag{10.12}$$

Returning to Eq. (10.9), we have

$$4.7068 \times 10^{-8}\left(77^4 - T_2^4\right) = hA_2(T_2 - T_\infty) + \varepsilon_2 \sigma A_2\left(T_2^4 - T_{surr}^4\right) \tag{10.13}$$

We can use Excel's *Goal Seek* program for the solution. Rearranging Eq. (10.13) for such a solution, we get function $f_1$:

$$f_1 = 4.7068 \times 10^{-8}\left(77^4 - T_2^4\right) - hA_2(T_2 - T_\infty) - \varepsilon_2 \sigma A_2\left(T_2^4 - T_{surr}^4\right) = 0 \tag{10.14}$$

Putting the various constants and parameters into this equation and getting convective coefficient $h$ from Eq. (7.40), we have a function that only depends on the unknown outer surface temperature $T_2$. If function $f_1$ equals zero, we have a solution. In the spreadsheet, we give a guess for the value of $T_2$. *Goal Seek* iterates on this value until a value that satisfies $f_1 = 0$ is hopefully obtained. We now give a detailed description of the Excel spreadsheet. Readers uninterested in the details can go directly to the end of this description for the results.

**Excel spreadsheet for Example 10.2 Part (a).** The general approach is to put labels in one column and their values/expressions in the adjacent column. Entries are in SI units and temperatures in Kelvin.

| | A | B | D | F |
|---|---|---|---|---|
| 1 | T2 | 300 (initial guess) | =0.589*B150.25 | =B19-B20-B21 |
| 2 | Tinf | 295.15 | =1+(0.469/B12)^(9/16) | |
| 3 | Tsurr | 295.15 | | |
| 4 | D1 | 1.5 | | |
| 5 | D2 | 1.6 | | |
| 6 | emiss | 0.2 | | |
| 7 | sigma | 5.67E-8 | | |
| 8 | g | 9.807 | | |
| 9 | Tfilm | =(B1+B2)/2 | | |
| 10 | k | =4.629E-12*B9^3-2.899E-8*B9^2+9.041E-5*B9+0.001009 | | |
| 11 | nu | =-1.725E-14*B9^3+1.027E-10*B9^2+3.725E-8*B9-4.196E-6 | | |
| 12 | Pr | =2.916E-13*B9^4-1.148E-9*B9^3+1.675E-6*B9^2-1.021E-3*B9+0.9126 | | |
| 13 | beta | =1/B9 | | |
| 14 | Gr | =B8*B13*ABS(B1-B2)*B5^3/(B11^2) | | |
| 15 | Ra | =B14*B12 | | |
| 16 | Nu | =2+D1/D2^(4/9) | | |
| 17 | h | =B10/B5*B16 | | |
| 18 | A | =PI()*B5^2 | | |
| 19 | q1 | =4.7068E-8*(77^4-B1^4) | | |
| 20 | q2 = qconv | =B17*B18*(B1-B2) | | |
| 21 | q3 = qrad | =B6*B7*B18*(B1^4-B3^4) | | |

Calling up *Goal Seek*, we tell it to change cell B1 until cell F1 is zero. Convergence is reached, and the value in B19 becomes −300.23. (Minus value means flow is *into* the liquid nitrogen.) Cells B15 and B12 were checked and found to be within the appropriate limits for Eq. (7.40).

The heat transfer into the liquid nitrogen is 300.2 W.

**Part (b).**

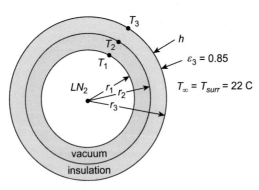

Preliminary: There is radiation between the two spheres and conduction through the insulation. There is natural convection between the outer surface of the insulation and the room air. There is also radiation between the outer surface of the insulation and the room's enclosure. We have steady state. We have series heat flow. The rate of radiative heat transfer from the outer sphere to the inner sphere is equal to the rate of conductive heat transfer through the insulation from the outer surface of the insulation to its inner surface. Both of these heat transfers are equal to the convection and radiation from the room air and enclosure to the outer surface of the insulation. Because radiation is involved, temperatures must be in absolute units. Eqs. (10.2), (10.4), (10.6) will be used for the properties of air.

$$r_1 = 0.75 \text{ m} \quad r_2 = 0.8 \text{ m} \quad r_3 = 0.8508 \text{ m} \quad \varepsilon_1 = \varepsilon_2 = 0.2 \quad \varepsilon_3 = 0.85$$

$$k_{ins} = 0.036 \text{ W/m C}$$

$$A_1 = 4\pi r_1^2 = 7.069 \text{ m}^2 \quad A_2 = 4\pi r_2^2 = 8.042 \text{ m}^2 \quad A_3 = 4\pi r_3^2 = 9.096 \text{ m}^2$$

$$T_\infty = T_{surr} = 22 + 273.15 = 295.15 \text{ K}$$

Let $q_1$ = the radiative transfer from the inner sphere to the outer sphere
$q_2$ = the conductive transfer through the insulation
$q_3$ = the convective transfer from the surface of the insulation to the room air
$q_4$ = the radiative transfer from the surface of the insulation to the room enclosure.

The energy conservation equation is $q_1 = q_2 = q_3 + q_4$        (10.15)

The heat flow between the spheres was discussed in Part (a), with the result given in Eq. (10.10):

$$q_1 = 4.7068 \times 10^{-8}\left(77^4 - T_2^4\right) \tag{10.16}$$

Eq. (3.73) for the conduction through a spherical shell can be used for the conduction through the insulation.

$$q_2 = \frac{4\pi k_{ins}}{\left(\dfrac{1}{r_2} - \dfrac{1}{r_3}\right)}(T_2 - T_3) \tag{10.17}$$

The above discussion in Part (a) regarding the convection and radiation from the outer surface of the sphere is applicable to the convection and radiation from the outer surface of the insulation for this Part (b). Eq. (7.40) still applies for the natural convection, and the corresponding equations for Eqs. (10.11) and (10.12) are

$$q_3 = hA_3(T_3 - T_\infty) \tag{10.18}$$

$$q_4 = \varepsilon_3 \sigma A_3\left(T_3^4 - T_{surr}^4\right) \tag{10.19}$$

Returning to Eq. (10.15), we have

$$q_1 = q_2 = q_3 + q_4 \tag{10.15}$$

Using Eqs. (10.16) through (10.19), Eq. (10.15) becomes

$$4.7068 \times 10^{-8}\left(77^4 - T_2^4\right) = \frac{4\pi k_{ins}}{\left(\dfrac{1}{r_2} - \dfrac{1}{r_3}\right)}(T_2 - T_3) = hA_3(T_3 - T_\infty) + \varepsilon_3 \sigma A_3\left(T_3^4 - T_{surr}^4\right) \tag{10.20}$$

There are two unknowns in this problem: $T_2$ and $T_3$. Excel's *Goal Seek* cannot be used for solution as that program is only suitable for one unknown. However, Excel's *Solver* add-in program can handle more than one unknown and we will use it for our solution. Looking at Eq. (10.15), we can create two functions that must be zero for a solution. First, we have $q_1 = q_2$, which leads to the function

$$f_1 = q_1 - q_2 = 0 \tag{10.21}$$

We can also create a second function $f_2$ from the equation $q_1 = q_3 + q_4$.

$$f_2 = q_1 - q_3 - q_4 = 0 \tag{10.22}$$

Both $f_1$ and $f_2$ need to be zero for a solution. This requirement will be satisfied if $f_1^2 + f_2^2 = 0$. In the Excel spreadsheet, one cell contains $f_1^2 + f_2^2$. The *Solver* program iterates on the values of $T_2$ and $T_3$ until that cell becomes zero.

***Excel spreadsheet for Example 10.2 Part (b).*** The general approach is to put labels in one column and their values/expressions in the adjacent column. Entries are in SI units and temperatures in Kelvin.

| | A | B | D | F |
|---|---|---|---|---|
| 1 | T1 | 77 | =0.589*B15^0.25 | =B24-B25 |
| 2 | T2 | 250 (initial guess) | =1+(0.469/B12)̂(9/16) | =B24-B26-B27 |
| 3 | T3 | 270 (initial guess) | | =F12+F22 |
| 4 | Tinf | 295.15 | | |
| 5 | Tsurr | 295.15 | | |
| 6 | r1 | 0.75 | | |
| 7 | r2 | 0.8 | | |
| 8 | r3 | 0.8508 | | |
| 9 | emiss 1,2 | 0.2 | | |
| 10 | emiss 3 | 0.85 | | |
| 11 | sigma | 5.67E-8 | | |
| 12 | g | 9.807 | | |
| 13 | Tfilm | =(B3+B4)/2 | | |
| 14 | k ins | 0.036 | | |
| 15 | k air | =4.629E-12*B13^3-2.899E-8*B13^2+9.041E-5*B13+0.001009 | | |
| 16 | nu air | =-1.725E-14*B13^3+1.027E-10*B13^2+3.725E-8*B13-4.196E-6 | | |
| 17 | Pr air | =2.916E-13*B13^4-1.148E-9*B13^3+1.675E-6*B13^2-1.021E-3*B13+0.9126 | | |
| 18 | beta | =1/B13 | | |
| 19 | Gr | =B12*B18*ABS(B3-B4)*(B8*2)^3/(B16^2) | | |
| 20 | Ra | =B19*B17 | | |
| 21 | Nu | =2+D1/D2^(4/9) | | |
| 22 | h | =B15/(2*B8)*B21 | | |
| 23 | A | =4*PI()*B8^2 | | |
| 24 | q1 | =4.7068E-8*(77^4-B1^4) | | |
| 25 | q2 | =4*PI()*B14/(1/B6-1/B7)*(B2-B3) | | |
| 26 | q3 = qconv | =B22*B23*(B3-B4) | | |
| 27 | q4 = qrad | =B10*B11*B23*(B3^4-B5^4) | | |

Calling up *Solver*, we tell it to change cells B2 and B3 until cell F3 is zero. F3 does not reach zero, but it is very small and we conclude that an acceptable solution is reached. Convergence is reached. The values in B24 and B25 become $-198.0$, and the sum of B26 and B27 is $-198.0$. (A minus value means the flow is *into* the liquid nitrogen.) Cells B20 and B17 were checked and found to be within the appropriate limits for Eq. (7.40).

With the insulation, the heat transfer into the liquid nitrogen is 198.0 W.

***Part (c).*** As noted in the problem statement, the heat of vaporization of liquid nitrogen is 200 kJ/kg. Without insulation, the heat input to the $LN_2$ is 300.2 W or 300.2 J/s. The evaporation rate of the $LN_2$ is

$$300.2 \text{ J} / \text{s} \times (1 \text{ kg} / 200,000 \text{ J}) \times (3600 \text{ s} / \text{h}) \times (24 \text{ h} / \text{day}) = 129.7 \text{ kg/day}$$

With insulation, the heat input drops to 198.0 W and the evaporation rate is

$$(198.0 / 300.2)( 129.7) = 85.5 \text{ kg/day}$$

The addition of the insulation reduces the evaporation rate of the nitrogen by about 1/3. We conclude that addition of the insulation was worthwhile.

## Example 10.3
### Cooling of a copper cube

*Problem*

A copper cube ($k = 375$ W/m C, $\rho = 8950$ kg/m³, $c = 385$ J/kg C) is 2 cm by 2 cm by 2 cm. It is initially uniform at 200 C in an oil bath. The cube is taken out of the bath and suspended by wires in the 20 C room air. The emissivity of the cube's surface is 0.05. How long will it take for the temperature of the cube to drop to 50 C?

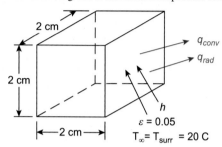

*Solution*

Preliminary: This is a transient problem. The temperature of the cube changes with time. The cube is of high conductivity and is also quite small. Therefore, it appears that the lumped method would be appropriate. We will assume this and then check the lumped Biot number at the end of the solution to see if our assumption was correct. There is convection and radiation from the surface of the cube. No information was given as to an air flow rate, so we will assume that the convection is natural convection. Lacking information about the room, we will assume it is a large room and that the temperature of the surroundings is the same as the temperature of the room air. As radiation is involved, temperatures must be in absolute units. Eqs. (10.2), (10.4), (10.6) will be used for the properties of air.

$S$ = side of the cube = 0.02 m   $A$ = surface area of the cube = $(6)(0.02)^2 = 0.0024$ m²

$V$ = volume of the cube = $(0.02)^3 = 8 \times 10^{-6}$ m³

$T_\infty = T_{surr} = 20 + 273.15 = 293.15$ K

$\varepsilon = 0.05$

We are assuming that the lumped technique is appropriate. The energy word equation for the cooling of the cube is

Rate of decrease in internal energy of the body = Rate of heat flow from the body by convection and radiation

This equation is presented mathematically by Eq. (4.3) for only convection. Modifying Eq. (4.3) to include radiation gives

$$-\rho c V \frac{dT}{dt} = hA(T - T_\infty) + \varepsilon \sigma A\left(T^4 - T_{surr}^4\right) \tag{10.23}$$

where $T$ is the temperature of the cube at time $t$

$\rho, c,$ and $V$ are the density, specific heat, and volume of the cube

$A$ is the surface area of the cube

Solution of the differential equation of Eq. (10.23) will give the temperature $T$ of the cube at a given time $t$. Unfortunately, the equation is nonlinear due to $h$ and the $T^4$ term, and it cannot be readily solved analytically. We will therefore solve it numerically, as follows:

Dividing both sides of Eq. (10.23) by $\rho c V$, we get

$$\frac{dT}{dt} = -\frac{hA(T - T_\infty)}{\rho c V} - \frac{\varepsilon \sigma A\left(T^4 - T_{surr}^4\right)}{\rho c V} \tag{10.24}$$

We can solve this numerically, working forward in time, by changing $dt$ to a finite time step $\Delta t$ and changing $dT$ to a finite temperature difference $\Delta T$. Eq. (10.24) then becomes

$$\Delta T = -\left[\frac{hA(T - T_\infty)}{\rho c V} + \frac{\varepsilon \sigma A(T^4 - T^4_{surr})}{\rho c V}\right]\Delta t \tag{10.25}$$

We pick a time step $\Delta t$ and then iterate using Eq. (10.25) over and over. Each time we iterate, we are moving forward in time by an amount $\Delta t$. The right side of Eq. (10.25) is recalculated and the temperature $T$ of the cube changes by an amount $\Delta T$. The value of $h$ changes at each iteration as $h$ is a function of $T$. For each iteration, $\Delta T = T_{new} - T_{old}$, so Eq. (10.25) could be written as

$$T_{new} = T_{old} - \left[\frac{hA(T - T_\infty)}{\rho c V} + \frac{\varepsilon \sigma A(T^4 - T^4_{surr})}{\rho c V}\right]\Delta t \tag{10.26}$$

The procedure starts with $T_{old}$ being the initial temperature of the cube, which is 200 $C = 473.15$ K. An appropriate equation for the convective coefficient for natural convection for a cube is Eq. (7.41):

$$\text{Nu} = \frac{hL^{**}}{k} = 5.748 + 0.752\left\{\frac{\text{Ra}_{L^{**}}}{\left[1 + (0.492/\text{Pr})^{9/16}\right]^{16/9}}\right\}^{0.252} \tag{7.41}$$

The characteristic length $L^{**}$ for a cube is $L^{**} = 3S\sqrt{\pi}$. For our cube, $L^{**} = (3)(0.02)\sqrt{\pi} = 0.1063$ m. This problem has been solved by Matlab and by Microsoft Excel. The Matlab program is given below.

*Matlab program for Example 10.3*

```
% Example Chpt 10 Example 10.3
% Cooling of a Copper Cube
clear, clc
% Properties for the cube
k=375; rho=8950; c=385;
% Kelvin temperatures since we have radiation
tsurr=293.15;
tinf=293.15;
% S = side of cube
S=0.02;
A=6*S2;
V=S3;
L=3*S*sqrt(pi);
g=9.807;
sigma=5.67e-8;
eps=0.05;
% Time step (seconds)
dtime=1;
% Set initial temperatures at 200 C = 473.15 K
t=473.15;
% Begin looping; Each pass moves time ahead by dtime
for loop=1:10000
    tfilm=(t+tinf)/2;
    kair=4.629e-12*tfilm^3-2.899e-8*tfilm^2+9.041e-5*tfilm+.001009;
    nuair=-1.725e-14*tfilm^3+1.027e-10*tfilm^2+3.725e-8*tfilm-4.196e-6;
    Pr=2.916e-13*tfilm^4-1.148e-9*tfilm^3+1.675e-6*tfilm^2-1.021e-3*tfilm+0.9126;
```

```
beta=1/tfilm;
Gr=g*beta*abs(t-tinf)*L^3/nuair^2;
Ra=Gr*Pr;
d1=(1+(.492/Pr)^(9/16))^(16/9);
Nu=5.748+0.752*(Ra/d1)^0.252;
h=Nu*kair/L;
qconv=h*A*(t-tinf);
qrad=eps*sigma*A*(t^4-tsurr^4);
rhocv=rho*c*V;
dt=-(qconv+qrad)/rhocv*dtime;
t=t+dt;
if t<=323.15
    t,h,time_seconds=loop*dtime,time_minutes=time_seconds/60
    break
end
end
```

The temperature of the cube dropped from 200 C to 50  C in 2189 iterations when the time step was 1  s. This corresponds to 36.48  min.

The temperature of the cube dropped from 200 to 50  C in 36.5  min.

The lumped Biot number will now be checked to see if the lumped assumption was appropriate.

$$\text{Bi}_{lumped} = \frac{h\left(\dfrac{V}{A}\right)}{k_{cube}} < 0.1 \text{ for a lumped analysis}$$

The *h-value* should be a combined *h* for convection and radiation. The largest coefficient for convection during the process was 10.6  W/m² C. The radiative heat transfer is much smaller than the convective heat transfer during the process, so the combined *h* is not much greater than the *h* for convection. We will conservatively say it is 20  W/m² C.

$$\text{Bi}_{lumped} = \frac{20\left(\dfrac{8 \times 10^{-6}}{0.0024}\right)}{375} = 0.00018$$

As $\text{Bi}_{lumped} \ll 0.1$, the assumption of a lumped system was indeed appropriate.

Excel can also be used to perform iterations, but it is a little tricky. First, we have to shutoff automatic calculation and change it to manual calculation. On the main screen, click the "Formulas" tab, then go to "Calculation Options" and check "Manual." This menu also has a "Calculate Now" button that we will use later to run the program.

We also have to enable iterative calculations. Click the Microsoft button in the upper left corner of the main screen and then click "Excel Options" in the bottom of the window. Click "Formulas" on the left panel, and then click "Enable iterative calculation." Enter the number of desired iterations in the "Maximum Iterations" box.

The Excel spreadsheet is given below. The operation of the program is controlled by the A1 cell in conjunction with the B2 cell. To run the spreadsheet, we first enter the number of desired iterations as outlined above. Then enter "1" in cell A1 and click "Calculate Now." Because of the IF statement in cell B2, this will cause the initial value of the cube's temperature (473.15) to be placed in cell B2. Then go back to cell A1 and place "=A1+1" in it and click "Calculate Now." This will cause the program to iterate the number of times set in the "Maximum Iterations" box. For each iteration, the time goes forward an amount dt. After the iterations are done, the ending temperature of the cube will be in cell B2. We can do another set of iterations, continuing working forward in time, by clicking "Calculate Now" again. (Note: During this process you will probably get a "Circular Reference" warning. This is due to cell A1 referencing itself. Nothing is wrong. We want the circular reference.)

### Excel spreadsheet for Example 10.3

The general approach is to put labels in one column and their values/expressions in the adjacent column. Cell A1 is an exception, described above. Entries are in SI units and temperatures in Kelvin.

| | A | B | C |
|---|---|---|---|
| 1 | (see above) | | =(1+(0.492/B17)^(9/16))^(16/9) |
| 2 | T | =IF(A1=1,473.15,B2+B28) | |
| 3 | Tinf | 293.15 | |
| 4 | Tsurr | 293.15 | |
| 5 | S | 0.02 | |
| 6 | V | =B53 | |
| 7 | L** | =3*B5*SQRT(PI()) | |
| 8 | rho | 8950 | |
| 9 | k cube | 375 | |
| 10 | c | 385 | |
| 11 | emiss | 0.05 | |
| 12 | g | 9.807 | |
| 13 | sigma | 5.67E-8 | |
| 14 | Tfilm | =(B2+B3)/2 | |
| 15 | k air | =4.629E-12*B14^3−2.899E-8*B14^2+9.041E-5*B14+0.001009 | |
| 16 | nu air | =-1.725E-14*B14^3+1.027E-10*B14^2 +3.725E-8*B14-4.196E-6 | |
| 17 | Pr air | =2.916E-13*B14^4−1.148E-9*B14^3+1.675E-6*B14^2-1.021E-3*B14+0.9126 | |
| 18 | beta | =1/B14 | |
| 19 | Gr | =B12*B18*ABS(B2−B3)*B7^3/(B16^2) | |
| 20 | Ra | =B19*B17 | |
| 21 | Nu | =5.748+0.752*(B20/C1)^0.252 | |
| 22 | h | =B21*B15/B7 | |
| 23 | A | =6*B5^2 | |
| 24 | qconv | =B22*B23*(B2−B3) | |
| 25 | q rad | =B11*B13*B23*(B2^4−B4^4) | |
| 26 | rho*c*V | =B8*B10*B6 | |
| 27 | dt (sec) | 1 | |
| 28 | dT (K) | =−(B24+B25)/B26*B27 | |
| 29 | time (sec) | =(A1−1)*B27 | |

Both Matlab and Excel gave the same results. The temperature of the cube dropped from 200 to 50 C in 36.5 min.

---

## Example 10.4

### Heat transfer from a pipe

*Problem*

Water at 90 C is flowing at a rate of 10 gpm through a 2″ Sch 40 galvanized steel pipe. The pipe is 50 m long. It is in a factory where the air temperature is 18 C.

**(a)** What is the rate of heat transfer from the pipe to the room?

**(b)** Rigid polyethylene insulation is installed on the pipe. The insulation is 3/4 inch thick. What is the new rate of heat transfer to the room? What is the % reduction in heat transfer due to the insulation?

*Solution*
**Part (a).**

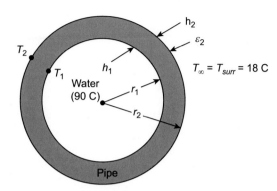

Preliminary: There is forced convection at the inner surface of the pipe. There is convection to the room air at the outer surface of the pipe. We will assume this is natural convection as no information was given as to any air flow across the pipe. We will assume that there is also radiation from the outer surface of the pipe to the room enclosure. The pipe is very long, so we will assume it is horizontal and we will assume that the room is large. As no information was given as to the temperature of the room enclosure, we will assume that the room surroundings are the same temperature as the room air. We have steady state. We need information as to the inside and outside diameters of the pipe, the conductivity of the pipe, and the emissivity of the outer surface of the pipe. We have series heat flow. Convection from the water to the pipe equals conduction through the pipe. In addition, these heat transfers are equal to the sum of the convection and radiation from the pipe's outer surface to the room. As radiation is involved, temperatures must be in absolute units. Eqs. (10.2), (10.4), (10.6) will be used for the properties of air.

From an internet search, the OD of a 2 inch Sch 40 pipe is 60.3 mm and the ID is 52.6 mm. Therefore, $r_1 = 0.0263$ m and $r_2 = 0.03015$ m. The conductivity of galvanized pipe is about 43  W/m C. A reasonable value for the emissivity of the outer surface of the pipe is 0.6.

$$T_{water} = 90 \text{ C} = 363.15 \text{ K} \quad T_\infty = T_{surr} = 18 \text{ C} = 291.15 \text{ K}$$

The properties of water at 90  C are $k = 0.675$ W/m C; $\rho = 965.3$ kg/m$^3$; $\mu = 0.315 \times 10^{-3}$ kg/m s; Pr $= 1.96$
The volumetric flow rate of the water is 10 gpm. The density of water at 90  C is 965.3  kg/m$^3$.
The mass flow rate of the water is therefore 10 gal/min x (1  m$^3$/264.17 gal) × (1  min/60 s) × (965.3  kg/m$^3$) $=$ 0.609  kg/s.
The mass flow rate is $\dot{m} = \rho A V$, where $\rho$ is the density of the water, $A$ is the flow area, and $V$ is the flow velocity.

$$A = \pi r_1^2 = \pi(0.0263)^2 = 0.002173 \text{ m}^2 \quad V = \frac{\dot{m}}{\rho A} = \frac{0.609}{(965.3)(0.002173)} = 0.290 \text{ m/s}$$

For the forced convection at the inner surface of the pipe, we will use Eq. (6.135):

$$\text{Nu}_D = \frac{hD}{k} = 0.023\text{Re}_D^{0.8}\text{Pr}^n \tag{10.27}$$

In this equation, $n = 0.3$ as the water is being cooled. Properties are at the water temperature.
The Reynolds number is $\text{Re}_D = \frac{VD}{v} = \frac{\rho VD}{\mu} = \frac{(965.3)(0.290)(0.0526)}{0.315\times10^{-3}} = 4.675 \times 10^4$.
Putting this in Eq. (10.27) and solving for $h_1$, we get

$$h_1 = \left(\frac{k}{D}\right)(0.023\text{Re}_D^{0.8}\text{Pr}^{0.3}) = \left(\frac{0.675}{0.0526}\right)(0.023)(4.675 \times 10^4)^{0.8}(1.96)^{0.3} = 1966 \text{ W/m}^2 \text{ C}$$

For the natural convection at the outside surface of the pipe, we will use Eq. (7.32):

$$\text{Nu} = \frac{hD}{k} = 0.36 + \frac{0.518\text{Ra}_D^{1/4}}{\left[1 + (0.559/\text{Pr})^{9/16}\right]^{4/9}} \quad \text{for} \quad 10^{-6} < \text{Ra}_D < 10^9 \tag{10.28}$$

where $\text{Ra}_D = \text{Gr}_D \text{ Pr} = \frac{g\beta(T_s - T_\infty)D^3}{\nu^2}\text{Pr}$.

Conduction through the pipe wall is given by Eq. (3.55):

$$q = \frac{2\pi k L}{\ln\left(\dfrac{r_o}{r_i}\right)}(T_i - T_o) \tag{10.29}$$

where $L$ is the length of the pipe.

Let $q_1$ = the convection from the water to the inner pipe surface

$q_2$ = the conduction through the pipe

$q_3$ = the convection and radiation from the outer pipe surface to the room.

$$\text{The energy conservation equation is } q_1 = q_2 = q_3 \tag{10.30}$$

$$\text{For } q_1 \text{ we have } q_1 = h_1 A_1 (T_{water} - T_1) \tag{10.31}$$

where $A_1 = \pi D_1 L$, $L$ = length of the pipe, and $h_1$ = 1966 W/m$^2$ C (found earlier).

$$\text{For } q_2 \text{ we have } \quad q_2 = \frac{2\pi k_{pipe} L}{\ln\left(\dfrac{r_2}{r_1}\right)}(T_1 - T_2) \tag{10.32}$$

$$\text{For } q_3 \text{ we have } q_3 = h_2 A_2 (T_2 - T_\infty) + \varepsilon_2 \sigma A_2 \left(T_2^4 - T_{surr}^4\right) \tag{10.33}$$

where $A_2 = \pi D_2 L$ and $h_2$ is obtained from Eq. (10.28).

There are two unknowns in this problem: $T_1$ and $T_2$. Once these are obtained, the heat flows can be calculated. We will use Excel's *Solver* program for the solution.

From Eq. (10.30), we create two functions which must be zero for the solution. These functions are

$$f_1 = q_1 - q_2 = 0 \text{ and } f_2 = q_1 - q_3 = 0 \tag{10.34}$$

(Note: There is flexibility in choosing the functions. For example, we could have used $q_2 - q_3$ as one of the functions. We just need two functions that include all three heat flows among them.)

*Solver* needs a single criterion for solution. Our criterion is $f_1^2 + f_2^2 = 0$. If this is zero, then both functions will be zero. We give initial guesses for $T_1$ and $T_2$. *Solver* performs iterations on these values until $f_1^2 + f_2^2 = 0$. Sometimes *Solver* cannot reach zero, but it can reach very close to zero. The Excel spreadsheet for the solution is as follows:

**Excel spreadsheet for Example 10.4 Part (a).** The general approach is to put labels in one column and their values/expressions in the adjacent column. Entries are in SI units and temperatures in Kelvin.

| | A | B | D | F |
|---|---|---|---|---|
| 1 | T water | =90+273.15 | =0.518*B19^0.25 | =B24−B25 |
| 2 | T1 | 360 (initial guess) | =1+(0.559/B16)^(9/16) | =B25−B26 |
| 3 | T2 | 358 (initial guess) | | =F1^2+F2^2 |
| 4 | Tinf | 291.15 | | |
| 5 | Tsurr | 291.15 | | |
| 6 | r1 | 0.0263 | | |
| 7 | r2 | 0.03015 | | |
| 8 | k pipe | 43 | | |
| 9 | h1 | 1966 | | |
| 10 | emiss | 0.6 | | |
| 11 | sigma | 5.67E-8 | | |
| 12 | g | 9.807 | | |
| 13 | Tfilm | =(B3+B4)/2 | | |
| 14 | k air | =4.629E-12*B13^3-2.899E-8*B13^2+9.041E-5*B13+0.001009 | | |
| 15 | nu air | =-1.725E-14*B13^3+1.027E-10*B13^2+3.725E-8*B13-4.196E-6 | | |
| 16 | Pr air | =2.916E-13*B13^4-1.148E-9*B13^3+1.675E-6*B13^2-1.021E-3*B13+0.9126 | | |

| 17 | beta | $=1/\text{B}13$ |
| 18 | Gr | $=\text{B}12*\text{B}17*(\text{B}3-\text{B}4)*(2*\text{B}7)\char`\^3/$ $(\text{B}15\char`\^2)$ |
| 19 | Ra | $=\text{B}18*\text{B}16$ |
| 20 | Nu | $=0.36+\text{D}1/\text{D}2\char`\^(4/9)$ |
| 21 | h2 | $=\text{B}14*\text{B}20/(2*\text{B}7)$ |
| 22 | A1/L | $=\text{PI}()*2*\text{B}6$ |
| 23 | A2/L | $=\text{PI}()*2*\text{B}7$ |
| 24 | q1/L | $=\text{B}9*\text{B}22*(\text{B}1-\text{B}2)$ |
| 25 | q2/L | $=2*\text{PI}()*\text{B}8/\text{LN}(\text{B}7/\text{B}6)*(\text{B}2-\text{B}3)$ |
| 26 | q3/L | $=\text{B}21*\text{B}23*(\text{B}3-$ $\text{B}4)+\text{B}10*\text{B}11*\text{B}23*(\text{B}3\char`\^4-\text{B}5\char`\^4)$ |

Calling up *Solver*, we tell it to change cells B2 and B3 until cell F3 is zero. The results were

$$T_1 = 362.71 \text{ K}, \quad T_2 = 362.63 \text{ K}, \quad q/L = 144.5 \text{ W/m}$$

$$q = (q/L)(L) = (144.5)(50) = 7225 \text{ W}$$

The rate of heat transfer from the pipe to the room is 7225 W.

**Part (b).**

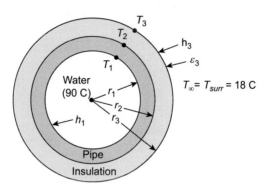

Preliminary: From an Internet search, the polyethylene insulation has a conductivity of about 0.037 W/m C. We will assume that the emissivity of the outer surface of the insulation is the same as that of the bare pipe, i.e., 0.6. We have steady-state and series heat flow. Convection from the water to the pipe equals conduction through the pipe, which equals the conduction through the insulation, which equals the convection plus radiation from the outer surface of the insulation. As radiation is involved, temperatures must be in absolute units. Eqs. (10.2), (10.4), (10.6) will be used for the properties of air.

$r_1 = 0.0263 \text{ m} \quad r_2 = 0.03015 \text{ m} \quad r_3 = r_2 + \frac{3}{4} \text{ inch} = 0.0492 \text{ m} k_{pipe} = 43 \text{ W/m C} \quad k_{ins} = 0.037 \text{ W/m C} \quad \varepsilon_3 = 0.6$

Let $q_1$ = the convection from the water to the inner pipe surface
$q_2$ = the conduction through the pipe wall
$q_3$ = the conduction through the insulation
$q_4$ = the convection and radiation from the outer surface of the insulation to the room.

The energy conservation equation is $q_1 = q_2 = q_3 = q_4$     (10.35)

For $q_1$ we have $q_1 = h_1 A_1 (T_{water} - T_1)$     (10.36)

where $A_1 = \pi D_1 L$, $L$ = length of the pipe, and $h_1$ = 1966 W/m² C (found earlier)

For $q_2$ we have $q_2 = \dfrac{2\pi k_{pipe} L}{\ln\left(\dfrac{r_2}{r_1}\right)} (T_1 - T_2)$     (10.37)

For $q_3$ we have $q_3 = \dfrac{2\pi k_{ins} L}{\ln\left(\dfrac{r_3}{r_2}\right)} (T_2 - T_3)$     (10.38)

For $q_4$ we have $q_4 = h_3 A_3 (T_3 - T_\infty) + \varepsilon_3 \sigma A_3 (T_3^4 - T_{surr}^4)$ (10.39)

where $A_3 = \pi D_3 L$ and $h_3$ is obtained from Eq. (10.28).

There are three unknowns in this problem: $T_1$, $T_2$, and $T_3$. Once these are obtained, the heat flows can be calculated. We will use Excel's *Solver* program for the solution.

The energy conservation equation is Eq. (10.35).

$$q_1 = q_2 = q_3 = q_4 \qquad (10.35)$$

We have three unknowns. From Eq. (10.35), we create three functions that must be zero for the solution. We chose the functions

$$f_1 = q_1 - q_2 = 0 \quad f_2 = q_3 - q_4 = 0 \quad \text{and} \quad f_3 = q_1 - q_4 = 0 \qquad (10.40)$$

(Note: There is flexibility in choosing the functions, but we need to include all four heat flows somewhere in the three functions.)

*Solver* needs a single criterion for solution. Our criterion is $f_1^2 + f_2^2 + f_3^2 = 0$. If this is zero, then all three functions will be zero.

We give initial guesses for $T_1$, $T_2$, and $T_3$. *Solver* performs iterations on these values until $f_1^2 + f_2^2 + f_3^2 = 0$. The Excel spreadsheet for the solution is as follows.

**Excel spreadsheet for Example 10.4 Part (b).** The general approach is to put labels in one column and their values/expressions in the adjacent column. Entries are in SI units and temperatures in Kelvin.

|    | A        | B                                                         | D                   | F                |
|----|----------|-----------------------------------------------------------|---------------------|------------------|
| 1  | T water  | =90+273.15                                                | =0.518*B22^0.25     | =B28−B29         |
| 2  | T1       | 360 (initial guess)                                       | =1+(0.559/B19)^(9/16) | =B30−B31       |
| 3  | T2       | 358 (initial guess)                                       |                     | =B28−B31         |
| 4  | T3       | 300 (initial guess)                                       |                     | =F1^2+F2^2+F3^2  |
| 5  | Tinf     | 291.15                                                    |                     |                  |
| 6  | Tsurr    | 291.15                                                    |                     |                  |
| 7  | r1       | 0.0263                                                    |                     |                  |
| 8  | r2       | 0.03015                                                   |                     |                  |
| 9  | r3       | 0.0492                                                    |                     |                  |
| 10 | k pipe   | 43                                                        |                     |                  |
| 11 | k ins    | 0.037                                                     |                     |                  |
| 12 | h1       | 1966                                                      |                     |                  |
| 13 | emiss    | 0.6                                                       |                     |                  |
| 14 | sigma    | 5.67E-8                                                   |                     |                  |
| 15 | g        | 9.807                                                     |                     |                  |
| 16 | Tfilm    | =(B4+B5)/2                                                |                     |                  |
| 17 | k air    | =4.629E-12*B16^3−2.899E-8*B16^2+9.041E-5*B16+0.001009     |                     |                  |
| 18 | nu air   | =-1.725E-14*B16^3+1.027E-10*B16^2+3.725E-8*B16-4.196E-6   |                     |                  |
| 19 | Pr air   | =2.916E-13*B16^4-1.148E-9*B16^3+1.675E-6*B16^2-1.021E-3*B16+0.9126 |            |                  |
| 20 | beta     | =1/B16                                                    |                     |                  |
| 21 | Gr       | =B15*B20*(B4−B5)*(2*B9)^3/(B18^2)                         |                     |                  |
| 22 | Ra       | =B21*B19                                                  |                     |                  |
| 23 | Nu       | =0.36+D1/D2^(4/9)                                         |                     |                  |
| 24 | h3       | =B17*B23/(2*B9)                                           |                     |                  |
| 25 | A1/L     | =PI()*2*B7                                                |                     |                  |
| 26 | A2/L     | =PI()*2*B8                                                |                     |                  |

| 27 | A3/L | =PI()*2*B9 |
|---|---|---|
| 28 | q1/L | =B12*B25*(B1−B2) |
| 29 | q2/L | =2*PI()*B10/LN(B8/B7)* (B2−B3) |
| 30 | q3/L | =2*PI()*B11/LN(B9/B8)*(B3  B4) |
| 31 | q4/L | =B24*B27*(B4− |
|  |  | B5)+B13*B14*B27*(B4^4−B6^4) |

Calling up *Solver*, we tell it to change cells B2, B3, and B4 until cell F4 is zero. The results were

$$T_1 = 363.064 \text{ K} \quad T_2 = 363.049 \text{ K} \quad T_3 = 304.005 \text{ K} \quad q/L = 28.05 \text{ W/m}$$

$$q = (q/L)(L) = (28.05)(50) = 1403 \text{ W}$$

$$\% \text{ Reduction} = \frac{7225 - 1403}{7225} \times 100 = 80.6 \%$$

With insulation, the new heat flow to the room is 1403  W. The insulation reduced the heat flow by 80.6%.

## Example 10.5
### Heat transfer through a building wall

*Problem*

A wall of a building is 8 feet high and 15 feet wide. It consists of 1/2 inch thick plasterboard ($k_1 = 0.173$ W/m C), 3½ inch thick fiberglass insulation ($k_2 = 0.039$ W/m C), and 9/16 inch thick T-111 siding ($k_3 = 0.115$ W/m C). The air inside the house is quiet at 72  F. Outside the house, the air is at 35  F, and it is blowing horizontally across the siding at 10 miles per hour. What is the rate of heat transfer through the wall?

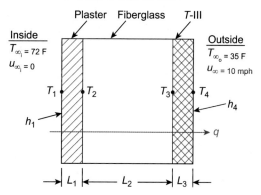

Preliminary: There is natural convection at the inside surface of the wall and forced convection on the outside surface. We will assume that radiation is insignificant. The heat transfer is steady state. Composite walls were considered in Chapter 3. We will use the resistance concept from that chapter to solve the problem. Eqs. (10.2), (10.4), (10.6) will be used for the properties of air. These equations are for air temperatures being in Kelvin degrees.
The applicable equation is Eq. (3.18):

$$q = \frac{\Delta T_{overall}}{\sum R} \tag{3.18}$$

For this problem, the overall temperature difference is the difference between the inside and outside air temperatures. There are five resistances: three conductive resistances for the wall layers and two convective resistances for the inner and outer surfaces of the wall. For a layer, the resistance is $\frac{L}{kA}$, where $L$ is the thickness of the layer, $k$ is the conductivity of the layer, and $A$ is the cross-sectional area through which the heat flows. For the convection, the resistance is $\frac{1}{hA}$, where $h$ is the convective coefficient. Hence, for this problem, the heat flow through the wall is, from Eq. (3.18),

$$q = \frac{T_{\infty_i} - T_{\infty_o}}{\left(\frac{1}{h_1 A}\right) + \left(\frac{L_1}{k_1 A}\right)_{plaster} + \left(\frac{L_2}{k_2 A}\right)_{fiberglass} + \left(\frac{L_3}{k_3 A}\right)_{T-111} + \left(\frac{1}{h_4 A}\right)} \tag{10.41}$$

$L_1 = 0.5$ inch $\times \left(\frac{0.0254 \text{ m}}{\text{inch}}\right) = 0.0127$ m. Similarly, $L_2 = 0.0889$ m and $L_3 = 0.01429$ m

$$T_{\infty_i} = 72 \text{ F} = (5/9)(72 - 32) = 22.22 \text{ C} \quad T_{\infty_o} = 35 \text{ F} = (5/9)(35 - 32) = 1.67 \text{ C}$$

Rearranging Eq. (10.41) and putting in values, we have

$$\frac{q}{A} = \frac{20.55}{2.477 + \left(\frac{1}{h_1}\right) + \left(\frac{1}{h_4}\right)} \tag{10.42}$$

Looking at Eq. (10.42), we see that we need convective coefficients $h_1$ and $h_4$ to determine the heat flow through the wall. These convective coefficients depend on their respective film temperatures. Hence, we need to determine $T_1$ and $T_4$. There is no need to determine interface temperatures $T_2$ and $T_3$. The resistance concept will be used to obtain equations for the surface temperatures as follows:

For $T_1$ we have 
$$\frac{q}{A} = \frac{T_{\infty_i} - T_1}{\left(\frac{1}{h_1}\right)} = \frac{T_1 - T_{\infty_o}}{\left(\frac{L_1}{k_1}\right) + \left(\frac{L_2}{k_2}\right) + \left(\frac{L_3}{k_3}\right) + \left(\frac{1}{h_4}\right)} \tag{10.43}$$

For $T_4$ we have 
$$\frac{q}{A} = \frac{T_{\infty_i} - T_4}{\left(\frac{1}{h_1}\right) + \left(\frac{L_1}{k_1}\right) + \left(\frac{L_2}{k_2}\right) + \left(\frac{L_3}{k_3}\right)} = \frac{T_4 - T_{\infty_o}}{\left(\frac{1}{h_4}\right)} \tag{10.44}$$

Putting values into Eq. (10.43), we get

$$\frac{q}{A} = \frac{22.22 - T_1}{\left(\frac{1}{h_1}\right)} = \frac{T_1 - 1.67}{2.477 + \left(\frac{1}{h_4}\right)} \tag{10.45}$$

Putting values into Eq. (10.44), we get

$$\frac{q}{A} = \frac{22.22 - T_4}{2.477 + \left(\frac{1}{h_1}\right)} = \frac{T_4 - 1.67}{\left(\frac{1}{h_4}\right)} \tag{10.46}$$

The convective coefficients $h_1$ and $h_4$ are functions of $T_1$ and $T_4$, respectively. Eqs. (10.45) and (10.46) are solved simultaneously for $T_1$ and $T_4$. In doing so, the values of $h_1$ and $h_4$ are obtained. The rate of heat transfer through the wall may then be determined from any one of several of the above equations.

An appropriate equation for the forced convection coefficient $h_4$ on the outside surface of the wall is Equation (6.92).

$$h = \left(0.037 \text{Re}_L^{0.8} - 871\right) \text{Pr}^{1/3} (k / L) \tag{10.47}$$

The characteristic length $L$ is the length of the wall in the direction of flow. We have $L = 15$ feet $= 4.572$ m. The Reynolds number in Eq. (10.47) is $\text{Re}_L = \frac{u_\infty L}{v}$, where $v$ is the kinematic viscosity of air at the film temperature $(T_4 + T_{\infty_o})/2$. The other properties of air in the equation are also taken at the film temperature. The free-stream velocity is $u_\infty = 10$ miles per hour $= 4.47$ m/s.

An appropriate equation for the natural convection coefficient $h_1$ at the inside surface of the wall is Eq. (7.11).

$$\text{Nu} = \frac{hL}{k} = \left\{ 0.825 + \frac{0.387 \text{Ra}_L^{1/6}}{\left[1 + (0.492/\text{Pr})^{9/16}\right]^{8/27}} \right\}^2 \tag{10.48}$$

The characteristic length is the height of the wall, $L = 8$ feet $= 2.438$ m. Eq. (10.48) is for a surface having both laminar and turbulent boundary layers. The results of the analysis will confirm that this is the case for this problem.

As mentioned above, there are two unknowns in this problem: $T_1$ and $T_4$. Once these are obtained, the heat flow through the wall can be calculated. We will use Excel's *Solver* program for the solution.

From Eqs. (10.45) and (10.46), we create two functions that must be zero for the solution. These functions are

$$f_1 = h_1(22.22 - T_1) - \frac{T_1 - 1.67}{2.477 + \left(\dfrac{1}{h_4}\right)} = 0 \tag{10.49}$$

$$f_2 = \frac{22.22 - T_4}{2.477 + \left(\dfrac{1}{h_1}\right)} - h_4(T_4 - 1.67) = 0 \tag{10.50}$$

*Solver* needs a single criterion for solution. Our criterion is $f_1^2 + f_2^2 = 0$. If this is zero, then both functions will be zero. We give initial guesses for $T_1$ and $T_4$. *Solver* performs iterations on these values until $f_1^2 + f_2^2 = 0$. Sometimes *Solver* cannot reach zero, but it can reach very close to zero. The Excel spreadsheet for the solution is as follows:

**Excel spreadsheet for Example 10.5.** The general approach is to put labels in one column and their values/expressions in the adjacent column. Entries are in SI units, film temperatures in Kelvin, and other temperatures in C.

| | A | B | C and D | F |
|---|---|---|---|---|
| 1 | Tinf in | 22.22 | (see below) | =B20*(22.22−B3)−(B3−1.67)/(2.477+1/B15) |
| 2 | Tinf out | 1.67 | | =(22.22−B4)/(2.477+1/B20)−B15*(B4−1.67) |
| 3 | T1 | 20 (initial guess) | | =F1^2+F2^2 |
| 4 | T4 | 3 (initial guess) | | |
| 5 | u inf | 4.47 | | |
| 6 | L1 | 2.438 | | |
| 7 | L4 | 4.572 | | |
| 8 | Tfilm 1 | =(B3+B1)/2+273.15 | | |
| 9 | Tfilm 4 | =(B4+B2)/2+273.15 | | |
| 10 | g | 9.807 | | |
| 11 | k air 1 | =4.629E-12*B8^3-2.899E-8*B8^2+9.041E-5*B8+0.001009 | | |
| 12 | nu air 1 | =-1.725E-14*B8^3+1.027E-10*B8^2+3.725E-8*B8-4.196E-6 | | |
| 13 | Pr air 1 | =2.916E-13*B8^4−1.148E-9*B8^3+1.675E-6*B8^2−1.021E-3*B8+0.9126 | | |
| 14 | Re | =B5*B7/D12 | | |
| 15 | h4 | =(0.037*B14^0.8−871)*D13^(1/3)*D11/B7 | | |
| 16 | beta | =1/B8 | | |
| 17 | Gr | =B10*B16*(B1−B3)*B6^3/B12^2 | | |
| 18 | Ra | =B17*B13 | | |
| 19 | Nu | =(0.825+B22/B23^(8/27))^2 | | |
| 20 | h1 | =B19*B11/B6 | | |
| 21 | q/A | =20.55/(2.477+1/B20+1/B15) | | |
| 22 | | =0.387*B18^(1/6) | | |
| 23 | | =1+(0.492/B13)^(9/16) | | |

| | C | D |
|---|---|---|
| 11 | k air 4 | =4.629E-12*B9^3−2.899E-8*B9^2 +9.041E-5*B9+0.001009 |
| 12 | nu air 4 | =-1.725E-14*B9^3+1.027E-10*B9^2+3.725E-8*B9-4.196E-6 |
| 13 | Pr air 4 | =2.916E-13*B9^4-1.148E-9*B9^3+1.675E-6*B9^2-1.021E-3*B9+0.9126 |

Calling up *Solver,* we tell it to change cells B3 and B4 until cell F3 is zero. The results were

$$T_1 = 18.998 \text{ C} \quad T_4 = 2.272 \text{ C} \quad h_1 = 2.096 \text{ W/m}^2 \text{ C} \quad h_4 = 11.219 \text{ W/m}^2 \text{ C}$$

$$\frac{q}{A} = 6.753 \text{ W/m}^2. \text{ The wall is 8 ft by 15 ft, so } A = 120 \text{ ft}^2 = 11.15 \text{ m}^2$$

$$q = (6.753)(11.15) = 75.3 \text{ W} = 257 \text{ Btu/h}$$

The rate of heat transfer through the wall is 75.3 W = 257 Btu/hr.

We also solved this problem using Matlab. We used the Gauss–Seidel method (see Appendix J). Eqs. (10.45) and (10.46) were rearranged to get the unknowns $T_1$ and $T_4$ on the left side of the equal signs.

From Eq. (10.45), $T_1 = \dfrac{22.22 + \dfrac{1.67}{h_1[2.477 + (1/h_4)]}}{1 + \dfrac{1}{h_1[2.477 + (1/h_4)]}}$

From Eq. (10.46) $T_4 = \dfrac{1.67 + \dfrac{22.22}{h_4[2.477 + (1/h_1)]}}{1 + \dfrac{1}{h_4[2.477 + (1/h_1)]}}$

The program loops update $T_1$ and $T_4$ after each loop. The program is run several times. For each run, we manually changed the range of the looping. We kept increasing the range until the values of $T_1$ and $T_4$ were the same as those of the previous run. It was found that convergence was quick. Only about five loops were needed. The Matlab results were the same as the Excel results.

**Matlab program for Example 10.5.** % Example 10.5

```
% Building Wall
clear, clc
tinf_in=22.22;
tinf_out=1.67;
uinf=4.47;
L1=2.438;
L4=4.572;
A=L1*L4;
g=9.807;
% Gauss Seidel looping for t1 and t2
% Initial guesses for t1 and t4
t1=20;
t4=2;

for loop = 1:100.
    tfilm1=(t1+tinf_in)/2+273.15;
    tfilm4=(t4+tinf_out)/2+273.15;
    kair1=4.629e-12*tfilm1^3-2.899e-8*tfilm1^2+9.041e-5*tfilm1+.001009;
    nuair1=-1.725e-14*tfilm1^3+1.027e-10*tfilm1^2+3.725e-8*tfilm1-4.196e-6;
    Pr1=2.916e-13*tfilm1^4-1.148e-9*tfilm1^3+1.675e-6*tfilm1^2-1.021e-
3*tfilm1+0.9126;
    kair4=4.629e-12*tfilm4^3-2.899e-8*tfilm4^2+9.041e-5*tfilm4+.001009;
    nuair4=-1.725e-14*tfilm4^3+1.027e-10*tfilm4^2+3.725e-8*tfilm4-4.196e-6;
    Pr4=2.916e-13*tfilm4^4-1.148e-9*tfilm4^3+1.675e-6*tfilm4^2-1.021e-
3*tfilm4+0.9126;
```

```
      Re=uinf*L4/nuair4;
      h4=(0.037*Re^0.8-871)*Pr4^(1/3)*kair4/L4;
      beta=1/tfilm1;
      Gr=g*beta*abs(t1-tinf_in)*L1^3/nuair1^2;
      Ra=Gr*Pr1;
      Nu=(0.825+0.387*Ra^(1/6)/(1+(0.492/Pr1)^(9/16))^(8/27))^2;
      h1=Nu*kair1/L1;
      a=22.22+1.67/h1/(2.477+1/h4);
      b=1+1/h1/(2.477+1/h4);
      c=1.67+22.22/h4/(2.477+1/h1);
      d=1+1/h4/(2.477+1/h1);
      t1=a/b;
      t4=c/d;
end
t1,t4
q_over_A=(t4-1.67)*h4
q=q_over_A*A
```

## Example 10.6
### Pin fin

*Problem*

An aluminum pin fin is attached to a wall. The fin is horizontal and has a diameter of 1.5 cm and a length of 50 cm. The wall is at 200 C and the surrounding air is at 15 C. To increase heat transfer the fin has a flat black coating of emissivity 0.9.

**(a)** What is the rate of heat transfer from the fin if the air is quiet?

**(b)** What is the rate of heat transfer from the fin if the air flows across the fin at a speed of 2 m/s?

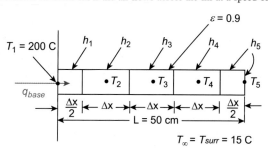

Preliminary: There is convection from the fin to the surrounding air. For Part (a), the convection is natural; for Part (b), it is forced. There is also radiation from the fin surface to the surroundings. Lacking any info to the contrary, we will assume that the temperature of the surroundings is the same as the room air. The heat transfer is steady state. We will do a numerical analysis with the nodes as shown in the accompanying figure. As radiation is involved, we will work with temperatures in absolute units, i.e., Kelvin. Eqs. (10.2), (10.4), (10.6) will be used for the properties of air. These equations are for air temperatures being in Kelvin degrees.

For the aluminum fin, $k = 210$ W/m C.

The nodal spacing is $\Delta x = L/4 = (50 \text{ cm})/4 = 12.5 \text{ cm} = 0.125$ m.

$T_\infty = T_{surr} = 15 \text{ C} = 288.15$ K

We have a steady-state problem. In developing nodal equations, we will assume that all heat flows are *into* the node. We will sum the heat flows and set the sum to zero.

Node 1 has a surface area in contact with the room air of $\pi D(\Delta x / 2)$. Summing up all the heat flows into Node 1 and setting the sum to zero, we have the nodal equation for Node 1:

$$q_{base} + \frac{kA(T_2 - T_1)}{\Delta x} + h_1[\pi D(\Delta x / 2)](T_\infty - T_1) + \varepsilon\sigma[\pi D(\Delta x / 2)]\left(T_{surr}^4 - T_1^4\right) = 0 \tag{10.51}$$

where

$q_{base} = $ rate of heat transfer from the base into the fin

$A = $ cross section area of the fin $= \dfrac{\pi}{4}D^2 = \dfrac{\pi}{4}(0.015)^2 = 0.0001767$ m$^2$

Summing up the heat flows for the four other nodes and setting each result to zero, we get

For Node 2, $\dfrac{kA(T_1 - T_2)}{\Delta x} + \dfrac{kA(T_3 - T_2)}{\Delta x} + h_2[\pi D(\Delta x)](T_\infty - T_2) + \varepsilon\sigma[\pi D(\Delta x)]\left(T_{surr}^4 - T_2^4\right) = 0$ \hfill (10.52)

For Node 3, $\dfrac{kA(T_2 - T_3)}{\Delta x} + \dfrac{kA(T_4 - T_3)}{\Delta x} + h_3[\pi D(\Delta x)](T_\infty - T_2) + \varepsilon\sigma[\pi D(\Delta x)]\left(T_{surr}^4 - T_3^4\right) = 0$ \hfill (10.53)

For Node 4, $\dfrac{kA(T_3 - T_4)}{\Delta x} + \dfrac{kA(T_5 - T_4)}{\Delta x} + h_4[\pi D(\Delta x)](T_\infty - T_4) + \varepsilon\sigma[\pi D(\Delta x)]\left(T_{surr}^4 - T_4^4\right) = 0$ \hfill (10.54)

For Node 5, $\dfrac{kA(T_4 - T_5)}{\Delta x} + h_5[\pi D(\Delta x / 2) + A](T_\infty - T_5) + \varepsilon\sigma[\pi D(\Delta x / 2) + A]\left(T_{surr}^4 - T_5^4\right) = 0$ \hfill (10.55)

There are five unknowns in these equations: $q_{base}, T_2, T_3, T_4,$ and $T_5$. But $q_{base}$ is only in one of the equations, Eq. (10.51). Hence, we can solve Eqs. (10.52) through (10.55) simultaneously for the four temperatures and then use Eq. (10.51) to get $q_{base}$.

The solution could be performed with Excel *Solver* as we have done for previous examples in this chapter. The expressions in Eqs. (10.52) through (10.55) each become functions $f_2$ through $f_5$. Guesses are made for the temperatures $T_2$ through $T_5$. *Solver* then changes the temperatures until $\sum_{2}^{5} f_i^2 = 0$. We have used *Solver* for this problem and have obtained a solution. For variety, we have decided to outline a different solution technique, the Gauss–Seidel method. This method was used in Chapter 5 for linear equations. We will use it here for our nonlinear equations. The process is as follows:

Rearrange Eqs. (10.52) through (10.55) so that one temperature is on the left side of the equal sign. For Eq. (10.52), $T_2$ will be on the left. For Eq. (10.53), $T_3$ will be on the left, similarly for $T_4$ and $T_5$. The $T^4$ terms can stay on the right side of the equal signs.

Doing this, Eqs. (10.52) through (10.55) become

$$T_2 = \frac{\dfrac{kA}{\Delta x}(T_1 + T_3) + h_2[\pi D(\Delta x)]T_\infty + \varepsilon\sigma[\pi D(\Delta x)](T_{surr}^4 - T_2^4)}{2\dfrac{kA}{\Delta x} + h_2[(\pi D(\Delta x)]} \tag{10.56}$$

$$T_3 = \frac{\dfrac{kA}{\Delta x}(T_2 + T_4) + h_3[\pi D(\Delta x)]T_\infty + \varepsilon\sigma[\pi D(\Delta x)]\left(T_{surr}^4 - T_3^4\right)}{2\dfrac{kA}{\Delta x} + h_3[(\pi D(\Delta x)]} \tag{10.57}$$

$$T_4 = \cfrac{\frac{kA}{\Delta x}(T_3 + T_5) + h_4[\pi D(\Delta x)]T_\infty + \varepsilon\sigma[\pi D(\Delta x)](T_{surr}^4 - T_4^4)}{2\frac{kA}{\Delta x} + h_4[(\pi D(\Delta x)]}$$

(10.58)

$$T_5 = \cfrac{\frac{kA}{\Delta x}(T_4) + h_5[\pi D(\Delta x/2) + A]T_\infty + \varepsilon\sigma[\pi D(\Delta x/2) + A](T_{surr}^4 - T_3^4)}{\frac{kA}{\Delta x} + h_5[\pi D(\Delta x/2) + A]}$$

(10.59)

A Matlab program was used for the Gauss–Seidel solution. In the program, initial guesses are given for the unknown temperatures $T_2$ through $T_5$. The program then loops through Eqs. (10.56)–(10.59). Each time a loop is executed, the values of the temperatures are updated. When the changes between the temperatures from one loop to the next become small enough, the program has converged to the solution. It was found that this method did indeed converge, and the convergence was quite quick. Having the solution for the temperatures, Eq. (10.51) above was used to get the rate of heat transfer $q_{base}$ from the fin to the room air and surroundings.

**Part (a)** deals with natural convection from the fin. An appropriate equation for the convective coefficient is Eq. (7.32):

$$Nu = \frac{hD}{k} = 0.36 + \frac{0.518 Ra_D^{1/4}}{\left[1 + (0.559/Pr)^{9/16}\right]^{4/9}}$$

(7.32)

**Part (b)** deals with forced convection from the fin. An appropriate equation for the convective coefficient is Eq. (6.96):

$$Nu = \frac{hD}{k} = 0.3 + \frac{0.62 Re_D^{1/2} Pr^{1/3}}{\left[1 + \left(\frac{0.4}{Pr}\right)^{2/3}\right]^{1/4}} \left[1 + \left(\frac{Re_D}{2.82 \times 10^5}\right)^{5/8}\right]^{4/5}$$

(6.96)

The solution for Part (a), natural convection, was

$$T_2 = 395.20 \text{ K}, \quad T_3 = 355.13 \text{ K}, \quad T_4 = 335.65 \text{ K}, \quad T_5 = 329.62 \text{ K}, \quad \text{and} \quad q_{base} = 35.37 \text{ W}$$

The solution for Part (b), forced convection, was

$$T_2 = 362.63 \text{ K}, \quad T_3 = 318.97 \text{ K}, \quad T_4 = 302.27 \text{ K}, \quad T_5 = 297.83 \text{ K}, \quad \text{and} \quad q_{base} = 59.85 \text{ W}$$

The heat transfer from the fin is 35.4 W for natural convection and 59.9 W for forced convection with an air velocity of 2 m/s across the fin. As expected, the heat transfer for forced convection is greater than that for natural convection. The Matlab programs for the two parts of the problem follow:

**Matlab program for Part (a)—Natural convection.**
```
% Chpt 10 Example 10.6 Part (a) Natural Convection
% Heat Transfer from a Pin Fin
clear, clc
% Properties for the fin
kfin=210; emiss=0.9; D=0.015; A=pi/4*D^2;dx=0.125;
% Kelvin temperatures since we have radiation
tsurr=288.15;
tinf=288.15;
Tbase=200+273.15;
g=9.807;
sigma=5.67e-8;
% Give initial guesses for T2 through T5
Told=[Tbase,450,400,375,300];
% Constants used in the loop
c1=kfin*A/dx; c2=pi*D*dx;
% Begin iterating
```

```
for loop=1:1000
        loop
        tfilm=(Told+tinf)/2;
        kair=4.629e-12*tfilm.^3-2.899e-8*tfilm.^2+9.041e-5*tfilm+.001009;
        nuair=-1.725e-14*tfilm.^3+1.027e-10*tfilm.^2+3.725e-8*tfilm-4.196e-6;
        Pr=2.916e-13*tfilm.^4-1.148e-9*tfilm.^3+1.675e-6*tfilm.^2-1.021e-
3*tfilm+0.9126;
        beta=1./tfilm;
        Gr=g*beta.*abs(Told-tinf)*D^3./nuair.^2;
        Ra=Gr.*Pr;
        d1=0.518*Ra.^(1/4);
        d2=1+(0.559./Pr).^(9/16);
        Nu=0.36+d1./d2.^(4/9);
        h=Nu.*kair/D;
        Tnew(1)=Tbase;
        Tnew(2)=(c1*(Told(1)+Told(3))+h(2)*c2*tinf+emiss*sigma*c2*(tsurr^4-    Told(2)^
4))/(2*c1+h(2)*c2);
        Tnew(3)=(c1*(Told(2)+Told(4))+h(3)*c2*tinf+emiss*sigma*c2*(tsurr^4-    Told(3)^
4))/(2*c1+h(3)*c2);
        Tnew(4)=(c1*(Told(3)+Told(5))+h(4)*c2*tinf+emiss*sigma*c2*(tsurr^4-    Told(4)^
4))/(2*c1+h(4)*c2);
        Tnew(5)=(c1*Told(4)+h(5)*(c2/2+A)*tinf+emiss*sigma*(c2/2+A)*(tsurr^4-    Told(5)^
4))/(c1+h(5)*(c2/2+A));
        if abs(Told(2)-Tnew(2))<.0001
                Tnew
                qbase=c1*(Tnew(1)-Tnew(2))+h(1)*c2/2*(Tnew(1)-tinf)+emiss*sigma*c2/
2*(Tnew(1)^4- tsurr^4)
                break
        else
                Told=Tnew;
        end
end
```

***Matlab program for Part (b)—Forced convection.*** ``% Chpt 10 Example 10.6 Part (b) Forced Convection``

```
% Heat Transfer from a Pin Fin
clear, clc
% Properties for the fin
kfin=210; emiss=0.9; D=0.015; A=pi/4*D^2;dx=0.125;
% Kelvin temperatures since we have radiation
tsurr=288.15;
tinf=288.15;
Tbase=200+273.15;
% Air flow velocity
uinf=2;
sigma=5.67e-8;
% Give initial guesses for T2 through T5
Told=[Tbase,450,400,375,300];
% Constants used in the loop
```

```
c1=kfin*A/dx; c2=pi*D*dx;
% Begin iterating
for loop=1:1000
    loop
    tfilm=(Told+tinf)/2;
    kair=4.629e-12*tfilm.^3-2.899e-8*tfilm.^2+9.041e-5*tfilm+.001009;
    nuair=-1.725e-14*tfilm.^3+1.027e-10*tfilm.^2+3.725e-8*tfilm-4.196e-6;
    Pr=2.916e-13*tfilm.^4-1.148e-9*tfilm.^3+1.675e-6*tfilm.^2-1.021e-
3*tfilm+0.9126;
    % Reynolds Number
    Re=uinf*D./nuair;
    d1=0.62*Re.^0.5.*Pr.^(1/3);
    d2=1+(Re./2.82e5).^(5/8);
    d3=1+(0.4./Pr).^(2/3);
    Nu=0.3+d1.*d2.^(4/5)./d3.^(.25);
    h=Nu.*kair/D;
    Tnew(1)=Tbase;
    Tnew(2)=(c1*(Told(1)+Told(3))+h(2)*c2*tinf+emiss*sigma*c2*(tsurr^4-    Told(2)^
4))/(2*c1+h(2)*c2);
    Tnew(3)=(c1*(Told(2)+Told(4))+h(3)*c2*tinf+emiss*sigma*c2*(tsurr^4-    Told(3)^
4))/(2*c1+h(3)*c2);
    Tnew(4)=(c1*(Told(3)+Told(5))+h(4)*c2*tinf+emiss*sigma*c2*(tsurr^4-    Told(4)^
4))/(2*c1+h(4)*c2);
    Tnew(5)=(c1*Told(4)+h(5)*(c2/2+A)*tinf+emiss*sigma*(c2/2+A)*(tsurr^4-    Told(5)^
4))/(c1+h(5)*(c2/2+A));
    if abs(Told(2)-Tnew(2))<0.0001
        Tnew
        qbase=c1*(Tnew(1)-Tnew(2))+h(1)*c2/2*(Tnew(1)-tinf)+emiss*sigma*c2/
2*(Tnew(1)^4- tsurr^4)
        break
    else
        Told=Tnew;
    end
end
```

## Example 10.7
### Temperature measurement

*Problem*
Air is flowing at a rate of 1000 cfm through a 16″ by 12″ duct. A sensor in the center of the flow measures the temperature of the air. The sensor is a sphere of 3 mm diameter. Its surface has an emissivity of 0.4. The measured temperature is 120 F. If the duct walls are at 250 F, what is the actual temperature of the air?

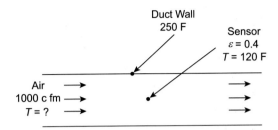

*Solution*

Preliminary: There is convection between the sensor and the flowing air. There is radiation between the sensor and the duct wall. The convection is forced convection over a sphere. The heat transfer is steady state.

The temperature of the temperature sensor is $T_{ts} = 120\,\text{F} = (120 - 32)(5/9) = 48.89\,\text{C} = 322.0\,\text{K}$.

The temperature of the duct wall is $T_{duct} = 250\,\text{F} = (250 - 32)(5/9) = 121.11\,\text{C} = 394.3\,\text{K}$.

An appropriate equation for the forced convection is Whitaker's equation, Eq. (6.111):

$$\text{Nu} = \frac{hD}{k} = 2 + \left(0.4\text{Re}_D^{1/2} + 0.06\text{Re}_D^{2/3}\right)\text{Pr}^{0.4}\left(\frac{\mu}{\mu_s}\right)^{1/4} \tag{6.111}$$

where fluid properties are at the fluid temperature except for $\mu_s$ which is at the surface temperature. We will assume that the fluid temperature is close to the sensor temperature. Then $\mu_s \approx \mu$ and the viscosity ratio factor in Eq. (6.111) can be deleted. (Note: The solution will confirm that this is appropriate.) Fluid properties will be taken at 50 C.

Air at 50 C: $\rho = 1.09\,\text{kg/m}^3$ $\mu = 1.96 \times 10^{-5}\,\text{kg/m s}$ $k = 0.027\,\text{W/m C}$ Pr $= 0.72$

The Reynolds number is $\text{Re}_D = \frac{\rho VD}{\mu}$. We need the velocity $V$ of the flow. We can get this from the volumetric flow rate and the density of the air.

$$Q = 1000\,\text{cfm} = 1000\frac{\text{ft}^3}{\text{min}} \times \frac{1\,\text{min}}{60\,\text{s}} \times \frac{1\,\text{m}^3}{(3.2808\,\text{ft})^3} = 0.472\,\text{m}^3/\text{s}$$

$$A = \text{flow area} = (16)(12) = 192\,\text{inch}^2 \times \left(\frac{0.0254\,\text{m}}{\text{inch}}\right)^2 = 0.1239\,\text{m}^2$$

$$V = \frac{Q}{A} = \frac{0.472}{0.1239} = 3.81\,\text{m/s}$$

Back to the Reynolds number: $\text{Re}_D = \frac{\rho VD}{\mu} = \frac{(1.09)(3.81)(0.003)}{1.96 \times 10^{-5}} = 636$

Putting values into Eq. (6.111), we get Nu $= 14.738$ and $h = 132.6\,\text{W/m}^2$ C

We now consider the energy balance on the temperature sensor.

Heat transfer by convection from the air to the sensor $=$ Heat transfer by radiation from the sensor to the duct wall

$$hA(T_{air} - T_{ts}) = \varepsilon\sigma A\left(T_{ts}^4 - T_{duct}^4\right) \tag{10.60}$$

where $T_{ts}$ is the measurement of the sensor.

Putting in values, we have $132.6(T_{air} - 322.0) = (0.4)\left(5.67 \times 10^{-8}\right)\left[(322.0)^4 - (394.3)^4\right]$ Solving this, we get $T_{air} = 319.7\,\text{K} = 115.8\,\text{F}$

The actual air temperature is 115.8 F. The sensor is reading 4.2 F too high.

## Example 10.8
### Environmental test chamber

*Problem*

An environmental test chamber is cylindrical in shape. The inside dimensions are as follows: The top and bottom are 0.5 m diameter and the height is 0.8 m. There is a vacuum in the chamber during tests. An electrical heater below the chamber keeps the bottom surface of the chamber at 150 C. The sidewall of the chamber is well-insulated. The top of the chamber is a door that can be opened to move specimens in and out of the chamber. The door is 1.5 cm thick and of a material whose thermal conductivity is 15 W/m C. All three interior surfaces of the chamber have an emissivity of 0.88. The outside surface of the door has an emissivity of 0.75. There is convection from the outside surface of the door to the 20 C room air. There is also radiation from the outside surface of the door to the 25 C surroundings.

For steady-state operation of the chamber,

**(a)** What are the temperatures of the surfaces of the door and the temperature of the sidewall?

**(b)** How much power must the electric heater deliver to the chamber to keep the bottom surface of the chamber at 150 C?

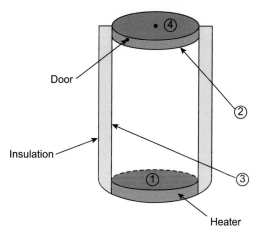

Door

Insulation

Heater

*Solution*

The bottom surface of the chamber is Surface 1. The top surface of the chamber (and bottom of the door) is Surface 2. The sidewall is Surface 3, and the top surface of the door is Surface 4.

Given

$$\varepsilon_1 = \varepsilon_2 = \varepsilon_3 = 0.88 \quad \varepsilon_4 = 0.75$$
$$D = 0.5 \text{ m} \quad H = 0.8 \text{ m}$$
$$\text{Areas}: A_1 = A_2 = A_4 = (\pi/4)D^2 = 0.1963 \text{ m}^2 \quad A_3 = \pi DH = 1.257 \text{ m}^2$$
$$k_{door} = 15 \text{ W/m C} \quad L_{door} = 0.015 \text{ m}$$
$$T_\infty = 20 \text{ C} = 293.15 \text{ K} \quad T_{surr} = 25 \text{ C} = 298.15 \text{ K}$$

Let us first get the shape factors. Fig. 9.6 gives the factor for parallel coaxial disks, Using the equations for Fig. 9.6, we have

$$x = H/r_1 = 0.8/0.25 = 3.2 \quad y = r_2/H = 0.25/0.8 = 0.3125$$
$$S = 1 + x^2(1 + y^2) = 12.24$$
$$F_{1-2} = (1/2)\left\{S - \left[S^2 - 4(xy)^2\right]^{1/2}\right\} = 0.0823$$
$$F_{1-1} + F_{1-2} + F_{1-3} = 1 \quad F_{1-3} = 1 - F_{1-1} - F_{1-2} = 1 - 0 - 0.0823 = 0.9177$$
$$\text{Reciprocity}: A_1 F_{1-3} = A_3 F_{3-1} \quad F_{3-1} = (A_1/A_3)F_{1-3} = (0.1963/1.257)(0.9177) = 0.1433$$

The circuit diagram for radiant exchange in the chamber is given in the figure below.

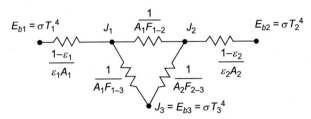

$$E_{b1} = \sigma T_1^4 = (5.67 \times 10^{-8})(150 + 273.15)^4 = 1817.9 \text{ W/m}^2$$

When values are put into the circuit diagram, it becomes

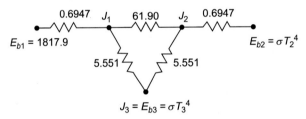

We use this diagram to write equations for nodes $J_1, J_2,$ and $J_3$.
Sum of the heat flows into each node $= 0$.

$$\text{For Node } J_1, \frac{1817.9 - J_1}{0.6947} + \frac{J_2 - J_1}{61.9} + \frac{J_3 - J_1}{5.551} = 0 \tag{1}$$

$$\text{For Node } J_2, \frac{J_1 - J_2}{61.90} + \frac{\sigma T_2^4 - J_2}{0.6947} + \frac{J_3 - J_2}{5.551} = 0 \tag{2}$$

$$\text{For Node} J_3, \frac{J_1 - J_3}{5.551} + \frac{J_2 - J_3}{5.551} = 0 \text{ or } J_1 + J_2 - 2J_3 = 0 \tag{3}$$

At the outer surface of the door, we have

$$\text{Conduction through door} = \text{Convection} + \text{Radiation to Environment}$$

$$q = \frac{k_{door}A_4}{L_{door}}(T_2 - T_4) = h_4 A_4(T_4 - T_\infty) + \varepsilon_4 \sigma A_4 \left(T_4^4 - T_{surr}^4\right)$$

Putting in values, this equation becomes

$$1000(T_2 - T_4) - h_4(T_4 - 293.15) - 4.253 \times 10^{-8}\left(T_4^4 - 298.15^4\right) = 0 \tag{4}$$

We also have a relation equating the radiant energy going to the bottom of the door with the convection and radiation from the outer surface of the door to the environment. Looking at the circuit diagram, we can write

$$\frac{J_2 - \sigma T_2^4}{0.6947} = h_4 A_4(T_4 - T_\infty) + \varepsilon_4 \sigma A_4 \left(T_4^4 - T_{surr}^4\right)$$

Putting in values and rearranging, we get

$$J_2 - \sigma T_2^4 - 0.13637 h_4(T_4 - 293.15) - 5.799 \times 10^{-9}\left(T_4^4 - 298.15^4\right) = 0 \tag{5}$$

Eqs. (4) and (5) have convective coefficient $h_4$. This is for natural convection from the upper surface of the door. The surface is a hot surface facing up, so an appropriate relation for $h_4$ is Eq. (7.24).

$$h_4 = 0.54 \text{Ra}^{1/4}(k_{air} / L_{conv}) \tag{7.24}$$

where $\text{Ra} = \frac{g\beta(T_4 - T_\infty)L_{conv}^3 \text{Pr}}{v_{air}^2}$

$$L_{conv} = 0.25 \quad D = 0.125 \text{ m} \quad g = 9.807 \text{ m/s}^2$$
$$\beta = 1/T_{film} \text{ where } T_{film} = (T_4 + T_\infty)/2$$

$k_{air}$, $v_{air}$, air Pr are given by Eqs. (10.2), (10.4), (10.6)

Eqs. (1) through (5) are solved simultaneously for the unknowns $J_1$, $J_2$, $J_3$, $T_2$, and $T_4$.

Unfortunately, the results of both Excel *Solver* and Matlab (Gauss−Seidel solution method) did not converge acceptably. Convergence for these methods sometimes depends on the initial guesses for the unknowns. For this problem, we were unsuccessful in choosing initial guesses leading to convergence. Due to the non-convergence, we took the following alternative approach:

First we assumed a temperature for the outer surface of the door, $T_4$. With this temperature and the given $T_\infty$, we calculated convective coefficient $h_4$ using Eq. (7.24). Then we calculated the heat flow $q$ to the environment using

$$q = h_4 A_4 (T_4 - T_\infty) + \varepsilon_4 \sigma A_4 (T_4^4 - T_{surr}^4) \tag{6}$$

But this heat flow is also the conduction through the door $q = \frac{k_{door} A_4}{L_{door}} (T_2 - T_4)$. Rearranging, we have $T_2 = T_4 + \frac{q L_{door}}{k_{door} A_4}$.

After we calculate $T_2$, we can calculate $E_{b2} = \sigma T_2^4$.

Looking back at the last circuit diagram, it is seen that the heat flow from the bottom surface of the chamber to the bottom of the door is

$$q = \frac{E_{b1} - E_{b2}}{\text{Resistance between } E_{b1} \text{ and } E_{b2}} = \frac{1817.9 - E_{b2}}{0.6947 + \dfrac{1}{[(1/61.9) + (1/11.102)]} + 0.6947}$$

$$q = \frac{1817.9 - E_{b2}}{10.803} \tag{7}$$

The heat flow in Eq. (7) equals the heat flow in Eq. (6). We try different values of $T_4$ until both heat flows are equal. When that happens, we have the solution. The following Matlab program loops for a range of $T_4$ values and gives a result of the $T_4$ value that makes the two heat flows as equal as possible. The program also gives $T_2$ and $q$.

The results for this problem were $q = 98.29$ W, $T_2 = 339.81$ K, and $T_4 = 339.31$ K

We also need the temperature of the sidewall, $T_3$. The calculations are as follows:

Using the circuit diagram,

$$q = \frac{J_2 - E_{b2}}{0.6947} = 98.29 \text{ W}$$

$$E_{b2} = \sigma T_2^4 = (5.67 \times 10^{-8})(339.81)^4 = 756.0 \text{ W/m}^2$$

$$J_2 = E_{b2} + (0.6947)(98.29) = 756.0 + (0.6947)(98.29) = 824.3 \text{ W/m}^2$$

$$q = \frac{E_{b1} - J_1}{0.6947} = 98.29 \text{ W}$$

$$J_1 = E_{b1} - (0.6947)(98.29) = 1817.9 - (0.6947)(98.29) = 1749.6 \text{ W/m}^2$$

From Eq. (3) above,

$$J_3 = \frac{J_1 + J_2}{2} = \frac{1749.6 + 824.3}{2} = 1287.0 \text{ W / m}^2 = \sigma T_3^4$$

$$T_3 = \left(\frac{1287.0}{5.67 \times 10^{-8}}\right)^{1/4} = 388.15 \text{ K}$$

The solution is

(a) Temperature of the top of the door $= T_4 = 339.31$ K $= 66.2$ C
   Temperature of the bottom of the door $= T_2 = 339.81$ K $= 66.7$ C
   Temperature of sidewall $= T_3 = 388.15$ K $= 115$ C
(b) Power needed to the electric heater $= 98.3$ W

**Matlab program for Example 10.8.** % Example 10.8

```
% Environmental Test Chamber
clear, clc
format compact
% Temperatures in Kelvin
D=0.5;
eps=0.88; eps4=0.75;
Lconv=0.25*D;
Ldoor=0.015;
kdoor=15;
tinf=293.15; tsurr=298.15;
A4=pi/4*D^2; A2=A4;
g=9.807;
sigma=5.67e-8;
min_diff=10;
for t4=330:.001:350
     tfilm=(t4+tinf)/2;
     kair=4.629e-12*tfilm^3-2.899e-8*tfilm^2+9.041e-5*tfilm+.001009;
     nuair=-1.725e-14*tfilm^3+1.027e-10*tfilm^2+3.725e-8*tfilm-4.196e-6;
     Pr=2.916e-13*tfilm^4-1.148e-9*tfilm^3+1.675e-6*tfilm^2-1.021e-3*tfilm+0.9126;
     beta=1/tfilm;
     Gr=g*beta*abs(t4-tinf)*Lconv^3/nuair^2;
     Ra=Gr*Pr;
     Nu=0.54*Ra^0.25;
     h4=Nu*kair/Lconv;
     q1=h4*A4*(t4-tinf)+eps4*sigma*A4*(t4^4-tsurr^4);
     t2=t4+q1*Ldoor/kdoor/A4;
     Eb2=sigma*t2^4;
     q2=(1817.9-Eb2)/10.803;
     diff=abs(q1-q2);
     if diff<min_diff
          min_diff=diff; t2_min=t2; t4_min=t4;q1_min=q1; q2_min=q2;
     end
end
q1_min, q2_min, t2_min, t4_min
```

## 10.4 Chapter summary and final remarks

This chapter provided several examples involving multimode heat transfer. The examples were designed to give the reader additional insight into solution set-up and execution. Solutions were obtained through the use of Microsoft Excel's *Solver* add-in and Matlab programs. The examples in this chapter incorporate material presented in earlier chapters of the book, thereby providing a useful review of previously gained knowledge.

## 10.5 Problems

Notes:

- If needed information is not given in the problem statement, then use the Appendix for material properties and other information. If the Appendix does not have sufficient information, then use the Internet or other reference sources. Some good reference sources are mentioned at the end of Chapter 1. If you use the Internet, double-check the validity of the information by using more than one source.
- Your solution should include a sketch of the problem.
- It is strongly recommended that the reader spend some time (in some cases perhaps as little as a few minutes) to give some thought to the problem before immediately jumping into its solution. Draw the sketch. Consider the questions of Section 10.2. Determine what additional information is needed and think about where it can be obtained. Determine what sections and equations of the text are relevant. Consider the possible solution techniques and choose one. This time in preparation will indeed be time well-spent and will lead to efficient and successful solution of the problem.
- Use Eqs. (10.1)–(10.6) for air properties as needed.

**10-1** The above-ground part of a basement wall is 10-inch-thick poured concrete (k $=$ 1.4 W/m C). The wall is 4 ft high and 15 ft wide. The outside air temperature is 5 C and the inside air temperature is 20 C. There is 3½ inches of fiberglass insulation (k $=$ 0.039 W/m C) on the inner surface of the concrete wall. The air is quiet on both the outside and inside surfaces of the wall. What is the rate of heat flow through this portion of the basement wall?

**10-2** Air at 100 C is blowing at a speed of 3 m/s over the top surface of a horizontal steel plate (k $=$ 40 W/m C). The plate is 0.2 m by 0.4 m by 4 mm thick. The air is blowing parallel to the 0.4 m side. There is stagnant air at 30 C on the bottom surface of the plate.
**(a)** What is the rate of heat flow (W) through the plate?
**(b)** What is the temperature of the bottom surface of the plate?

**10-3** A sandwich of four square plates is on a hot plate. The plates are all 15 cm by 15 and 1 cm thick. Starting at the bottom plate that is in contact with the hot plate, we have: carbon steel (k $=$ 50 W/m C), insulating board (k $=$ 0.02 W/m C), aluminum alloy (k $=$ 180 W/m C), and stainless steel (k $=$ 18 W/m C). The bottom of the carbon steel plate is at 800 C. The exposed surface of the stainless steel plate has natural convection to the 20 C room air.
What is the temperature at the interface between the insulating board and the aluminum?

**10-4** Water at 60 C is flowing at a rate of 5 gpm through a horizontal 1/2 inch Type L copper tube. The air in the factory is quiet and at 15 C. The walls of the factory are also at 15 C. The outer surface of the tube has an emissivity of 0.4. Including both convective and radiative heat transfer,
**(a)** What is the rate of heat flow from the hot water to the room per meter length of tube (W/m)?

(b) What are the temperatures of the inner and outer surfaces of the tube?

(c) Fiberglass insulation ($k$ = 0.035 W/m C), 1.27 cm thick, is placed on the tube. What is the new rate of heat flow to the room? Assume that the emissivity at the outer surface of the fiberglass is 0.4.

10-5 A thin-walled cylindrical container is 61 cm diameter and 150 cm high. It is filled with liquid helium (LHe). The temperature of the LHe is 4.2 K. The bottom of the container is on the floor of a factory. The container's outer surfaces are insulated with 1/2 inch thick multilayer insulation ($k$ = 1 × $10^{-4}$ W/m K). The room air is at 20 C and the emissivity of the outer surface of the insulation is 0.2. What is the rate at which the helium vents from the container (kg/day)? The heat of vaporization of liquid helium is 0.0845 kJ/mol.

10-6 A hollow copper sphere has an inside diameter of 6 cm and an outside diameter of 12 cm. The inner surface of the sphere is at 300 C, and the outer surface convects to the surrounding 25 C air that is quiet. Assume that the outer surface of the sphere is highly polished and radiation heat transfer is negligible.

(a) What is the rate of heat flow through the wall of the sphere?

(b) If insulation (k = 0.035 W/m C) is added to the outer surface of the sphere, what thickness of insulation is needed to reduce the heat flow of Part (a) by 75%?

10-7 A copper sphere has a diameter of 2 cm. It is initially heated to a uniform temperature of 200 C in an oven. It is then removed from the oven and hung by wires in the 20 C room air. Air is blowing over the sphere at a speed of 3 m/s.

(a) How long will it take for the center of the sphere to drop to 60 C if the sphere is highly polished and its emissivity is 0.05 ?

(b) How long will it take for the center of the sphere to drop to 60 C if the sphere is painted black and its emissivity is 0.9 ?

10-8 A horizontal, uninsulated #8 AWG copper wire ($k$ = 385 W/m C) carries a current of 30 A. The resistivity of the wire is 1.68 × $10^{-8}$ Ω·m. The surrounding air is quiet at 22 C. What is the temperature of the surface of the wire?

10-9 A long cylinder of 1.5 cm radius has a conductivity of 5 W/m C and uniform internal heat generation of 8 × $10^{5}$ W/m³. The cylinder is horizontal, and air flows across it at a velocity of 2.6 m/s. The air temperature is 30 C. What is the surface temperature of the cylinder?

10-10 A sphere of conductivity $k$ = 1.5 W/m C has a radius of 4 cm and a heat generation rate of 6 × $10^{4}$ W/m³. The surface of the sphere convects to the surrounding quiet air that is at 25 C.

(a) What is the convective coefficient at the surface of the sphere (W/m² C)?

(b) What is the rate of heat flow (W) to the air?

10-11 A horizontal aluminum pin fin has a square cross section of 0.4 cm by 0.4 cm and a length of 25 cm. The fin is attached to a wall at 150 C. Air at 10 C blows across the fin at a speed of 1.5 m/s. The fin's surface has an emissivity of 0.7. Determine, numerically, the rate of heat transfer to the air and the surroundings.

10-12 The horizontal roof of a house is a 15 cm thick concrete slab ($k$ = 2 W/m C) that is 10 m wide and 20 m long. Wind at 10 C blows at a speed of 15 miles per hour across the roof parallel to the 20  m side. Inside the house, the room air is quiet at 22  C. The emissivity of both surfaces of the roof is 0.9 and the night sky temperature is 100  K.

**(a)** What is the rate of heat transfer through the roof?

**(b)** What is the temperature of the inside surface of the roof?

**10-13**  A double-pane window is 1.5  m high and 2  m wide. The window has two 2.5  mm thick layers of glass ($k = 0.8$  W/m K) separated by a 12  mm thick stagnant layer of argon ($k = 0.016$  W/m K). The air both inside and outside the house is quiet. The inside air is at 22  C, and the outside air is at 4  C. Neglecting radiation, what is the rate of heat transfer through the window?

**10-14**  Consider Figure P 10-14. A transparent film is to be bonded to the top surface of a square plate that is 50  cm by 50  cm. The film ($k = 0.05$  W/m C) is 0.8  mm thick and the plate ($k = 1.5$  W/m C) is 0.5  cm thick. The bottom surface of the plate is maintained at 55  C. In order for proper bonding of the film to the plate, the interface between the film and plate must be maintained at 72  C for a period of time. Hot air is blown across the top of the film at a velocity of 5  m/s. How hot does the air have to be to result in the needed 72  C interface temperature?

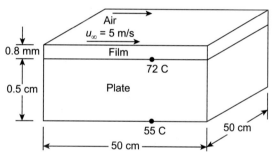

**10-15**  The thin, vertical electric heater shown in Figure P 10-15 produces a heat output of 500  W. The heater is square, 1m by 1m, and is sandwiched between two layers of different materials. Layer A has a conductivity of 0.4  W/m C and a thickness of 1.5  cm. Layer B has a conductivity of 0.03  W/m C and a thickness of 4.5  cm. There is forced convection and radiation from the exposed surfaces of Layers A and B. Air at 20  C and a velocity of 1  m/s flows across the exposed surface of Layer A. Air at 100  C and a velocity of 4  m/s flows across the exposed surface of Layer B. The emissivity of the surface of Layer A is 0.2. The emissivity of the surface of Layer B is 0.8. The surroundings on the Layer A side are at 20  C. The surroundings on the Layer B side are at 40  C. What are the temperatures of the heater and the exposed surfaces of both layers?

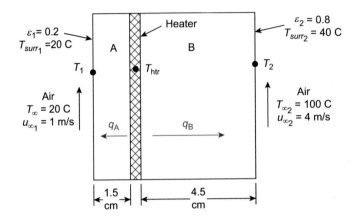

**10-16**  Air at 15 C flows across a 35-cm-diameter cylinder at a speed of 20 m/s. The surface of the cylinder is at 150 C and its emissivity is 0.7. The surroundings are at 30 C. What is the heat transfer per meter length from the cylinder to the air and the surroundings?

**10-17**  A 12-ounce can of beer is initially at a temperature of 85 F. It is placed in a refrigerator having an air temperature of 37 F. How long will it take for the beer to reach the optimum drinking temperature of 44 F?

**10-18**  A cylinder has a diameter of 40 mm and a length of 200 mm. It is being heated in a large furnace whose walls are at 1000 K. Air at 450 K blows across the cylinder at a velocity of 4 m/s. The emissivity of the cylinder's surface is 0.6. What is the temperature of the surface of the cylinder at steady state?

**10-19**  Do Problem 10-18 with the air flow being parallel to the cylinder's surface rather than being across the surface.

**10-20**  A flat horizontal plate receives solar insolation at a rate of 650 W/m$^2$. The plate is square, 1.2 m by 1.2 m, and is perfectly insulated on its back side. The plate's exposed surface has the spectral emissivity shown in Figure P 10-20. The air and sky temperatures are 25 and 10 C, respectively. What is the equilibrium temperature of the plate?

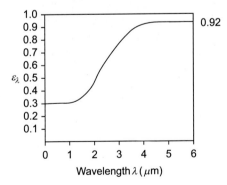

**10-21**  A thick concrete slab $\left(\rho = 2100 \text{ kg}/\text{m}^3, \ c = 910 \text{ J}/\text{kg C}, \ k = 1.4 \text{ W}/\text{m C}\right)$ is at a uniform initial temperature of 280 C. It suddenly starts cooling from its surface to the room air which is quiet at 20 C.

(a) What is the temperature at the surface of the slab after 30 min of cooling?

(b) What is the temperature at a depth of 5 cm from the surface after 30 min of cooling?

**10-22**  A long stainless steel cylinder $\left(\rho = 7900 \text{ kg}/\text{m}^3, \ c = 480 \text{ J}/\text{kg C}, \ k = 15 \text{ W}/\text{m C}\right)$ of 50 cm diameter is initially uniform at 450 C and is being cooled by an air stream that is blowing across it. The air is at 30 C and its velocity is 6 m/s.

(a) How long does it take for the center of the cylinder to reach 250 C?

(b) What is the surface temperature of the cylinder at the time found in Part (a)?

**10-23**  Do Problem 10-22, but with the air quiet at 30 C. The cylinder is horizontal.

**10-24**  A person living in northern New York forgot to drain his sprinkler system. The region is experiencing a cold snap, and the water pipes are only buried 1 foot into the ground. He is worrying about the pipes freezing. The cold snap lasts for 4 days. During this period, the air is blowing over the ground at an average temperature of 14 F and an average speed of 8 mph. The effective sky temperature is 5 F. At the beginning of the cold snap, the ground is at a uniform temperature of 50 F. The ground has a thermal conductivity of 0.7 Btu/h ft F, a density of 115 lbm/ft$^3$, a specific heat of 0.45 Btu/lbm F, and an emissivity of 0.9. Will the pipes freeze during the cold snap?

**10-25**  A vacuum chamber is cubical, with inner dimensions 0.5 × 0.5 × 0.5 m. An electrical heater below the chamber provides a heat flux from the bottom surface into the chamber of 650 W/m$^2$. The sidewalls of the chamber are at a temperature of 120 C. The top of the chamber is a door that can be opened to move specimens in and out of the chamber. The door is 1 cm thick and of a material whose thermal conductivity is 4 W/m C. The bottom surface of the chamber and the sidewalls have an emissivity of 0.8. The door has an emissivity of 0.6 on both sides. There is convection from the outside surface of the door to the 20 C room air. The air is flowing across the door at a speed of 2 m/s. There is also radiation from the outside surface of the door to the 25 C surroundings.

For steady-state operation of the chamber,

(a) What are the temperatures of the top and bottom surfaces of the door and the bottom surface of the chamber?

(b) What is the rate of heat transfer through the door and to the environment?

# Mass transfer

## 11.1 Introduction

This chapter deals with diffusion mass transfer. There are several types of processes that involve mass transfer. Flow of a fluid through a pipe certainly transfers mass. Blowing a gas over a surface for cooling or heating purposes entails mass transfer of the gas. And, currents created near a surface due to natural convection is also mass transfer. However, these types of mass transfer are macroscopic and are not the focus of this chapter. In this chapter, we discuss several topics relevant to diffusion mass transfer. This is mass transfer caused by concentration gradients in a material. We will start the chapter by presenting terminology relevant to concentrations in gas mixtures.

Heat Transfer Principles and Applications. https://doi.org/10.1016/B978-0-12-802296-2.00011-1

## 11.2 Concentrations in a gas mixture

Let us consider a box that contains two different species of gas. One gas, Gas A, is in the left half of the box. The other gas, Gas B, is in the right half of the box. The gases are at the same pressure and temperature and are separated by an impervious partition. When the partition is removed, Gas A starts to move into the Gas B part of the box, and Gas B starts to move into the Gas A part of the box. The diffusion of the gases continues until the mixture of the gases is homogeneous and there are no concentration gradients in the mixture.

Let us now look at a mixture of two or more species in a container. The concentration of each species may be expressed as a mass concentration or a molar concentration. The mass concentration of species $i$ is the mass density $\rho_i$ with units of $kg/m^3$. The molar concentration is $C_i$ with the units of $kmol/m^3$. A "kmol" is a kilogram-mole, which is the amount of gas in kilograms that is numerically equal to the molecular weight $M_i$ of the gas. For example, 1 kmol of oxygen $(O_2)$ is 32 kg; 1 kmol of nitrogen $(N_2)$ is 28 kg; and 1 kmol of argon $(Ar)$ is 39.95 kg.

From the definition of kilogram-mole, the density of species $i$ of the mixture is

$$\rho_i = M_i C_i \tag{11.1}$$

For a gaseous mixture, the mass fraction of species $i$, $mf_i$, is

$$\text{mass fraction} = mf_i = \frac{m_i}{m} = \frac{\rho_i}{\rho} \tag{11.2}$$

where

  $m_i$ is the mass of species $i$

  $m$; is the total mass of the mixture.

The mole fraction of species $i$ in a gaseous mixture is

$$\text{mole fraction} = y_i = \frac{n_i}{n} \tag{11.3}$$

where

  $n_i$ is the number of kmoles of species $i$

  $n$ is the total number of kmoles in the mixture.

Many common gases can be considered to be ideal gases. If all the gases in a mixture are ideal gases, and the mixture itself is an ideal gas, we can use Dalton's law to relate the partial pressure fractions of the individual gases to the mole fractions. Let us assume that each gas is in its individual container of volume $V$ and that all the gases are at the same pressure $P$ and temperature $T$. The gases are then combined into a container of volume $V$. The resulting mixture of gases is at pressure $P$ and temperature $T$.

$$\text{For the mixture, we have } PV = n\overline{R}T \tag{11.4}$$

where

  $n = \sum n_i$ = number of kmoles in the mixture
  $P$ = pressure of the mixture $(kPa)$
  $V$ = volume of the mixture $(m^3)$
  $T$ = absolute temperature of the mixture $(K)$
  $\overline{R}$ = universal gas constant = $8.31446 \text{ kJ}/\text{kmol K}$

Eq. (11.4) can be written for each of the individual species. That is,

$$P_i V = n_i \bar{R} T \tag{11.5}$$

where $P_i$ is the partial pressure of species $i$.

Summing Eq. (11.5) for all the gases in the mixture, we have

$$\left( \sum P_i \right) V = \left( \sum n_i \right) \bar{R} T \tag{11.6}$$

Comparing Eq. (11.6) with Eq. (11.4), we have

$$P = \sum P_i \tag{11.7}$$

That is, the pressure of the mixture is the sum of the partial pressures of the individual species. And, using Eq. (11.3), (11.4), (11.5), we see that the mole fraction $y_i$ of species $i$ is the ratio of the partial pressure of species $i$ to the pressure of the mixture.

$$\text{That is, } y_i = \frac{n_i}{n} = \frac{P_i}{P} \tag{11.8}$$

Let us now use Amagat's law and look at the volume fraction $V_i/V$ of the different gases that form the mixture. Each individual gas is in its own container of volume $V_i$, and all of the gases are at the same pressure $P$ and temperature $T$. The gases are then combined into a container of volume $V$. The gas mixture is at the same pressure and temperature as the individual gases had before they were combined.

$$\text{For each species, before combination, we have } PV_i = n_i \bar{R} T \tag{11.9}$$

$$\text{For the mixture, we have } PV = n \bar{R} T \tag{11.10}$$

Dividing Eq. (11.9) by Eq. (11.10), we have

$$\frac{V_i}{V} = \frac{n_i}{n} \tag{11.11}$$

In summary, for each species of an ideal gas mixture, the volume fraction, the mole fraction, and the ratio of the partial pressure to the total pressure of the mixture are equal.

## Example 11.1
### Converting from volumetric to mass analysis

*Problem*

A volumetric analysis of a gaseous mixture gave the following results:

| | |
|---|---|
| $N_2$ | 82.0% |
| $CO_2$ | 11.0% |
| $O_2$ | 5.0% |
| Ar | 2.0% |

What is
**(a)** the analysis on a mass basis?
**(b)** the molecular weight of the mixture?

(c) the gas constant $R$ of the mixture?

*Solution*

We will assume that all the gases are ideal gases (a good assumption).

**Gas 1: N₂.** Percentage by volume (%) = 82

Molecular weight: $M_1 = 28$ kg/kmol

Mole fraction = $n_1/n$ = volumetric fraction = 0.82

Mass per kmol of mixture = 0.82 (28) = 22.96 kg

**Gas 2: CO₂.** Percentage by volume (%) = 11

Molecular weight: $M_2 = 44$ kg/kmol

Mole fraction = $n_2/n$ = volumetric fraction = 0.11

Mass per kmol of mixture = 0.11 (44) = 4.84 kg

**Gas 3: O₂.** Percentage by volume (%) = 5

Molecular weight: $M_3 = 32$ kg/kmol

Mole fraction = $n_3/n$ = volumetric fraction = 0.05

Mass per kmol of mixture = 0.05 (32) = 1.60 kg

**Gas 4: Ar.** Percentage by volume (%) = 2

Molecular weight: $M_4 = 39.95$ kg/kmol

Mole fraction = $n_4/n$ = volumetric fraction = 0.02

Mass per kmol of mixture = 0.02 (39.95) = 0.80 kg

(a) Total mass in 1 kmol of mixture = 22.96 + 4.84 + 1.60 + 0.80 = 30.20 kg

Percentage by mass

N₂: (22.96/30.20) × 100 = 76.03%

CO₂: (4.84/30.20) × 100 = 16.03%

O₂: (1.60/30.20) × 100 = 5.30%

Ar: (0.80/30.20) × 100 = 2.65%

(b) Molecular weight of mixture = 30.2

(c) Gas constant of mixture = $R = \frac{\bar{R}}{M} = \frac{8.31446}{30.2} = 0.275$ kJ/kg K

## 11.3 Fick's law of diffusion

At the beginning of the previous section, we considered a partitioned box with Gas A in the left side of the box and Gas B in the right side. When the partition is removed, Gas A migrates toward the right side and Gas B migrates toward the left side. For Gas A, a concentration gradient occurs, which causes the gas to diffuse toward the right. This gradient, during the process after removal of the partition, is shown in Fig. 11.1. It is seen that the gradient curve has a negative slope. That is, $\partial C_A / \partial x$ is negative.

At a location $x$, the flux of Gas A, on a molar basis is

$$\frac{\dot{n}_A}{A} = -D_{AB}\frac{\partial C_A}{\partial x} \tag{11.12}$$

where $\frac{\dot{n}_A}{A}$ is the molar flux of species A (kmol/s m²), $D_{AB}$ is the binary diffusion coefficient (m²/s), $C_A$ is the molar concentration of species A (kmol/m³).

Eq. (11.12) is **Fick's law of diffusion.** The law can be expressed on a mass basis by multiplying both sides of Eq. (11.12) by molecular weight $M_A$ and using Eq. (11.1). This results in

$$\frac{\dot{m}_A}{A} = \left(\frac{\dot{n}_A}{A}\right)M_A = -D_{AB}\frac{\partial C_A}{\partial x}(M_A) = -D_{AB}\frac{\partial \rho_A}{\partial x} \tag{11.13}$$

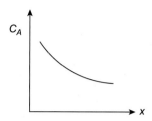

**FIGURE 11.1**

Concentration gradient in Gas A.

where $\frac{\dot{m}_A}{A}$ is the mass flux of species A (kg/s m$^2$), $D_{AB}$ is the binary diffusion coefficient (m$^2$/s), $\rho_A$ is the density of species A (kg/m$^3$).

The minus sign in Eqs. (11.12) and (11.13) makes the molar and mass flows positive for flow in the $+x$ direction.

## 11.3.1 Binary gas diffusion coefficient

Many experiments have been performed to determine binary diffusion coefficients. And, several correlation equations have been proposed to fit experimental data and to predict the coefficient for pairs of gases for which experimental data are unavailable [1—6]. One such equation, by **Fuller, Schettler, and Giddings** [5], was developed by correlating data for 340 experiments. The equation has a very good average error of only 4.3%. The equation is

$$D_{AB} = \frac{10^{-7} T^{1.75} \left( \dfrac{1}{M_A} + \dfrac{1}{M_B} \right)^{1/2}}{P \left[ (V_A)^{1/3} + (V_B)^{1/3} \right]^2}$$  (11.14)

where $D_{AB}$ = binary gas diffusion coefficient (m$^2$/s),

$P$ = total pressure (atm),

$T$ = temperature (K),

$M$ = molecular weight of the gas (kg/kmol),

$V_A$ and $V_B$ = diffusion molecular volumes of the two gases.

The diffusion molecular volumes of some simple gases are given in Table 11.1. For organic vapors, the diffusion volume increments given in Table 11.2 may be appropriately summed to obtain an equivalent volume to use in Eq. (11.14). For example, if methane (CH$_4$) is Gas A, then, using the values in Table 11.2, $V_A = (16.5) + 4(1.98) = 24.42$.

It is seen from Eq. (11.14) that the diffusion coefficient is directly proportional to the temperature to the 1.75 power and inversely proportional to the pressure. Hence, if $D_{AB}$ is known at State 1, then $D_{AB}$ at State 2 can be approximated by using the relation

$$(D_{AB})_2 = (D_{AB})_1 \left( \frac{P_1}{P_2} \right) \left( \frac{T_2}{T_1} \right)^{1.75}$$  (11.15)

**Table 11.1 Diffusion volumes of simple molecules.**

| | | | | | |
|---|---|---|---|---|---|
| $H_2$ | 7.07 | Kr | 22.8 | $(Xe)^a$ | 37.9 |
| He | 2.88 | CO | 18.9 | $(SF_6)$ | 69.7 |
| $N_2$ | 17.9 | $CO_2$ | 26.9 | $(Cl_2)$ | 37.7 |
| $O_2$ | 16.6 | $N_2O$ | 35.9 | $(Br_2)$ | 67.2 |
| Air | 20.1 | $NH_3$ | 14.9 | $(SO_2)$ | 41.1 |
| Ne | 5.59 | $H_2O$ | 12.7 | | |
| Ar | 16.1 | | | | |

$^a$*Items with parentheses are based on only a few data points and may be less accurate than the other items.*

**Table 11.2 Atomic diffusion volume increments.**

| | | | |
|---|---|---|---|
| C | 16.5 | $(N)^a$ | 5.69 |
| H | 1.98 | (Cl) | 19.5 |
| O | 5.48 | (S) | 17.0 |

$^a$*Items with parentheses are based on only a few data points and may be less accurate than the other items.*

(Note: Temperatures in Eq. (11.15) must be in absolute units, e.g., Kelvin.)

Experimental values of gas diffusion coefficients are given in Table 11.3. Most of the experimental values in the table are from a listing in Ref. [5], which can be consulted for the original data sources. If the listed coefficient value was for a temperature and/or pressure different from 300 K and 1 atm, then Eq. (11.15) was used to modify the value to the 300 K, 1 atm reference values.

**Table 11.3 Binary gas diffusion coefficients (at 300 K and 1 atm).**

| Gases | $D_{AB} \times 10^4$ (m²/s) | Gases | $D_{AB} \times 10^4$ (m²/s) | Gases | $D_{AB} \times 10^4$ (m²/s) |
|---|---|---|---|---|---|
| $H_2$–air | 0.73 | $H_2$–$N_2$ | 0.81 | $N_2$–CO | 0.22 |
| He–air | 0.71 | $H_2$–He | 1.14 | $N_2$–$CO_2$ | 0.17 |
| $CO_2$–air | 0.17 | $H_2$–Ar | 0.89 | $N_2$–$NH_3$ | 0.26 |
| $O_2$–air | 0.21 | $H_2$–CO | 0.76 | Ar–CO | 0.19 |
| $H_2O$–air | 0.26 | $H_2$–$CO_2$ | 0.66 | Ar–$CO_2$ | 0.15 |
| $NH_3$–air | 0.25 | $H_2$–$NH_3$ | 0.86 | Ar–$O_2$ | 0.21 |
| $SO_2$–air | 0.13 | He–Ar | 0.75 | Ar–$NH_3$ | 0.24 |
| $H_2$–$H_2O$ | 0.97 | He–CO | 0.72 | $O_2$–$NH_3$ | 0.26 |
| He–$H_2O$ | 0.88 | He–$CO_2$ | 0.61 | $O_2$–$N_2$ | 0.21 |
| $CO_2$–$H_2O$ | 0.19 | He–$N_2$ | 0.74 | CO–$CO_2$ | 0.16 |
| $N_2$–$H_2O$ | 0.25 | He–$NH_3$ | 0.85 | CO–$NH_3$ | 0.25 |
| $O_2$–$H_2O$ | 0.27 | He–$O_2$ | 0.73 | $CO_2$–$N_2O$ | 0.12 |

## Example 11.2

**Diffusion coefficient of oxygen in air**

*Problem*

What is the binary diffusion coefficient of oxygen in air at 500 K and 1.5 atm?

*Solution*

From Table 11.3, the binary diffusion coefficient of oxygen in air at 300 K and 1 atm is $0.21 \times 10^{-4}$ m²/s. Let State 1 be 300 K and 1 atm and State 2 be 500 K and 1.5 atm. Then, from Eq. (11.15),

$$(D_{AB})_2 = (D_{AB})_1 \left(\frac{P_1}{P_2}\right)\left(\frac{T_2}{T_1}\right)^{1.75} = 0.21 \times 10^{-4} \left(\frac{1}{1.5}\right)\left(\frac{500}{300}\right)^{1.75} = 3.4 \times 10^{-5} \text{ m}^2/s$$

The binary diffusion coefficient of oxygen in air at 500 K and 1.5 atm is $\mathbf{3.4 \times 10^{-5} \text{ m}^2/\text{s}}$.

## 11.3.2 Binary gas–liquid diffusion coefficient

The previous section provided binary diffusion coefficients for several different gas pairs. There can also be diffusion of gases through liquids. Table 11.4 gives the diffusion coefficients for several gases in water [7].

## 11.4 Diffusion in gases

In this section, we consider two topics. The first is Stefan's law, which involves evaporation of a liquid into and through a stagnant gas. A couple of relevant processes are the drying of fabrics and the drying of paint. The second topic is equimolar counterdiffusion, which is relevant to distillation processes.

**Table 11.4 Gas–water diffusion coefficient $D_{AB}$.**

| Gas | T(C) | $D_{AB}$ (m²/s) | Gas | T(C) | $D_{AB}$ (m²/s) |
|-----|------|-----------------|-----|------|-----------------|
| $H_2$ | 15 | $4.1 \times 10^{-9}$ | $N_2$ | 25 | $2.0 \times 10^{-9}$ |
| $H_2$ | 20 | $4.6 \times 10^{-9}$ | $NO_2$ | 20 | $1.2 \times 10^{-9}$ |
| $H_2$ | 25 | $5.1 \times 10^{-9}$ | $NO_2$ | 25 | $1.4 \times 10^{-9}$ |
| $H_2$ | 30 | $5.7 \times 10^{-9}$ | $NO_2$ | 30 | $1.6 \times 10^{-9}$ |
| $O_2$ | 15 | $1.7 \times 10^{-9}$ | $N_2O$ | 25 | $2.6 \times 10^{-9}$ |
| $O_2$ | 20 | $2.0 \times 10^{-9}$ | $SO_2$ | 20 | $1.6 \times 10^{-9}$ |
| $O_2$ | 25 | $2.4 \times 10^{-9}$ | $SO_2$ | 25 | $1.8 \times 10^{-9}$ |
| He | 15 | $6.2 \times 10^{-9}$ | $SO_2$ | 30 | $2.1 \times 10^{-9}$ |
| He | 20 | $6.7 \times 10^{-9}$ | $Cl_2$ | 25 | $1.9 \times 10^{-9}$ |
| He | 25 | $7.3 \times 10^{-9}$ | Ar | 25 | $2.5 \times 10^{-9}$ |
| He | 30 | $7.9 \times 10^{-9}$ | $CH_4$ | 25 | $1.8 \times 10^{-9}$ |

## 11.4.1 Stefan's law

Let's consider the open container shown in Fig. 11.2. The container contains Liquid A and a stagnant column of Gas B.

There is isothermal evaporation of liquid species A from the free surface of the liquid and the diffusion of the gaseous species A through the stagnant column of Gas B. We will assume that the process is steady, the gas pressure $P$ in the container is constant, and that Gas B is not soluble in liquid A. We will also assume that the two gases, A and B, are ideal gases.

There is movement of Gas B across the opening of the container to sweep away any amount of Gas A reaching the opening. That is, the concentration of Gas A at the opening is negligible. The movement of Gas B is gentle and has negligible effect on the gas movement inside the container.

Species A is evaporated from the liquid and moves upward by diffusion through the column of Gas B. Gas B moves downward by diffusion. However, Gas B is not soluble in liquid A so there is no downward movement of B at the liquid A surface. To have steady diffusion, there must be a bulk flow of A and B upward against the downward diffusion of B.

The downward mass diffusion flux of species B at a location $x$ is, by Eq. (11.13),

$$\frac{\dot{m}_B}{A} = -D_{BA}\frac{\partial \rho_B}{\partial x} \tag{11.16}$$

where $A$ is the cross-sectional area of the container.

From the ideal gas law, $\rho_B = \left(\frac{P_B M_B}{RT}\right)$. Therefore Eq. (11.16) becomes

$$\frac{\dot{m}_B}{A} = -D_{BA}\left(\frac{M_B}{RT}\right)\frac{\partial P_B}{\partial x} \tag{11.17}$$

The bulk mass flux of B upward at a location $x$ is

$$\frac{\dot{m}_{B,bulk}}{A} = \rho_B V \tag{11.18}$$

where $V$ is the bulk upward mass velocity.

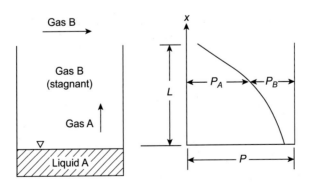

**FIGURE 11.2**

Evaporation through a stagnant gas layer.

Using the ideal gas law, Eq. (11.18) becomes

$$\frac{\dot{m}_{B,bulk}}{A} = \left(\frac{P_B M_B}{\overline{R}T}\right)V \tag{11.19}$$

We have steady state, and Gas B is stagnant in the container. Therefore, at any location $x$ in the container, the downward diffusion of B is counteracted by the upward bulk flow of B. We can use Eqs. (11.17) and (11.19) to solve for the bulk upward mass velocity, which is

$$V = D_{BA}\left(\frac{1}{P_B}\right)\frac{\partial P_B}{\partial x} \tag{11.20}$$

We now look at the upward diffusion of Gas A and the upward bulk flow of Gas A. Following the equations obtained for Gas B, we have:

The upward mass diffusion flux of species A at location $x$ is

$$\frac{\dot{m}_A}{A} = -D_{AB}\left(\frac{M_A}{\overline{R}T}\right)\frac{\partial P_A}{\partial x} \tag{11.21}$$

The bulk mass flux of A upward at location $x$ is

$$\frac{\dot{m}_{A,bulk}}{A} = \left(\frac{P_A M_A}{\overline{R}T}\right)V \tag{11.22}$$

Summing the mass fluxes from Eqs. (11.21) and (11.22), the total upward mass flux of A at location $x$ is

$$\left(\frac{\dot{m}_A}{A}\right)_{total} = -D_{AB}\left(\frac{M_A}{\overline{R}T}\right)\frac{\partial P_A}{\partial x} + \left(\frac{P_A M_A}{\overline{R}T}\right)V \tag{11.23}$$

Using $V$ from Eq. (11.20), Eq. (11.23) becomes

$$\left(\frac{\dot{m}_A}{A}\right)_{total} = -D_{AB}\left(\frac{M_A}{\overline{R}T}\right)\frac{\partial P_A}{\partial x} + \left(\frac{P_A M_A}{\overline{R}T}\right)\frac{D_{BA}}{P_B}\frac{\partial P_B}{\partial x} \tag{11.24}$$

The total pressure $P$ in the container is the sum of partial pressures $P_A$ and $P_B$. It is also constant, i.e., does not vary with location $x$. Differentiating the total pressure, we get

$$\frac{\partial P}{\partial x} = \frac{\partial(P_A + P_B)}{\partial x} = \frac{\partial P_A}{\partial x} + \frac{\partial P_B}{\partial x} = 0 \tag{11.25}$$

$$\text{And, } \frac{\partial P_B}{\partial x} = -\frac{\partial P_A}{\partial x} \tag{11.26}$$

Using this result and recognizing that $D_{AB} = D_{BA}$, Eq. (11.24) becomes

$$\left(\frac{\dot{m}_A}{A}\right)_{total} = D_{AB}\left(\frac{M_A}{\overline{R}T}\right)\left(1 - \frac{P_A}{P_B}\right)\frac{\partial P_A}{\partial x} \tag{11.27}$$

As $P_B = P - P_A$, this equation may be written as

$$\left(\frac{\dot{m}_A}{A}\right)_{total} = -D_{AB}\left(\frac{M_A}{\overline{R}T}\right)\left(\frac{P}{P - P_A}\right)\frac{\partial P_A}{\partial x} \tag{11.28}$$

Eq. (11.28) is known as **Stefan's law**.

With diffusion solely in the $x$ direction, $\partial P_A / \partial x = dP_A / dx$. Both sides of Eq. (11.28) may be integrated over the gas height $L$ of the container to get the mass flow of species A out of the container:

$$\dot{m}_A = D_{AB} A \left( \frac{M_A}{\overline{R}T} \right) \left( \frac{P}{L} \right) \ln \left[ \frac{(P - P_A)_{x=L}}{(P - P_A)_{x=0}} \right] \qquad (11.29)$$

where

$\dot{m}_A$ = evaporation rate of liquid A (kg/s)

$D_{AB}$ = binary gas diffusion coefficient $(m^2 / s)$

$A$ = cross − sectional area of container $(m^2)$

$M_A$ = molecular weight of species A (kg/kmol)

$\overline{R}$ = universal gas constant = 8.31446 kJ/kmol K

$P$ = total pressure of gases in container (kPa)

$P_A$ = partial pressure of Gas A (kPa)

$L$ = gas height in container (m)

This equation can be applied in an experiment to determine binary diffusion coefficients. The experimental setup is essentially identical to the open container discussed in this section. A vertical tube, closed at the bottom, contains the liquid of one of the gas species. The other species blows gently across the open upper end of the tube. The liquid in the tube is maintained at a constant temperature, and the evaporation of the liquid is measured by the drop of liquid level in the tube. Knowledge of the evaporation rate $\dot{m}_A$ of the liquid affords the determination of the gas diffusion coefficient $D_{AB}$. The tube used in the experiment is called a **Stefan tube**.

## Example 11.3
### Evaporation of water from a pail

*Problem*

A five-gallon pail from a home improvement store contains 1 inch of water. If the water is at 20 C and the room air is dry and at 20 C and 1 atm pressure, estimate the time it will take for all of the water in the pail to evaporate.

*Solution*

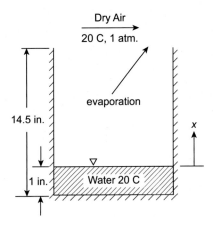

From the website of a home improvement store, a 5-gallon pail has an inside diameter of 12 inches and a height of 14.5 inches.

Diameter of pail $= D = 12$ in $\times$ (0.0254 m/inch) $= 0.3048$ m

$$\text{Cross}-\text{sectional area of pail}=A = \frac{\pi}{4}D^2 = \frac{\pi}{4}(0.3048)^2 = 0.0730 \text{ m}^2$$

$$\text{Volume of water in pail} = (A)(\text{water height}) = (0.0730 \text{ m}^2)(1 \text{ inch})\left(\frac{0.0254 \text{ m}}{\text{inch}}\right) = 0.001853 \text{ m}^3$$

$$\text{Mass of water in pail} = \rho_{water} V = (998 \text{ kg}/\text{m}^3)(0.001853 \text{ m}^3) = 1.85 \text{ kg}$$

Evaporation rate from Eq. (11.29):

$$\dot{m}_A = D_{AB}A\left(\frac{M_A}{RT}\right)\left(\frac{P}{L}\right)\ln\left[\frac{(P-P_A)_{x=L}}{(P-P_A)_{x=0}}\right] \tag{11.30}$$

Species A is the water vapor. Species B is the air.

At $x = 0$ (water surface), $P_A = P_{sat}$ at 20 C $= 2.339$ kPa (from steam tables)

At $x = L$ (pail top), $P_A = 0$ kPa (dry air)

$L =$ average height of air above water surface $= 14$ inch x (0.0254 m/inch) $= 0.3556$ m.

$P = 1$ atm $= 101.325$ kPa. $M_A =$ molecular weight of water $= 18$ kg/kmol.

From Table 11.3, $D_{AB}$ for a water–air mixture at 300 K and 1 atm is $0.26 \times 10^{-4}$ m²/s. Using Eq. (11.15), $D_{AB}$ at 293.15 K and 1 atm is $0.26 \times 10^{-4}\left(\frac{293.15}{300}\right)^{1.75} = 0.25 \times 10^{-4}$ m²/s.

From Eq. (11.30), the mass flow of water vapor out of the pail is

$$\dot{m}_A = D_{AB}A\left(\frac{M_A}{RT}\right)\left(\frac{P}{L}\right)\ln\left[\frac{(P-P_A)_{x=L}}{(P-P_A)_{x=0}}\right]$$

$$= (0.25 \times 10^{-4})(0.0730)\left(\frac{18}{(8.31446)(293.15)}\right)\left(\frac{101.325}{0.3556}\right)\ln\left(\frac{101.325-0}{101.325-2.339}\right)$$

$$= 8.97 \times 10^{-8} \text{ kg/s}$$

The time to evaporate the 1.85 kg of water in the pail is therefore

$$\Delta t = \frac{1.85 \text{ kg}}{8.97 \times 10^{-8} \text{ kg/s}} = (2.06 \times 10^7 \text{ s})\left(\frac{1 \text{ h}}{3600 \text{ s}}\right)\left(\frac{1 \text{ day}}{24 \text{ h}}\right) = 238 \text{ days}$$

It will take about 8 months to evaporate the 1 inch of water in the pail.

## 11.4.2 Equimolar counterdiffusion

Let us now look at Fig. 11.3 that shows two reservoirs connected by a tube. The tube has flow area $A$ and length $L$. The reservoirs contain gas mixtures of species A and B, with the left reservoir having a larger concentration of A than B and the right reservoir having a larger concentration of B than A. Both reservoirs are at the same pressure $P$ and temperature $T$. When the valve is opened, Gas A starts to diffuse to the right and Gas B starts to diffuse to the left.

The diffusion of the gases follows Fick's law. At steady state in the tube, we have, for the molar fluxes, from Eq. (11.12),

$$\frac{\dot{n}_A}{A} = -D_{AB}\frac{\partial C_A}{\partial x} \quad \text{and} \quad \frac{\dot{n}_B}{A} = -D_{AB}\frac{\partial C_B}{\partial x} \tag{11.31}$$

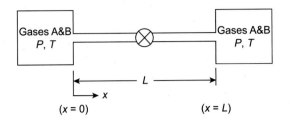

**FIGURE 11.3**

Equimolar counterdiffusion.

For the mass fluxes, we have, from Eq. (11.13),

$$\frac{\dot{m}_A}{A} = -D_{AB}\frac{\partial \rho_A}{\partial x} \quad \text{and} \quad \frac{\dot{m}_B}{A} = -D_{AB}\frac{\partial \rho_B}{\partial x} \tag{11.32}$$

Using Eq. (11.1), Eq. (11.31) may be written as

$$\frac{\dot{n}_A}{A} = -D_{AB}\left(\frac{1}{M_A}\right)\frac{\partial \rho_A}{\partial x} \quad \text{and} \quad \frac{\dot{n}_B}{A} = -D_{AB}\left(\frac{1}{M_B}\right)\frac{\partial \rho_B}{\partial x} \tag{11.33}$$

We will assume that the two gases are ideal. Then, for each species $i$, $\rho_i = \frac{P_i M_i}{RT}$ and Eq. (11.33) can be written in terms of pressure rather than density:

$$\frac{\dot{n}_A}{A} = -D_{AB}\left(\frac{1}{RT}\right)\frac{\partial P_A}{\partial x} \quad \text{and} \quad \frac{\dot{n}_B}{A} = -D_{AB}\left(\frac{1}{RT}\right)\frac{\partial P_B}{\partial x} \tag{11.34}$$

The total pressure $P$ in the reservoirs and tube is the sum of partial pressures $P_A$ and $P_B$ of the two gases. This pressure is constant and does not vary with location. Differentiating the total pressure with $x$, we have

$$\frac{\partial P}{\partial x} = \frac{\partial (P_A + P_B)}{\partial x} = \frac{\partial P_A}{\partial x} + \frac{\partial P_B}{\partial x} = 0 \tag{11.35}$$

From Eq. (11.35), we have $\frac{\partial P_A}{\partial x} = -\frac{\partial P_B}{\partial x}$. Putting this result into Eq. (11.34), we see that $\dot{n}_A$ and $\dot{n}_B$ have the same magnitude. However, the flows are in opposite directions. In the tube, Gas A flows to the right ($+x$ direction) while Gas B flows to the left ($-x$ direction).

Integrating Eq. (11.34) over the length $L$ of the tube, we get the molar fluxes (kmol/s m$^2$) of the two gas species:

$$\frac{\dot{n}_A}{A} = -D_{AB}\left(\frac{1}{RT}\right)\left[\frac{(P_A)_{x=L} - (P_A)_{x=0}}{L}\right] \quad \text{and} \quad \frac{\dot{n}_B}{A} = -D_{AB}\left(\frac{1}{RT}\right)\left[\frac{(P_B)_{x=L} - (P_B)_{x=0}}{L}\right] \tag{11.36}$$

The mass fluxes (kg/s m$^2$) of the two gas species are

$$\frac{\dot{m}_A}{A} = -D_{AB}\left(\frac{M_A}{RT}\right)\left[\frac{(P_A)_{x=L} - (P_A)_{x=0}}{L}\right] \quad \text{and} \quad \frac{\dot{m}_B}{A} = -D_{AB}\left(\frac{M_B}{RT}\right)\left[\frac{(P_B)_{x=L} - (P_B)_{x=0}}{L}\right] \tag{11.37}$$

These results are very useful in determining the rate of venting of a gas from a tank. This is illustrated in Example 11.4.

## Example 11.4
### Venting of helium to the atmosphere

*Problem*

A large tank of helium at 1 atm and 30 C vents to the atmosphere, which is also at 1 atm and 30 C. Venting is through a tube that is 1.5 cm diameter and 30 cm long.
**(a)** What is the rate (g/hr) at which the helium is venting from the tank?
**(b)** What is the rate (g/hr) at which the air in the atmosphere is going into the tank?

*Solution*

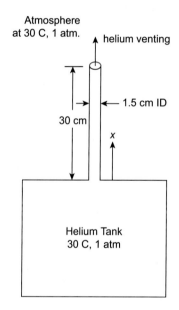

Let the helium be Gas A and the air be Gas B.
Then $M_A = 4$ and $M_B = 29$.
For the helium–air mixture: From Table 11.3, $D_{AB} = 0.71 \times 10^{-4}$ m$^2$/s at 1 atm and 300 K.
We have $T = 30$ C $= 303.15$ K. Correcting the diffusion coefficient to this temperature, using Eq. (11.15), we have

$$D_{AB} = 0.71 \times 10^{-4}(303.15/300)^{1.75} = 0.723 \times 10^{-4} \text{ m}^2/\text{s}.$$

The flow area of the tube is $A = \frac{\pi}{4}(0.015)^2 = 0.0001767$ m$^2$.
The universal gas constant is $\bar{R} = 8.31446$ kJ/kmol K.
The pressure in the atmosphere and in the tank is 1 atm $= 101.325$ kPa.
It is assumed that the concentration of helium in the atmospheric air is negligible and that the concentration of air in the tank of helium is negligible. Therefore,

at the tank end of the tube $(x = 0)$: $\begin{aligned} P_A &= P_{He} = 101.325 \text{ kPa} \\ P_B &= P_{air} \approx 0 \end{aligned}$

at the open end of the tube ($x = L = 0.3$ m): $\quad P_A = P_{He} \approx 0$
$$P_B = P_{air} = 101.325 \text{ kPa}$$
(a) From Eq. (11.37),

$$\frac{\dot{m}_{He}}{A} = \frac{\dot{m}_A}{A} = -D_{AB}\left(\frac{M_A}{RT}\right)\left[\frac{(P_A)_{x=L} - (P_A)_{x=0}}{L}\right] = -0.723 \times 10^{-4}\left(\frac{4}{(8.31446)(303.15)}\right)\left(\frac{0 - 101.325}{0.3}\right)$$

$$= 3.875 \times 10^{-5} \text{ kg/s m}^2$$

$$\dot{m}_{He} = (3.875 \times 10^{-5})A = (3.875 \times 10^{-5})(0.0001767) = 6.848 \times 10^{-9} \text{ kg/s}$$

$$\dot{m}_{He} = 6.848 \times 10^{-9} \text{ kg/s} \left(\frac{3600 \text{ s}}{1 \text{ h}}\right)\left(\frac{1000 \text{ g}}{1 \text{ kg}}\right) = 0.025 \text{ g/h}$$

The helium vents from the tank at a rate of 0.025 g/hr.
(b) From Eq. (11.37),

$$\frac{\dot{m}_{air}}{A} = \frac{\dot{m}_B}{A} = -D_{AB}\left(\frac{M_B}{RT}\right)\left[\frac{(P_B)_{x=L} - (P_B)_{x=0}}{L}\right] = -0.723 \times 10^{-4}\left(\frac{29}{(8.31446)(303.15)}\right)\left(\frac{101.325 - 0}{0.3}\right)$$

$$= -2.809 \times 10^{-4} \text{ kg/s m}^2$$

$$\dot{m}_{air} = -(2.809 \times 10^{-4})A = (2.809 \times 10^{-4})(0.0001767) = -4.964 \times 10^{-8} \text{ kg/s}$$

$$\dot{m}_{air} = -4.964 \times 10^{-8} \text{ kg/s} \left(\frac{3600 \text{ s}}{1 \text{ h}}\right)\left(\frac{1000 \text{ g}}{1 \text{ kg}}\right) = -0.179 \text{ g/h}$$

The air diffuses into the tank at a rate of 0.179 g/h.
**Note:** The mass flow rates of the helium and air are different. The air, having the greater molecular weight, flows at a greater rate than the helium. The molar flow rates of both gases are the same. Use of Eq. (11.36) gives the molar flow rate as $1.712 \times 10^{-9}$ kmol/s.

## 11.5 The mass-heat analogy

In Chapter 3, we discussed the heat-electric analogy, which was very useful in deriving equations for heat flow through layered systems such as composite walls. In this section, we will discuss another analogy, the mass-heat analogy.

Let us first go back to Fick's law of diffusion, Eq. (11.12):

$$\frac{\dot{n}_A}{A} = -D_{AB}\frac{\partial C_A}{\partial x} \tag{11.38}$$

On the left side of the equation, we have the molar flux and on the right side we have the binary gaseous diffusion coefficient and the concentration gradient. It is the concentration gradient that causes the molar flux. Fick's law is analogous to Fourier's law of heat conduction, stated in Eq. (1.3):

$$\frac{q_x}{A} = -k\frac{\partial T}{\partial x} \tag{11.39}$$

On the left side of the equation is the heat flux and on the right side is the thermal conductivity and the temperature gradient which is the driving force for the heat flux. By comparing Eqs. (11.38) and (11.39), it is seen that the molar flux corresponds to the heat flux, the gaseous diffusion coefficient corresponds to the thermal conductivity, and the concentration gradient corresponds to the temperature gradient.

### 11.5.1 Mass transfer through walls and membranes

In Chapter 3, we integrated Fourier's law and applied appropriate boundary conditions to obtain relations for the heat transfer rate $q$ through a variety of walls: flat walls, cylindrical walls, and spherical walls. These relations were given in Eqs. (3.10), (3.55), and (3.73) and were the following:

$$\text{For flat walls}: \quad q = \frac{kA}{L}(T_1 - T_2) \tag{11.40}$$

$$\text{For cylindrical walls}: \quad q = \frac{2\pi kL}{\ln\left(\dfrac{r_o}{r_i}\right)}(T_i - T_o) \tag{11.41}$$

$$\text{For spherical walls}: \quad q = \frac{4\pi k}{\left(\dfrac{1}{r_i} - \dfrac{1}{r_o}\right)}(T_i - T_o) \tag{11.42}$$

In Eq. (11.40), $L$ is the thickness of the wall, $A$ is the cross-sectional area of the wall, and $T_1$ and $T_2$ are the temperatures of the two surfaces of the wall.

In Eq. (11.41), $L$ is the length of the cylinder, $r_i$ and $r_o$ are the inner and outer radii of the wall, and $T_i$ and $T_o$ are the inner and outer surface temperatures of the wall.

In Eq. (11.42), $r_i$ and $r_o$ are the inner and outer radii of the wall and $T_i$ and $T_o$ are the inner and outer surface temperatures of the wall.

The analogous relations for molar flow of gas species A through walls are given in Eqs. (11.43), (11.44), and (11.45).

$$\text{For flat walls}: \dot{n}_A = \frac{D_{AB}A}{L}(C_{A_1} - C_{A_2}) \tag{11.43}$$

$$\text{For cylindrical walls}: \dot{n}_A = \frac{2\pi D_{AB}L}{\ln\left(\dfrac{r_o}{r_i}\right)}(C_{A_i} - C_{A_o}) \tag{11.44}$$

$$\text{For spherical walls}: \dot{n}_A = \frac{4\pi D_{AB}}{\left(\dfrac{1}{r_i} - \dfrac{1}{r_o}\right)}(C_{A_i} - C_{A_o}) \tag{11.45}$$

Table 11.5 gives some binary diffusion coefficients in solids [8−10].

From Eqs. (11.43)−(11.45), we see that the gas concentrations at the boundaries of the solid wall or membrane are needed for determination of the molar flux. At each boundary, there is a solid side and a gas side. We need the gas concentration on the solid side of the boundary. This concentration is related to the partial pressure of the diffusing gas by the **Solubility $S$**. That is

$$C_{A, \text{ solid side}} = SP_{A, \text{ gas side}} \tag{11.46}$$

**Table 11.5 Binary diffusion coefficient $D_{AB}$ in solids.**

| A | B | T(C) | $D_{AB}$ $(m^2/s)$ | A | B | T(C) | $D_{AB}$ $(m^2/s)$ |
|---|---|---|---|---|---|---|---|
| $H_2$ | Fe | 20 | $2.59 \times 10^{-13}$ | $O_2$ | Rubber | 25 | $1.58 \times 10^{-10}$ |
| $H_2$ | Fe | 100 | $1.24 \times 10^{-11}$ | $O_2$ | Lexan | 25 | $2.1 \times 10^{-12}$ |
| $H_2$ | Ni | 85 | $1.16 \times 10^{-12}$ | C | Fe | 800 | $1.5 \times 10^{-12}$ |
| $H_2$ | Ni | 165 | $10.5 \times 10^{-12}$ | C | Fe | 1100 | $4.5 \times 10^{-11}$ |
| $H_2$ | $SiO_2$ | 20 | $6.5 \times 10^{-14}$ | O | Fe | 1000 | $1.0 \times 10^{-14}$ |
| $H_2$ | $SiO_2$ | 500 | $1.3 \times 10^{-12}$ | N | Fe | 1100 | $4.0 \times 10^{-11}$ |
| $H_2$ | Lexan | 25 | $0.64 \times 10^{-10}$ | He | Pyrex | 20 | $4.5 \times 10^{-15}$ |
| $H_2$ | Rubber | 25 | $10.2 \times 10^{-10}$ | He | Pyrex | 500 | $2 \times 10^{-12}$ |
| $H_2$ | Polystyrene | 25 | $4.36 \times 10^{-10}$ | He | $SiO_2$ | 20 | $4 \times 10^{-14}$ |
| $O_2$ | Polystyrene | 25 | $0.11 \times 10^{-10}$ | He | $SiO_2$ | 500 | $7.8 \times 10^{-12}$ |
| $CO_2$ | Rubber | 25 | $1.1 \times 10^{-10}$ | He | Rubber | 25 | $2.16 \times 10^{-9}$ |
| $CH_4$ | Rubber | 25 | $0.89 \times 10^{-10}$ | $N_2$ | Rubber | 25 | $1.1 \times 10^{-10}$ |

where

$C_{A, \text{ solid side}}$ = concentration of gas species A at the solid side of the boundary $(kmol/m^3)$
$S$ = solubility for the gas–solid combination $(kmol/m^3 \text{ bar})$
$P_{A, \text{ gas side}}$ = partial pressure of gas species A on the gas side of the boundary (bar).

Some solubilities are given in Table 11.6 [8].

Another parameter often used in flow through walls and membranes is the **Permeability P.** This is the product of the diffusion coefficient and the solubility. That is,

$$P = D_{AB}S \qquad (11.47)$$

where

$P$ = permeability $(kmol/m \text{ s bar})$
$S$ = solubility for the gas-solid combination $(kmol/m^3 \text{ bar})$
$D_{AB}$ = diffusion coefficient $(m^2/s)$.

**Table 11.6 Solubility of gas–solid combinations.**

| Gas | Solid | T (K) | $S$ (kmol/m³ bar) |
|---|---|---|---|
| $O_2$ | Rubber | 298 | $3.12 \times 10^{-3}$ |
| $N_2$ | Rubber | 298 | $1.56 \times 10^{-3}$ |
| $H_2$ | Rubber | 298 | $2.1 \times 10^{-3}$ |
| $CO_2$ | Rubber | 298 | $4.015 \times 10^{-2}$ |
| He | $SiO_2$ | 293 | $4.5 \times 10^{-4}$ |
| $H_2$ | Ni | 358 | $9.01 \times 10^{-3}$ |

The next example, Example 11.5, shows that gases can indeed pass through solid containers. However, the leakage rate is usually very, very small.

---

## Example 11.5
### Diffusion of helium through a fused silica container

*Problem*

A spherical container of fused silica has a diameter of 2 m and a wall thickness of 2.5 cm. It contains helium at 25 C and a pressure of 2 atm. What is the diffusion rate of the helium through the container wall?

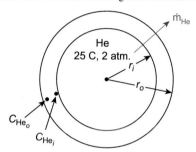

*Solution*

Let us assume that the diameter given is the inside diameter of the container. Then $r_i = 1$ m and $r_o = 1.025$ m. The molar rate of flow of the helium through the container wall is given by Eq. (11.45):

$$\dot{n}_A = \frac{4\pi D_{AB}}{\left(\dfrac{1}{r_i} - \dfrac{1}{r_o}\right)}(C_{A_i} - C_{A_o}) \tag{11.48}$$

We need the diffusion coefficient $D_{AB}$ for the helium-fused silica system and the concentrations $C_{A_i}$ and $C_{A_o}$ of the helium at the inner and outer surfaces of the container.

From Table 11.5, $D_{AB} = 4 \times 10^{-14}$ m$^2$/s.

From Table 11.6, the solubility for the helium-fused silica system $S = 4.5 \times 10^{-4}$ kmol/m$^3$ bar.

The helium pressure is 2 atm. Converting this to bars, we have

$P = 2$ atm x (101.325 kPa/1 atm) x (1 bar/100 kPa) = 2.0265 bar.

The container has only helium, so the partial pressure of the helium is 2.0265 bar.

From Eq. (11.46), the concentration of the helium on the silica side of the inner surface is

$$C_{A, \text{ solid–side}} = SP_{A, \text{ gas–side}} = (4.5 \times 10^{-4})(2.0265) = 0.000912 \text{ kmol/m}^3$$

On the silica side of the outer surface, the concentration of helium is essentially zero as there is negligible helium in the adjacent atmospheric air. Putting values into Eq. (11.48), we have, for the molar diffusion of helium through the container wall,

$$\dot{n}_A = \frac{4\pi D_{AB}}{\left(\dfrac{1}{r_i} - \dfrac{1}{r_o}\right)}(C_{A_i} - C_{A_o}) = \frac{4\pi(4 \times 10^{-14})}{\left(1 - \dfrac{1}{1.025}\right)}(0.000912 - 0) = 1.88 \times 10^{-14} \text{ kmol/s}$$

On a mass basis, the diffusion is

$$\dot{m}_A = \dot{n}_A M_A = (1.88 \times 10^{-14} \text{ kmol / s})(4 \text{ kg / kmol}) = 7.52 \times 10^{-14} \text{ kg/s} = 0.0024 \text{ g/year}$$

Yes, there will be leakage of helium through container wall, but the leakage will be very, very small!

Example 11.6 illustrates diffusion through a flat membrane having different gases on the two sides.

## Example 11.6

### Diffusion of hydrogen and oxygen through a flat membrane

*Problem*

A rubber membrane, 0.5 mm thick, separates hydrogen and oxygen gases. The hydrogen is at 25 C and 2 atm pressure and the oxygen is at 25 C and 0.9 atm pressure. What are the initial rates of mass flux of the hydrogen and oxygen through the membrane?

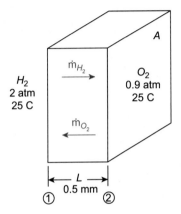

*Solution*

The molar flow through the membrane is given by Eq. (11.43):

$$n_A = \frac{D_{AB}A}{L}(C_{A_1} - C_{A_2}) \tag{11.49}$$

The diffusion coefficients are in Table 11.5. For the $H_2$–rubber system, we have $D_{AB} = 10.2 \times 10^{-10}$ m²/s, and for the $O_2$–rubber system, we have $D_{AB} = 1.58 \times 10^{-10}$ m²/s.

The concentrations of the gases in the membrane at the left and right boundaries can be determined from the solubilities of the gases and the pressure of the gases using Eq. (11.46):

$$C_{A, \text{ solid side}} = SP_{A, \text{ gas side}} \tag{11.50}$$

For the hydrogen on the left side of the membrane (location 1): $S = 2.1 \times 10^{-3}$ kmol/m³ bar from Table 11.6 and the partial pressure of the hydrogen is 2 atm × (101.325 kPa/atm) × (1 bar/100 kPa) = 2.0265 bar. Hence, from Eq. (11.50), $C_{H_2} = 0.00426$ kmol/m³ at location 1. On the right side of the membrane (location 2), $C_{H_2} = 0$.

For the oxygen on the right side of the membrane (location 2): $S = 3.12 \times 10^{-3}$ kmol/m³ bar from Table 11.6 and the partial pressure of the oxygen is 0.9 atm × (101.325 kPa/atm) × (1 bar/100 kPa) = 0.9119 bar. Hence, from Eq. (11.50), $C_{O_2} = 0.00285$ kmol/m³. On the left side of the membrane (location 1), $C_{O_2} = 0$.

The molar fluxes of the gases through the membrane are, from Eq. (11.49),

$$\frac{n_{H_2}}{A} = \frac{D_{AB}}{L}(C_{H_2}|_1 - C_{H_2}|_2) = \frac{10.2 \times 10^{-10}}{0.0005}(0.00426 - 0) = 8.69 \times 10^{-9} \text{ kmol / s m}^2$$

$$\frac{n_{O_2}}{A} = \frac{D_{AB}}{L}(C_{O_2}|_1 - C_{O_2}|_2) = \frac{1.58 \times 10^{-10}}{0.0005}(0 - 0.00285) = -9.01 \times 10^{-10} \text{ kmol / s m}^2$$

Multiplying these results by the respective molecular weights of the hydrogen and oxygen, we get the mass fluxes of the gases:

$$\frac{m_{H_2}}{A} = \frac{n_{H_2}}{A}M_{H_2} = (8.69 \times 10^{-9} \text{ kmol / s m}^2)(2 \text{ kg } H_2 \text{ / kmol } H_2) = 1.74 \times 10^{-8} \text{ kg/s m}^2$$

$$\frac{m_{O_2}}{A} = \frac{n_{O_2}}{A}M_{O_2} = (-9.01 \times 10^{-10} \text{ kmol / s m}^2)(32 \text{ kg } O_2 \text{ / kmol } O_2) = -2.88 \times 10^{-8} \text{ kg/s m}^2$$

The initial hydrogen mass flux through the membrane is to the right at $1.74 \times 10^{-8}$ kg/s per m$^2$ of membrane. The initial oxygen mass flux through the membrane is to the left at $2.88 \times 10^{-8}$ kg/s per m$^2$ of membrane.

## 11.5.2 Transient diffusion

The previous section dealt with steady-state diffusion. The mass-heat analogy is also useful for transient applications. One such application is the carburization of steel.

Carburization is used in the heat treatment of steel objects. In this process, a carbon-bearing material such as charcoal or carbon monoxide gas is applied to the surface of the object. The carbon diffuses into the object and produces a surface layer of higher carbon concentration than the base metal. This surface layer has increased hardness, toughness, and wear-resistance. Carburization is often used for machine parts such as bearings, gears, and axles.

The carburization process is often used on low-carbon and alloy steels with initial carbon content of 0.2%–0.3%. The carbon concentration in the carburized layer is usually in the range of 0.8%–1%, and the case depth generally ranges from about 0.1 to 1.5 mm.

The mass-heat analogy is useful for predicting the required time for the carburization process. In Chapter 4, we discussed transient conduction in a semi-infinite object. The object was initially at a uniform temperature and then the surface temperature was abruptly changed and we determined the resulting temperature distribution in the object for different times after the change of surface temperature. This situation is analogous to the carburization process. The object initially has a uniform carbon concentration and then the carbon concentration at the surface of the object is abruptly increased. We wish to determine the carbon concentration at a specific depth in the object at a specific time after the change in surface concentration.

The heat transfer equation for the transient conduction in a semi-infinite object with a surface temperature boundary condition is Equation (4.108):

$$\frac{T(x,t) - T_i}{T_o - T_i} = \text{erfc}\left(\frac{x}{2\sqrt{\alpha t}}\right) \tag{11.51}$$

where $T(x,t)$ = temperature at depth $x$ at time $t$
$T_i$ = initial uniform temperature of the object
$T_o$ = temperature at surface for $t \geq 0$
$\alpha$ = thermal diffusivity of object
The analogous mass diffusion equation to Eq. (11.51) is

$$\frac{C_A(x,t) - C_{A,i}}{C_{A,o} - C_{A,i}} = \text{erfc}\left(\frac{x}{2\sqrt{D_{AB}t}}\right). \tag{11.52}$$

where $C_A(x,t)$ = concentration of species A at depth $x$ at time $t$
$C_{A,\ i}$ = initial uniform concentration of species A in the object
$C_{A,\ o}$ = concentration of species A at surface for $t \geq 0$
$D_{AB}$ = diffusion coefficient
The concentrations on the left side of Eq. (11.52) are molar concentrations of species A. For dilute solutions, we could have used densities, mole fractions, or mass (weight) fractions in place of the molar concentrations.

Example 11.7 illustrates the application of Eq. (11.52) to a problem involving the carburization process.

## Example 11.7
### Carburization of a steel machine part

*Problem*
A steel machine part is to be carburized in a furnace. The steel initially has a carbon content of 0.3% by weight. The carburizing agent produces a surface carbon concentration of 1.4% by weight. It is desired that the carbon concentration at a depth of 0.75 mm from the surface be 1% by weight. How long should the part be kept in the carburizing furnace? Assume that the diffusion coefficient of carbon in steel is $3 \times 10^{-11}$ m²/s for the furnace temperature.

*Solution*
The appropriate equation for solution is Eq. (11.52).

$$\frac{C_A(x, t) - C_{A,i}}{C_{A,o} - C_{A,i}} = \text{erfc}\left(\frac{x}{2\sqrt{D_{AB}t}}\right) \tag{11.52}$$

Putting values in the equation, we have

$$\frac{0.01 - 0.003}{0.014 - 0.003} = 0.6364 = \text{erfc}\left(\frac{0.00075}{2\sqrt{(3 \times 10^{-11})t}}\right) \tag{11.53}$$

The argument of the erfc function that gives a result of 0.6364 is 0.3343. Therefore, from Eq. (11.53), we have

$$\frac{0.00075}{2\sqrt{(3 \times 10^{-11})t}} = 0.3343 \tag{11.54}$$

Solving Eq. (11.54) for time $t$, we have $t = 41{,}944$ s $= 11.7$ h.
To achieve the desired treatment, the part should be left in the furnace for 11.7 h.

## 11.6 Gas–liquid diffusion

Thus far, we have looked at diffusion through gases and diffusion of a gas through a solid. Let us now look at diffusion of a gas in a liquid. Consider Fig. 11.4 that shows a gas adjacent to the surface of a liquid.

The gas (species A) adjacent to the liquid surface has a partial pressure of $P_{A-\text{gas}}$. At the surface, some of Gas A dissolves in the liquid. The mole fraction of the gas on the liquid side of the interface is $y_{A-\text{liquid}}$. For dilute solutions, **Henry's law** relates the gas pressure to the mole fraction of the gas in the liquid. This law is

$$y_{A-\text{liquid}} = \frac{P_{A-\text{gas}}}{H} \tag{11.55}$$

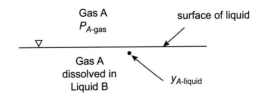

**FIGURE 11.4**

Interface between a gas and a liquid.

where $H$ = Henry's constant.

Table 11.7 gives Henry's constant for a variety of species in water. Although it is called a "constant," it is seen that $H$ varies significantly with temperature.

Examples 11.8 and 11.9 illustrate the use of Henry's law.

## Example 11.8
### Dissolved carbon dioxide in sparkling water

*Problem*

A one-liter bottle of sparkling water is charged with carbon dioxide at 2 atm pressure and 20 C. The gas space above the liquid contains a saturated mixture of water vapor and carbon dioxide. What is the amount of dissolved $CO_2$ in the water?

*Solution*

From the steam tables: For water at 20 C, the saturation pressure is 2.34 kPa. Therefore, above the liquid, $P_{H_2O} = 2.34$ kPa and $P_{CO_2} = 2$ atm $\times$ (101.325 kPa /atm) $- 2.34 = 200.3$ kPa $= 2.003$ bar.

$$\text{From Henry's law, at the liquid - gas interface,} \quad y_{A-\text{liquid}} = \frac{P_{A-\text{gas}}}{H} \tag{11.56}$$

We are concerned with the amount of carbon dioxide in the water, so species A in Eq. (11.56) is the carbon dioxide. Linearly interpolating from Table 11.7 for carbon dioxide at 20 C, we have $H = 1426$ bar. Hence, from Eq. (11.56), the mole fraction of $CO_2$ in the water is

$$y_{CO_2-\text{liquid}} = \frac{P_{CO_2-\text{gas}}}{H} = \frac{2.003}{1426} = 0.001405 \tag{11.57}$$

The volume of the water is 1 L x (1 m³/1000 L) = 0.001 m³.

The mass of water is $m_{H_2O} = \rho_{H_2O}V = (998 \text{ kg}/\text{m}^3)(0.001 \text{ m}^3) = 0.998$ kg.

The number of kmols of the water is $n_{H_2O} = \frac{0.998 \text{ kg}}{18 \text{ kg/kmol}} = 0.0554$ kmol.

From above, $y_{CO_2-\text{liquid}} = 0.001405$.

From the definition of mole fraction, we have $\frac{n_{CO_2}}{n_{CO_2}+n_{H_2O}} = \frac{n_{CO_2}}{n_{CO_2}+0.0554} = 0.001405$.

Solving this for the number of kmols of $CO_2$, we get $n_{CO_2} = 7.795 \times 10^{-5}$ kmol.

The mass of $CO_2$ dissolved in the water is

$$n_{CO_2}M_{CO_2} = (7.795 \times 10^{-5} \text{ kmol})(44 \text{ kg}/\text{kmol}) = 0.00343 \text{ kg} = 3.43 \text{ g}$$

There are 3.43 g of $CO_2$ dissolved in the water.

The following example is a transient diffusion problem incorporating Henry's law.

**Table 11.7 Henry's constant H for gases in water. [The table entries are pressure (bar)].**

|        | 290 K | 300 K | 310 K | 320 K |
|--------|-------|-------|-------|-------|
| Air    | $6.2 \times 10^4$ | $7.4 \times 10^4$ | $8.4 \times 10^4$ | $9.2 \times 10^4$ |
| $H_2$  | $6.7 \times 10^4$ | $7.2 \times 10^4$ | $7.6 \times 10^4$ | $7.9 \times 10^4$ |
| $N_2$  | $7.6 \times 10^4$ | $8.9 \times 10^4$ | $1.0 \times 10^5$ | $1.1 \times 10^5$ |
| $O_2$  | $3.8 \times 10^4$ | $4.5 \times 10^4$ | $5.2 \times 10^4$ | $5.7 \times 10^4$ |
| CO     | $5.1 \times 10^4$ | $6.0 \times 10^4$ | $6.7 \times 10^4$ | $7.4 \times 10^4$ |
| $CO_2$ | $1.3 \times 10^3$ | $1.7 \times 10^3$ | $2.2 \times 10^3$ | $2.7 \times 10^3$ |

## Example 11.9

### Oxygen diffusion in a pool

*Problem*

A homeowner fills his pool after cleaning it. The water initially has no oxygen in it. However, oxygen is gained from the atmospheric air by diffusion. What will be the mole fraction of oxygen at a depth of 1 cm from the surface 10 hours after the pool filling?

*Solution*

Atmospheric air contains about 21% oxygen. We will assume the air and water are at 300 K and the air is at standard atmospheric pressure of 101.325 kPa. The saturation pressure of water at 300 K is 3.57 kPa, so the partial pressure of the air is 101.325−3.57 = 97.76 kPa. The partial pressure of the oxygen in the air is

$$P_{O_2} = 0.21(97.76 \text{ kPa}) = 20.53 \text{ kPa} = 0.2053 \text{ bar} \qquad (11.58)$$

Using Eq. (11.55) and Henry's constant from Table 11.7, we can get the mole fraction of oxygen on the liquid side of the pool surface.

$$y_{O_2-\text{liquid}} = \frac{P_{O_2-\text{gas}}}{H} = \frac{0.2053}{45000} = 4.56 \times 10^{-6} \qquad (11.59)$$

This is the mole fraction at the surface of the water. To determine the mole fraction at a depth, we can use Eq. (11.52):

$$\frac{y_{O_2}(x,t) - y_{O_2,i}}{y_{O_{2,o}} - y_{O_{2,i}}} = \text{erfc}\left(\frac{x}{2\sqrt{D_{AB}t}}\right) \qquad (11.60)$$

$D_{AB}$ for gases in water is given in Table 11.4. For oxygen in water at about 300 K, $D_{AB}$ is about $2.4 \times 10^{-9}$ m²/s. The time is $t = 10 \text{ h} \times (3600 \text{ s/hour}) = 36000$ s.

Putting values in Eq. (11.60), we have

$$\frac{y_{O_2}(1 \text{ cm}, 10 \text{ hour}) - 0}{4.56 \times 10^{-6} - 0} = \text{erfc}\left(\frac{0.01}{2\sqrt{(2.4 \times 10^{-9})(36000)}}\right)$$

which reduces to $y_{O_2}(1 \text{ cm}, 10 \text{ hour}) = (4.56 \times 10^{-6})[\text{erfc}(0.538)] = 2.03 \times 10^{-6}$

The mole fraction of oxygen at a depth of 1 cm after 10 hours is $2.03 \times 10^{-6}$.

## 11.7 Mass transfer coefficient

A mass transfer coefficient can be defined similarly to how we defined the convective coefficient $h$. The equation for the heat transfer coefficient was

$$q = hA(T_1 - T_2) \qquad (11.61)$$

For the mass transfer, we have

$$\dot{m}_A = M_{coef}A(C_{A_1} - C_{A_2}) \qquad (11.62)$$

where $\dot{m}_A$ = mass flow rate of species A (kg/s)

$A$ = area through which species A flows (m²)

$M_{coef}$ = mass transfer coefficient (m/s)

$C_A$ = mass concentration of species A (kg/m³).

As used in Eq. (11.62), the concentration of species A is actually the density $\rho_A$ of species A.

### 11.7.1 Dimensionless parameters

In heat transfer, several dimensionless parameters were used to correlate experimental data. These parameters included the Reynolds number, Re; the Prandtl number, Pr; the Nusselt number, Nu; and the Grashof number, Gr. Mass transfer also has its dimensionless parameters, which include

$$\text{Schmidt Number Sc} = \frac{v}{D} = \frac{\mu}{\rho D}$$

where

$v =$ kinematic viscosity

$D =$ the diffusion coefficient

$\mu =$ absolute viscosity

$\rho =$ density

$$\text{Lewis Number Le} = \frac{\alpha}{D}$$

where

$$\alpha = \text{thermal diffusivity}$$

The Lewis number is also the Schmidt number divided by the Prandtl number. That is,

$$\text{Le} = \frac{\text{Sc}}{\text{Pr}}$$

$$\text{Sherwood Number Sh} = \frac{M_{coef} L}{D}$$

where

$D =$ the diffusion coefficient

$L =$ a characteristic length of the problem

In a previous section, we discussed some useful analogies between heat and mass transfer. For example, we used equations for heat transfer through a wall to obtain equations for mass diffusion of a gas through the wall of a container. Another heat-mass analogy is the **Chilton–Colburn Analogy**, which is useful when both heat transfer and mass transfer are occurring simultaneously. The analogy results in the following relation between the convective heat transfer coefficient $h$ and the mass transfer coefficient $M_{coef}$:

$$\frac{h}{M_{coef}} = \rho c_p (Le)^{2/3} \tag{11.63}$$

where $c_p =$ specific heat at constant pressure.

This relation will be used in the following section on the wet-bulb and dry-bulb psychrometer.

### 11.7.2 Wet-bulb and dry-bulb psychrometer

The wet-bulb and dry-bulb psychrometer measures the wet-bulb and dry-bulb temperatures of air. Knowing these temperatures, the moisture content of the air can be determined. The psychrometer consists of two bulb thermometers. One is uncovered (the dry bulb). The other (the wet bulb) has a wick covering that is kept wet from a small water container in the device.

Fig. 11.5 shows the wet-bulb thermometer portion of the psychrometer.

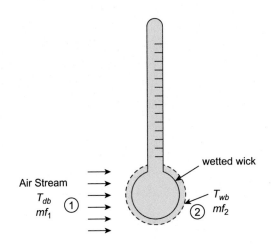

**FIGURE 11.5**

Wet-bulb thermometer.

The air, at temperature $T_{db}$, flows over the wet-bulb thermometer, evaporating water from the wick. The temperature measured by the wet-bulb thermometer is $T_{wb}$. Heat from the air stream causes evaporation of water from the wick. At steady state, we have

$$hA(T_{db} - T_{wb}) = \dot{m}_w h_{fg} \tag{11.64}$$

where

$\dot{m}_w$ = evaporation rate of the water (kg/s)
$h_{fg}$ = enthalpy of vaporization of water (kJ/kg)

For the water being evaporated, we have, from Eq. (11.62),

$$\dot{m}_w = M_{coef} A \rho (mf_2 - mf_1) \tag{11.65}$$

where

$\rho$ = density of the air stream

$mf_1$ = mass fraction of water vapor in the air stream

$mf_2$ = mass fraction of water vapor in the air immediately adjacent to the wick

Combining Eqs. (11.64) and (11.65), we have

$$h(T_{db} - T_{wb}) = M_{coef}(mf_2 - mf_1)\rho h_{fg} \tag{11.66}$$

Using Eq. (11.63), Eq. (11.66) modifies to

$$T_{db} - T_{wb} = \frac{h_{fg}}{c_p}\left(\frac{Sc}{Pr}\right)^{-2/3}(mf_2 - mf_1) \tag{11.67}$$

Example 11.10 illustrates the use of Eq. (11.67) in determining the relative humidity of air in a room when the wet-bulb and dry-bulb temperatures are known.

## Example 11.10
### Relative humidity of air

*Problem*

The air in a room has a dry-bulb temperature of 35 C and a wet-bulb temperature of 24 C. What is the relative humidity of the air?

*Solution*

We will assume that the atmospheric pressure is 101.325 kPa.
We will evaluate the properties $c_p$, $\alpha$, and $D$ at the film temperature $T_{film}$.

$$T_{film} = \frac{T_{db} + T_{wb}}{2} = \frac{35 + 24}{2} = 29.5 \text{ C} = 303 \text{ K}$$

Air at 303 K: $\alpha = 2.21 \times 10^{-5} \text{ m}^2/\text{s}$ and $c_p \doteq 1.007 \text{ kJ/kg C}$.
For water vapor and air, from Table 11.3, $D = 0.26 \times 10^{-4} \text{ m}^2/\text{s}$.
At the wet-bulb temperature of 24 C: From steam tables, $P_{sat}$ for the water vapor $= 3.003$ kPa and $h_{fg} = 2440$ kJ/kg.
Also, $\bar{R} = 8.31446 \text{ kJ/kmol K}$, $M_{H_2O} = 18 \text{ kg/kmol}$, and. $M_{air} = 29 \text{ kg/kmol}$.
The applicable equation is Eq. (11.67)

$$T_{db} - T_{wb} = \frac{h_{fg}}{c_p} \left( \frac{\text{Sc}}{\text{Pr}} \right)^{-2/3} (mf_2 - mf_1) \tag{11.67}$$

To get the relative humidity in the air stream, we need to find $mf_1$. Rearranging Eq. (11.67), we get

$$mf_1 = mf_2 - \frac{c_p}{h_{fg}} \left( \frac{\text{Sc}}{\text{Pr}} \right)^{2/3} (T_{db} - T_{wb}) \tag{11.68}$$

The mole fraction of water vapor in the air by the wick is

$$y_{H_2O}\big|_2 = \frac{P_{sat}}{P} = \frac{3.003}{101.325} = 0.02964 \tag{11.69}$$

The mass fraction of water vapor in the air by the wick is

$$mf_2 = \frac{\left(y_{H_2O}\big|_2\right)(M_{H_2O})}{\left(y_{H_2O}\big|_2\right)(M_{H_2O}) + \left(1 - y_{H_2O}\big|_2\right)(M_{air})} = \frac{(0.02964)(18)}{(0.02964)(18) + (1 - 0.02964)(29)} = 0.01861$$

The ratio (Sc/Pr) in Eq. (11.68) is the Lewis number, which is also $(\alpha/D)$ For this problem,
$\left( \frac{\text{Sc}}{\text{Pr}} \right) = \left( \frac{\alpha}{D} \right) = \frac{2.21 \times 10^{-5}}{0.26 \times 10^{-4}} = 0.85$.

Putting values into Eq. (11.68), we can now solve for mass fraction $mf_1$ :

$$mf_1 = mf_2 - \frac{c_p}{h_{fg}} \left( \frac{\text{Sc}}{\text{Pr}} \right)^{2/3} (T_{db} - T_{wb}) = 0.01861 - \frac{1.007}{2440}(0.85)^{2/3}(35 - 24) = 0.01454$$

The mole fraction of water vapor in the air stream is

$$y_{H_2O}\big|_1 = \frac{(mf_1/18)}{(mf_1/18) + (1 - mf_1)/29} = \frac{(0.01454/18)}{(0.01454/18) + (1 - 0.01454)/29} = 0.02322$$

The partial pressure of water vapor in the air stream is

$$P_{H_2O}\big|_1 = \left(y_{H_2O}\big|_1\right)P = (0.02322)(101.325) = 2.353 \text{ kPa}$$

By definition, the relative humidity $RH$ is the ratio of the partial pressure of water vapor in the air stream divided by the saturation pressure at the air stream temperature. From steam tables, the saturation pressure of water at $T_{db} = 35$ C is 5.625 kPa.
The relative humidity of the air streams is therefore

$$RH(\%) = \frac{P_{H_2O}\big|_1}{P_{sat} \text{ at } T_{db}} \times 100 = \frac{2.353}{5.628} \times 100 = 41.8\%$$

The relative humidity of the air stream is 41.8%.

## 11.8 Chapter summary and final remarks

This chapter was an introduction to mass transfer, a topic of major importance to chemical engineers. We first discussed the terminology relevant to concentrations in gas mixtures. We then discussed mass diffusion, including diffusion of a gas in another gas, diffusion of gas in a liquid, and diffusion in a solid. We covered leakage of gases through container walls and carburization of steel for case hardening. We discussed the heat-mass analogy, whereby equations from heat transfer, with minor modification, can be applied to applications in mass transfer. Finally, we discussed the dry-bulb and wet-bulb psychrometer whose operation is based on simultaneous heat and mass transfer. But, we have only touched the surface of mass transfer. Additional information can be found in Ref. [11−17].

## 11.9 Problems

Notes:

- If needed information is not given in the problem statement, then use the Appendix for material properties and other information. If the Appendix does not have sufficient information, then use the Internet or other reference sources. Some good reference sources are mentioned at the end of Chapter 1. If you use the Internet, double-check the validity of the information by using more than one source.
- Your solution should include a sketch of the problem.

**11-1** Moist air has a volumetric analysis of 78% nitrogen, 20% oxygen, and 2% water vapor.
   **(a)** What is the analysis on a mass basis?
   **(b)** What is the molecular weight of the moist air?

**11-2** The mass analysis of a gas mixture is 60% nitrogen, 25% carbon dioxide, and 15% oxygen. The mixture is at a temperature of 310 K and a pressure of 280 kPa.
   **(a)** What is the molal analysis of the mixture?
   **(b)** What are the partial pressures of the gases?
   **(c)** What is the molecular weight of the gas mixture?
   **(d)** What is the gas constant of the mixture?

**11-3** A gas mixture consists of 6 kmol of $H_2$, 3 kmol of $N_2$, and 2 kmol of $CO_2$.
   **(a)** What is the mass fraction of each component?
   **(b)** What is the mole fraction of each component?
   **(c)** If the mixture is at a temperature of 420 K and a pressure of 110 kPa, what are the partial pressures of the individual gases?
   **(d)** What is the molecular weight of the gas mixture?

**11-4** A gas mixture is made up of the following components: 2.6 kg of $O_2$, 6 kg of $N_2$, 4.3 kg of $SO_2$, and 2 kg of $N_2O$.
   **(a)** What is the percentage of each component on a mass basis?
   **(b)** What is the percentage of each component on a volumetric basis?
   **(c)** What is the mass fraction of each component?
   **(d)** What is the mole fraction of each component?
   **(e)** What is the molecular weight of the mixture?
   **(f)** What is the gas constant of the mixture?

**11-5** What is the binary diffusion coefficient of carbon dioxide in air at 400 K and 0.8 atm?

**11-6** What is the binary diffusion coefficient of helium in air at 500 K and 100 kPa?

**11-7** What is the binary diffusion coefficient of water vapor in air at 320 K and 0.9 atm?

**11-8** What is the binary diffusion coefficient of ethane in carbon dioxide at 320 K and 95 kPa?

**11-9** A vertical test tube has a diameter of 1 cm and a height of 20 cm. It contains liquid benzene at 20 C. The distance from the surface of the benzene to the open top of the tube is 16 cm. Dry air at 20 C blows gently across the open end of the tube. What is the evaporation rate of the benzene (g/hr)?

**11-10** A pail of diameter 15 cm and height 20 cm contains 5 cm of water at 35 C. What is the evaporation rate of the water if (a) the air in the room is dry at 35 C and (b) the air in the room is 40% relative humidity and 15 C?

**11-11** It is desired to determine the binary diffusion coefficient for a gas species A in air. Species A has a molecular weight of 60, a liquid density of 850 kg/m³, and a vapor pressure of 25 kPa at 30 C. The diffusion coefficient is to be determined by using a Stefan tube. The tube has a diameter of 2 cm and a height of 6 cm. Eight cc of liquid A is put into the tube. The tube is open to dry air at 30 C. The tube is in operation for 10 h. During this time, the level of Liquid A drops by 5 mm. What is the diffusion coefficient for the A−air pair of gases?

**11-12** A large tank of methane at 27 C and 1 atm is vented to the atmosphere through a tube that is 0.8 cm diameter and 20 cm long. The atmospheric air is also at 27 C and 1 atm.
  **(a)** At what rate is the methane leaking out of the tank though the vent tube?
  **(b)** At what rate is air intruding into the tank through the vent tube?

**11-13** Carbon dioxide is in a large tank at 30 C and 95 kPa. There is a vent on the tank that has a diameter of 1.5 cm and a length of 35 cm. The atmospheric air is also at 30 C and 95 kPa.
  **(a)** At what rate does the $CO_2$ vent from the tank?
  **(b)** At what rate does the air diffuse into the tank through the vent tube?

**11-14** Hydrogen gas at a pressure of 250 kPa and a temperature of 85 C is being stored in a cylindrical nickel container. The container has an inside diameter of 20 cm and a length of 40 cm. The wall thickness of the container is 5 mm.
  **(a)** What is the rate of mass loss from the container?
  **(b)** What is the pressure decrease of the hydrogen in 10 days?

**11-15** Oxygen at 25 C is being stored in a spherical Lexan container that has an inside diameter of 1.5 m and a wall thickness of 1 cm. The molar concentration of $O_2$ in the Lexan is 0.00055 kmol/m³ at the inner surface of the container and negligible at the outer surface.
  **(a)** What is the mass flow rate of $O_2$ through the container wall?
  **(b)** How much oxygen is lost from the container in a month?

**11-16** A spherical balloon having a diameter of 0.3 m contains helium at 110 kPa and 25 C. The rubber wall of the balloon is 0.15 mm thick. The permeability of rubber to helium is $9.4 \times 10^{-13}$ kmol/m s bar.
  **(a)** What is the rate of helium loss from the balloon (kg/s)?
  **(b)** How long does it take for the helium pressure to decrease to the atmospheric pressure of 101.325 kPa? Assume that the volume of the balloon remains constant and neglect any diffusion of air into the balloon.

**11-17** Nitrogen at 1 atm and 300 K is flowing through a rubber pipe. The pipe is 25 m long. It has an inside diameter of 2.5 cm and a wall thickness of 2 mm.

    **(a)** What is the rate of $N_2$ leakage from the pipe if the surroundings are a vacuum?

    **(b)** What is the rate of $N_2$ leakage if the surroundings are air with 79% nitrogen and 21% oxygen?

**11-18** A steel machine part is to be case hardened in a furnace. The steel initially has a uniform carbon content of 0.2% by weight. It is desired that the carbon concentration at a depth of 0.5 mm from the surface be 0.8% by weight and that the carburization time be no longer than 5 hours. The diffusion coefficient of carbon in steel is $5 \times 10^{-11}$ m$^2$/s at the furnace temperature.

What is the needed carbon surface concentration (% by weight) to achieve the desired heat treatment?

**11-19** A steel plate is to be carburized in a furnace to harden its surface. The steel initially has a uniform carbon content of 0.24% by weight. It is desired that the carbon concentration at a depth of 0.7 mm from the surface be 1% by weight after 3 hours of treatment in the furnace. The carburizing agent keeps the surface carbon concentration at 1.8%. The diffusion coefficient of carbon in the steel is $D_{AB} = 0.67 \times 10^{-4} e^{-\left(\frac{19000}{T}\right)}$, where $D_{AB}$ is in m$^2$/s and $T$ is in K. What is the needed temperature of the furnace to achieve the desired heat treatment?

**11-20** A steel part undergoes a carburization process in a furnace. The steel initially has a uniform carbon content of 0.25% by weight. The diffusion coefficient of carbon in the steel is $2 \times 10^{-10}$ m$^2$/s and the carburizing agent holds the surface at 2% carbon concentration by weight. What is the carbon concentration at a depth of 1.5 mm after the part has been in the furnace for 6 hours?

**11-21** A two-liter bottle of soda is charged with carbon dioxide at 3 atm pressure and 40 C. Assume the soda has the properties of water. The gas space above the soda is a saturated mixture of water vapor and $CO_2$.

    **(a)** What is the mole fraction of water vapor in the $CO_2$ gas above the surface of the soda?

    **(b)** What is the mole fraction of $CO_2$ on the liquid side of the surface of the soda?

    **(c)** What is the amount of dissolved carbon dioxide (grams) in the soda?

**11-22** The atmospheric pressure at the surface of a pond is 96 kPa and the temperature of the surface of the pond is 15 C.

    **(a)** What is the mole fraction of the water vapor in the air at the surface of the pond?

    **(b)** What is the mole fraction of air in the water near the surface of the pond?

**11-23** For the problem of Example 11.9, what will be the mole fraction of the oxygen at a depth of 2 cm after a long period of time has passed after pool filling?

**11-24** The water in a pond initially has no air in it. The water and the atmospheric air are at 290 K and the atmospheric air has a pressure of 95 kPa. Assume that the air is 78% nitrogen and 21% oxygen. What are the mole fractions of nitrogen and oxygen at a depth of 1.5 cm from the surface after 5 hours?

**11-25** The air in a room has a dry-bulb temperature of 85 F and a wet-bulb temperature of 75 F. The atmospheric pressure is 14.1 psia. What is the relative humidity of the air?

**11-26** The outdoor air has a dry-bulb temperature of 25 C and a wet-bulb temperature of 11 C. The atmospheric pressure is 90 kPa.

(a) What is the relative humidity of the air?

(b) Go on the Internet and use a search engine to find an online relative humidity calculator. Use the calculator with the inputs of dry- and wet-bulb temperatures and compare the relative humidity result with that of Part (a). If there is a difference, is it significant? Which is correct—the result of Part (a) or that from the online calculator?

---

# References

[1] J.H. Arnold, Studies in diffusion, I − estimation of diffusivities in gaseous systems, Ind. Eng. Chem. 22 (1930) 1091−1095.

[2] E.R. Gilliland, Diffusion coefficients in gaseous systems, Ind. Eng. Chem. 26 (1934) 681−685.

[3] L. Andrussow, Z. Elektrochem. 54 (1950) 566.

[4] N.H. Chen, D.F. Othmer, New generalized equation for gas diffusion coefficient, J. Chem. Eng. Data 7 (1962) 37−41.

[5] E.N. Fuller, P.D. Schettler, J.C. Giddings, A new method for prediction of binary gas-phase diffusion coefficients, Ind. Eng. Chem. 58 (1966) 19−27.

[6] P.C. Singh, S. Singh, Development of a new correlation for binary gas phase diffusion coefficients, Int. Commun. Heat Mass Transf. 10 (1983) 123−140.

[7] Handbook of Chemistry and Physics, 85th Edition, CRC Press, 2004.

[8] R.M. Barrer, Diffusion in and Through Solids, Cambridge University Press, Macmillan, 1951.

[9] J.L. Plavsky, Transport Phenomena Fundamentals, Marcel Dekker, 2001.

[10] E.L. Cussler, Diffusion Mass Transfer in Fluid Systems, second ed., Cambridge University Press, 1997.

[11] A.H.P. Skelland, Diffusional Mass Transfer, Wiley, 1974.

[12] C.J. Geankoplis, Mass Transport Phenomena, Holt, Rinehart, and Winston, 1972.

[13] W. Jost, Diffusion in Solids, Liquids, and Gases, Academic Press, 1952.

[14] R.B. Bird, W.E. Stewart, E.N. Lightfoot, Transport Phenomena, Wiley, 1960.

[15] C.O. Bennett, J.E. Myers, Momentum, Heat, and Mass Transfer, McGraw-Hill, 1962.

[16] W.M. Rohsenow, H.Y. Choi, Heat, Mass and Momentum Transfer, Prentice-Hall, 1961.

[17] A.F. Mills, Heat and Mass Transfer, R. D. Irwin, 1995.

# Special topics

## Chapter outline

## 12.1  Introduction

This chapter contains special topics that may be included in a heat transfer course if time permits and the instructor so desires. Two of these topics are internal heat generation and heat transfer in buildings. They are extensions of material covered in earlier chapters. The other two topics, contact resistance and condensation/boiling, are unique to this chapter.

## 12.2  Internal heat generation

Section 3.4 considered heat generation in a cylinder. In this section, we will consider heat generation in plane walls and spheres.

Heat Transfer Principles and Applications. https://doi.org/10.1016/B978-0-12-802296-2.00012-3

## 12.2.1 Heat generation in a plane wall

Fig. 12.1 shows a plane wall. It has a constant, uniformly distributed internal heat generation source of strength $q_{gen}$ W/m$^3$. The wall has a thickness of $L$ in the $x$ direction. Its lengths in the $y$ and $z$ directions are large so that the heat flow can be assumed to be one-dimensional. The heat flow is also steady, so the temperature in the wall is $T = T(x)$. As shown in the figure, there are temperature boundary conditions at the two wall surfaces.

The appropriate heat conduction equation is Equation (3.2), repeated here:

$$\frac{\partial}{\partial x}\left(k\frac{\partial T}{\partial x}\right) + q_{gen} = 0 \tag{12.1}$$

We will assume that the conductivity and internal heat generation are uniform and constant and that the heat flow is one-dimensional. Then Eq. (12.1) becomes

$$\frac{d^2 T}{dx^2} + \frac{q_{gen}}{k} = 0 \tag{12.2}$$

Integrating Eq. (12.2), we get

$$\frac{dT}{dx} = \frac{-q_{gen}}{k}x + C_1 \tag{12.3}$$

Integrating again, we get

$$T(x) = \frac{-q_{gen}}{2k}x^2 + C_1 x + C_2 \tag{12.4}$$

where $C_1$ and $C_2$ are constants. These constants can be obtained from the boundary conditions, which are

$$\text{At } x=0, \ T = T_1 \ \text{and At } x=L, \ T = T_2 \tag{12.5}$$

From Eq. (12.4) and the first boundary condition, we get

$$T_1 = \frac{-q_{gen}}{2k}(0)^2 + C_1(0) + C_2, \text{or}$$

$$C_2 = T_1 \tag{12.6}$$

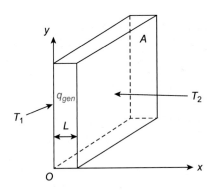

**FIGURE 12.1**

Plane wall with uniform heat generation.

From Eq. (12.4) and the second boundary condition, we get

$$T_2 = \frac{-q_{gen}}{2k}L^2 + C_1L + C_2 \tag{12.7}$$

We found that $C_2 = T_1$. Solving Eq. (12.7) for $C_1$, we get

$$C_1 = \frac{T_2 - T_1}{L} + \frac{q_{gen}L}{2k} \tag{12.8}$$

Putting Eqs. (12.6) and (12.8) into Eq. (12.4), we get the temperature distribution $T(x)$ in the wall.

$$T(x) = T_1 + \frac{q_{gen}}{2k}\left(Lx - x^2\right) + \frac{T_2 - T_1}{L}x \tag{12.9}$$

Let us now determine the location of the maximum temperature in the wall and its value. The maximum temperature will be at the location where $\frac{dT}{dx} = 0$. Differentiating Eq. (12.9) and setting the result to zero, we get

$$\frac{dT}{dx} = 0 = \frac{q_{gen}}{2k}\left(L - 2x\right) + \left(\frac{T_2 - T_1}{L}\right) \tag{12.10}$$

Solving Eq. (12.10) for $x$, we get the location $x_{max}$ where the temperature in the wall is a maximum:

$$x_{max} = \frac{L}{2} + \frac{k}{q_{gen}}\left(\frac{T_2 - T_1}{L}\right) \tag{12.11}$$

If we use this $x$-value in Eq. (12.9), we can get the maximum temperature in the wall.

**Caution**: Looking at the temperature distribution given in Eq. (12.9), it is seen that there is a heat generation term and a linear term. If the heat generation rate is small, the linear term will dominate, and there will be no maximum in the wall where $\frac{dT}{dx} = 0$. Eq. (12.11) will then give a location $x_{max}$, which lies outside the wall. That is, $x_{max}$ will be less than zero or greater than $L$. For this situation, the location of the maximum temperature in the wall will be at a wall surface. This is illustrated by Example 12.1 which follows.

---

## Example 12.1
### Wall with uniform internal heat generation

*Problem*

A wall shown below has uniform internal heat generation at a rate of $6 \times 10^7$ W/m$^3$. The wall has a thickness of 3 cm and a conductivity of 25 W/m C. The left surface of the wall is at 20 C and the right surface is at 150 C.
(a)  What is the maximum temperature in the wall and where does it occur?
(b)  What is the heat flux through the left surface of the wall?
(c)  What is the heat flux through the right surface of the wall?
(d)  Do Parts (a), (b), and (c) for a heat generation rate of $4 \times 10^6$ W/m$^3$.

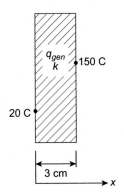

20 C

3 cm

$x$

## Solution

For this problem, $q_{gen} = 6 \times 10^7$ W/m$^3$, $T_1 = 20$ C, $T_2 = 150$ C, and $L = 0.03$ m.

**(a)** The location of the maximum temperature is given by Eq. (12.11):

$$x_{max} = \frac{L}{2} + \frac{k}{q_{gen}}\left(\frac{T_2 - T_1}{L}\right)$$

Putting in values, we have $x_{max} = \frac{0.03}{2} + \frac{25}{6 \times 10^7}\left(\frac{150-20}{0.03}\right) = 0.0168$ m.

The temperature at this location is found from Eq. (12.9):

$$T(x) = T_1 + \frac{q_{gen}}{2k}\left(Lx - x^2\right) + \frac{T_2 - T_1}{L}x$$

Putting in values, we have

$$T(x_{max}) = 20 + \frac{6 \times 10^7}{2(25)}\left[(0.03)(0.0168) - (0.0168)^2\right] + \frac{150-20}{0.03}(0.0168) = 359.8 \text{ C}$$

The maximum temperature in the wall is 359.8 C, and it occurs at x = 1.68 cm, which is 0.18 cm to the right of the center plane.

**(b)** The heat flux at the left surface is

$$\frac{q_{x=0}}{A} = -k\frac{dT}{dx} \qquad (12.12)$$

From Eq. (12.3), $\frac{dT}{dx} = \frac{-q_{gen}}{k}x + C_1$, and $C_1 = \frac{T_2-T_1}{L} + \frac{q_{gen}L}{2k}$ from Eq. (12.8).

$$C_1 = \frac{T_2 - T_1}{L} + \frac{q_{gen}L}{2k} = \frac{150-20}{0.03} + \frac{6 \times 10^7(0.03)}{2(25)} = 40333$$

$$\frac{dT}{dx} = \frac{-q_{gen}}{k}x + C_1 = \frac{-6 \times 10^7}{25}(0) + 40333 = 40333 \text{ C/m}$$

From Eq. (12.12), $\frac{q_{x=0}}{A} = -k\frac{dT}{dx} = -25(40333) = -1.008 \times 10^6$ W/m$^2$.

The heat flux at the left surface is $1.008 \times 10^6$ W/m$^2$ toward the left.

**(c)** Following the procedure of Part (b), we have, for the right surface:

$$\frac{dT}{dx} = \frac{-q_{gen}}{k}x + C_1 = \frac{-6 \times 10^7}{25}(0.03) + 40333 = -31667 \text{ C/m}$$

The heat flux at the right surface is $\frac{q_{x=L}}{A} = -k\frac{dT}{dx} = -25(-31667) = 7.92 \times 10^5$ W/m$^2$.

The heat flux at the right surface is $7.92 \times 10^5$ W/m$^2$ toward the right.

Note: As a check we can sum these values: $1.008 \times 10^6 + 0.792 \times 10^6 = 1.8 \times 10^6$ W/m$^2$.
For steady state, the heat generation in the wall equals the heat flows out the two surfaces. The heat generation in the wall is $q_{gen}$(Volume) $= q_{gen}(LA) = 6 \times 10^7(0.03)A = 1.8 \times 10^6(A)$ W, which is indeed equal to the heat flows out the two surfaces.

**(d)** For a heat generation rate of $4 \times 10^6$ W/m$^3$, Eq. (12.11) gives the location of the maximum temperature as

$$x_{max} = \frac{L}{2} + \frac{k}{q_{gen}}\left(\frac{T_2 - T_1}{L}\right) = \frac{0.03}{2} + \frac{25}{4 \times 10^6}\left(\frac{150 - 20}{0.03}\right) = 0.042 \text{ m}$$

This value is greater than 0.03 m and outside the domain of the wall.
The maximum temperature is at the right surface of the wall and is 150 C.
The heat flux at the left surface is

$$\frac{q_{x=0}}{A} = -k\frac{dT}{dx} \tag{12.12}$$

Following the calculations of Part (b), we have

$$C_1 = \frac{T_2 - T_1}{L} + \frac{q_{gen}L}{2k} = \frac{150 - 20}{0.03} + \frac{4 \times 10^6(0.03)}{2(25)} = 6733.3$$

$$\frac{dT}{dx} = \frac{-q_{gen}}{k}x + C_1 = \frac{-4 \times 10^6}{25}(0) + 6733.3 = 6733.3 \text{ C/m}$$

From Eq. (12.12), $\frac{q_{x=0}}{A} = -k\frac{dT}{dx} = -25(6733.3) = -1.683 \times 10^5$ W/m$^2$.
The heat flux at the left surface is $1.683 \times 10^5$ W/m$^2$ toward the left.
The heat flux at the right surface is

$$\frac{q_{x=L}}{A} = -k\frac{dT}{dx} \tag{12.12}$$

Following the calculations of Part (c), we have

$$\frac{dT}{dx} = \frac{-q_{gen}}{k}x + C_1 = \frac{-4 \times 10^6}{25}(0.03) + 6733.3 = 1933.3 \text{ C/m}$$

From Eq. (12.12), $\frac{q_{x=0}}{A} = -k\frac{dT}{dx} = -25(1933.3) = -0.483 \times 10^5$ W/m$^2$.
The heat flux at the right surface is $0.483 \times 10^5$ W/m$^2$ toward the left.
Note: For this lower heat generation rate, the conduction transfer through the wall overcomes the heat generation. The heat fluxes at both surfaces are to the left. Let us check our results. The input to the wall is the heat generation plus the heat flowing into the wall through the right surface. The heat generation is $q_{gen}$(Volume) $= (4 \times 10^6)(0.03)(A) = 1.2 \times 10^5 A$ W. The flow into the wall through the right surface is $0.483 \times 10^5 A$ W, giving a total input of $1.683 \times 10^5 A$ W. From above, the heat flow from the wall through the left surface is $1.683 \times 10^5 A$ W, which confirms our calculations.

Let us now look at the situation shown in Fig. 12.2, where both wall surfaces are at the same temperature $T_0$. That is, $T_1 = T_2 = T_0$.

Eq. (12.9) gives the temperature distribution in the wall. For $T_1 = T_2 = T_0$, we have

$$T(x) = T_0 + \frac{q_{gen}}{2k}\left(Lx - x^2\right) \tag{12.13}$$

From Eq. (12.11), the maximum temperature occurs at the midplane $x = L/2$. This makes sense as the situation is symmetrical.

Using Eq. (12.9) with $x = L/2$, we can get the maximum temperature in the wall:

$$T_{max} = T_0 + \frac{q_{gen}L^2}{8k} \tag{12.14}$$

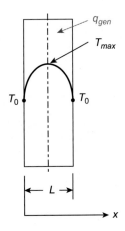

**FIGURE 12.2**

Wall with heat generation and equal surface temperatures.

Let us now look at the heat flows in the wall. For a wall with cross-sectional area $A$, the heat flow at location $x$ is

$$q = -kA\frac{dT}{dx} \qquad (12.15)$$

Differentiating Eq. (12.13), we get

$$\frac{dT}{dx} = \frac{q_{gen}}{2k}(L - 2x) \qquad (12.16)$$

Putting this into Eq. (12.15), the heat flow at location $x$ is

$$q = q_{gen}A\left(x - \frac{L}{2}\right) \qquad (12.17)$$

Let us now look at the heat flows at the two surfaces. At the left surface ($x = 0$), the heat flow, from Eq. (12.17), is

$$q_{x=0} = -q_{gen}\left(\frac{AL}{2}\right) \qquad (12.18)$$

At the right surface ($x = L$), the heat flow is

$$q_{x=L} = q_{gen}\left(\frac{AL}{2}\right) \qquad (12.19)$$

The heat flows at the two surfaces both have the same magnitude. However, the heat flow at the left surface is negative, which means that the flow is in the minus $x$ direction. The heat flow at the right surface is positive and in the plus $x$ direction. Regarding the magnitude of the heat flows: $q_{gen}$ is the generation rate per unit volume and $(AL/2)$ is the volume of half of the wall. Therefore, the magnitudes of the heat flows at the wall surfaces are equal to the heat generated in half of the wall. Finally, at the center plane of the wall ($x = L/2$), $dT/dx = 0$ and the heat flow is zero from Eq. (12.15). There is no heat flow across the center plane of the wall.

Summarizing, we see that the heat generated in the left half of the wall flows out the left surface of the wall. And, the heat generated in the right half of the wall flows out the right surface of the wall.

The above is for temperature boundary conditions. Let us now look at convection boundary conditions at the two surfaces. Fig. 12.3 shows this situation.

The fluid on the left surface of the wall has a temperature $T_{\infty_1}$ and a convective coefficient $h_1$. The fluid on the right surface is at $T_{\infty_2}$ with a convective coefficient $h_2$.

The general solution for constant conductivity and uniform heat generation is still given by Eq. (12.4).

$$T(x) = \frac{-q_{gen}}{2k}x^2 + C_1x + C_2 \qquad (12.20)$$

Constants $C_1$ and $C_2$ are obtained from the boundary condition equations, as follows:

At the left surface, we have heat flow continuity and the convection equals the conduction. That is,

$$\text{At } x=0, \ h_1A(T_{\infty_1} - T) = -kA\frac{dT}{dx} \qquad (12.21)$$

Differentiating Eq. (12.20), we have

$$\frac{dT}{dx} = \frac{-q_{gen}x}{k} + C_1 \qquad (12.22)$$

Therefore, at $x = 0$, $\frac{dT}{dx} = C_1$. Also, from Eq. (12.20) it is seen that, at $x = 0$, $T = C_2$. Eq. (12.21) then becomes $h_1(T_{\infty_1} - C_2) = -kC_1$. Rearranging this, we get

$$kC_1 - h_1C_2 = -h_1T_{\infty_1} \qquad (12.23)$$

At the right surface, continuity of heat flow gives

$$\text{At } x = L - kA\frac{dT}{dx} = h_2A(T - T_{\infty_2}) \qquad (12.24)$$

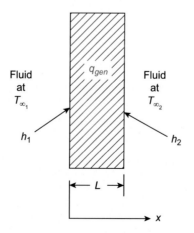

**FIGURE 12.3**

Wall with heat generation and convection boundary conditions.

Using Eqs. (12.20) and (12.22), Eq. (12.24) becomes

$$-k\left(-\frac{q_{gen}L}{k}+C_1\right) = h_2\left(-\frac{q_{gen}L^2}{2k}+C_1L+C_2-T_{\infty_2}\right)$$

This reduces to $(k+h_2L)C_1+h_2C_2 = q_{gen}L+h_2\left(\frac{q_{gen}L^2}{2k}+T_{\infty_2}\right)$ (12.25)

Eqs. (12.23) and (12.25) are simultaneous linear equations that can be easily solved for constants $C_1$ and $C_2$, as follows: Let us look at the coefficients of the unknowns and the constants on the right sides of the equations. For Eq. (12.23), we have $a_1C_1 + b_1C_2 = d_1$, where $a_1 = k$, $b_1 = -h_1$, and $d_1 = -h_1T_{\infty_1}$.

For Eq. (12.25), we have $a_2C_1 + b_2C_2 = d_2$, where $a_2 = k+h_2L$ $b_2 = h_2$ and $d_2 = q_{gen}L + h_2\left(\frac{q_{gen}L^2}{2k}+T_{\infty_2}\right)$.

Through algebra, it can be shown that $C_2 = \frac{a_2d_1-a_1d_2}{a_2b_1-a_1b_2}$ and $C_1 = \frac{d_1-b_1C_2}{a_1}$.

A special case is when the fluids on both sides of the wall have the same temperature and same convective coefficient. That is, $T_{\infty_1} = T_{\infty_2} = T_\infty$ and $h_1 = h_2 = h$. For this case, there is symmetry and the surface temperatures are equal. From the boundary conditions, it can be shown that the surface temperature $T_0$ is

$$T_0 = T_\infty + \frac{q_{gen}L}{2h}$$ (12.26)

The maximum temperature in the wall is at $x = L/2$ and is given by Eq. (12.14):

$$T_{max} = T_0 + \frac{q_{gen}L^2}{8k}$$ (12.27)

In terms of the fluid temperature, using Eqs. (12.26) and (12.27), the maximum temperature in the wall is at $x = L/2$ and is

$$T_{max} = T_\infty + q_{gen}\left(\frac{L}{2h}+\frac{L^2}{8k}\right)$$ (12.28)

---

### Example 12.2

**Wall with uniform heat generation, insulated on one side**

*Problem*
The wall shown below is 4 cm thick and has uniform internal heat generation at a strength of $10^5$ W/m$^3$. The thermal conductivity of the wall is 15 W/m C. One surface of the wall is perfectly insulated. The other surface convects to an adjacent fluid that is at 20 C. The convective coefficient is 25 W/m$^2$ C. What is the maximum temperature in the wall, and where does it occur?

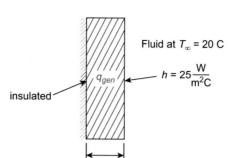

## Solution

The general solution for the temperature distribution in the wall is given by Eq. (12.4).

$$T(x) = \frac{-q_{gen}}{2k}x^2 + C_1 x + C_2 \tag{12.29}$$

We get constants $C_1$ and $C_2$ by applying the boundary conditions. The left surface is insulated and $q = 0$. The boundary condition equation is

$$\text{At } x = 0, \frac{dT}{dx} = 0 \tag{12.30}$$

Differentiating Eq. (12.29), we get

$$\frac{dT}{dx} = \frac{-q_{gen}x}{k} + C_1 \tag{12.31}$$

Applying the boundary condition of Eq. (12.30) to Eq. (12.31) it is seen that $C_1 = 0$.
At the right surface, the conduction heat flux equals the convection heat flux. The boundary condition equation is

$$\text{At } x = L, \ -k\frac{dT}{dx} = h(T - T_\infty) \tag{12.32}$$

Using Eqs. (12.29) and (12.31) and remembering that $C_1 = 0$, Eq. (12.32) becomes

$$-k\left(\frac{-q_{gen}L}{k}\right) = h\left(\frac{-q_{gen}L^2}{2k} + C_2 - T_\infty\right) \tag{12.33}$$

Solving for $C_2$, we get    $C_2 = T_\infty + \frac{q_{gen}L^2}{2k} + \frac{q_{gen}L}{h} \tag{12.34}$

From Eq. (12.29), the temperature distribution in the wall is

$$T(x) = T_\infty + \frac{q_{gen}}{2k}\left(L^2 - x^2\right) + \frac{q_{gen}L}{h} \tag{12.35}$$

From Eq. (12.35), it is seen that the maximum temperature occurs at $x = 0$, the insulated surface. This is consistent with $dT/dx$ being zero at $x = 0$. A maximum or minimum is indicated for a zero $dT/dx$. The maximum temperature is

$$T_{max} = T_\infty + \frac{q_{gen}L^2}{2k} + \frac{q_{gen}L}{h} \tag{12.36}$$

Putting the values of this problem into Eq. (12.36), we have

$$T_{max} = T_\infty + \frac{q_{gen}L^2}{2k} + \frac{q_{gen}L}{h} = 20 + \frac{10^5(0.04)^2}{2(15)} + \frac{10^5(0.04)}{25} = 185.33 \text{ C}$$

The maximum temperature in the wall is 185.33 C, and it occurs at the insulated surface of the wall.

Note: This problem could have been solved a lot quicker by recognizing the symmetry of problem. A wall having perfect insulation on one surface is really equivalent to that of a wall twice as thick having the same convection at both surfaces. For this latter wall, $dT/dx$ is zero and the heat flow is zero at the center plane.

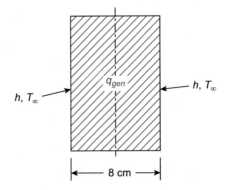

The maximum temperature is at the center plane of this wall and is given by Equation (12.28).

$$T_{max} = T_\infty + q_{gen}\left(\frac{L}{2h} + \frac{L^2}{8k}\right) = 20 + 10^5\left(\frac{0.08}{2(25)} + \frac{(0.08)^2}{8(15)}\right) = 185.33\ C$$

Thus far, we have discussed temperature and convection boundary conditions at the wall surfaces. Let us now look at the situation shown in Fig. 12.4 where there is radiation at the surfaces.

The left surface of the wall has an emissivity of $\varepsilon_1$ and surroundings at absolute temperature $T_{surr_1}$. The right surface has an emissivity of $\varepsilon_2$ and surroundings at absolute temperature $T_{surr_2}$.

The general solution for constant conductivity and uniform heat generation is still given by Eq. (12.4).

$$T(x) = \frac{-q_{gen}}{2k}x^2 + C_1 x + C_2 \tag{12.37}$$

Constants $C_1$ and $C_2$ are obtained from the boundary condition equations as follows:
At the left surface, we have heat flow continuity and the radiation equals the conduction. That is,

$$\text{At } x = 0, \quad \varepsilon_1 \sigma A\left(T_{surr_1}^4 - T^4\right) = -kA\frac{dT}{dx} \tag{12.38}$$

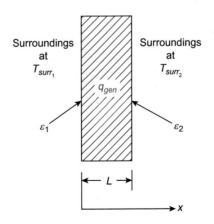

**FIGURE 12.4**

Wall with heat generation and radiation boundary conditions.

Differentiating Eq. (12.37), we have

$$\frac{dT}{dx} = \frac{-q_{gen}x}{k} + C_1 \tag{12.39}$$

From Eqs. (12.37) and (12.39), it is seen that the temperature at $x = 0$ is $C_2$ and the derivative $dT/dx$ at $x = 0$ is $C_1$. Using these results, Eq. (12.38) becomes

$$\varepsilon_1 \sigma \left( T_{surr_1}^4 - C_2^4 \right) + kC_1 = 0 \tag{12.40}$$

At the right surface, the radiation equals the conduction. The boundary equation is

$$\text{At } x = L, \quad \varepsilon_2 \sigma \left( T^4 - T_{surr_2}^4 \right) = -k\frac{dT}{dx} \tag{12.41}$$

Using Eqs. (12.37) and (12.39), Eq. (12.41) becomes

$$\varepsilon_2 \sigma \left[ \left( \frac{-q_{gen}}{2k}L^2 + C_1 L + C_2 \right)^4 - T_{surr_2}^4 \right] + kC_1 - q_{gen}L = 0 \tag{12.42}$$

Eqs. (12.40) and (12.42) may be solved simultaneously for $C_1$ and $C_2$. The equations are nonlinear and must be solved numerically. In Example 12.3 below, we use the Excel *Solver* program for their solution. Once the two constants are known, Eq. (12.37) gives the temperature distribution in the wall.

The maximum temperature in the wall will be at the location where $dT/dx = 0$. From Eq. (12.39), the location is

$$x_{max} = \frac{kC_1}{q_{gen}} \tag{12.43}$$

---

## Example 12.3
### Wall with uniform heat generation and radiation boundary conditions

*Problem*
The wall shown below is 5 cm thick. It has uniform internal heat generation at a rate of $3 \times 10^5$ W/m$^3$. The thermal conductivity of the wall is 30 W/m C. The wall is in a vacuum and its surfaces radiate to the surroundings. One surface has an emissivity of 0.2 and the other surface has an emissivity of 0.9. The surroundings on both sides of the wall are at 0 C.
**(a)** What is the maximum temperature in the wall, and where does it occur?
**(b)** What are the temperatures of the surfaces of the wall?

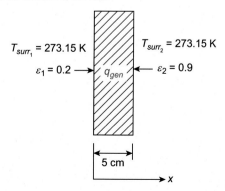

*Solution*

An Excel spreadsheet was used for the solution of the problem. The values in this problem, i.e., values for $q_{gen}$, $L$, $k$, $\varepsilon_1$, $\varepsilon_2$, $\sigma$, $T_{surr_1}$, and $T_{surr_2}$, were entered in cells. The unknown constants $C_1$ and $C_2$ were also assigned cells. The functions on the left sides of Eqs. (12.40) and (12.42) were assigned cells. Let us say these last 2 cells were cells F1 and F2. Finally, a cell F3 was defined as F1²+F2². We put starting values of 100 in the cells for $C_1$ and $C_2$ and called up the *Solver* add-in program. We told *Solver* to vary the values in the cells for $C_1$ and $C_2$ until cell F3 was a minimum. (Ideally we want F3 to be zero, but we are satisfied if F3 is very small.)

When we did this for the parametric values of the problem, we got a very small value for F3 and the values for $C_1$ and $C_2$ were $C_1 = 94.31$ and $C_2 = 710.36$. With these values, we used Eq. (12.37) to get the surface temperatures and Eq. (12.43) to get the location of the maximum temperature. Using this location, we used Eq. (12.37) to get the maximum temperature in the wall.

Summarizing: $C_1 = 94.31$ and $C_2 = 710.36$

$$T(x) = \frac{-q_{gen}}{2k}x^2 + C_1 x + C_2$$

At the left surface, $x = 0$, and $T = C_2 = 710.36$ K $= 437.2$ C.

At the right surface, $x = 0.05$ m and

$$T = \frac{-3 \times 10^5}{2(30)}(0.05)^2 + 94.31(0.05) + 710.36 = 702.58 \text{ K} = 429.4 \text{ C}$$

The maximum occurs at $x_{max} = \frac{kC_1}{q_{gen}} = \frac{(30)(94.31)}{3 \times 10^5} = 0.00943$ m and the maximum temperature is $T_{max} = \frac{-3 \times 10^5}{2(30)}(0.00943)^2 + 94.31(0.00943) + 710.36 = 710.80$ K $= 437.7$ C

The left surface of the wall is at 437.2 C and the right surface is at 429.4 C. The maximum temperature in the wall is 437.7 C and it is located 9.43 mm from the left surface.

Note: The equations for $C_1$ and $C_2$ are nonlinear. Successful solution of the equations often depends on the guessed starting values for $C_1$ and $C_2$. Several attempts with different starting values may be necessary for success.

We have covered situations for temperature, convection, and radiation boundary conditions at the surfaces of the wall. We can also have both convection and radiation occurring simultaneously at the wall surfaces. The treatment of this situation is left as a problem in the problems section of the chapter.

## 12.2.2 Heat generation in a sphere

We now wish to find the temperature distribution in a sphere with uniform heat generation. The sphere is shown in Fig. 12.5. It has volumetric heat generation of strength $q_{gen}$ W/m³, thermal conductivity $k$, and radius $r_o$. The temperature at the surface of the sphere is $T_o$.

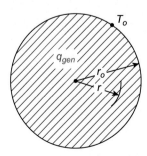

**FIGURE 12.5**

Sphere with uniform heat generation.

The differential equation for the temperature distribution in the sphere is Equation (3.62).

$$\frac{1}{r^2}\frac{d}{dr}\left(kr^2\frac{dT}{dr}\right)+q_{gen}=0 \tag{12.44}$$

Multiplying the equation by $r^2$, we get

$$\frac{d}{dr}\left(kr^2\frac{dT}{dr}\right)+q_{gen}r^2=0 \tag{12.45}$$

Integrating Eq. (12.45) gives

$$kr^2\frac{dT}{dr}+\frac{q_{gen}r^3}{3}=C_1 \tag{12.46}$$

Dividing the terms of this equation by $kr^2$ and integrating the result gives

$$T(r)=C_2-\frac{C_1}{kr}-\frac{r^2}{6k}q_{gen} \tag{12.47}$$

The temperature, however, is finite at the center ($r=0$) of the sphere. Therefore, $C_1$ must be zero and Eq. (12.47) becomes

$$T(r)=C_2-\frac{r^2}{6k}q_{gen} \tag{12.48}$$

[Note: If the sphere were hollow, there would be no material at $r=0$ and we would retain the $C_1$ term in Eq. (12.47)].

We have a temperature boundary condition: At $r=r_o$, $T=T_o$. Putting this into Eq. (12.48), we have $T(r_o)=T_o=C_2-\frac{r_o^2}{6k}q_{gen}$ and $C_2=T_o+\frac{r_o^2}{6k}q_{gen}$.

Putting $C_2$ into Eq. (12.48), we finally get the temperature distribution in the sphere:

$$T(r)=T_o+\frac{q_{gen}}{6k}\left(r_o^2-r^2\right) \tag{12.49}$$

It is seen from Eq. (12.49) that the maximum temperature is at the center ($r=0$) of the sphere and its value is

$$T_{max}=T_o+\frac{q_{gen}r_o^2}{6k} \tag{12.50}$$

Let us now consider the situation where the sphere has a convection boundary condition as shown in Fig. 12.6.

The temperature distribution in the sphere is given by Eq. (12.48) and the constant $C_2$ is obtained from the boundary condition, as follows:

$$T(r)=C_2-\frac{r^2}{6k}q_{gen} \tag{12.51}$$

At the surface of the sphere, the conduction equals the convection. That is,

$$\text{At } r=r_o, \quad -k\frac{dT}{dr}=h(T-T_\infty) \tag{12.52}$$

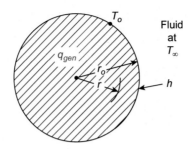

**FIGURE 12.6**

Sphere with uniform heat generation and convection.

Differentiating Eq. (12.51), we get

$$\frac{dT}{dr} = -\frac{r}{3k}q_{gen} \tag{12.53}$$

Eq. (12.52) then becomes $-k\left(\frac{-r_o}{3k}q_{gen}\right) = h\left(C_2 - \frac{r_o^2}{6k}q_{gen} - T_\infty\right)$.

Solving for $C_2$, we get $C_2 = T_\infty + q_{gen}\left(\frac{r_o^2}{6k} + \frac{r_o}{3h}\right)$. Putting $C_2$ into Eq. (12.51), we get the temperature distribution in the sphere:

$$T(r) = T_\infty + \frac{(r_o^2 - r^2)}{6k}q_{gen} + \frac{r_o}{3h}q_{gen} \tag{12.54}$$

The temperature at the surface of the sphere $(r = r_o)$ is

$$T(r_o) = T_o = T_\infty + \frac{r_o}{3h}q_{gen} \tag{12.55}$$

The maximum temperature in the sphere is at the center $(r = 0)$ and is

$$T_{max} = T_\infty + q_{gen}\left(\frac{r_o^2}{6k} + \frac{r_o}{3h}\right) \tag{12.56}$$

Let us now consider a sphere with uniform heat generation that has convection and radiation at its surface. This is shown in Fig. 12.7.

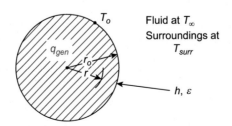

**FIGURE 12.7**

Sphere with uniform heat generation, convection, and radiation.

The temperature distribution in the sphere is given by Eq. (12.48) and the constant $C_2$ is obtained from the boundary condition, as follows:

$$T(r) = C_2 - \frac{r^2}{6k}q_{gen} \qquad (12.57)$$

At the surface of the sphere, the conduction equals the convection plus the radiation. That is,

$$\text{At } r = r_o, \quad -k\frac{dT}{dr} = h(T - T_\infty) + \varepsilon\sigma(T^4 - T^4_{surr}) \qquad (12.58)$$

Differentiating Eq. (12.57), we get

$$\frac{dT}{dr} = -\frac{r}{3k}q_{gen} \qquad (12.59)$$

Eq. (12.58) then becomes

$$-k\left(\frac{-r_o}{3k}q_{gen}\right) = h\left(C_2 - \frac{r_o^2}{6k}q_{gen} - T_\infty\right) + \varepsilon\sigma\left[\left(C_2 - \frac{r_o^2}{6k}q_{gen}\right)^4 - T^4_{surr}\right] \qquad (12.60)$$

Rearranging Eq. (12.60), we get

$$\varepsilon\sigma\left[\left(C_2 - \frac{r_o^2}{6k}q_{gen}\right)^4 - T^4_{surr}\right] + h\left(C_2 - \frac{r_o^2}{6k}q_{gen} - T_\infty\right) - \frac{r_o q_{gen}}{3} = 0 \qquad (12.61)$$

Eq. (12.61) is nonlinear and may be solved for $C_2$ by Excel's *Goal Seek* or *Solver* programs or by using Matlab's *fzero* function.

Once $C_2$ is determined, the surface temperature and the maximum temperature in the sphere are given by

$$T(r_o) = T_o = C_2 - \frac{r_o^2}{6k}q_{gen} \qquad (12.62)$$

$$T_{max} = T(0) = C_2 \qquad (12.63)$$

We end this section with two examples that incorporate material from earlier sections of this text.

---

### Example 12.4
### Buried sphere with internal heat generation

*Problem*

As shown below, a sphere is buried in the ground. The soil has a thermal conductivity of 0.45 W/m C. The center of the sphere is 2 m below the ground surface. The sphere is 1 m diameter and has a conductivity of 15 W/m C. It generates heat uniformly at a rate of 3000 W/m³. The surface of the ground is at 30 C. What is the maximum temperature in the sphere?

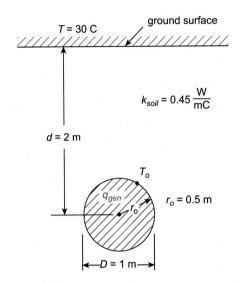

## Solution

This problem involves the conduction shape factor discussed in Section 3.7. The appropriate shape factor, from Table 3.3, is

$$S = \frac{2\pi D}{1 - \dfrac{D}{4d}} \tag{12.64}$$

where $D$ is the diameter of the sphere and $d$ is the depth of the center of the sphere. For our problem, $D = 1$ m and $d = 2$ m. Therefore, the shape factor is

$$S = \frac{2\pi D}{1 - \dfrac{D}{4d}} = \frac{2\pi(1)}{1 - \dfrac{1}{4(2)}} = 7.181 \text{ m}$$

The heat transfer is given by Equation (3.106), which for this problem is

$$q = k_{soil} S (T_o - T_{ground}) \tag{12.65}$$

For our problem, $S = 7.181$ m, $T_{ground} = 30$ C, and $k_{soil} = 0.45$ W/m C. Putting these values into Eq. (12.65), we have

$$q = 0.45(7.181)(T_o - 30) = 3.231(T_o - 30) \tag{12.66}$$

Now let us look at the heat generation in the sphere. The heat generation is the heat generation rate times the volume of the sphere or

$$q = q_{gen}(Volume) = q_{gen}\left(\frac{4}{3}\pi r_o^3\right) = (3000)(4/3)\pi(0.5)^3 = 1570.8 \text{ W} \tag{12.67}$$

For steady state, the heat flows given by Eqs. (12.66) and (12.67) are equal. That is $3.231(T_o - 30) = 1570.8$ and the surface temperature of the sphere is $T_o = 516.2$ C

The maximum temperature in the sphere is given by Eq. (12.50):

$$T_{max} = T_o + \frac{q_{gen} r_o^2}{6k} = 516.2 + \frac{3000(0.5)^2}{6(15)} = 524.5 \text{ C}$$

The maximum temperature in the sphere is at the center and is 525 C.

## Example 12.5

### Sphere with heat generation and convection

*Problem*

Air at 25 C flows over a sphere at a velocity of 2 m/s. The sphere has a diameter of 20 cm and internal heat generation at a rate of 5000 W/m³. The thermal conductivity of the sphere is 25 W/m C.
**(a)** What is the surface temperature of the sphere?
**(b)** What is the center temperature of the sphere?

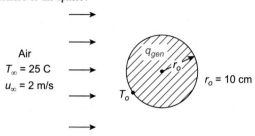

Air
$T_\infty = 25$ C
$u_\infty = 2$ m/s

$q_{gen}$
$r_o$
$r_o = 10$ cm
$T_o$

*Solution*

The first thing to do is to determine the convective coefficient $h$ at the surface of the sphere. An appropriate equation for forced convection over a sphere is Equation (6.111):

$$\text{Nu} = \frac{hD}{k} = 2 + \left(0.4\text{Re}_D^{1/2} + 0.06\text{Re}_D^{2/3}\right)\text{Pr}^{0.4}\left(\frac{\mu}{\mu_s}\right)^{1/4} \tag{12.68}$$

The properties in this equation, except for $\mu_s$, are taken at the fluid temperature, which is 25 C.
For air at 25 C,

$$k = 0.026 \text{ W/m C}, \quad v = 1.55 \times 10^{-5} \text{ m}^2/\text{s}, \quad \text{Pr} = 0.729, \quad \text{and} \quad \mu = 1.84 \times 10^{-5} \text{ kg/m s}$$

$$\text{Re}_D = \frac{u_\infty D}{v} = \frac{(2)(0.2)}{1.55 \times 10^{-5}} = 25800$$

We will assume that $\mu_s \approx \mu$. Then, Eq. (12.68) is

$$\text{Nu} = \frac{hD}{k} = 2 + \left(0.4\text{Re}_D^{1/2} + 0.06\text{Re}_D^{2/3}\right)\text{Pr}^{0.4}$$

Putting values into this equation, we get Nu = 104.78 and

$$h = \frac{k}{D}\text{Nu} = \frac{0.026}{0.2}(104.78) = 13.6 \text{ W/m}^2 \text{ C}$$

The surface temperature of the sphere is given by Eq. (12.55):

$$T(r_o) = T_o = T_\infty + \frac{r_o}{3h}q_{gen} = 25 + \frac{0.1}{(3)(13.6)}(5000) = 37.3 \text{ C}$$

The center temperature of the sphere is given by Eq. (12.56):

$$T_{max} = T(0) = T_\infty + q_{gen}\left(\frac{r_o^2}{6k} + \frac{r_o}{3h}\right) = 25 + (5000)\left(\frac{(0.1)^2}{6(25)} + \frac{0.1}{(3)(13.6)}\right) = 37.6 \text{ C}$$

**(a)** The surface temperature of the sphere is 37.3 C.
**(b)** The center temperature of the sphere is 37.6 C.

Note: Our assumption that $\mu_s \approx \mu$ was appropriate. There is only a 12 C difference between fluid and surface temperatures and $\mu/\mu_s = 0.99$.

## 12.3 Contact resistance

Let us consider two large plates of area $A$ which are in contact. Plate 1 has thermal conductivity $k_1$ and Plate 2 has conductivity $k_2$. This is shown in Fig. 12.8.

The outside surface of Plate 1 is at temperature $T_1$ and the outside surface of Plate 2 is at a lower temperature $T_2$. Ideally, the contact surfaces of the two plates are perfectly smooth and there is no resistance to heat flow at the contact interface. Then, as the plates are large, the heat flow will essentially be one-dimensional and will be

$$q = \frac{\Delta T_{overall}}{\sum R} = \frac{T_1 - T_2}{\dfrac{L_1}{k_1 A} + \dfrac{L_2}{k_2 A}} \tag{12.69}$$

The temperature profile is shown in Fig. 12.9.

But, real surfaces are not perfectly smooth. They have peaks and valleys, as shown in Fig. 12.10, and the two plates are not actually in contact at all points of the interface.

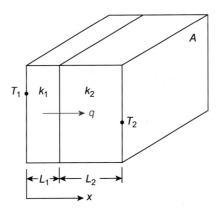

**FIGURE 12.8**

Two large plates in contact.

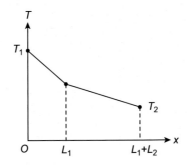

**FIGURE 12.9**

Temperature profile for plates in contact.

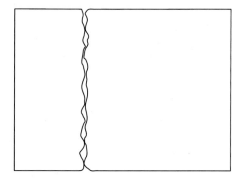

**FIGURE 12.10**

Real surfaces in contact.

This lack of perfect contact creates a thermal resistance at the interface, which is manifested by a temperature difference between the two plates in contact. This temperature difference is shown in Fig. 12.11.

The heat flow through the interface is

$$q = h_c A (T_a - T_b) \tag{12.70}$$

where $h_c$ is the contact conductance.

If the contact resistance $\left( \frac{1}{h_c A} \right)$ is included in the series system of the two plates, then Eq. (12.69) becomes

$$q = \frac{T_1 - T_2}{\dfrac{L_1}{k_1 A} + \dfrac{1}{h_c A} + \dfrac{L_2}{k_2 A}} \tag{12.71}$$

Much experimentation has been performed to measure contact conductance. A typical experimental apparatus is shown in Fig. 12.12. The two specimens are cylindrical and are placed, one

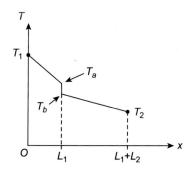

**FIGURE 12.11**

Temperature profile for plates in contact with contact resistance.

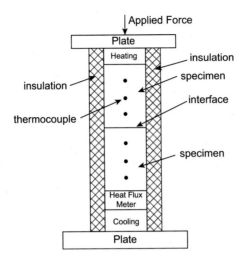

**FIGURE 12.12**

Test apparatus for contact conductance.

above the other, in the apparatus. Below the lower specimen is a heat flux meter. There is a heating section above the upper specimen and a cooling section below the heat flux meter. Insulation is provided around the specimens so that heat flow is axial. Thermocouples are provided in the test specimens to determine the temperature profiles. Finally, there is a mechanism provided to vary the axial force on the specimens so that results can be obtained for different contact pressures.

Heat transfer between surfaces in contact consists of conduction through the solid areas in contact and conduction through the fluid that is trapped in the spaces created by the roughness of the surfaces. (If the surfaces are in a vacuum, the latter item, of course, does not apply.) For surfaces at very high temperatures, radiation heat transfer between the surfaces is also a factor. The contact conductance depends on several parameters, including the material(s) of the surfaces; surface roughness; surface flatness; surface condition (e.g., conductance is decreased if the surfaces are oxidized); the contact pressure; and the fluid between the surfaces.

In general, contact conductance increases with an increase in contact pressure as a pressure increase results in an increased area of surface contact. The conductance also increases with smoother and flatter surfaces, with trapped fluids of higher conductivity, and for higher contact temperatures.

One way to increase contact conductance is to insert a very thin foil of soft metal between the surfaces. Another way is to use thermal greases, many of which are silicon-based.

Much experimentation has been performed to determine contact conductance. Some experimental results are given in Table 12.1 [1–4]. Other experimental and theoretical studies on contact resistance are in Refs. [5–11].

We conclude this section with two examples involving contact resistance.

### Table 12.1  Contact conductance.

| Material | Roughness μm (rms) | Gap material | Contact temperature (C) | Contact pressure (kPa) | $h_c$ (W/m² C) |
|---|---|---|---|---|---|
| 2024-T3 (Al) | 1.2–1.7 | Vacuum | 43 | 20 | 140 |
| 2024-T3 (Al) | 1.2–1.7 | Vacuum | 43 | 200 | 455 |
| 2024-T3 (Al) | 0.2–0.46 | Vacuum | 43 | 20 | 210 |
| 2024-T3 (Al) | 0.2–0.46 | Vacuum | 43 | 200 | 680 |
| 75S-T6 (Al) | 3 | Air | 93 | 200 | 3400 |
| 75S-T6 (Al) | 3 | Air | 93 | 2000 | 9650 |
| 75S-T6 (Al) | 1.7 | Air | 93 | 200 | 5700 |
| 75S-T6 (Al) | 1.7 | Air | 93 | 2000 | 12,000 |
| 75S-T6 (Al) | 0.3 | Air | 93 | 200 | 10,200 |
| 75S-T6 (Al) | 0.3 | Air | 93 | 2000 | 17,600 |
| 304 (SS) | 1.1–1.5 | Vacuum | 30 | 200 | 280 |
| 304 (SS) | 1.1–1.5 | Vacuum | 30 | 2000 | 650 |
| 304 (SS) | 0.3–0.4 | Vacuum | 30 | 200 | 520 |
| 304 (SS) | 0.3–0.4 | Vacuum | 30 | 2000 | 2900 |
| 416 (SS) | 2.5 | Air | 93 | 200 | 2300 |
| 416 (SS) | 2.5 | Air | 93 | 2000 | 2900 |
| Copper (OFHC) | 0.2 | Vacuum | 46 | 200 | 7400 |
| Copper (OFHC) | 0.2 | Vacuum | 46 | 2000 | 9950 |

### Example 12.6
#### Cylinders in contact

*Problem*

Two cylinders of 416 stainless steel ($k = 25$ W/m C) are in contact, as shown below. The cylinders are in an air environment. Each cylinder is 8 cm diameter and 25 cm long. The contact surfaces of both cylinders have a roughness of 2.5 μm (rms) and the contact pressure is 1000 kPa. The outside ends of the cylinders are at 20 and 250 C, respectively. The side surfaces of the cylinders are perfectly insulated so all heat flow is axial.
**(a)** What is the rate of heat flow through the cylinders?
**(b)** What is the temperature drop across the contact interface?

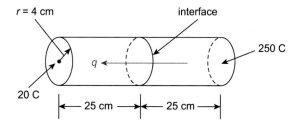

*Solution:*

$$L = 0.25 \text{ m} \quad A = \frac{\pi}{4}(0.08)^2 = 0.00503 \text{ m}^2$$

From Table 12.1, $h_c \approx 2500 \text{ W/m}^2 \text{ C}$.

**(a)** $q = \dfrac{\Delta T_{overall}}{\sum R} = \dfrac{\Delta T_{overall}}{2\left(\dfrac{L}{kA}\right)_{cylinder} + \left(\dfrac{1}{h_cA}\right)_{contact}}$

Putting in values, we have

$$q = \frac{250 - 20}{2\left(\dfrac{0.25}{(25)(0.00503)}\right)_{cylinder} + \left(\dfrac{1}{(2500)(0.00503)}\right)_{contact}} = 56.7 \text{ W}$$

**(b)** At the interface, we have $q = h_c A (\Delta T)_{contact}$

$$(\Delta T)_{contact} = \frac{q}{h_c A} = \frac{56.7}{(2500)(0.00503)} = 4.51 \text{ C}$$

The rate of heat flow through the cylinders is 56.7 W, and the temperature drop at the contact interface is 4.51 C.

---

## Example 12.7
### Aluminum–Insulator sandwich

*Problem*

As shown below, an insulating material ($k = 0.3$ W/m C) is sandwiched between two large aluminum plates ($k = 200$ W/m C). The insulator is 5 mm thick, and the aluminum plates are both 3 cm thick. The temperatures at the outer surfaces of the assembly are maintained at 20 and 180 C, respectively.
**(a)** If we assume there is no thermal contact resistance between the contacting surfaces, what is the heat flux through the assembly?
**(b)** If the contact resistance is actually $2.5 \times 10^{-4}$ m$^2$ C/W at the insulator–aluminum interfaces, what is the heat flux through the assembly?

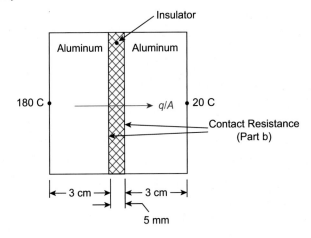

*Solution*

**(a)** With no contact resistance:

$$q = \frac{\Delta T_{overall}}{\sum R} = \frac{\Delta T_{overall}}{2\left(\dfrac{L}{kA}\right)_{plates} + \left(\dfrac{L}{kA}\right)_{insulator}}$$

$$\frac{q}{A} = \frac{180 - 20}{2\left(\dfrac{0.03}{200}\right)_{plates} + \left(\dfrac{0.005}{0.3}\right)_{insulator}} = 9430 \text{ W/m}^2$$

**(b)** Including the contact resistance:

$$q = \frac{\Delta T_{overall}}{\sum R} = \frac{\Delta T_{overall}}{2\left(\dfrac{L}{kA}\right)_{plates} + \left(\dfrac{L}{kA}\right)_{insulator} + 2\left(\dfrac{1}{h_c A}\right)_{interfaces}}$$

$$\frac{q}{A} = \frac{180 - 20}{2\left(\dfrac{0.03}{200}\right)_{plates} + \left(\dfrac{0.005}{0.3}\right)_{insulator} + 2(2.5 \times 10^{-4})_{interfaces}} = 9160 \text{ W/m}^2$$

The heat flux without contact resistance is 9430 W/m². With contact resistance, the heat flux is 9160 W/m². Including the contact resistance results in a 2.9% decrease in the heat flux.

# 12.4 Condensation and boiling

Up to now, we have only considered heat transfer involving a single phase. This section considers heat transfer involving a change of phase for a substance. When the temperature of a subcooled liquid is raised above its saturation temperature, vaporization or boiling will occur. And, when the temperature of a superheated vapor is lowered below its saturation temperature, condensation will occur. Such changes in phase result in relatively large heat transfer rates. Applications include power and refrigeration cycles and compact heat exchangers.

Condensation and boiling heat transfer are significantly more complex than single-phase heat transfer. We will consider only the fundamentals of the phenomena. The references at the end of this chapter may be consulted for more detailed treatments of the topics.

## 12.4.1 Condensation heat transfer

As mentioned above, condensation takes place when a condensable vapor is cooled to a temperature lower than its saturation temperature $T_{sat}$. This is usually done by bringing the vapor into contact with a surface that is at a temperature lower than $T_{sat}$. The vapor will condense, and the resulting condensate will flow down the surface under the action of gravity. The condensation will be either **dropwise** or **filmwise**. In dropwise condensation, drops of condensate form on the surface and flow downward in random paths over the surface. This will happen if the condensate does not "wet" the surface. On the other hand, if the condensate "wets" the surface, then a condensate film will form on the surface and the condensate will flow downwards as a film over all of the surface. At the upper end of the surface, the thickness of the film will be zero or minimal. As the condensate flows downward, the film thickness increases.

The layer of condensate in film condensation causes a resistance to heat flow between the vapor and the surface. Therefore, dropwise condensation results in significantly higher heat transfer rates than film condensation and is to be preferred if such higher heat rates are desired. But dropwise condensation is much harder to sustain than film condensation, and film condensation is the predominant mode of condensation occurring in practice. For this reason, design of heat transfer devices usually assumes film condensation rather than dropwise condensation.

We will look at film condensation for vertical and inclined plates. We will also look at external film condensation for vertical and horizontal cylinders and spheres.

### 12.4.1.1 Film condensation for vertical and inclined plates

#### 12.4.1.1.1 Vertical plates

Fig. 12.13 shows film condensation on a vertical flat surface. This phenomenon was first analyzed in 1916 by Nusselt [12]. The surface is at temperature $T_s$ and the vapor adjacent to the film is at saturation temperature $T_v$. The thickness of the film is $\delta(x)$. It is zero at $x = 0$, and it increases with $x$.

An element of the condensate film between $x$ and $x + dx$ is shown in Fig. 12.14. The figure also shows the forces acting on the element.

The element has a thickness of $\delta - y$, a height of $dx$, and a width of unity (in the $z$ direction). Its volume is therefore $(\delta - y)(dx)(1)$. It is assumed that the condensate flow down the surface is laminar, the temperature distribution from the plate surface to the vapor is linear, and there is no shear stress at the interface between the condensate and the vapor. Three forces act on the element as follows: There is the weight of the condensate in the element, the buoyant force on the element (which is the weight of the displaced vapor), and the shear force on the left face of the element. The first force acts downward in the $+x$ direction. The other two act upwards in the $-x$ direction. We assume that the acceleration of the element is negligible. Newton's Law then gives

$$\sum F_x = \rho g(\delta - y)dx - \mu \frac{du}{dy}(dx) - \rho_v g(\delta - y)dx = 0 \qquad (12.72)$$

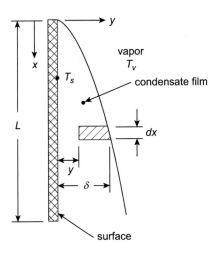

**FIGURE 12.13**

Film condensation on a vertical flat surface.

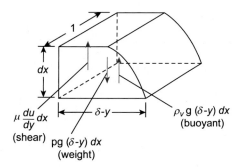

**FIGURE 12.14**

Forces on an element of the condensate film.

where $g$ = acceleration of gravity
$\rho$ = density of the condensate
$\mu$ = absolute viscosity of the condensate
$u$ = velocity of the element in the $x$ direction
$\rho_v$ = density of the vapor.
Rearranging Eq. (12.72), we have

$$\frac{du}{dy} = \frac{(\rho - \rho_v)g}{\mu}(\delta - y) \tag{12.73}$$

The condensate sticks to the surface, so $u = 0$ at $y = 0$. Integrating Eq. (12.73) with respect to $y$, we get

$$u = \frac{(\rho - \rho_v)g}{\mu}\left(\delta y - \frac{y^2}{2}\right) \tag{12.74}$$

The mass flow rate of condensate (per unit width of film) at any location $x$ is

$$\dot{m}(x) = \int_0^\delta \rho u \, dy = \int_0^\delta \frac{\rho(\rho - \rho_v)g}{\mu}\left(\delta y - \frac{y^2}{2}\right)dy = \frac{\rho(\rho - \rho_v)g\delta^3}{3\mu} \tag{12.75}$$

The increase in mass flow rate of condensate from $x = x$ to $x = x + dx$ is

$$\dot{m}_{increase} = \dot{m}(x + dx) - \dot{m}(x) = \frac{d\dot{m}(x)}{dx}dx = \frac{d}{dx}\left(\frac{\rho(\rho - \rho_v)g\delta^3}{3\mu}\right)dx = \frac{\rho(\rho - \rho_v)g\delta^2}{\mu}d\delta \tag{12.76}$$

This increase in condensate results from condensing of the vapor through conduction heat transfer from the vapor to the wall. With $h_{fg}$ being the latent heat of vaporization and $k$ being the thermal conductivity of the condensate, the heat transfer balance is

$$\dot{m}_{increase}h_{fg} = kdx\frac{T_v - T_s}{\delta} \tag{12.77}$$

Using Eq. (12.76) for $\dot{m}_{increase}$, Eq. (12.77) is

$$\frac{\rho(\rho - \rho_v)g\delta^2 d\delta}{\mu}h_{fg} = kdx\frac{T_v - T_s}{\delta} \tag{12.78}$$

The film thickness $\delta$ at the top of the surface (i.e., at $x = 0$) is zero. Integrating Eq. (12.78) with respect to $x$, we get the following relation for the film thickness:

$$\delta(x) = \left[\frac{4\mu k(T_v - T_s)x}{\rho(\rho - \rho_v)gh_{fg}}\right]^{1/4} \tag{12.79}$$

For unit width, the local heat transfer flux from the vapor to the plate's surface at location $x$ is

$$\frac{q}{dx} = h_x(T_v - T_s) = k\frac{T_v - T_s}{\delta} \tag{12.80}$$

The local heat transfer coefficient $h_x$ is then

$$h_x = \frac{k}{\delta(x)} \tag{12.81}$$

Using Eq. (12.79), the local coefficient is

$$h_x = \left[\frac{\rho(\rho - \rho_v)gh_{fg}k^3}{4\mu(T_v - T_s)x}\right]^{1/4} \tag{12.82}$$

The local heat transfer coefficient is integrated over the height $L$ of the surface to get the average heat transfer coefficient $h$. That is,

$$h = \frac{1}{L}\int_0^L \left[\frac{\rho(\rho - \rho_v)gh_{fg}k^3}{4\mu(T_v - T_s)x}\right]^{1/4} dx = 0.943\left[\frac{\rho(\rho - \rho_v)gh_{fg}k^3}{\mu L(T_v - T_s)}\right]^{1/4} \tag{12.83}$$

The heat transfer coefficient is based on laminar flow in the film. It also assumes that the temperature profile is linear across the film thickness. However, the profile is actually nonlinear, and the condensate film is being cooled to below the saturation temperature. These two factors can be compensated for by replacing $h_{fg}$ in Eq. (12.83) by $h'_{fg}$, which is defined by

$$h'_{fg} = h_{fg} + 0.68c_p(T_v - T_s) \tag{12.84}$$

where $c_p$ = specific heat of the condensate.

Therefore, Eq. (12.83) becomes

$$h = 0.943\left[\frac{\rho(\rho - \rho_v)gh'_{fg}k^3}{\mu L(T_v - T_s)}\right]^{1/4} \tag{12.85}$$

For low condensate flow rates, the flow is laminar. For high flow rates, the flow is turbulent. The Reynolds number Re indicates the type of flow. For the condensate flow, the Reynolds number is defined as

$$\text{Re} = \frac{\rho V D_H}{\mu} = \frac{\rho V(4A/P)}{\mu} \tag{12.86}$$

where $D_H$ = hydraulic diameter

$A$ = flow area

$P$ = wetted perimeter = width of the film = unity for our model

$V$ = average flow velocity.

The mass flow rate is $\dot{m} = \rho A V$, so Eq. (12.86) becomes

$$\text{Re} = \frac{4\dot{m}}{P\mu} \tag{12.87}$$

Let us look at the heat transfer between the wall and the saturated vapor. The heat flow rate $q$ is

$$q = hA_s(T_v - T_s) = \dot{m}h'_{fg} \tag{12.88}$$

where $A_s$ = area of surface in contact with condensate = $L \times 1$ for our model.

Putting $\dot{m}$ from Eq. (12.88) into Eq. (12.87), the Reynolds number is

$$\text{Re} = \frac{4hL(T_v - T_s)}{\mu h'_{fg}} \tag{12.89}$$

where $h$ is given by Eq. (12.85).

It was found by experimentation that the critical Reynolds number, where the flow turns turbulent, is about 1800. It turns out that Eq. (12.85) for the heat transfer coefficient gives accurate results if Re is less than about 30. However, between Reynolds numbers of 30 and 1800, the surface of the condensate becomes wavy and the heat transfer coefficient in this region is somewhat greater than that given by Eq. (12.85). McAdams [13] suggested that the heat transfer coefficient given by Eq. (12.85) be increased by 20% in the wavy region for design purposes. If this is done, Eq. (12.85) for vertical surfaces becomes

$$h = 1.13 \left[ \frac{\rho(\rho - \rho_v)gh'_{fg}k^3}{\mu L(T_v - T_s)} \right]^{1/4} \tag{12.90}$$

The properties of the condensate in the previous equations should be taken at the film temperature $T_f = \frac{T_s + T_v}{2}$. The latent heat of vaporization $h_{fg}$ and the density of the saturated vapor $\rho_v$ are taken at the saturation temperature $T_v$. The above equations can also be used to determine the heat transfer coefficient in cases where the vapor is superheated. For those cases, the heat flow should be calculated using the difference between the surface temperature and the saturation temperature corresponding to the pressure of the vapor.

An alternative approach to Eq. (12.90) for the wavy surface region of $30 < \text{Re} < 1800$ was developed by Kutateladze [14] who correlated experimental data and recommended the following equation for the heat transfer coefficient on a vertical flat surface:

$$h = \frac{\text{Re}\ k}{1.08\ \text{Re}^{1.22} - 5.2} \left( \frac{g}{v^2} \right)^{1/3} \quad \text{for} \quad 30 < \text{Re} < 1800 \text{ and } \rho_v \ll \rho \tag{12.91}$$

where: $v$ = kinematic viscosity

The Reynolds number to be used in Eq. (12.91) is

$$\text{Re} = \left[ 4.81 + \frac{3.70 kL(T_v - T_s)}{\mu h'_{fg}} \left(\frac{g}{v^2}\right)^{1/3} \right]^{0.82} \quad \text{for} \quad \rho_v \ll \rho \tag{12.92}$$

For the turbulent region of Re > 1800, Labuntsov [15] recommends

$$h = \frac{\text{Re } k}{8750 + 58 \text{Pr}^{-0.5}(\text{Re}^{0.75} - 253)} \left(\frac{g}{v^2}\right)^{1/3} \quad \text{for} \quad \text{Re} > 1800 \text{ and } \rho_v \ll \rho \tag{12.93}$$

The Reynolds number to be used in Eq. (12.93) is

$$\text{Re} = \left[ \frac{0.0690 kLPr^{0.5}(T_v - T_s)}{\mu h'_{fg}} \left(\frac{g}{v^2}\right)^{1/3} - 151 \text{ Pr}^{0.5} + 253 \right]^{4/3} \quad \text{for} \quad \rho_v \ll \rho \tag{12.94}$$

---

## Example 12.8
### Film condensation on a vertical flat plate

*Problem*

A vertical flat plate is 25 cm wide and 80 cm high. The plate is at 40 C and is exposed to saturated steam at 1 atm pressure. What is the rate of heat transfer and the amount of steam condensed in 1 hour?

*Solution*

Plate width = 0.25 m

Plate height = $L = 0.8$ m

Plate surface area = $(0.25)(0.8) = 0.2 \text{ m}^2$

$$T_s = 40 \text{ C } T_v = \text{saturation temperature at 1 atm} = 100 \text{ C}$$

Condensate properties are at film temperature

$T_f = \frac{T_s + T_v}{2} = \frac{40 + 100}{2} = 70 \text{ C. } \rho = 978 \text{ kg/m}^3$

$c_p = 4190 \text{ J/kg C}$

Properties of water at 70 C are $k = 0.659 \text{ W/m C}$

$\mu = 0.404 \times 10^{-3} \text{ kg/m s}$

$v = \mu/\rho = 4.13 \times 10^{-7} \text{ m}^2/\text{s}$

For the saturated steam at 100 C, we have $\rho_v = 0.60 \text{ kg/m}^3$

The corrected heat of vaporization is $h_{fg} = 2260 \text{ kJ/kg}$

$$h'_{fg} = h_{fg} + 0.68 c_p (T_v - T_s) = 2260 \times 10^3 + 0.68(4190)(100 - 40) = 2430 \times 10^3 \text{ J/kg}$$

Let us first use Eq. (12.90) for the heat transfer coefficient.

$$h = 1.13 \left[ \frac{\rho(\rho - \rho_v)gh'_{fg}k^3}{\mu L(T_v - T_s)} \right]^{1/4} = 1.13 \left[ \frac{978(978 - 0.60)(9.807)(2430 \times 10^3)(0.659)^3}{(0.404 \times 10^{-3})(0.8)(100 - 40)} \right]^{1/4}$$

$$h = 4840 \text{ W/m}^2 \text{ C}$$

$$q = hA(T_v - T_s) = (4840)(0.2)(100 - 40) = 58080 \text{ W}$$

For the condensate flow rate, we have

$$q = \dot{m} h'_{fg} \quad \dot{m} = \frac{q}{h'_{fg}} = \frac{58080}{2430 \times 10^3} = 0.024 \text{ kg/s} \times (3600 \text{ s}/\text{h}) = 86.4 \text{ kg/h}$$

Let us check the Reynolds number from Eq. (12.89).

$$\text{Re} = \frac{4hL(T_v - T_s)}{\mu h'_{fg}} = \frac{4(4840)(0.8)(100 - 40)}{(0.404 \times 10^{-3})(2430 \times 10^3)} = 947$$

This is for the wavy surface region, so let us see the results using Eqs. (12.91) and (12.92). Using Eq. (12.92), the Reynolds number is

$$\text{Re} = \left[ 4.81 + \frac{3.70kL(T_v - T_s)}{\mu h'_{fg}} \left( \frac{g}{v^2} \right)^{1/3} \right]^{0.82}$$

$$= \left[ 4.81 + \frac{3.70(0.659)(0.8)(100 - 40)}{(0.404 \times 10^{-3})(2430 \times 10^3)} \left( \frac{9.807}{(4.13 \times 10^{-7})^2} \right)^{1/3} \right]^{0.82} = 1009$$

Using this Reynolds number in Eq. (12.91), we have

$$h = \frac{\text{Re } k}{1.08\text{Re}^{1.22} - 5.2} \left( \frac{g}{v^2} \right)^{1/3} = \frac{(1009)(0.659)}{1.08(1009)^{1.22} - 5.2} \left( \frac{9.807}{(4.13 \times 10^{-7})^2} \right)^{1/3} = 5147 \text{ W/m}^2 \text{ C}$$

$$q = hA(T_v - T_s) = (5147)(0.2)(100 - 40) = 61765 \text{ W}$$

For the condensate flow rate, we have

$$\dot{m} = \frac{q}{h'_{fg}} = \frac{61765}{2430 \times 10^3} = 0.0254 \text{ kg/s} \times (3600 \text{ s}/\text{h}) = 91.5 \text{ kg/h}$$

The rate of heat transfer is 61765 W, and the rate of steam condensation is 91.5 kg/h.
Note: There is only a 6% difference between the results of Eqs. (12.90) and (12.91). The laminar analysis with a 20% increase for the wavy surface region did a pretty good job of determining the heat transfer coefficient.

### 12.4.1.1.2 Inclined plates
Fig. 12.15 shows film condensation on an inclined surface. For the inclined surface, the acceleration $g$ in the above equations should be replaced by the component of gravity parallel to the surface. If $\theta$ is the angle of the surface from the vertical, then $g$ should be replaced with $g \cos \theta$.

### 12.4.1.2 Film condensation for vertical cylinders
The above equations for vertical flat surfaces may be used for vertical cylinders if the diameter of the cylinder is large compared with the condensate film thickness.

### 12.4.1.3 Film condensation for horizontal cylinders and for spheres
The heat transfer coefficient for a horizontal cylinder of diameter $D$ is

$$h_{horizontal \ cylinder} = 0.729 \left[ \frac{\rho(\rho - \rho_v)gh'_{fg}k^3}{\mu D(T_v - T_s)} \right]^{1/4} \tag{12.95}$$

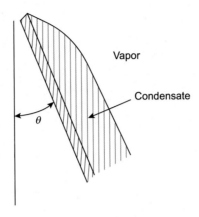

**FIGURE 12.15**

Film condensation on an inclined surface.

The heat transfer coefficient for a sphere of diameter $D$ is

$$h_{sphere} = 0.815 \left[ \frac{\rho(\rho - \rho_v)gh'_{fg}k^3}{\mu D(T_v - T_s)} \right]^{1/4} \tag{12.96}$$

Tubes in a condenser are often horizontal and arranged in tube banks. If $N$ tubes are in a vertical stack as shown in Fig. 12.16, then the average heat transfer coefficient for the $N$ tubes is given by

$$h_{N \ horizontal \ cylinders} = 0.729 \left[ \frac{\rho(\rho - \rho_v)gh'_{fg}k^3}{\mu ND(T_v - T_s)} \right]^{1/4} = N^{-1/4} h_{horizontal \ cylinder} \tag{12.97}$$

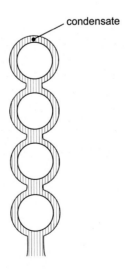

condensate

**FIGURE 12.16**

Film condensation on stacked horizontal tubes.

## Example 12.9
### Steam condenser

*Problem*

A steam condenser has 60 horizontal tubes arranged in a tube bundle 10 tubes wide by 6 tubes high. Each tube has an outer diameter of 2 cm and a length of 10 m. The outer surfaces of the tubes are at 20 C. The steam is saturated and at a pressure of 20 kPa. What is the heat transfer rate for the condenser and the rate of condensation of the steam?

*Solution:*

Tube diameter $= D = 0.02$ m
Tube length $= L = 10$ m

$$\text{Surface area of a single tube} = \pi DL = \pi(0.02)(10) = 0.6283 \text{ m}^2$$

$$T_s = 20 \text{ C} \quad T_v = \text{saturation temperature at 20 kPa} = 60 \text{ C}$$

Condensate properties are at film temperature $\quad T_f = \frac{T_s + T_v}{2} = \frac{20 + 60}{2} = 40$ C.

Properties of water at 40 C are
$$\rho = 992 \text{ kg/m}^3$$
$$c_p = 4179 \text{ J/kg C}$$
$$k = 0.629 \text{ W/m C}$$
$$\mu = 0.653 \times 10^{-3} \text{ kg/m s}$$

For the saturated steam at 60 C, we have
$$\rho_v = 0.13 \text{ kg/m}^3$$
$$h_{fg} = 2360 \text{ kJ/kg}$$

The corrected heat of vaporization is

$$h'_{fg} = h_{fg} + 0.68c_p(T_v - T_s) = 2360 \times 10^3 + 0.68(4179)(60 - 20) = 2474 \times 10^3 \text{ J/kg}$$

From Eq. (12.95), the heat transfer coefficient for a single tube is

$$h_{horizontal\ cylinder} = 0.729 \left[\frac{\rho(\rho - \rho_v)gh'_{fg}k^3}{\mu D(T_v - T_s)}\right]^{1/4}$$

$$= 0.729 \left[\frac{(992)(992 - 0.13)(9.807)(2474 \times 10^3)(0.629)^3}{(0.653 \times 10^{-3})(0.02)(60 - 20)}\right]^{1/4} = 7528 \text{ W/m}^2 \text{ C}$$

The tubes are in stacks of 6 tubes high. Therefore, by Eq. (12.97), we have

$$h_{N\ horizontal\ cylinders} = N^{-1/4}h_{horizontal\ cylinder} = (6)^{-1/4}(7528) = 4810 \text{ W/m}^2 \text{ C}$$

There are 60 tubes, so the total heat transfer rate is

$$q = hA(T_v - T_s) = (4810)[(60)(0.6283)](60 - 20) = 7.253 \times 10^6 \text{ W}$$

The condensate flow is $\dot{m} = \frac{q}{h'_{fg}} = \frac{7.253 \times 10^6}{2474 \times 10^3} = 2.94 \text{ kg/s} = 10550 \text{ kg/h}$

The heat transfer rate for the steam condenser is $7.273 \times 10^6$ W, and the rate of condensate flow is 10550 kg/h.

## 12.4.2 Boiling heat transfer

Boiling occurs when a solid surface is in contact with a liquid, and the temperature of the solid surface is sufficiently greater than the saturation temperature of the liquid. Let us consider a surface immersed in a liquid. The boiling heat transfer causes some of the liquid to vaporize, resulting in vapor bubbles and films. If the liquid is initially stagnant, motion of the vapor and heated water will occur due to buoyancy effects. There will be natural convection or **pool boiling**. If the liquid is initially in motion, then the boiling will affect the initial fluid motion. This boiling is **flow boiling.** We will discuss pool boiling. References [16−18] may be consulted for information on flow boiling.

### 12.4.2.1 Regions of pool boiling

Fig. 12.17 shows a typical boiling curve for water at 1 atm pressure. Other liquids have similarly-shaped curves. The boiling curve gives heat flux $(q/A)$ at the surface of the immersed object versus the excess temperature $\Delta T_{excess}$, which is the difference between the surface temperature $T_s$ and the saturated water temperature $T_{sat}$. That is, $\Delta T_{excess} = T_s - T_{sat}$.

The figure shows four regions, a maximum, and a minimum. Let us first discuss the regions.

Region I is for an excess temperature less than about 5 C. In this region, there is natural convection heat transfer.

Region II is the nucleate boiling region that is for excess temperatures between about 5 C and 30 C. At the lower end of this range, bubbles start to appear on the surface of the solid. As the excess temperature increases, these bubbles rise to the free surface of the liquid.

Region III is the transition boiling region, covering excess temperatures between about 30 C and 120 C. The creation of bubbles is quite rapid and a vapor film starts to form on the solid surface. This film provides a resistance to heat transfer, which results in the decrease in heat flux as the excess temperature increases. This region is also called the unstable film region as there is fluctuation between nucleate and film boiling on the surface.

Region IV is the film boiling region. The excess temperature is high, greater than about 120 C, and the heating surface is covered by a film, or blanket, of vapor. As the excess temperature increases, radiation heat transfer through the vapor film becomes significant and increases the heat flux.

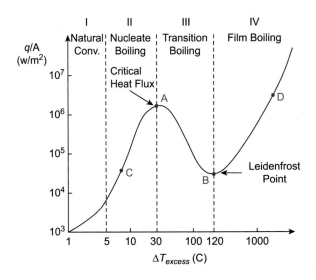

**FIGURE 12.17**

Regions of pool boiling for water at 1 atm pressure.

The maximum heat flux (Point A) is the critical heat flux (CHF). Let us say we start at a point on the boiling curve (e.g., Point C) in the nucleate boiling region to the left of Point A and slowly increase the heat flux. Following the curve, the excess temperature also increases. However, an important and unexpected thing happens when the CHF point is reached. With further increase in heat flux, there is an abrupt increase of surface temperature and the system no longer follows the boiling curve. There is a jump to Point D in the film boiling region. And, this new surface temperature may indeed be greater than the melting point of the solid, thereby causing failure of the device. For this reason, Point A is often called the burnout point. This characteristic was first observed by Nukiyama [19] in his experiments on heated nichrome and platinum wires in water.

To avoid the possibility of burnout, devices are generally designed to operate in the higher heat flux area of the nucleate boiling region, but below the CHF point. The heat flux for this operating area is high and the excess temperature is small.

The minimum heat flux (Point B) is called the Leidenfrost point. At this point, the surface is completely covered by a film of vapor.

### 12.4.2.2 Nucleate pool boiling

In 1952, Rohsenow [20] proposed a boiling curve correlation for the nucleate pool boiling region. This correlation, which finds often use today, is

$$\frac{c_p \Delta T_{excess}}{h_{fg} \mathrm{Pr}^s} = C_{sf} \left[ \frac{(q/A)}{\mu h_{fg}} \sqrt{\frac{\sigma}{g(\rho - \rho_v)}} \right]^{1/3} \tag{12.98}$$

Rearranging this equation, the heat flux ($q/A$) is given by

$$\frac{q}{A} = \mu h_{fg} \left[ \frac{(\rho - \rho_v)g}{\sigma} \right]^{1/2} \left[ \frac{c_p \Delta T_{excess}}{C_{sf} h_{fg} \mathrm{Pr}^s} \right]^3 \tag{12.99}$$

In these two equations,

Pr = Prandtl number of saturated liquid (dimensionless)

$c_p$ = specific heat of saturated liquid (J/kg C)

$\mu$ = absolute viscosity of saturated liquid (kg/m s)

$\rho$ = density of saturated liquid $(kg/m^3)$

$\rho_v$ = density of saturated vapor $(kg/m^3)$

$h_{fg}$ = latent heat of vaporization (J/kg)

g = acceleration of gravity $(m/s^2)$

$\sigma$ = surface tension of liquid − vapor interface (N/m)

$\Delta T_{excess}$ = excess temperature = $T_s - T_{sat}$ (C)

$C_{sf}$ = experimental constant depending on the surface − fluid combination

$s$ = experimental constant depending on the fluid ($s = 1$ for water and 1.7 for other liquids)

**Table 12.2 Liquid−vapor surface tension for water.**

| T (C) | σ (N/m) | T (C) | σ (N/m) |
|-------|---------|-------|---------|
| 0.01 | 0.0756 | 200 | 0.0377 |
| 20 | 0.0727 | 220 | 0.0331 |
| 40 | 0.0696 | 240 | 0.0284 |
| 60 | 0.0662 | 260 | 0.0237 |
| 80 | 0.0627 | 280 | 0.0189 |
| 100 | 0.0589 | 300 | 0.0143 |
| 120 | 0.0550 | 320 | 0.0098 |
| 140 | 0.0509 | 340 | 0.0056 |
| 160 | 0.0466 | 360 | 0.0019 |
| 180 | 0.0422 | 374 | 0 |

The properties of the liquid and the vapor should be evaluated at the saturation temperature $T_{sat}$.

Surface tension values for water are given in Table 12.2 [21]. Values for other liquids may be found in Ref. [22].

Values of $C_{sf}$ for various surfaces immersed in water are given in Table 12.3 [23].

As noted above, the CHF (Point A in Fig. 12.17), at the end of the nucleate boiling region, is very important. The CHF is given by

$$CHF = (q/A)_{max} = Ch_{fg}\rho_v \left[ \frac{\sigma g(\rho - \rho_v)}{\rho_v^2} \right]^{1/4} \tag{12.100}$$

Several researchers have worked to determine the value of constant $C$ in this equation [14,24−28]. Some values of $C$ are in Table 12.4. Values for other geometries are given in the references.

**Table 12.3 Coefficient $C_{sf}$ for various surfaces immersed in water.**

| | $C_{sf}$ |
|---|---------|
| Copper, polished with emery paper | 0.0120 |
| Copper, polished with mirror finish | 0.0091 |
| Copper, mirror finish scored by steel | 0.0070 |
| Stainless steel, polished | 0.0100 |
| Stainless steel, ground | 0.0079 |
| Stainless steel, surface milled | 0.0215 |
| Stainless steel, Teflon sprayed | 0.0058 |
| Stainless steel, polished with emery paper | 0.0133 |
| Stainless steel, etched with HCl acid | 0.0134 |

**Table 12.4 Constant $C$ for Eq. (12.100).**

| | $L$ | $C$ |
|---|---|---|
| Infinitely long horizontal flat heater | Width | 0.149 for $L' \geq 27$ |
| Horizontal cylindrical heater | Radius | 0.118 for $L' \geq 1.2$ |
| Horizontal cylindrical heater | Radius | $0.123 / (L')^{1/4}$ for $0.15 \leq L' \leq 1.2$ |
| Spherical heater | Radius | 0.11 for $L' \geq 4.26$ |
| Spherical heater | Radius | $0.227 / (L')^{1/2}$ for $0.15 \leq L' \leq 4.26$ $L' = L \left[ \frac{g(\rho - \rho_v)}{\sigma} \right]^{1/2}$ |

## Example 12.10

### Boiling water in a pot

*Problem*

A 10-inch stainless steel pot is boiling water at 1 atm pressure on the top of a gas stove. The gas burner has an output of 13,000 BTU, and it is assumed that 80% of the output goes into the water. What is the temperature of the inner surface of the bottom of the pot?

*Solution:*

$$D = 10 \text{ inch} \times 0.0254 \text{ m/inch} = 0.254 \text{ m}$$

$$A = \frac{\pi}{4}D^2 = \frac{\pi}{4}(0.254)^2 = 0.05067 \text{ m}^2$$

$$q = 13000 \frac{\text{BTU}}{\text{h}} \times \frac{1 \text{ W}}{3.412 \text{ BTU/h}} \times 0.8 = 3048 \text{ W}$$

$$\frac{q}{A} = \frac{3048}{0.05067} = 60154 \text{ W/m}^2$$

(Note: "13,000 BTU" in the problem statement means 13,000 BTU/h)

The equation relating excess temperature and heat flux in the nucleate boiling region are Eqs. (12.98) and (12.99). Using Eq. (12.98), we have

$$\frac{c_p \Delta T_{excess}}{h_{fg} \text{Pr}^s} = C_{sf} \left[ \frac{(q/A)}{\mu h_{fg}} \sqrt{\frac{\sigma}{g(\rho - \rho_v)}} \right]^{1/3} \qquad (12.98)$$

For 1 atm pressure, $T_{sat} = 100$ C. At this temperature, for water,

$$\sigma = 0.0589 \text{ N/m}$$
$$\text{Pr} = 1.75$$
$$\mu = 0.282 \times 10^{-3} \text{ kg/m s}$$
$$\rho = 958 \text{ kg/m}^3$$
$$\rho_v = 0.60 \text{ kg/m}^3$$
$$h_{fg} = 2257 \times 10^3 \text{ J/kg}$$
$$c_p = 4217 \text{ J/kg C}$$
$$\text{also, } g = 9.807 \text{ m/s}^2 \quad \text{and} \quad s = 1 \text{ for water}$$

Assuming the pot is polished, we have, from Table 12.3, that $C_{sf} = 0.01$.
Rearranging Eq. (12.98), we have

$$\Delta T_{excess} = C_{sf} \frac{h_{fg} \mathrm{Pr}^s}{c_p} \left[ \frac{(q/A)}{\mu h_{fg}} \sqrt{\frac{\sigma}{g(\rho - \rho_v)}} \right]^{1/3} \qquad (12.101)$$

Putting in values, we have

$$\Delta T_{excess} = 0.01 \frac{(2257 \times 10^3)(1.75)}{4217} \left[ \frac{60154}{(0.282 \times 10^{-3})(2257 \times 10^3)} \sqrt{\frac{0.0589}{9.807(958 - 0.60)}} \right]^{1/3}$$

$$= 5.79 \text{ C}$$

Looking at Fig. 12.17, it is seen that $\Delta T_{excess}$ lies in the nucleate boiling region. The use of Eq. (12.98) was indeed appropriate.

$$T_s = T_{sat} + \Delta T_{excess} = 100 + 5.79 = 105.79 \text{ C}$$

The temperature of the inner surface of the bottom of the pot is 105.8 C.

## Example 12.11
### Critical heat flux for horizontal cylindrical heater

*Problem*

Water is being boiled at 1 atm pressure. The heater surface is cylindrical, 4 cm diameter, and 2 m long.
**(a)** What is the CHF for the heater?
**(b)** If the surface is polished stainless steel, what is the temperature of the heater's surface at the CHF point?

*Solution*
**(a)** Eq. (12.100) gives the CHF.

$$CHF = (q/A)_{max} = C h_{fg} \rho_v \left[ \frac{\sigma g(\rho - \rho_v)}{\rho_v^2} \right]^{1/4} \qquad (12.100)$$

The water is at 100 C, the same temperature as Example 12.10, so we can use the property values from Example 12.10 for this example.
Constant $C$ in Eq. (12.100) is from Table 12.4. For the cylindrical heater, $L = 0.02$ m and

$$L' = L \left[ \frac{g(\rho - \rho_v)}{\sigma} \right]^{1/2} = 0.02 \left[ \frac{9.807(958 - 0.6)}{0.0589} \right]^{1/2} = 7.99$$

Therefore, assuming the heater is horizontal, $C = 0.118$ from Table 12.4.
Putting values into Eq. (12.100), we have

$$CHF = (q/A)_{max} = (0.118)(2257 \times 10^3)(0.6) \left[ \frac{(0.0589)(9.807)(958 - 0.6)}{(0.6)^2} \right]^{1/4} = 1.000 \times 10^6 \text{ W/m}^2$$

The CHF is $10^6$ W/m$^2$.
**(b)** The CHF point is in the nucleate boiling region. Eq. (12.101) gives the excess temperature for a given heat flux $q/A$.

$$\Delta T_{excess} = C_{sf} \frac{h_{fg} \mathrm{Pr}^s}{c_p} \left[ \frac{(q/A)}{\mu h_{fg}} \sqrt{\frac{\sigma}{g(\rho - \rho_v)}} \right]^{1/3} \qquad (12.101)$$

From Example 12.10, $C_{sf} = 0.01$ and $s = 1$.

Putting in values, we have

$$\Delta T_{excess} = 0.01 \frac{(2257 \times 10^3)(1.75)}{4217} \left[ \frac{10^6}{(0.282 \times 10^{-3})(2257 \times 10^3)} \sqrt{\frac{0.0589}{9.807(958 - 0.60)}} \right]^{1/3}$$

$$= 14.8\,C$$

$$T_s = T_{sat} + \Delta T_{excess} = 100 + 14.8 = 114.8\,C$$

The surface temperature of the heater is 114.8 C for the CHF.

Please take special note of this answer! The heat flux is extremely high ($10^6$ W/m$^2$), and the surface temperature is only 14.8 C above the water temperature!

### 12.4.2.3 Film boiling

For the film boiling region of Fig. 12.17, Bromley [29] recommended the following relation for the boiling heat transfer coefficient for a horizontal cylinder of diameter $D$:

$$h_{boil} = 0.62 \left[ \frac{\rho_v(\rho - \rho_v)gh'_{fg}k_v^3}{\mu_v D(T_s - T_{sat})} \right]^{1/4} \tag{12.102}$$

where

$T_s$ = surface temperature  (C)

$T_{sat}$ = saturation temperature at the given pressure  (C)

$\rho$ = liquid density at $T_{sat}$  (kg / m$^3$)

$\rho_v$ = vapor density  (kg / m$^3$)

$k_v$ = thermal conductivity of vapor  (W / m C)

$\mu_v$ = absolute viscosity of vapor  (kg / m s)

$h'_{fg}$ = corrected latent heat of vaporization  (J / kg)

$D$ = diameter  (m)

The corrected latent heat of vaporization is

$$h'_{fg} = h_{fg} + 0.4c_{pv}(T_s - T_{sat}) \tag{12.103}$$

where
$\begin{array}{rcl} h_{fg} & = & \text{latent heat of vaporization at } T_{sat} \text{ (J / kg)} \\ c_{pv} & = & \text{specific heat of vapor (J / kg C)} \end{array}$

The vapor properties should be evaluated at the film temperature $T_f = \frac{T_s + T_{sat}}{2}$.

Eq. (12.102) can also be used for a **sphere** in the film boiling region if the 0.62 value is replaced with 0.67.

For high surfaces temperatures (say, above 300 C), radiation across the film may be significant. If so, a radiation coefficient is defined is

$$h_{rad} = \frac{\varepsilon\sigma\left(T_s^4 - T_{sat}^4\right)}{T_s - T_{sat}} \tag{12.104}$$

where $\sigma$ = Stefan–Boltzmann constant = $5.67 \times 10^{-8}$ W/m$^2$ K$^4$ and $\varepsilon$ = emissivity of the surface. In Eq. (12.104), the temperatures must be in absolute temperature units, i.e., degrees K.

The total heat transfer coefficient is given by

$$h_{total} = h_{boil}\left(\frac{h_{boil}}{h_{total}}\right)^{1/3} + h_{rad} \tag{12.105}$$

Looking at Eq. (12.105), it is seen that iteration is required to obtain the total heat transfer coefficient $h_{total}$.

The heat flux at the surface of the cylinder is

$$\left(\frac{q}{A}\right) = h_{total}(T_s - T_{sat}) \tag{12.106}$$

---

## Example 12.12
### Cylindrical heater in the film boiling region

*Problem*

A horizontal cylindrical heater is boiling water at 1 atm pressure. The heater has a diameter of 2 cm and a length of 0.8 m. Its surface temperature is 460 C, and the emissivity of its surface is 0.6. What is the heat transfer rate to the water?

*Solution:*

Looking at Fig. 12.17, with an excess temperature of 360 C, the boiling is certainly in the film boiling region. The boiling heat transfer coefficient is given by Eq. (12.102).

$$h_{boil} = 0.62\left[\frac{\rho_v(\rho - \rho_v)gh'_{fg}k_v^3}{\mu_v D(T_s - T_{sat})}\right]^{1/4} \tag{12.102}$$

$T_s = 460$ C and $T_{sat} = 100$ C
The vapor properties are taken at the film temperature $T_f = \frac{T_s + T_{sat}}{2} = (460 + 100)/2 = 280$ C.
For water vapor at 280 C,

$\rho_v = 33.15$ kg / m$^3$
$k_v = 0.0605$ W / m C
$\mu_v = 1.870 \times 10^{-5}$ kg / m s
$c_{pv} = 4835$ J / kg C
For water at 100 C, $\quad \rho = 958$ kg / m$^3$
$\qquad\qquad\qquad h_{fg} = 2257$ kJ / kg

From Eq. (12.103), the corrected latent heat of vaporization is

$$h'_{fg} = h_{fg} + 0.4c_{pv}(T_s - T_{sat}) = 2257 \times 10^3 + 0.4(4835)(460 - 100) = 2953 \times 10^3 \text{ J/kg}$$

Putting values into Eq. (12.102), we have

$$h_{boil} = 0.62 \left[ \frac{\rho_v(\rho - \rho_v)gh'_{fg}k_v^3}{\mu_v D(T_s - T_{sat})} \right]^{1/4}$$

$$= 0.62 \left[ \frac{(33.15)(958 - 33.15)(9.807)(2953 \times 10^3)(0.0605)^3}{(1.870 \times 10^{-5})(0.02)(460 - 100)} \right]^{1/4} = 682 \text{ W/m}^2 \text{ C}$$

For the radiation, from Eq. (12.104),

$$h_{rad} = \frac{\varepsilon\sigma(T_s^4 - T_{sat}^4)}{T_s - T_{sat}} = \frac{(0.6)(5.67 \times 10^{-8})\left[(460 + 273.15)^4 - (100 + 273.15)^4\right]}{460 - 100} = 25.5 \text{ W/m}^2 \text{ C}$$

The total heat transfer coefficient is given by Eq. (12.105).

$$h_{total} = h_{boil}\left(\frac{h_{boil}}{h_{total}}\right)^{1/3} + h_{rad} \tag{12.105}$$

Putting in values, we get

$$h_{total} = 682\left(\frac{682}{h_{total}}\right)^{1/3} + 25.5 \tag{12.107}$$

Rearranging Eq. (12.107) for a solution by Excel's *Goal Seek* program, we have

$$h_{total} - 682\left(\frac{682}{h_{total}}\right)^{1/3} - 25.5 = 0 \tag{12.108}$$

Solving Eq. (12.108) with Excel's *Goal Seek*, we got $h_{total} = 701$ W / m$^2$ C.
The heat flow to the water is $q = h_{total}A(T_s - T_{sat}) = (701)[(\pi)(0.02)(0.8)](460 - 100) = 12685$ W.
The heat flow to the water is 12700 W.

# 12.5 Energy usage in buildings

In this section, we discuss energy usage in buildings. In particular, we discuss the **degree-day method**. The method could be applied to various types of buildings and for both heating and cooling. However, it was originally developed for heating of residences, and it is for this application that it has the highest accuracy. Hence, we restrict our discussion to the heating of residences. Using the degree-day method, we will estimate the energy needed to provide adequate heating of residences.

Let's first consider the definition of "degree-day" If we look at a given day, the day has a maximum temperature $T_{max}$ and a minimum temperature $T_{min}$. If we average these temperatures, we get $T_{avg}$, the average temperature for the day. That is, $T_{avg} = \frac{T_{min} + T_{max}}{2}$. Another temperature of interest is the balance point temperature. This is the outdoor temperature above which no heating is needed. When the degree-day method originated, the balance point temperature was about 65 F. If the temperature on a day was 65 F or higher, no heating would be needed on that day. However, if the temperature was lower than 65 F, heating would indeed be needed. The amount of needed heating was postulated to be proportional to the difference between 65 F and the average temperature $T_{avg}$ for the day.

The number of degree days for a particular day is then the difference between the balance point temperature (originally 65 F) and the average temperature of the day if the day's average temperature is below the balance point temperature. If the average temperature is above the balance point temperature, the day has zero degree days. For example, let us say that the balance point temperature for a building is 65 F, and the average temperature of a day is 43 F. Then the number of degree days for the day is 65 − 43 or 22 degree days. But if the balance point temperature is 65 F and the average temperature for the day is 74 F, no degree-days will be obtained for the day. Weather bureaus keep track of annual degree days and often monthly degreedays for given locations. As shown below, these values can be used to predict energy usage for heating a building.

Over the years, buildings have become better insulated. They have also become "tighter" and infiltration of cold air has been reduced. Finally, the use of heat-generating appliances has increased. All these factors have caused the balance point temperature to drop a bit from the original 65 F. Degree-day data for the base of 65 F is still often used in determining energy usage. However, use of the 65 F base tends to overpredict the energy usage, and a correction factor less than unity is applied to the result so it is in better agreement with the actual energy usage of the building.

The equation for predicting heating energy usage of a building by the degree-day method is

$$F = \frac{24(q_L)(DD)}{\eta(HV)(T_i - T_o)}C_D \qquad (12.109)$$

where

$F$ = fuel consumption for the desired period  (units of fuel; e. g., gallons, ft$^3$, kWh)

$q_L$ =   the total calculated heat load based on design temperatures $T_i$ and  $T_o$  (Btu / hr)

$DD$ = the number of degree days for the desired period (F − day)

$\eta$ = efficiency of heating system; usually between about 0.6 and 0.95.  For electricity, $\eta$  = 1

$HV$ = heating value of fuel  (Btu / unit of fuel)

For natural gas, $HV$ = 1050 Btu/ft$^3$
For propane, $HV$ = 90,000 Btu/gal.
For #2 fuel oil, $HV$ = 140,000 Btu/gal.
For electricity, $HV$ = 3412 Btu/kWh.

$T_i$ = inside design temperature  (F)

$T_o$ = outside design temperature  (F)

$C_D$ = correction factor = 1  if  $DD$ is based on the actual balance point temperature

of the building

= 0.77  if  $DD$ is based on a 65 F  balance point temperature

Let us now discuss some of the items in Eq. (12.109).

The inside design temperature $T_i$ is arbitrary and determined from the desires of the occupants. A temperature of 72 F is often used. The outside design temperature depends on the location of the residence and the historical outside temperatures at that location. A location in the northern part of the United States will have a lower design temperature than locations further south. Table 12.5 gives ASHRAE recommended outside design temperatures and approximate annual heating degree days (65 F base) for some locations in the United States [30].

The parameter $q_L$ in Eq. (12.109) is the calculated heat load of the building based on the inside and outside design temperatures. The two major components of this parameter are the heat transfers through the building envelope (walls, roof, floor, foundation) and the infiltration of cold air into the building. Heat is needed to bring the infiltrating air up to the inside temperature.

**Table 12.5 Outside design temperatures and annual heating degree days for locations in the United States.**

| Location | Temp (F) | Degree days (F-day) | Location | Temp (F) | Degree days (F-day) |
|---|---|---|---|---|---|
| Anchorage, AK | −23 | 10360 | Minneapolis, MN | −7.6 | 7565 |
| Flagstaff, AZ | 9.4 | 6912 | St. Louis, MO | 10.2 | 4504 |
| San Francisco, CA | 40.8 | 2708 | Albuquerque, NM | 31.2 | 4069 |
| Denver, CO | 6.9 | 5942 | New York, NY | 17.3 | 4603 |
| Washington, DC | 15.7 | 4735 | Syracuse, NY | 2.9 | 6635 |
| Miami, FL | 49.0 | 172 | Fargo, ND | −15.2 | 8793 |
| Atlanta, GA | 23.2 | 3045 | Pittsburgh, PA | 9.4 | 5624 |
| Chicago, IL | 4.3 | 5930 | Norfolk, VA | 25.6 | 3244 |
| Louisville, KY | 14.5 | 4168 | Spokane, WA | 9.6 | 6687 |
| Boston, MA | 12.4 | 5621 | Charleston, WV | 14.7 | 4443 |
| Portland, ME | 4.2 | 7082 | Green Bay, WI | −4.3 | 7684 |

In Chapter 3, we discussed steady-state heat transfer through walls and included convection at the wall surfaces in our calculations. This information can be used to determine the heat transfers through the building envelope. We defined an overall heat transfer coefficient $U$ by the relation

$$q = UA \ \Delta T \tag{12.110}$$

where $q$ is the heat flow through the wall, $U$ is the overall heat transfer coefficient, $A$ is the cross-sectional area of the wall, and $\Delta T$ is the temperature difference across the wall.

The infiltration part of the heating load can be estimated from the relation

$$q = \rho_o c_{po}(ACH)(V)(T_i - T_o) \tag{12.111}$$

where

$\rho_o$ = density of the air at the outside design condition $\left(\text{lbm} / \text{ft}^3\right)$

$c_{po}$ = specific heat of the infiltrating air (Btu / lbm F)

$V$ = air volume of the building $\left(\text{ft}^3\right)$

$T_i$ and $T_o$ = inside and outside design temperatures (F)

$ACH$ = assumed number of air changes per hour

The infiltration $ACH$ value ranges from about 0.2 for a tight new construction house to about 2 for an old leaky house. Perhaps, a good value to use is 0.9 if details on the particular house are not known.

The degree-day method is useful for estimating the energy needed for heating a house and also for evaluating the energy impact of potential remodels. This is illustrated in Example 12.13.

---

## Example 12.13
### Estimated energy usage of a house

*Problem*

A one-story house located in Syracuse, NY, is rectangular and 30 feet by 60 feet. The walls are 8 feet high. The house has eight vinyl, double-pane insulated windows. Three are 24″ by 60″, three are 30″ by 60″, and two are 30″ by 48″. There are three exterior steel doors. One is 30″ by 80″, one is 36″ by 80″, and one is a patio door 72″ by 80″. The exterior walls are T-111 wood siding, 5/8″ thick. The floor is 3/4″ thick plywood with wood laminate flooring. The roof is asphalt shingles. The walls are 2 × 4 with R-11 insulation. The ceiling has 2 × 6 joists with R-19 insulation The floor has 2 × 8 joists with R-19 insulation. The floor is elevated from the ground and open to the outside environment. The furnace is oil-fired and has an efficiency of about 0.75.

**(a)** If the cost of oil is $ 3.10/gallon, estimate the annual heating cost for the house.

**(b)** The homeowner is considering adding R-19 of insulation to the ceiling to bring it to R-38. The cost of the insulation is $ 30 for a 15″ × 39.2 ft roll. Estimate the energy cost savings and payback period if this is done.

*Solution*

**(a)** From Table 12.5, the recommended outside design temperature for Syracuse is 2.9 F. We will use an inside design temperature of 72 F.

We first look at the heat load of the windows and doors. From an Internet search, a typical $U$ value for vinyl double-pane insulating windows is 0.3. For the steel exterior doors, a typical $U$ value is 0.27.

The total area for the windows is

$$A_{windows} = (3)(24 \times 60) \ + \ (24 \times 60)(30 \times 48) + (3)(30 \times 60)$$
$$= 12600 \text{ in}^2 \times \left(1 \text{ ft}^2/144 \text{ in}^2\right) = 87.5 \text{ ft}^2$$

The design heat load for the windows is

$$q = UA_{windows}(T_i - T_o) = (0.3)(87.5)(72 - 2.9) = 1814 \text{ Btu/h}$$

The total area for the doors

$$A_{doors} = (30 \times 80) + (36 \times 80) + (72 \times 80) = 11040 \text{ in}^2 \times \left(1 \text{ ft}^2/ \ 144 \text{ in}^2\right) = 76.7 \text{ ft}^2$$

The design heat load for the doors is

$$q = UA_{doors}(T_i - T_o) = (0.27)(76.7)(72 - 2.9) = 1431 \text{ Btu/h}$$

Now let us look at the walls.

The gross wall area is $(8)(30 \ + 60 \ + 30 \ + 60) \ = \ 1440 \text{ ft}^2$. Subtracting the window and door areas, we have the net wall area.

$$A_{walls} = 1440 - 87.5 - 76.7 = 1276 \text{ ft}^2$$

For a quick estimate of the design heat load for the walls, we will neglect the details of the wall construction and assume that the wall area consists only of the R-11 insulation. (Note: This is really not a bad initial approach as the insulation is, usually by far, the major component of the thermal resistance. We are neglecting the conduction through the studs and assuming that there is insulation in the stud area. The actual heat flow through the wall will be a little larger than we calculate as the conduction through the studs is greater than the conduction through the insulation. On the other hand, we are neglecting the decrease in heat flow due to the added resistance of the siding and the convection. We thus have two opposing factors that somewhat cancel each other out.) We will make the same assumption for the floor and ceiling. For the walls,

$$q = UA_{wall}(T_i - T_o) = \frac{A_{wall}}{R_{wall}}(T_i - T_o) = \frac{1276}{11}(72 - 2.9) = 8016 \text{ Btu/h}$$

For the floor,

$$q = UA_{floor}(T_i - T_o) = \frac{A_{floor}}{R_{floor}}(T_i - T_o) = \frac{1800}{19}(72 - 2.9) = 6546 \text{ Btu/h}$$

The ceiling has the same heat transfer as the floor.

The total heat transfer through the building's envelope is thus

$$q_{envelope} = 1814 + 1431 + 8016 + 6546 + 6546 = 24353 \text{ Btu/h}$$

The other major heat usage is due to the infiltration. This is calculated from Eq. (12.111).

$$q = \rho_o c_{po} (ACH)(V)(T_i - T_o) \tag{12.111}$$

We will assume that $ACH = 0.9$. The volume of the building is $V = (60)(30)(8) = 14{,}400 \text{ ft}^3$. At the outside temperature of 2.9 F, $\rho_o = 0.086 \text{ lbm / ft}^3$ and $c_{po} = 0.239 \text{Btu / lbm F}$. The heat usage for the infiltration is then

$$q = \rho_o c_{po}(ACH)(V)(T_i - T_o) = (0.086)(0.239)(0.9)(14400)(72 - 2.9) = 18407 \text{ Btu/h}$$

The total heat loss $q_L = 24353 + 18407 = 42760 \text{ Btu / h}$.

Using the degree-day equation, Eq. (12.109), the annual consumption of oil is estimated to be

$$F = \frac{24(q_L)(DD)}{\eta(HV)(T_i - T_o)} C_D = \frac{24(42760)(6635)}{(0.75)(140000)(72 - 2.9)} (0.77) = 723 \text{ gallons/yr}$$

Annual heating cost $= 723 \text{ gallons/yr} \times \$3.10/\text{gallon} = \$2241/\text{yr}$

The annual heating cost is estimated at $ 2240.

**(b)** If R-19 insulation is added to the ceiling to get R-38, the new heat transfer for the ceiling will be

$$q = \frac{A_{ceiling}}{R_{ceiling}}(T_i - T_o) = \frac{1800}{38}(72 - 2.9) = 3273 \text{ Btu / hr}$$

Adding the insulation will reduce the heating load for the ceiling by $6546 - 3273 = 3273$ Btu/hr. The new load will be $42{,}760 - 3273 = 39{,}487$ Btu/hr. The new oil consumption, using Eq. (12.109), is

$$F = \frac{24(q_L)(DD)}{\eta(HV)(T_i - T_o)} C_D = \frac{24(39487)(6635)}{(0.75)(140000)(72 - 2.9)} 0.77) = 667 \text{ gallons / yr}$$

New annual heating cost $= 667 \text{ gallons/yr} \times \$ 3.10 / \text{ gallon} = \$ 2068/\text{yr}$

Savings due to the new insulation $= 2240 - 2068 = \$172/\text{yr}$

Cost per square foot of insulation $= \dfrac{\$30}{(15/12)(39.2) \text{ ft}^2} = \$0.612/\text{ft}^2$

Cost of insulation $= (1800 \text{ ft}^2)(\$0.612 / \text{ ft}^2) = \$1102$

Payback period $= \dfrac{\text{Cost}}{\text{Savings per year}} = \dfrac{1102}{172} = 6.4 \text{ years}$

It is estimated that adding the ceiling insulation will result in an energy cost savings of about $ 172 per year and that the payback period will be about 6.4 years.

(Note: The given $ 30 cost of the package of insulation is material cost alone. No installation cost is included. Thus, the 6.4 year payback period is based on the homeowner adding the insulation by him/herself. If someone else does it and there are installation costs, then the payback period will increase.)

The degree-day method can also be used to determine energy needed for cooling a building [31]. Many different methods are now available to estimate energy usage of buildings [32]. Commercial software is also available to compute energy usage [33].

# 12.6 Chapter summary and final remarks

This chapter contains four topics that could be included in the heat transfer course if time permits and the instructor so desires.

## 12.7 Problems

Notes:

- If needed information is not given in the problem statement, then use the Appendix for material properties and other information. If the Appendix does not have sufficient information, then use the Internet or other reference sources. Some good reference sources are mentioned at the end of Chapter 1. If you use the Internet, double-check the validity of information by using more than one source.
- Your solution should include a sketch of the problem.

**12-1** A wall has uniform internal heat generation at a rate of $5 \times 10^6$ W/m$^3$. The wall has a thickness of 2 cm and a conductivity of 30 W/m C. The left surface of the wall is at 100 C and the right surface is at 300 C.

(a) What is the maximum temperature in the wall and where does it occur?

(b) What is the heat flux through the left surface of the wall?

(c) What is the heat flux through the right surface of the wall?

**12-2** A wall has uniform internal heat generation at a rate of $10^8$ W/m$^3$. The wall has a thickness of 4 cm and a conductivity of 35 W/m C. The left surface of the wall is at 100 C and the right surface is at 150 C.

(a) What is the maximum temperature in the wall and where does it occur?

(b) If the wall has a surface area of 5 m$^2$, what is the rate of heat flow through the left surface of the wall?

(c) If the wall has a surface area of 5 m$^2$, what is the rate of heat flow through the right surface of the wall?

**12-3** A wall has uniform internal heat generation. It has a thickness of 3.5 cm and a thermal conductivity of 25 W/m C. Both wall surfaces are at 150 C and the maximum temperature in the wall is 225 C. What is the heat generation rate (W/m$^3$)?

**12-4** A wall is $L$ thick and has uniform internal heat generation at a rate of $q_{gen}$ W/m$^3$. Its thermal conductivity is $k$. There is convection and radiation at both surfaces of the wall. At one surface, the convective coefficient is $h_1$ and the emissivity is $\varepsilon_1$. The surroundings are at $T_{surr_1}$ and the adjacent fluid is at $T_{\infty_1}$. For the other surface, the convective coefficient is $h_2$ and the emissivity is $\varepsilon_2$. The surroundings are at $T_{surr_2}$ and the adjacent fluid is at $T_{\infty_2}$.

(a) Give the differential equation for the temperature distribution $T(x)$ in the wall.

(b) Give the general solution for the temperature distribution $T(x)$ in the wall.

(c) Give the boundary equations for the two surfaces of the wall.

**12-5** A wall is 5 cm thick and has internal heat generation of $2 \times 10^6$ W/m$^3$ and a conductivity of 15 W/m C. Both surfaces have convection and radiation. At one surface, $h = 20$ W/m$^2$ C and $\varepsilon = 0.1$. The adjacent air is at 20 C and the surroundings are at 50 C. For the other surface, $h = 100$ W / m$^2$ C and $\varepsilon = 0.8$. The adjacent air is at 100 C and the surroundings are at 150 C.

(a) What are the surface temperatures of the wall?

(b) What is the maximum temperature in the wall and where does it occur?

**12-6** A wall is 8 cm thick and has an internal heat generation of $8 \times 10^5$ W/m$^3$ and a conductivity of 25 W/m C. One surface is kept at 150 C. The other surface convects to the adjacent 20 C air with a convective coefficient of 120 W/m$^2$ C.
   **(a)** What is the temperature of the convecting surface?
   **(b)** What is the maximum temperature in the wall and where does it occur?

**12-7** A sphere has a diameter of 15 cm, an internal heat generation rate of $10^5$ W/m$^3$, and a thermal conductivity of 12 W/m C. It is in still air at 300 K. Radiation from the surface of the sphere is negligible.
   **(a)** What is the temperature of the surface of the sphere?
   **(b)** What is the temperature of the center of the sphere?

**12-8** A sphere has a diameter of 10 cm, an internal heat generation rate of 25000 W/m$^3$, and a thermal conductivity of 25 W/m C. The sphere is in a vacuum and has radiation to the surroundings, which are at 30 C. The surface of the sphere has an emissivity of 0.5.
   **(a)** What is the temperature of the surface of the sphere?
   **(b)** What is the temperature of the center of the sphere?

**12-9** A hollow sphere has an inside diameter of 10 cm and an outside diameter of 15 cm. It is made of a material that uniformly generates heat at a rate of 15000 W/m$^3$. The thermal conductivity of the sphere's material is 14 W/m C. The inside surface of the sphere is perfectly insulated and the outside surface convects to the surrounding 25 C air with a convective coefficient of 5 W/m$^2$ C.
   **(a)** What are the temperatures of the inner and outer surfaces of the sphere?
   **(b)** What is the temperature at the midpoint radius of the sphere?

**12-10** A hollow sphere has an inside diameter of 15 cm and an outside diameter of 20 cm. It is made of a material that uniformly generates heat at a rate of $10^5$ W/m$^3$. The thermal conductivity of the sphere's material is 5 W/m C. The inside surface of the sphere convects to 20 C air with a convective coefficient of 12 W/m$^2$ C. The outside surface has an emissivity of 0.7 and radiates to the surroundings, which are at 0 C. What are the temperatures of the inner and outer surfaces of the sphere?

**12-11** A sphere of 0.5 m diameter generates heat uniformly at a rate of 3000 W/m$^3$ and has a thermal conductivity of 20 W/m C. It is buried at a depth of 5 m in soil that has a thermal conductivity of 0.6 W/m C. The surface of the ground is at 20 C. What is the center temperature of the sphere?

**12-12** Air at a temperature of 20 C flows over a sphere that is 0.5 m diameter. The sphere generates heat at a rate of 2500 W/m$^3$ and has a thermal conductivity of 15 W/m C.
What speed of air flow is needed to keep the maximum temperature in the sphere below 60 C?

**12-13** Air at 27 C is blowing over a transistor. The transistor's surface area in contact with the air is 4 cm$^2$ and the convective coefficient is 15 W/m$^2$ C. The transistor is mounted on a substrate that is maintained at 30 C. The contact resistance between the transistor and the substrate is $0.6 \times 10^{-4}$ m$^2$ C/W and the area of contact is 0.8 cm$^2$. If the transistor's temperature cannot exceed 70 C, what is the maximum allowed power generation by the transistor?

**12-14** Circumferential fins were discussed in Chapter 3. In calculating the heat transfer, we assumed that there was no thermal resistance between the fin and the tube it was attached to. However, such fins are often installed by a press-fit process and there indeed is some contact resistance. Consider an aluminum circumferential fin that is attached to a 1 inch OD aluminum tube.

The fin has a thickness of 1.25 mm and a length of 1.5 cm. It convects to the surrounding 25 C air with a convective coefficient of 60 W/m$^2$ C. The outside surface of the aluminum tube is at 70 C.

(a) What is the heat transfer from the fin to the air if there is no contact resistance at the fin–tube interface?

(b) What is the heat transfer from the fin to the air if the fin is press-fit on the tube and the contact resistance is $0.5 \times 10^{-4}$ m$^2$ C/W?

**12-15** Two metal cylinders ($k = 125$ W/m C) of 10 cm diameter and 20 cm length are stacked vertically. The temperature at the top of the upper cylinder is 200 C and the temperature at the bottom of the lower cylinder is 25 C. The lateral surfaces of the cylinders are perfectly insulated so that all heat flow is axial. Force is applied to the ends of the assembly to give an appreciable contact pressure at the interface between the cylinders. For this contact pressure, it is found that the contact conductance is 2500 W/m$^2$ C.

(a) What is the rate of heat transfer through the cylinders?

(b) What is the temperature drop at the interface between the cylinders?

(c) What would be the rate of heat transfer through the cylinders if the contact resistance at the interface was zero?

**12-16** A vertical plate is 40 cm wide and 60 cm high. One surface is at 50 C and is exposed to saturated steam at 90 C. What is the rate of heat transfer from the steam to the surface and the mass of steam condensed per hour at the surface?

**12-17** A square vertical plate is 0.5 m by 0.5 m. The plate is at 30 C. One surface is exposed to saturated steam at 1 atm pressure. What is the rate of heat transfer from the steam to the surface and the mass of steam condensed per hour at the surface?

**12-18** If the plate of Problem 12-17 is inclined with its surface 30 degrees from the vertical, what is the mass of steam condensed per hour?

**12-19** A horizontal tube has an outer diameter of 1.5 cm and a length of 2 m. Its outer surface is at 35 C and is exposed to saturated steam at 25 kPa. What is the rate of heat transfer and the rate at which steam is condensed by the tube?

**12-20** A vertical cylinder has a diameter of 10 cm and a length of 35 cm. The cylinder's surface is at 20 C. Saturated steam at 100 C condenses on the cylinder's surface. What is the condensation rate (kg/hr)?

**12-21** A vertical plate is 5 cm wide and 5 cm high. One surface is at 90 C and has adjacent saturated steam at 100 C. What is the rate of heat transfer from the steam to the surface and the mass condensed per hour at the surface?

**12-22** A horizontal cold water pipe has an outer surface temperature of 12 C. The pipe is 3/4 inch Type L copper and it is 5 m long. The pipe is in a room where the air is at 25 C with 80% relative humidity. What is the rate at which condensate will drip from the pipe? (You may assume that the pipe is exposed to saturated vapor at the partial pressure of the water vapor in the air.)

**12-23** A rectangular tube bundle in a steam condenser has 200 horizontal tubes arranged 20 tubes wide by 10 tubes high. Each tube has an outer diameter of 3 cm and a length of 20 m. The outer surfaces of the tubes are at 15 C. The steam is saturated and at a pressure of 12 kPa. What is the heat transfer rate for the condenser and the rate of condensate flow?

**12-24** A cylindrical 8 mm diameter polished copper heater is boiling water at 1 atm pressure. The excess temperature is 12 C. What is the heat transfer rate per meter length of the heater?

**12-25** Water is boiling in a stainless steel pot on a stove. The atmospheric pressure is 85 kPa. If the heat flux to the water is 75000 $W/m^2$, what is the temperature of the bottom of the pot?

**12-26** Water is boiled at 1 atm pressure by a polished cylindrical copper heater of 5 mm diameter.
(a) What is the critical heat flux (CHF)?
(b) What is the surface temperature of the heater for the CHF?

**12-27** Water is boiled at 1 atm pressure by a horizontal cylindrical heater. The heater is 1 cm diameter and 30 cm long. The emissivity of the heater's surface is 0.4. The surface temperature of the heater is 350 C.
(a) What is the rate of heat transfer to the water?
(b) What is the rate of evaporation of the water?

**12-28** An immersion electric heater is used to boil water in an uncovered pot. The atmospheric pressure is 1 atm. When the water boils, it is observed that 0.5 liter of water evaporates in 30 min. The heater is cylindrical with an outer diameter of 0.5 cm and a length of 25 cm. Its surface is roughly polished stainless steel.
(a) What is the power rating of the heater (W)?
(b) What is the surface temperature of the heater?

**12-29** Water is being boiled in a stainless steel pot that has a bottom heating surface of 35 cm diameter. The water is at 120 C and the bottom of the pot is at 130 C.
(a) What is the power required to boil the water?
(b) What is the evaporation rate of the water?

**12-30** For the house of Example 12.13, determine the annual heating cost if it were heated by electricity rather than oil, and the cost of electricity were 18¢ per kWh.

**12-31** A house in Minneapolis, MN, has a design heating load of 84,000 Btu/hr. Determine the annual cost of heating if
(a) The house has an oil-fired furnace of 75% efficiency and the cost of oil is $ 2.50 per gallon.
(b) The house is heated by electric baseboard units and the cost of electricity is 20¢ per kWh.

**12-32** A home in Chicago has a design heating load of 112,000 Btu/hr. The inside design temperature is 73 F. The house has a propane furnace of 88% efficiency.
(a) What is the annual cost of heating if the propane is $ 1.80/gallon?
(b) If the owner lowers the house temperature to 70 F, what is the annual cost of heating?

**12-33** A 2000 square foot residence in St. Louis has a gas-fired furnace of 85% efficiency. The design heating load is 75,000 Btu/hr. How many therms of natural gas are needed for the heating season?

**12-34** A small 400 $ft^2$ cabin in Green Bay, WI, has a design heating load of 13,000 Btu/hr. If the cabin is heated by a portable kerosene heater, how many gallons of kerosene are needed for the heating season?

**12-35** Design Problem: A skiing enthusiast wants to build a small (16 ft by 20 ft) cabin in the woods near Gore Mountain. The cabin will be rustic and simple. It will be one-storey on treated wood posts. The wall studs will be 2 × 4's, the floor joists will be 2 × 8's, and the roof rafters for the shed roof will be 2 × 8's, all on 16 inch centers. There will be two vinyl double-pane insulating windows and one exterior door. The walls will have R-13 insulation and the floor will have R-19 insulation. There will be an attic space for storage, and the attic floor will have R-38 insulation.

**(a)** Estimate the design heat load for the building, considering heat flows through the building envelope and infiltration.

**(b)** Study possible heating systems for the building. Consider portable propane, kerosene, and electric heaters. Using the Internet, determine current fuel prices and estimate the annual heating costs for the three possible options. Select the heating option you will use.

**(c)** Using material costs from local box stores, estimate the cost of constructing the cabin, assuming that you will be building the cabin yourself. Also estimate the cost of the cabin if you use a contractor for the construction.

# References

[1] M.E. Barzelay, K.N. Tong, G.F. Holloway, Effect of Pressure on Thermal Conductance of Contact Joints, NACA TN - 3295 (Syracuse University), Washington, D. C., May 1955.

[2] E. Fried, Study of Interface Thermal Contact Conductance, General Electric Company Document No. 64SD652, May 1964.

[3] H. Fenech, W.M. Rohsenow, Prediction of thermal conductance of metallic surfaces in contact, J. Heat Transf. 85 (1963) 15−24.

[4] P.J. Schneider, Conduction, in: W.M. Rohsenow, J.P. Hartnett (Eds.), Handbook of Heat Transfer, 1973, pp. 3-14−3-18.

[5] E. Fried, F.A. Costello, Interface thermal contact resistance problem in space vehicles, ARS J. 32 (2) (1962) 237−243.

[6] H.J. Sauer Jr., C.R. Remington, W.E. Stewart Jr., J.T. Lin, Thermal Contact Conductance with Several Interstitial Materials, 11th Intl. Thermal Conductivity Conference, Albuquerque, NM, Sept. 28 - Oct. 1, 1971.

[7] C.V. Madhusudana, L.S. Fletcher, Contact heat transfer − the last decade, AIAA J. 24 (1986) 510−523.

[8] L.S. Fletcher, Recent developments in contact conductance heat transfer, J. Heat Transf. 110 (1988) 1059−1070.

[9] E. Fried, H.L. Atkins, Interface thermal conductance in a vacuum, J. Spacecr. Rocket. 2 (4) (1965) 591−593.

[10] S. Song, M.W. Yovanovich, F.O. Goodman, Thermal gap conductance of conforming surfaces in contact, J. Heat Transf. 115 (1993) 533−540.

[11] M. Mittelbach, C. Vogd, L.S. Fletcher, G.P. Peterson, The interfacial pressure distribution and thermal conductance of bolted joints, J. Heat Transf. 116 (1994) 823−828.

[12] W. Nusselt, Die Oberflachenkondensation des Wasserdampfes, Z. Des. Vereines Dtsch. Ingenieure 60 (1916) 541−569.

[13] W.H. McAdams, Heat Transmission, third ed., McGraw-Hill, 1954.

[14] S.S. Kutateladze, Fundamentals of Heat Transfer, Academic Press, 1963.

[15] D.A. Labuntsov, Heat transfer in film condensation of pure steam on vertical surface and horizontal tubes, Teploenergetika 4 (1957) 72−80.

[16] R.W. Bjorge, G.R. Hall, W.M. Rohsenow, Correlation of forced convection boiling heat transfer data, Int. J. Heat Mass Transf. 25 (1982) 753−757.

[17] W.M. Rohsenow, Boiling, in: W.M. Rohsenow, J.P. Hartnett (Eds.), Handbook of Heat Transfer, Chpt. 13, McGraw-Hill, 1973.

[18] A.E. Bergles, W.M. Rohsenow, The determination of forced convection surface boiling heat transfer, J. Heat Transf. 86 (1964) 365−372.

[19] S. Nukiyama, The maximum and minimum values of the heat transmitted from metal to boiling water at atmospheric pressure, J. Soc. Mech. Engrs. Japan 37 (1934) 367–374. Translated in Int. J. Heat Mass Transfer, vol. 9, pp. 1419–1433, 1966.

[20] W.M. Rohsenow, A method of correlating heat transfer data for surface boiling of liquids, ASME Transactions 74 (1952) 969–975.

[21] J. Pellicer, V. Garcia-Morales, L. Guanter, M.J. Hernandez, M. Dolz, On the experimental values of the water surface tension used in some textbooks, Am. J. Phys. 70 (2002) 705–709.

[22] Handbook of Chemistry and Physics, CRC Press.

[23] R.I. Vachon, G.H. Nix, G.E. Tanger, Evaluation of constants for the Rohsenow pool-boiling correlation, J. Heat Transf. 90 (1968) 239–246.

[24] N. Zuber, On the stability of boiling heat transfer, ASME Trans. 80 (1958) 711–720.

[25] K. Sun, J.H. Lienhard, The peak pool boiling heat flux on horizontal cylinders, Int. J. Heat Mass Transf. 13 (1970) 1425–1430.

[26] J.S. Ded, J.H. Lienhard, The peak pool boiling heat flux from a sphere, AIChE J. 18 (1972) 337–342.

[27] J.H. Lienhard, V.K. Dhir, Hydrodynamic predictions of peak pool-boiling heat fluxes from finite bodies, J. Heat Transf. 95 (1973) 152–158.

[28] J.H. Lienhard, V.K. Dhir, D.M. Riherd, Peak pool boiling heat-flux measurements on finite horizontal flat plates, J. Heat Transf. 95 (1973) 477–482.

[29] L.A. Bromley, Heat transfer in stable film boiling, Chem. Eng. Prog. 46 (1950) 221–227.

[30] American Society of Heating, Refrigerating and Air-Conditioning Engineers, Inc., 2009. ASHRAE Handbook - Fundamentals, Chpt. 14, ASHRAE, Atlanta, GA, 2009.

[31] R.H. Howell, H.J. Sauer Jr., W.J. Coad, Principles of Heating Ventilating and Air Conditioning, ASHRAE, Atlanta, GA, 2005.

[32] American Society of Heating, Refrigerating and Air-Conditioning Engineers, Inc., 2009. ASHRAE Handbook - Fundamentals, Chpt. 19, ASHRAE, Atlanta, GA, 2017.

[33] Carrier Corporation, HAP (Hourly Analysis Program), www.carrier.com/commercial/.

# Appendices

| | |
|---|---|
| A | Properties of metals |
| B | Properties of nonmetals and building materials |
| C | Properties of liquids |
| D | Properties of gases |
| E | Properties of saturated steam |
| F | Hemispherical emissivity of surfaces |
| G | Bessel function of the first kind |
| H | Modified Bessel functions |
| I | Complementary error function |
| J | Computations using Excel and Matlab |
| K | Computer programs |

# Appendix A
## Properties of metals
### (a) Properties at 20 C

| Metal | Density $\rho$ (kg/m$^3$) | Specific heat $c_p$ (J/kg C) | Thermal conductivity $k$ (W/m C) |
|---|---|---|---|
| Pure aluminum | 2700 | 900 | 237 |
| Al alloy 2024-T6 | 2770 | 875 | 174 |
| Pure copper | 8933 | 385 | 401 |
| Commercial bronze | 8800 | 420 | 52 |
| Brass | 8530 | 380 | 111 |
| Gold | 19300 | 130 | 317 |
| Pure iron | 7870 | 447 | 80 |
| **Carbon steels** | | | |
| AISI 1010 | 7830 | 434 | 64 |
| AISI 1042 | 7840 | 460 | 50 |
| AISI 4130 | 7840 | 460 | 43 |
| **Stainless steels** | | | |
| AISI 304 | 7900 | 477 | 15 |
| AISI 316 | 8240 | 468 | 13 |
| Lead | 11340 | 29 | 35 |
| Pure nickel | 8900 | 444 | 91 |
| Silver | 10500 | 235 | 429 |
| Titanium | 4500 | 522 | 22 |
| Zinc | 7140 | 390 | 116 |

### (b) Thermal conductivity (W/m C) variation with temperature

| Metal | 20 C | 200 C | 400 C | 600 C |
|---|---|---|---|---|
| Pure aluminum | 237 | 238 | 240 | 233 |
| Al alloy 2024-T6 | 174 | 189 | | |
| Pure copper | 401 | 387 | 374 | 362 |
| Commercial Bronze | 52 | 53 | | |
| Brass | 111 | 141 | 145 | |

| | | | | |
|---|---|---|---|---|
| Gold | 317 | 306 | 293 | 280 |
| Pure iron | 80 | 64 | 51 | 40 |
| **Carbon steels** | | | | |
| AISI 1010 | 64 | 55 | 46 | 37 |
| AISI 1042 | 50 | 48 | 42 | 33 |
| AISI 4130 | 43 | 41 | 39 | 34 |
| **Stainless steels** | | | | |
| AISI 304 | 15 | 18 | 21 | 24 |
| AISI 316 | 13 | 16 | 19 | 20 |
| Lead | 35 | 33 | | |
| Pure nickel | 91 | 74 | 67 | 69 |
| Silver | 429 | 421 | 406 | 390 |
| Titanium | 22 | 20 | 19 | 20 |
| Zinc | 116 | 108 | 100 | |

# Appendix B
## Properties of nonmetals and building materials

| Properties are at 300 K unless otherwise noted | | | |
|---|---|---|---|
| **Material** | **Density ρ (kg/m³)** | **Specific heat c_p (J/kg C)** | **Thermal conductivity k (W/m C)** |
| **Plastics** | | | |
| Neoprene | 1250 | 1930 | 0.19 |
| Polyethylene | 960 | 2090 | 0.33 |
| Polypropylene | 1170 | 1930 | 0.17 |
| PVC | 1714 | 1050 | 0.092 |
| Teflon | 2200 | 1050 | 0.35 |
| Plexiglass | 1180 | 1500 | 0.2 |
| Lexan | 1200 | 1300 | 0.2 |
| **Brick** | | | |
| Common | 1920 | 835 | 0.72 |
| Face | 2083 | | 1.3 |
| Fireclay | 2645 | 960 | 0.9 (125 C) |
| | | | 1.4 (525 C) |
| | | | 1.8 (1325 C) |
| **Concrete** | | | |
| 1:2:4 mix | 2100 | 880 | 1.4 |
| **Glass** | | | |
| Fused silica | 2220 | 745 | 1.38 |
| Pyrex | 2640 | 800 | 1.09 |
| Soda lime | 2400 | 840 | 0.88 |
| Fiberglass batt | 40 | 835 | 0.035 |
| Glass wool | 200 | 670 | 0.04 |
| Fiber insulating board | 240 | | 0.048 |
| **Loose fill** | | | |
| Cellulose (wood or paper pulp) | 45 | | 0.039 |
| Vermiculite (expanded) | 122 | | 0.069 |
| **Rigid foam** | | | |
| Polyurethane | 70 | 1045 | 0.026 |
| Styrofoam | | | 0.033 |

| Soil | | | |
|---|---|---|---|
| Dry | 1500 | 1900 | 1.0 |
| Wet | 1900 | 2200 | 2.0 |
| Plywood | 545 | 1215 | 0.12 |
| **Gypsum plaster** | | | |
| Lightweight aggregate | 720 | | 0.23 |
| Sand aggregate | 1680 | 840 | 0.81 |
| Plasterboard | 800 | 1090 | 0.176 |
| Acoustic tile | 290 | 1340 | 0.058 |
| **Particle board** | | | |
| Low density | 590 | 1300 | 0.078 |
| High density | 1010 | 1300 | 0.17 |
| Wood subfloor | | | 0.12 |
| **Woods** | | | |
| Hardwood | 720 | 1255 | 0.16 |
| Softwood | 510 | 1380 | 0.12 |
| **Snow** | | | |
| Loose | 110 | | 0.05 |
| Packed | 500 | | 0.19 |

# Appendix C
## Properties of liquids
### (a) Saturated water

| T (C) | $\rho$ (kg/m³) | $c_p$ (J/kg C) | k (W/mC) | $\mu$ (kg/m s) | $\nu$ (m²/s) | $\beta$ (1/K) | Pr |
|---|---|---|---|---|---|---|---|
| 0 | 999.8 | 4217 | 0.562 | $1.792 \times 10^{-3}$ | $1.792 \times 10^{-6}$ | $-0.085 \times 10^{-3}$ | 13.5 |
| 5 | 999.9 | 4205 | 0.572 | $1.519 \times 10^{-3}$ | $1.519 \times 10^{-6}$ | $0.005 \times 10^{-3}$ | 11.2 |
| 10 | 999.7 | 4194 | 0.582 | $1.307 \times 10^{-3}$ | $1.307 \times 10^{-6}$ | $0.082 \times 10^{-3}$ | 9.45 |
| 15 | 999.1 | 4186 | 0.591 | $1.138 \times 10^{-3}$ | $1.139 \times 10^{-6}$ | $0.148 \times 10^{-3}$ | 8.09 |
| 20 | 998.0 | 4182 | 0.600 | $1.002 \times 10^{-3}$ | $1.004 \times 10^{-6}$ | $0.207 \times 10^{-3}$ | 7.01 |
| 25 | 997.0 | 4180 | 0.608 | $0.891 \times 10^{-3}$ | $0.894 \times 10^{-6}$ | $0.259 \times 10^{-3}$ | 6.14 |
| 30 | 996.0 | 4178 | 0.615 | $0.798 \times 10^{-3}$ | $0.801 \times 10^{-6}$ | $0.306 \times 10^{-3}$ | 5.42 |
| 35 | 994.0 | 4178 | 0.622 | $0.720 \times 10^{-3}$ | $0.724 \times 10^{-6}$ | $0.349 \times 10^{-3}$ | 4.83 |
| 40 | 992.1 | 4179 | 0.629 | $0.653 \times 10^{-3}$ | $0.658 \times 10^{-6}$ | $0.389 \times 10^{-3}$ | 4.32 |
| 45 | 990.1 | 4180 | 0.635 | $0.596 \times 10^{-3}$ | $0.602 \times 10^{-6}$ | $0.427 \times 10^{-3}$ | 3.91 |
| 50 | 988.1 | 4181 | 0.641 | $0.547 \times 10^{-3}$ | $0.554 \times 10^{-6}$ | $0.462 \times 10^{-3}$ | 3.55 |
| 55 | 985.2 | 4183 | 0.646 | $0.504 \times 10^{-3}$ | $0.512 \times 10^{-6}$ | $0.496 \times 10^{-3}$ | 3.25 |
| 60 | 983.3 | 4185 | 0.651 | $0.467 \times 10^{-3}$ | $0.475 \times 10^{-6}$ | $0.529 \times 10^{-3}$ | 2.99 |
| 65 | 980.4 | 4187 | 0.655 | $0.433 \times 10^{-3}$ | $0.442 \times 10^{-6}$ | $0.560 \times 10^{-3}$ | 2.75 |
| 70 | 977.5 | 4190 | 0.659 | $0.404 \times 10^{-3}$ | $0.413 \times 10^{-6}$ | $0.590 \times 10^{-3}$ | 2.55 |
| 75 | 974.7 | 4193 | 0.663 | $0.378 \times 10^{-3}$ | $0.388 \times 10^{-6}$ | $0.619 \times 10^{-3}$ | 2.38 |
| 80 | 971.8 | 4197 | 0.667 | $0.355 \times 10^{-3}$ | $0.365 \times 10^{-6}$ | $0.647 \times 10^{-3}$ | 2.22 |
| 85 | 968.1 | 4201 | 0.670 | $0.333 \times 10^{-3}$ | $0.344 \times 10^{-6}$ | $0.675 \times 10^{-3}$ | 2.08 |
| 90 | 965.3 | 4206 | 0.673 | $0.315 \times 10^{-3}$ | $0.326 \times 10^{-6}$ | $0.702 \times 10^{-3}$ | 1.96 |
| 95 | 961.5 | 4212 | 0.675 | $0.297 \times 10^{-3}$ | $0.309 \times 10^{-6}$ | $0.728 \times 10^{-3}$ | 1.85 |
| 100 | 957.9 | 4217 | 0.678 | $0.282 \times 10^{-3}$ | $0.294 \times 10^{-6}$ | $0.755 \times 10^{-3}$ | 1.75 |
| 120 | 943.4 | 4244 | 0.683 | $0.232 \times 10^{-3}$ | $0.246 \times 10^{-6}$ | $0.859 \times 10^{-3}$ | 1.44 |
| 140 | 921.7 | 4286 | 0.685 | $0.197 \times 10^{-3}$ | $0.214 \times 10^{-6}$ | $0.966 \times 10^{-3}$ | 1.24 |
| 160 | 907.4 | 4340 | 0.682 | $0.170 \times 10^{-3}$ | $0.187 \times 10^{-6}$ | $1.084 \times 10^{-3}$ | 1.09 |
| 180 | 887.3 | 4410 | 0.675 | $0.150 \times 10^{-3}$ | $0.169 \times 10^{-6}$ | $1.216 \times 10^{-3}$ | 0.98 |
| 200 | 864.3 | 4500 | 0.663 | $0.134 \times 10^{-3}$ | $0.155 \times 10^{-6}$ | $1.350 \times 10^{-3}$ | 0.91 |

### (b) Ethylene glycol

| T (C) | $\rho$ (kg/m³) | $c_p$ (J/kg C) | k (W/mC) | $\mu$ (kg/m s) | $\nu$ (m²/s) | Pr |
|---|---|---|---|---|---|---|
| 0 | 1132 | 2295 | 0.254 | $65.1 \times 10^{-3}$ | $57.5 \times 10^{-6}$ | 588 |
| 10 | 1123 | 2341 | 0.255 | $36.5 \times 10^{-3}$ | $32.5 \times 10^{-6}$ | 335 |

| 20 | 1115 | 2386 | 0.257 | $21.4 \times 10^{-3}$ | $19.2 \times 10^{-6}$ | 199 |
| 30 | 1108 | 2431 | 0.258 | $13.7 \times 10^{-3}$ | $12.4 \times 10^{-6}$ | 129 |
| 40 | 1102 | 2476 | 0.259 | $9.57 \times 10^{-3}$ | $8.68 \times 10^{-6}$ | 91 |
| 50 | 1096 | 2520 | 0.261 | $7.02 \times 10^{-3}$ | $6.41 \times 10^{-6}$ | 68 |
| 60 | 1090 | 2565 | 0.262 | $5.17 \times 10^{-3}$ | $4.74 \times 10^{-6}$ | 51 |
| 70 | 1083 | 2610 | 0.263 | $3.87 \times 10^{-3}$ | $3.57 \times 10^{-6}$ | 39 |
| 80 | 1076 | 2656 | 0.265 | $3.19 \times 10^{-3}$ | $2.96 \times 10^{-6}$ | 32 |
| 90 | 1068 | 2702 | 0.266 | $2.94 \times 10^{-3}$ | $2.75 \times 10^{-6}$ | 29 |
| 100 | 1060 | 2750 | 0.267 | $2.11 \times 10^{-3}$ | $1.99 \times 10^{-6}$ | 22 |

### (c)  Unused engine oil

| T (C) | $\rho$ (kg/m$^3$) | $c_p$ (J/kg C) | k (W/mC) | $\mu$ (kg/m s) | $\nu$ (m$^2$/s) | Pr |
|---|---|---|---|---|---|---|
| 0 | 899 | 1797 | 0.147 | $3850 \times 10^{-3}$ | $4280 \times 10^{-6}$ | 47100 |
| 10 | 894 | 1838 | 0.146 | $1826 \times 10^{-3}$ | $2043 \times 10^{-6}$ | 23600 |
| 20 | 888 | 1879 | 0.145 | $800 \times 10^{-3}$ | $901 \times 10^{-6}$ | 10400 |
| 30 | 882 | 1921 | 0.144 | $363 \times 10^{-3}$ | $412 \times 10^{-6}$ | 4450 |
| 40 | 876 | 1963 | 0.143 | $212 \times 10^{-3}$ | $242 \times 10^{-6}$ | 2870 |
| 50 | 870 | 2005 | 0.142 | $123 \times 10^{-3}$ | $141 \times 10^{-6}$ | 1720 |
| 60 | 864 | 2047 | 0.141 | $72.9 \times 10^{-3}$ | $84.4 \times 10^{-6}$ | 1050 |
| 70 | 858 | 2090 | 0.140 | $46.0 \times 10^{-3}$ | $53.6 \times 10^{-6}$ | 680 |
| 80 | 852 | 2113 | 0.139 | $31.8 \times 10^{-3}$ | $37.3 \times 10^{-6}$ | 480 |
| 90 | 846 | 2176 | 0.138 | $23.8 \times 10^{-3}$ | $28.1 \times 10^{-6}$ | 360 |
| 100 | 840 | 2219 | 0.137 | $18.3 \times 10^{-3}$ | $21.8 \times 10^{-6}$ | 277 |
| 110 | 834 | 2262 | 0.136 | $13.9 \times 10^{-3}$ | $16.7 \times 10^{-6}$ | 210 |
| 120 | 829 | 2306 | 0.135 | $10.3 \times 10^{-3}$ | $12.4 \times 10^{-6}$ | 175 |
| 130 | 823 | 2350 | 0.134 | $8.2 \times 10^{-3}$ | $10.0 \times 10^{-6}$ | 141 |
| 140 | 817 | 2394 | 0.133 | $6.5 \times 10^{-3}$ | $8.0 \times 10^{-6}$ | 116 |
| 150 | 812 | 2439 | 0.133 | $5.5 \times 10^{-3}$ | $6.8 \times 10^{-6}$ | 98 |
| 160 | 806 | 2484 | 0.132 | $4.5 \times 10^{-3}$ | $5.6 \times 10^{-6}$ | 84 |

# Appendix D
## Properties of gases
### (a) Gases at 20 C and 1 atm (101.325 kPa)

| Gas | ρ (kg/m³) | $c_p$ (J/kg C) | k (W/mC) | μ (kg/m s) | ν (m²/s) | Pr |
|---|---|---|---|---|---|---|
| Air | 1.204 | 1007 | 0.0251 | $18.20 \times 10^{-6}$ | $15.11 \times 10^{-6}$ | 0.730 |
| Argon | 1.682 | 521 | 0.0174 | $22.27 \times 10^{-6}$ | $13.24 \times 10^{-6}$ | 0.668 |
| Carbon dioxide | 1.853 | 847 | 0.0161 | $14.69 \times 10^{-6}$ | $7.93 \times 10^{-6}$ | 0.773 |
| Helium | 0.1685 | 5200 | 0.146 | $19.56 \times 10^{-6}$ | $116.0 \times 10^{-6}$ | 0.695 |
| Hydrogen | 0.0849 | 14280 | 0.178 | $8.78 \times 10^{-6}$ | $103.4 \times 10^{-6}$ | 0.706 |
| Nitrogen | 1.164 | 1041 | 0.0253 | $17.50 \times 10^{-6}$ | $15.03 \times 10^{-6}$ | 0.719 |
| Oxygen | 1.331 | 919 | 0.256 | $20.32 \times 10^{-6}$ | $15.27 \times 10^{-6}$ | 0.729 |

### (b) Dry air at 1 atm (101.325 kPa)

| T (C) | ρ (kg/m³) | $c_p$ (J/kg C) | k (W/mC) | μ (kg/m s) | ν (m²/s) | Pr |
|---|---|---|---|---|---|---|
| −30 | 1.451 | 1004 | 0.0214 | $15.51 \times 10^{-6}$ | $10.69 \times 10^{-6}$ | 0.748 |
| −20 | 1.394 | 1005 | 0.0221 | $16.08 \times 10^{-6}$ | $11.54 \times 10^{-6}$ | 0.744 |
| −10 | 1.341 | 1006 | 0.0229 | $16.63 \times 10^{-6}$ | $12.40 \times 10^{-6}$ | 0.741 |
| 0 | 1.292 | 1006 | 0.0236 | $17.17 \times 10^{-6}$ | $13.29 \times 10^{-6}$ | 0.737 |
| 10 | 1.246 | 1006 | 0.0244 | $17.69 \times 10^{-6}$ | $14.19 \times 10^{-6}$ | 0.734 |
| 20 | 1.204 | 1007 | 0.0251 | $18.20 \times 10^{-6}$ | $15.11 \times 10^{-6}$ | 0.730 |
| 30 | 1.164 | 1007 | 0.0259 | $18.69 \times 10^{-6}$ | $16.05 \times 10^{-6}$ | 0.728 |
| 40 | 1.127 | 1007 | 0.0266 | $19.17 \times 10^{-6}$ | $17.01 \times 10^{-6}$ | 0.725 |
| 50 | 1.092 | 1007 | 0.0273 | $19.64 \times 10^{-6}$ | $17.98 \times 10^{-6}$ | 0.722 |
| 60 | 1.059 | 1007 | 0.0281 | $20.09 \times 10^{-6}$ | $18.97 \times 10^{-6}$ | 0.720 |
| 70 | 1.028 | 1007 | 0.0288 | $20.54 \times 10^{-6}$ | $19.98 \times 10^{-6}$ | 0.717 |
| 80 | 1.000 | 1008 | 0.0295 | $21.01 \times 10^{-6}$ | $21.01 \times 10^{-6}$ | 0.715 |
| 90 | 0.9718 | 1008 | 0.0302 | $21.43 \times 10^{-6}$ | $22.02 \times 10^{-6}$ | 0.713 |
| 100 | 0.9458 | 1009 | 0.0309 | $21.85 \times 10^{-6}$ | $23.11 \times 10^{-6}$ | 0.711 |
| 120 | 0.8977 | 1011 | 0.0323 | $22.69 \times 10^{-6}$ | $25.27 \times 10^{-6}$ | 0.708 |
| 140 | 0.8542 | 1013 | 0.0337 | $23.49 \times 10^{-6}$ | $27.50 \times 10^{-6}$ | 0.704 |
| 160 | 0.8148 | 1016 | 0.0350 | $24.28 \times 10^{-6}$ | $29.80 \times 10^{-6}$ | 0.702 |
| 180 | 0.7788 | 1019 | 0.0364 | $25.05 \times 10^{-6}$ | $32.16 \times 10^{-6}$ | 0.700 |
| 200 | 0.7459 | 1023 | 0.0377 | $25.80 \times 10^{-6}$ | $34.59 \times 10^{-6}$ | 0.698 |
| 250 | 0.6746 | 1033 | 0.0409 | $27.61 \times 10^{-6}$ | $40.92 \times 10^{-6}$ | 0.695 |

| 300 | 0.6158 | 1045 | 0.0441 | $29.33 \times 10^{-6}$ | $47.63 \times 10^{-6}$ | 0.693 |
| 350 | 0.5664 | 1055 | 0.0471 | $30.99 \times 10^{-6}$ | $54.71 \times 10^{-6}$ | 0.693 |
| 400 | 0.5243 | 1066 | 0.0500 | $32.58 \times 10^{-6}$ | $62.14 \times 10^{-6}$ | 0.694 |
| 450 | 0.4880 | 1077 | 0.0528 | $34.11 \times 10^{-6}$ | $69.91 \times 10^{-6}$ | 0.696 |
| 500 | 0.4565 | 1088 | 0.0555 | $35.61 \times 10^{-6}$ | $78.00 \times 10^{-6}$ | 0.699 |
| 550 | 0.4282 | 1100 | 0.0581 | $37.00 \times 10^{-6}$ | $86.41 \times 10^{-6}$ | 0.701 |
| 600 | 0.4042 | 1111 | 0.0607 | $38.45 \times 10^{-6}$ | $95.12 \times 10^{-6}$ | 0.704 |
| 650 | 0.3824 | 1122 | 0.0631 | $39.81 \times 10^{-6}$ | $104.1 \times 10^{-6}$ | 0.707 |
| 700 | 0.3627 | 1133 | 0.0655 | $41.12 \times 10^{-6}$ | $113.4 \times 10^{-6}$ | 0.710 |
| 750 | 0.3450 | 1144 | 0.0678 | $42.41 \times 10^{-6}$ | $122.9 \times 10^{-6}$ | 0.712 |
| 800 | 0.3289 | 1154 | 0.0700 | $43.64 \times 10^{-6}$ | $132.7 \times 10^{-6}$ | 0.715 |
| 850 | 0.3142 | 1163 | 0.0722 | $44.84 \times 10^{-6}$ | $142.7 \times 10^{-6}$ | 0.717 |
| 900 | 0.3008 | 1171 | 0.0743 | $46.01 \times 10^{-6}$ | $153.0 \times 10^{-6}$ | 0.720 |
| 950 | 0.2886 | 1178 | 0.0763 | $47.16 \times 10^{-6}$ | $163.4 \times 10^{-6}$ | 0.723 |
| 1000 | 0.2772 | 1183 | 0.0782 | $48.24 \times 10^{-6}$ | $174.0 \times 10^{-6}$ | 0.726 |

## (c) Argon at 1 atm (101.325 kPa)

| T (C) | $\rho$ (kg/m³) | $c_p$ (J/kg C) | k (W/mC) | $\mu$ (kg/m s) | $\nu$ (m²/s) | Pr |
|---|---|---|---|---|---|---|
| −50 | 2.197 | 521 | 0.0137 | $17.64 \times 10^{-6}$ | $8.03 \times 10^{-6}$ | 0.670 |
| 0 | 1.810 | 521 | 0.0164 | $20.99 \times 10^{-6}$ | $11.60 \times 10^{-6}$ | 0.669 |
| 50 | 1.516 | 521 | 0.0188 | $24.11 \times 10^{-6}$ | $15.91 \times 10^{-6}$ | 0.668 |
| 100 | 1.297 | 521 | 0.0211 | $27.03 \times 10^{-6}$ | $20.84 \times 10^{-6}$ | 0.667 |
| 150 | 1.136 | 521 | 0.0233 | $29.76 \times 10^{-6}$ | $26.21 \times 10^{-6}$ | 0.666 |
| 200 | 1.016 | 521 | 0.0253 | $32.34 \times 10^{-6}$ | $31.82 \times 10^{-6}$ | 0.665 |
| 250 | 0.926 | 521 | 0.0273 | $34.77 \times 10^{-6}$ | $37.53 \times 10^{-6}$ | 0.664 |
| 300 | 0.856 | 521 | 0.0292 | $37.08 \times 10^{-6}$ | $43.33 \times 10^{-6}$ | 0.665 |
| 350 | 0.796 | 521 | 0.0309 | $39.29 \times 10^{-6}$ | $49.34 \times 10^{-6}$ | 0.661 |
| 400 | 0.742 | 521 | 0.0327 | $41.40 \times 10^{-6}$ | $55.81 \times 10^{-6}$ | 0.660 |
| 450 | 0.689 | 521 | 0.0344 | $43.43 \times 10^{-6}$ | $63.01 \times 10^{-6}$ | 0.659 |
| 500 | 0.637 | 521 | 0.0360 | $45.39 \times 10^{-6}$ | $71.22 \times 10^{-6}$ | 0.657 |
| 550 | 0.587 | 521 | 0.0376 | $47.28 \times 10^{-6}$ | $80.55 \times 10^{-6}$ | 0.656 |
| 600 | 0.542 | 521 | 0.0391 | $49.11 \times 10^{-6}$ | $90.66 \times 10^{-6}$ | 0.655 |
| 650 | 0.507 | 521 | 0.0405 | $50.89 \times 10^{-6}$ | $100.3 \times 10^{-6}$ | 0.654 |
| 700 | 0.492 | 521 | 0.0419 | $52.60 \times 10^{-6}$ | $107.0 \times 10^{-6}$ | 0.653 |

### (d) Carbon dioxide at 1 atm (101.325 kPa)

| T (C) | $\rho$ (kg/m$^3$) | $c_p$ (J/kg C) | k (W/mC) | $\mu$ (kg/m s) | $\nu$ (m$^2$/s) | Pr |
|-------|----------|----------|--------|--------------|-------------|-------|
| −50 | 2.421 | 781 | 0.0111 | $11.32 \times 10^{-6}$ | $4.68 \times 10^{-6}$ | 0.799 |
| 0 | 1.994 | 828 | 0.0146 | $13.75 \times 10^{-6}$ | $6.90 \times 10^{-6}$ | 0.779 |
| 50 | 1.670 | 874 | 0.0183 | $16.06 \times 10^{-6}$ | $9.62 \times 10^{-6}$ | 0.766 |
| 100 | 1.429 | 917 | 0.0222 | $18.27 \times 10^{-6}$ | $12.79 \times 10^{-6}$ | 0.756 |
| 150 | 1.251 | 958 | 0.0261 | $20.39 \times 10^{-6}$ | $16.30 \times 10^{-6}$ | 0.747 |
| 200 | 1.119 | 996 | 0.0302 | $22.42 \times 10^{-6}$ | $20.03 \times 10^{-6}$ | 0.740 |
| 250 | 1.020 | 1030 | 0.0342 | $24.36 \times 10^{-6}$ | $23.87 \times 10^{-6}$ | 0.733 |
| 300 | 0.943 | 1062 | 0.0383 | $26.23 \times 10^{-6}$ | $22.82 \times 10^{-6}$ | 0.727 |
| 350 | 0.877 | 1090 | 0.0423 | $28.02 \times 10^{-6}$ | $31.95 \times 10^{-6}$ | 0.722 |
| 400 | 0.817 | 1116 | 0.0463 | $29.75 \times 10^{-6}$ | $36.40 \times 10^{-6}$ | 0.717 |
| 450 | 0.759 | 1139 | 0.0502 | $31.42 \times 10^{-6}$ | $41.37 \times 10^{-6}$ | 0.713 |
| 500 | 0.702 | 1159 | 0.0539 | $33.03 \times 10^{-6}$ | $47.04 \times 10^{-6}$ | 0.710 |
| 550 | 0.647 | 1177 | 0.0574 | $34.58 \times 10^{-6}$ | $53.47 \times 10^{-6}$ | 0.709 |
| 600 | 0.597 | 1194 | 0.0607 | $36.09 \times 10^{-6}$ | $60.44 \times 10^{-6}$ | 0.710 |
| 650 | 0.559 | 1210 | 0.0638 | $37.55 \times 10^{-6}$ | $67.13 \times 10^{-6}$ | 0.713 |
| 700 | 0.542 | 1227 | 0.0665 | $38.96 \times 10^{-6}$ | $71.85 \times 10^{-6}$ | 0.718 |

### (e) Helium at 1 atm (101.325 kPa)

| T (C) | $\rho$ (kg/m$^3$) | $c_p$ (J/kg C) | k (W/mC) | $\mu$ (kg/m s) | $\nu$ (m$^2$/s) | Pr |
|-------|----------|----------|--------|--------------|-------------|-------|
| −50 | 0.2202 | 5200 | 0.124 | $16.17 \times 10^{-6}$ | $73.44 \times 10^{-6}$ | 0.679 |
| 0 | 0.1813 | 5200 | 0.140 | $18.62 \times 10^{-6}$ | $102.7 \times 10^{-6}$ | 0.692 |
| 50 | 0.1519 | 5200 | 0.156 | $20.93 \times 10^{-6}$ | $137.8 \times 10^{-6}$ | 0.698 |
| 100 | 0.1300 | 5200 | 0.172 | $23.13 \times 10^{-6}$ | $178.0 \times 10^{-6}$ | 0.700 |
| 150 | 0.1138 | 5200 | 0.188 | $25.23 \times 10^{-6}$ | $221.8 \times 10^{-6}$ | 0.698 |
| 200 | 0.1018 | 5200 | 0.204 | $27.25 \times 10^{-6}$ | $267.7 \times 10^{-6}$ | 0.694 |
| 250 | 0.0928 | 5200 | 0.220 | $29.20 \times 10^{-6}$ | $314.6 \times 10^{-6}$ | 0.689 |
| 300 | 0.0857 | 5200 | 0.236 | $31.08 \times 10^{-6}$ | $362.6 \times 10^{-6}$ | 0.684 |
| 350 | 0.0797 | 5200 | 0.252 | $32.91 \times 10^{-6}$ | $412.7 \times 10^{-6}$ | 0.678 |
| 400 | 0.0743 | 5200 | 0.268 | $34.69 \times 10^{-6}$ | $467.0 \times 10^{-6}$ | 0.673 |
| 450 | 0.0690 | 5200 | 0.284 | $36.44 \times 10^{-6}$ | $527.9 \times 10^{-6}$ | 0.668 |
| 500 | 0.0638 | 5200 | 0.299 | $38.16 \times 10^{-6}$ | $597.9 \times 10^{-6}$ | 0.663 |
| 550 | 0.0588 | 5200 | 0.314 | $39.86 \times 10^{-6}$ | $677.9 \times 10^{-6}$ | 0.660 |
| 600 | 0.0543 | 5200 | 0.329 | $41.53 \times 10^{-6}$ | $765.1 \times 10^{-6}$ | 0.657 |
| 650 | 0.0508 | 5200 | 0.343 | $43.18 \times 10^{-6}$ | $849.3 \times 10^{-6}$ | 0.656 |
| 700 | 0.0493 | 5200 | 0.356 | $44.82 \times 10^{-6}$ | $909.2 \times 10^{-6}$ | 0.655 |

## (f) Hydrogen at 1 atm (101.325 kPa)

| T (C) | $\rho$ (kg/m$^3$) | $c_p$ (J/kg C) | k (W/mC) | $\mu$ (kg/m s) | $\nu$ (m$^2$/s) | Pr |
|---|---|---|---|---|---|---|
| −50 | 0.1109 | 13810 | 0.142 | $7.32 \times 10^{-6}$ | $66.00 \times 10^{-6}$ | 0.713 |
| 0 | 0.0913 | 14190 | 0.168 | $8.37 \times 10^{-6}$ | $91.6 \times 10^{-6}$ | 0.707 |
| 50 | 0.0765 | 14390 | 0.191 | $9.37 \times 10^{-6}$ | $122.6 \times 10^{-6}$ | 0.706 |
| 100 | 0.0654 | 14470 | 0.211 | $10.34 \times 10^{-6}$ | $158.0 \times 10^{-6}$ | 0.708 |
| 150 | 0.0573 | 14500 | 0.230 | $11.26 \times 10^{-6}$ | $196.5 \times 10^{-6}$ | 0.709 |
| 200 | 0.0513 | 14500 | 0.248 | $12.14 \times 10^{-6}$ | $236.8 \times 10^{-6}$ | 0.711 |
| 250 | 0.0468 | 14500 | 0.265 | $12.99 \times 10^{-6}$ | $277.9 \times 10^{-6}$ | 0.712 |
| 300 | 0.0432 | 14520 | 0.281 | $13.81 \times 10^{-6}$ | $319.8 \times 10^{-6}$ | 0.712 |
| 350 | 0.0402 | 14540 | 0.298 | $14.60 \times 10^{-6}$ | $363.3 \times 10^{-6}$ | 0.712 |
| 400 | 0.0374 | 14590 | 0.315 | $15.37 \times 10^{-6}$ | $410.4 \times 10^{-6}$ | 0.711 |
| 450 | 0.0348 | 14640 | 0.332 | $16.11 \times 10^{-6}$ | $463.2 \times 10^{-6}$ | 0.710 |
| 500 | 0.0322 | 14700 | 0.350 | $16.84 \times 10^{-6}$ | $523.8 \times 10^{-6}$ | 0.707 |
| 550 | 0.0296 | 14750 | 0.368 | $17.56 \times 10^{-6}$ | $593.3 \times 10^{-6}$ | 0.704 |
| 600 | 0.0273 | 14800 | 0.386 | $18.28 \times 10^{-6}$ | $669.1 \times 10^{-6}$ | 0.702 |
| 650 | 0.0256 | 14840 | 0.403 | $19.00 \times 10^{-6}$ | $742.3 \times 10^{-6}$ | 0.700 |
| 700 | 0.0248 | 14880 | 0.419 | $19.72 \times 10^{-6}$ | $794.3 \times 10^{-6}$ | 0.700 |

## (g) Nitrogen at 1 atm (101.325 kPa)

| T (C) | $\rho$ (kg/m$^3$) | $c_p$ (J/kg C) | k (W/mC) | $\mu$ (kg/m s) | $\nu$ (m$^2$/s) | Pr |
|---|---|---|---|---|---|---|
| −50 | 1.523 | 1042 | 0.0201 | $14.12 \times 10^{-6}$ | $9.28 \times 10^{-6}$ | 0.732 |
| 0 | 1.253 | 1041 | 0.0239 | $16.57 \times 10^{-6}$ | $13.23 \times 10^{-6}$ | 0.722 |
| 50 | 1.049 | 1041 | 0.0275 | $18.85 \times 10^{-6}$ | $17.97 \times 10^{-6}$ | 0.714 |
| 100 | 0.897 | 1043 | 0.0309 | $20.96 \times 10^{-6}$ | $23.37 \times 10^{-6}$ | 0.708 |
| 150 | 0.785 | 1047 | 0.0342 | $22.94 \times 10^{-6}$ | $29.22 \times 10^{-6}$ | 0.704 |
| 200 | 0.703 | 1053 | 0.0373 | $24.80 \times 10^{-6}$ | $35.30 \times 10^{-6}$ | 0.701 |
| 250 | 0.641 | 1060 | 0.0402 | $26.56 \times 10^{-6}$ | $41.45 \times 10^{-6}$ | 0.700 |
| 300 | 0.592 | 1069 | 0.0431 | $28.22 \times 10^{-6}$ | $47.67 \times 10^{-6}$ | 0.701 |
| 350 | 0.551 | 1080 | 0.0458 | $29.80 \times 10^{-6}$ | $54.10 \times 10^{-6}$ | 0.702 |
| 400 | 0.513 | 1091 | 0.0484 | $31.31 \times 10^{-6}$ | $61.01 \times 10^{-6}$ | 0.705 |
| 450 | 0.477 | 1103 | 0.0510 | $32.77 \times 10^{-6}$ | $68.71 \times 10^{-6}$ | 0.709 |
| 500 | 0.441 | 1115 | 0.0535 | $34.17 \times 10^{-6}$ | $72.51 \times 10^{-6}$ | 0.712 |
| 550 | 0.406 | 1128 | 0.0560 | $35.53 \times 10^{-6}$ | $87.52 \times 10^{-6}$ | 0.716 |
| 600 | 0.375 | 1140 | 0.0584 | $36.84 \times 10^{-6}$ | $98.35 \times 10^{-6}$ | 0.719 |
| 650 | 0.351 | 1152 | 0.0609 | $38.11 \times 10^{-6}$ | $108.7 \times 10^{-6}$ | 0.721 |
| 700 | 0.340 | 1162 | 0.0633 | $39.34 \times 10^{-6}$ | $115.7 \times 10^{-6}$ | 0.722 |

### (h) Oxygen at 1 atm (101.325 kPa)

| T (C) | $\rho$ (kg/m³) | $c_p$ (J/kg C) | k (W/mC) | $\mu$ (kg/m s) | $\nu$ (m²/s) | Pr |
|-------|------|------|--------|----------------------|----------------------|-------|
| −50 | 1.740 | 914 | 0.0203 | $16.21 \times 10^{-6}$ | $9.32 \times 10^{-6}$ | 0.729 |
| 0 | 1.432 | 917 | 0.0241 | $19.19 \times 10^{-6}$ | $13.40 \times 10^{-6}$ | 0.730 |
| 50 | 1.199 | 924 | 0.0278 | $21.95 \times 10^{-6}$ | $18.31 \times 10^{-6}$ | 0.728 |
| 100 | 1.025 | 935 | 0.0315 | $24.51 \times 10^{-6}$ | $23.92 \times 10^{-6}$ | 0.726 |
| 150 | 0.897 | 948 | 0.0352 | $26.91 \times 10^{-6}$ | $30.00 \times 10^{-6}$ | 0.726 |
| 200 | 0.803 | 963 | 0.0387 | $29.16 \times 10^{-6}$ | $36.33 \times 10^{-6}$ | 0.727 |
| 250 | 0.732 | 979 | 0.0421 | $31.29 \times 10^{-6}$ | $42.76 \times 10^{-6}$ | 0.728 |
| 300 | 0.676 | 995 | 0.0453 | $33.32 \times 10^{-6}$ | $49.27 \times 10^{-6}$ | 0.731 |
| 350 | 0.629 | 1010 | 0.0485 | $35.25 \times 10^{-6}$ | $56.02 \times 10^{-6}$ | 0.734 |
| 400 | 0.586 | 1024 | 0.0515 | $37.10 \times 10^{-6}$ | $63.28 \times 10^{-6}$ | 0.737 |
| 450 | 0.545 | 1037 | 0.0545 | $38.89 \times 10^{-6}$ | $71.38 \times 10^{-6}$ | 0.740 |
| 500 | 0.504 | 1049 | 0.0574 | $40.61 \times 10^{-6}$ | $80.64 \times 10^{-6}$ | 0.742 |
| 550 | 0.464 | 1059 | 0.0603 | $42.28 \times 10^{-6}$ | $91.16 \times 10^{-6}$ | 0.743 |
| 600 | 0.428 | 1069 | 0.0632 | $43.88 \times 10^{-6}$ | $102.6 \times 10^{-6}$ | 0.742 |
| 650 | 0.401 | 1077 | 0.0662 | $45.43 \times 10^{-6}$ | $113.4 \times 10^{-6}$ | 0.740 |
| 700 | 0.389 | 1086 | 0.0693 | $46.92 \times 10^{-6}$ | $120.7 \times 10^{-6}$ | 0.735 |

# Appendix E
## Properties of saturated steam

| T (C) | Saturation pressure (kPa) | $h_{fg}$ (kJ/kg) | Liquid $\rho$ (kg/m$^3$) | Vapor $\rho_v$ (kg/m$^3$) | $c_{pv}$ (J/kg C) | $k_v$ (W/mC) | $\mu_v$ (kg/m s) | Pr |
|---|---|---|---|---|---|---|---|---|
| 0.01 | 0.6113 | 2501 | 999.8 | 0.0049 | 1854 | 0.0171 | $0.922 \times 10^{-5}$ | 1.00 |
| 5 | 0.8721 | 2490 | 999.9 | 0.0068 | 1857 | 0.0173 | $0.934 \times 10^{-5}$ | 1.00 |
| 10 | 1.2276 | 2478 | 999.7 | 0.0094 | 1862 | 0.0176 | $0.946 \times 10^{-5}$ | 1.00 |
| 15 | 1.7051 | 2466 | 999.1 | 0.0129 | 1863 | 0.0179 | $0.959 \times 10^{-5}$ | 1.00 |
| 20 | 2.339 | 2454 | 998.0 | 0.0173 | 1867 | 0.0182 | $0.973 \times 10^{-5}$ | 1.00 |
| 25 | 3.169 | 2442 | 997.0 | 0.0231 | 1870 | 0.0186 | $0.987 \times 10^{-5}$ | 1.00 |
| 30 | 4.246 | 2431 | 996.0 | 0.0304 | 1875 | 0.0189 | $1.001 \times 10^{-5}$ | 1.00 |
| 35 | 5.628 | 2419 | 994.0 | 0.0397 | 1880 | 0.0192 | $1.016 \times 10^{-5}$ | 1.00 |
| 40 | 7.384 | 2407 | 992.1 | 0.0512 | 1885 | 0.0196 | $1.031 \times 10^{-5}$ | 1.00 |
| 45 | 9.593 | 2395 | 990.1 | 0.0655 | 1892 | 0.0200 | $1.046 \times 10^{-5}$ | 1.00 |
| 50 | 12.35 | 2383 | 988.1 | 0.0831 | 1900 | 0.0204 | $1.062 \times 10^{-5}$ | 1.00 |
| 55 | 15.76 | 2371 | 985.2 | 0.1045 | 1908 | 0.0208 | $1.077 \times 10^{-5}$ | 1.00 |
| 60 | 19.94 | 2359 | 983.3 | 0.1304 | 1916 | 0.0212 | $1.093 \times 10^{-5}$ | 1.00 |
| 65 | 25.03 | 2346 | 980.4 | 0.1614 | 1926 | 0.0216 | $1.110 \times 10^{-5}$ | 1.00 |
| 70 | 31.19 | 2334 | 977.5 | 0.1983 | 1936 | 0.0221 | $1.126 \times 10^{-5}$ | 1.00 |
| 75 | 38.58 | 2321 | 974.7 | 0.2421 | 1948 | 0.0225 | $1.142 \times 10^{-5}$ | 1.00 |
| 80 | 47.39 | 2309 | 971.8 | 0.2935 | 1962 | 0.0230 | $1.159 \times 10^{-5}$ | 1.00 |
| 85 | 57.83 | 2296 | 968.1 | 0.3536 | 1977 | 0.0235 | $1.176 \times 10^{-5}$ | 1.00 |
| 90 | 70.14 | 2283 | 965.3 | 0.4235 | 1993 | 0.0240 | $1.193 \times 10^{-5}$ | 1.00 |
| 95 | 84.55 | 2270 | 961.5 | 0.5045 | 2010 | 0.0246 | $1.210 \times 10^{-5}$ | 1.00 |
| 100 | 101.33 | 2257 | 957.9 | 0.5978 | 2029 | 0.0251 | $1.227 \times 10^{-5}$ | 1.00 |
| 120 | 198.5 | 2203 | 943.4 | 1.121 | 2120 | 0.0275 | $1.296 \times 10^{-5}$ | 1.00 |
| 140 | 361.3 | 2145 | 925.9 | 1.965 | 2244 | 0.0301 | $1.365 \times 10^{-5}$ | 1.02 |
| 160 | 617.8 | 2083 | 907.4 | 3.256 | 2420 | 0.0331 | $1.434 \times 10^{-5}$ | 1.05 |
| 180 | 1002.1 | 2015 | 887.3 | 5.153 | 2590 | 0.0364 | $1.502 \times 10^{-5}$ | 1.07 |
| 200 | 1553.8 | 1941 | 864.3 | 7852 | 2840 | 0.0401 | $1.571 \times 10^{-5}$ | 1.11 |
| 220 | 2318 | 1859 | 840.3 | 11.602 | 3110 | 0.0442 | $1.641 \times 10^{-5}$ | 1.15 |
| 240 | 3344 | 1767 | 813.7 | 16.734 | 3520 | 0.0487 | $1.712 \times 10^{-5}$ | 1.24 |
| 260 | 4688 | 1663 | 783.7 | 23.691 | 4070 | 0.0540 | $1.788 \times 10^{-5}$ | 1.35 |
| 280 | 6412 | 1544 | 750.8 | 33.145 | 4835 | 0.0605 | $1.870 \times 10^{-5}$ | 1.49 |
| 300 | 8581 | 1405 | 712.3 | 46.147 | 5980 | 0.0695 | $1.965 \times 10^{-5}$ | 1.69 |

# Appendix F
## Hemispherical emissivity of surfaces
### (a) Metals

|                     | 200 K | 300 K | 400 K | 600 K | 800 K | 1000 K |
|---------------------|-------|-------|-------|-------|-------|--------|
| Aluminum, polished  | 0.03  | 0.04  | 0.05  | 0.06  | 0.07  |        |
| Aluminum, oxidized  |       |       | 0.10  | 0.14  | 0.18  |        |
| Brass, polished     |       | 0.04  |       |       |       |        |
| Brass, oxidized     |       | 0.6   |       |       |       |        |
| Copper, polished    |       | 0.04  |       | 0.06  | 0.09  | 0.12   |
| Copper, oxidized    |       | 0.56  | 0.6   | 0.67  | 0.87  |        |
| Gold, polished      | 0.02  |       | 0.03  | 0.05  | 0.06  | 0.07   |
| Iron, polished      | 0.05  |       | 0.09  | 0.14  | 0.20  | 0.25   |
| Iron, rusted        |       | 0.83  |       |       |       |        |
| Nickel, polished    | 0.08  |       | 0.10  | 0.11  | 0.12  | 0.15   |
| Silver, polished    | 0.02  |       | 0.02  | 0.03  | 0.03  | 0.03   |

### (b) Nonmetals (at 300 K)

| | |
|---|---|
| Asbestos board | 0.93 |
| Brick | 0.90 |
| Concrete | 0.91 |
| Glass | 0.9 |
| Ice | 0.96 |
| **Paint** | |
| Flat black | 0.88 |
| Flat white | 0.91 |
| Aluminum | 0.45 |
| Sand | 0.76 |
| **Wood** | |
| Oak | 0.9 |
| Spruce | 0.8 |

## References for Appendices A to F

**Properties in Appendices A to F were from a variety of sources, including, but not limited to, the following:**

[1] Touloukian, Y. S. and Ho, C. Y., Eds., Thermophysical Properties of MatterI, Plenum Press, 1972.
[2] Bejan, A., Heat Transfer, Wiley, 1993.
[3] Cengel, Y. A., Heat Transfer, A Practical Approach, McGraw-Hill, 1998.
[4] Incropera, F. P. and DeWitt, D. P., Introduction to Heat Transfer, 4th Ed., Wiley, 1996.
[5] Mills, A. F., Basic Heat and Mass Transfer, Irwin, 1995.
[6] Ozisik, M. N., Heat Transfer, A Basic Approach, McGraw-Hill, 1985.
[7] Suryanarayana, N. V., Engineering Heat Transfer, West, 1995.

# Appendix G
## Bessel functions of the first kind

| $\alpha$ | $J_0(\alpha)$ | $J_1(\alpha)$ | $\alpha$ | $J_0(\alpha)$ | $J_1(\alpha)$ | $\alpha$ | $J_0(\alpha)$ | $J_1(\alpha)$ |
|---|---|---|---|---|---|---|---|---|
| 0.0 | 1.0000 | 0.0000 | | | | | | |
| 0.2 | 0.9900 | 0.0995 | 5.2 | −0.1103 | −0.3432 | 10.2 | −0.2496 | −0.0066 |
| 0.4 | 0.9604 | 0.1960 | 5.4 | −0.0412 | −0.3453 | 10.4 | −0.2434 | −0.0555 |
| 0.6 | 0.9120 | 0.2867 | 5.6 | 0.0270 | −0.3343 | 10.6 | −0.2276 | −0.1012 |
| 0.8 | 0.8463 | 0.3688 | 5.8 | 0.0917 | −0.3110 | 10.8 | −0.2032 | −0.1422 |
| 1.0 | 0.7652 | 0.4401 | 6.0 | 0.1506 | −0.2767 | 11.0 | −0.1712 | −0.1768 |
| 1.2 | 0.6711 | 0.4983 | 6.2 | 0.2017 | −0.2329 | 11.2 | −0.1330 | −0.2039 |
| 1.4 | 0.5669 | 0.5419 | 6.4 | 0.2433 | −0.1816 | 11.4 | −0.0902 | −0.2224 |
| 1.6 | 0.4554 | 0.5699 | 6.6 | 0.2740 | −0.1250 | 11.6 | −0.0446 | −0.2320 |
| 1.8 | 0.3400 | 0.5815 | 6.8 | 0.2931 | −0.0652 | 11.8 | 0.0020 | −0.2323 |
| 2.0 | 0.2239 | 0.5767 | 7.0 | 0.3001 | −0.0047 | 12.0 | 0.0477 | −0.2234 |
| 2.2 | 0.1104 | 0.5560 | 7.2 | 0.2951 | 0.0543 | 12.2 | 0.0908 | −0.2060 |
| 2.4 | 0.0025 | 0.5202 | 7.4 | 0.2786 | 0.1096 | 12.4 | 0.1296 | −0.1807 |
| 2.6 | −0.0968 | 0.4708 | 7.6 | 0.2516 | 0.1592 | 12.6 | 0.1626 | −0.1487 |
| 2.8 | −0.1850 | 0.4097 | 7.8 | 0.2154 | 0.2014 | 12.8 | 0.1887 | −0.1114 |
| 3.0 | −0.2601 | 0.3391 | 8.0 | 0.1717 | 0.2346 | 13.0 | 0.2069 | −0.0703 |
| 3.2 | −0.3202 | 0.2613 | 8.2 | 0.1222 | 0.2580 | 13.2 | 0.2167 | −0.0271 |
| 3.4 | −0.3643 | 0.1792 | 8.4 | 0.0692 | 0.2708 | 13.4 | 0.2177 | 0.0166 |
| 3.6 | −0.3918 | 0.0955 | 8.6 | 0.0146 | 0.2727 | 13.6 | 0.2101 | 0.0590 |
| 3.8 | −0.4026 | 0.0128 | 8.8 | −0.0392 | 0.2641 | 13.8 | 0.1943 | 0.0984 |
| 4.0 | −0.3971 | −0.0660 | 9.0 | −0.0903 | 0.2453 | 14.0 | 0.1711 | 0.1334 |
| 4.2 | −0.3766 | −0.1386 | 9.2 | −0.1367 | 0.2174 | 14.2 | 0.1414 | 0.1626 |
| 4.4 | −0.3423 | −0.2028 | 9.4 | −0.1768 | 0.1816 | 14.4 | 0.1065 | 0.1850 |
| 4.6 | −0.2961 | −0.2566 | 9.6 | −0.2090 | 0.1395 | 14.6 | 0.0679 | 0.1999 |
| 4.8 | −0.2404 | −0.2985 | 9.8 | −0.2323 | 0.0928 | 14.8 | 0.0271 | 0.2066 |
| 5.0 | −0.1776 | −0.3276 | 10.0 | −0.2459 | 0.0435 | 15.0 | −0.0142 | 0.2051 |

# Appendix H
## Modified Bessel functions

| x | $I_0 (x)$ | $I_1 (x)$ | $I_2 (x)$ | $K_0 (x)$ | $K_1 (x)$ |
|---|---|---|---|---|---|
| 0.0 | 1 | 0 | 0 | — | — |
| 0.1 | 1.003 | 0.050 | 0.001 | 2.427E+00 | 9.854E+00 |
| 0.5 | 1.063 | 0.258 | 0.032 | 9.244E-01 | 1.656E+00 |
| 1.0 | 1.266 | 0.565 | 0.136 | 4.210E-01 | 6.019E-01 |
| 1.5 | 1.647 | 0.982 | 0.338 | 2.138E-01 | 2.774E-01 |
| 2.0 | 2.280 | 1.591 | 0.689 | 1.139E-01 | 1.399E-01 |
| 2.5 | 3.290 | 2.517 | 1.276 | 6.235E-02 | 7.389E-02 |
| 3.0 | 4.881 | 3.953 | 2.245 | 3.474E-02 | 4.016E-02 |
| 3.5 | 7.378 | 6.206 | 3.832 | 1.960E-02 | 2.224E-02 |
| 4.0 | 11.302 | 9.759 | 6.422 | 1.116E-02 | 1.248E-02 |
| 4.5 | 17.48 | 15.39 | 10.64 | 6.400E-03 | 7.078E-03 |
| 5.0 | 27.24 | 24.34 | 17.51 | 3.691E-03 | 4.045E-03 |
| 5.5 | 42.69 | 38.59 | 28.66 | 2.139E-03 | 2.326E-03 |
| 6.0 | 67.23 | 61.34 | 46.79 | 1.244E-03 | 1.344E-03 |
| 6.5 | 106.29 | 97.74 | 76.22 | 7.259E-04 | 7.799E-04 |
| 7.0 | 168.59 | 156.04 | 124.01 | 4.248E-04 | 4.542E-04 |
| 7.5 | 268.16 | 249.58 | 201.61 | 2.492E-04 | 2.653E-04 |
| 8.0 | 427.56 | 399.87 | 327.60 | 1.465E-04 | 1.554E-04 |
| 8.5 | 683.16 | 641.62 | 532.19 | 8.626E-05 | 9.120E-05 |
| 9.0 | 1093.59 | 1030.91 | 864.50 | 5.088E-05 | 5.364E-05 |
| 9.5 | 1753.48 | 1658.45 | 1404.33 | 3.006E-05 | 3.160E-05 |
| 10.0 | 2815.72 | 2670.99 | 2281.52 | 1.778E-05 | 1.865E-05 |

# Appendix I
## Complementary error function

| $\alpha$ | erfc $(\alpha)$ | $\alpha$ | erfc $(\alpha)$ | $\alpha$ | erfc $(\alpha)$ |
|---|---|---|---|---|---|
| 0 | 1 | | | | |
| 0.05 | 0.94363 | 1.05 | 0.13756 | 2.05 | 0.00374 |
| 0.1 | 0.88754 | 1.1 | 0.11979 | 2.1 | 0.00298 |
| 0.15 | 0.83200 | 1.15 | 0.10388 | 2.15 | 0.00236 |
| 0.2 | 0.77730 | 1.2 | 0.08969 | 2.2 | 1.863E-03 |
| 0.25 | 0.72367 | 1.25 | 0.07710 | 2.25 | 1.463E-03 |
| 0.3 | 0.67137 | 1.3 | 0.06599 | 2.3 | 1.143E-03 |
| 0.35 | 0.62062 | 1.35 | 0.05624 | 2.35 | 8.893E-04 |
| 0.4 | 0.57161 | 1.4 | 0.04771 | 2.4 | 6.885E-04 |
| 0.45 | 0.52452 | 1.45 | 0.04030 | 2.45 | 5.306E-04 |
| 0.5 | 0.47950 | 1.5 | 0.03389 | 2.5 | 4.070E-04 |
| 0.55 | 0.43668 | 1.55 | 0.02838 | 2.55 | 3.107E-04 |
| 0.6 | 0.39614 | 1.6 | 0.02365 | 2.6 | 2.360E-04 |
| 0.65 | 0.35797 | 1.65 | 0.01962 | 2.65 | 1.785E-04 |
| 0.7 | 0.32220 | 1.7 | 0.01621 | 2.7 | 1.343E-04 |
| 0.75 | 0.28884 | 1.75 | 0.01333 | 2.75 | 1.006E-04 |
| 0.8 | 0.25790 | 1.8 | 0.01091 | 2.8 | 7.501E-05 |
| 0.85 | 0.22933 | 1.85 | 0.00889 | 2.85 | 5.566E-05 |
| 0.9 | 0.20309 | 1.9 | 0.00721 | 2.9 | 4.110E-05 |
| 0.95 | 0.17911 | 1.95 | 0.00582 | 2.95 | 3.020E-05 |
| 1 | 0.15730 | 2 | 0.00468 | 3 | 2.209E-05 |
| | | | | $\infty$ | 0 |

# Appendix J
## Computations using Excel and Matlab

Throughout this book, we have been using Microsoft *Excel* and MathWorks *Matlab* for calculations. The author has found these software packages to be excellent in serving the needs of the subject. In this Appendix, we will again discuss the features of the software for solving nonlinear equations and systems of simultaneous nonlinear equations. The use of the software for solving systems of linear equations was discussed in detail in, Chapter 5.

Nonlinear equations often arise in heat transfer when radiation is involved as terms contain the $T^4$ factor. Let us consider a heat transfer problem that involves radiation. We will derive the governing equations and solve them using Excel and Matlab.

Consider the figure below. It is a plane wall with an incoming heat flux on its left surface and convection and radiation to the surroundings on its right surface. The heat transfer is steady state. Therefore, the incoming heat flux to the wall on the left surface equals the heat conduction through the wall and also equals the heat convection and radiation to the surroundings from the right surface. The conduction, convection, and radiation heat transfers are

$$\left(\frac{q}{A}\right)_{cond} = \frac{k}{L}(T_1 - T_2) \tag{J.1}$$

$$\left(\frac{q}{A}\right)_{conv} = h(T_2 - T_\infty) \tag{J.2}$$

$$\left(\frac{q}{A}\right)_{rad} = \varepsilon\sigma\left(T_2^4 - T_{surr}^4\right) \tag{J.3}$$

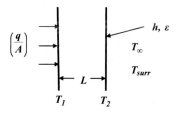

The energy equation for this system is

$$\left(\frac{q}{A}\right)_{in} = \left(\frac{q}{A}\right)_{cond} = \left(\frac{q}{A}\right)_{conv} + \left(\frac{q}{A}\right)_{rad} \tag{J.4}$$

Let us say that the incoming heat flux is known and we would like to determine the temperature $T_2$ of the right surface of the wall. Using the above equations, we have

$$\left(\frac{q}{A}\right)_{in} = h(T_2 - T_\infty) + \varepsilon\sigma\left(T_2^4 - T_{surr}^4\right) \tag{J.5}$$

Let us put some numbers into our problem:

$$\left(\frac{q}{A}\right)_{in} = 1500 \text{ W/m}^2$$

$$h = 80 \text{ W/m}^2 \text{ C}$$

$$\varepsilon = 0.85$$

$$\sigma = 5.67 \times 10^{-8} \text{ W/m}^2 \text{ K}^4$$

$$T_\infty = 20 \text{ C} = 293.15 \text{ K}$$

$$T_{surr} = 20 \text{ C} = 293.15 \text{ K}$$

Using these values in Eq. (J.5), we get the following equation to solve for $T_2$:

$$80T_2 + 4.8195 \times 10^{-8}T_2^4 - 25308 = 0 \qquad (\text{J.6})$$

Let us solve this equation using Excel and Matlab.

With Excel, we open a new spreadsheet. We will use cell A1 for the value of $T_2$. We start by putting a guess in the cell. We know that $T_2$ must be greater than 293.15, so let us guess 400 and put that value in cell A1. In cell A2, we will put the left side of Eq. (J.6). That is, in cell A2 we put = 80*A1 + 4.8195e-8 * A1^4 - 25308. We then call up the *Goal Seek* function of Excel and tell it to make cell A2 zero by changing cell A1. We do this and get the solution $T_2 = 310.73$. The temperature of the right surface of the wall is 310.73 K.

With Matlab, we can solve Eq. (J.6) interactively using the *fzero* function. The command is >>fzero ('80*x + 4.8195e-8 * x.^4 - 25308', 400). Like *Goal Seek*, we gave the function an initial guess of 400 for the solution. We need to use "x" as the variable in the *fzero* function. And we need to use a period before the ^ character as we are doing element-by-element operations instead of matrix operations. The Matlab solution for $T_2$ was the same as the Excel result. $T_2 = 310.73$ K.

Let us now look at a situation where we have simultaneous nonlinear equations. The below figure shows the same wall, but now, we have convection and radiation boundary conditions on both surfaces.

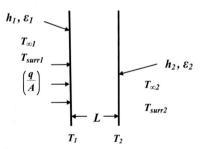

For steady heat transfer, the heat flux coming into the wall by convection and radiation at the left surface equals the conduction through the wall and also equals the convection and radiation from the right surface of the wall to the surroundings. That is

$$\left(\frac{q}{A}\right)_{\text{left surface}} = \left(\frac{q}{A}\right)_{\text{through wall}} = \left(\frac{q}{A}\right)_{\text{right surface}} \qquad (\text{J.7})$$

As all three heat fluxes are equal, we will call them all $\left(\frac{q}{A}\right)$ Eq. (J.7) is then equivalent to the following three simultaneous equations:

$$\left(\frac{q}{A}\right) = h_1(T_{\infty_1} - T_1) + \varepsilon_1\sigma\left(T_{surr_1}^4 - T_1^4\right) \tag{J.8}$$

$$\left(\frac{q}{A}\right) = \frac{k}{L}(T_1 - T_2) \tag{J.9}$$

$$\left(\frac{q}{A}\right) = h_2(T_2 - T_{\infty_2}) + \varepsilon_2\sigma\left(T_2^4 - T_{surr_2}^4\right) \tag{J.10}$$

The three unknowns are the heat flux $\left(\frac{q}{A}\right)$ and the surface temperatures $T_1$ and $T_2$. All other parameters are given in the problem statement. Let us put some values into Eqs. (J.8)–(J.10):

$$h_1 = 12 \text{ W/m}^2 \text{ C}, \quad h_2 = 80 \text{ W/m}^2 \text{ C}$$
$$\varepsilon_1 = 0.4, \quad \varepsilon_2 = 0.85$$
$$\sigma = 5.67 \times 10^{-8} \text{ W/m}^2 \text{ K}^4$$
$$T_{\infty_1} = 250 \text{ C} = 523.15 \text{ K}, \quad T_{\infty_2} = 20 \text{ C} = 293.15 \text{ K}$$
$$T_{surr_1} = 250 \text{ C} = 523.15 \text{ K}, \quad T_{surr_2} = 20 \text{ C} = 293.15 \text{ K}$$
$$k = 15 \text{ W/m C}$$
$$L = 2.5 \text{ cm} = 0.025 \text{ m}$$

With these values, Eqs. (J.8)–(J.10) become

$$\left(\frac{q}{A}\right) = 7976.6 - 12T_1 - 2.268 \times 10^{-8}T_1^4 \tag{J.11}$$

$$\left(\frac{q}{A}\right) = 600(T_1 - T_2) \tag{J.12}$$

$$\left(\frac{q}{A}\right) = 80T_2 - 23808 + 4.8195 \times 10^{-8}T_2^4 \tag{J.13}$$

Eqs. (J.11)–(J.13) are our three equations to solve for $\left(\frac{q}{A}\right), T_1,$ and $T_2$.

Let us first do this using Excel. We move all the terms in the equations to the left side and give them names. That is,

$$f_1 = \left(\frac{q}{A}\right) - 7976.6 + 12T_1 + 2.268 \times 10^{-8}T_1^4 = 0 \tag{J.14}$$

$$f_2 = \left(\frac{q}{A}\right) - 600(T_1 - T_2) = 0 \tag{J.15}$$

$$f_3 = \left(\frac{q}{A}\right) - 80T_2 + 23808 - 4.8195 \times 10^{-8}T_2^4 = 0 \tag{J.16}$$

To reach a solution for the three unknowns, the functions $f_1, f_2,$ and $f_3$ must all be zero. We will be using the Excel *Solver* program. With this program, the values of the unknowns are varied until one criterion is fulfilled. For $f_1, f_2,$ and $f_3$ to be zero, the appropriate criterion is that the sum of the squares of the functions is zero. That is, if $f_1^2 + f_2^2 + f_3^2 = 0$, then all three of the functions must be zero. This is the criterion that we will use with *Solver*.

To proceed with the Excel solution, we open a new spreadsheet. We will use cell A1 for the value of $\left(\frac{q}{A}\right)$, cell A2 for the value of $T_1$, and cell A3 for the value of $T_2$. We start by putting initial guesses in the cells. From the above values for the environment on the left surface, we know that $T_1$ must be less than 523.15. So, let us guess 400 and put that value in cell A2. From the values for the environment on the right surface, we know that $T_2$ must be greater than 293.15. It also must be less than $T_1$. So, let us guess 350 and put that value in cell A3. For a guess of $\left(\frac{q}{A}\right)$, we use Eq. (J.15) with the initial guesses for $T_1$ and $T_2$ and put 30000 in cell A1. We will put the three functions in cells A4, A5, and A6. That is, in cell A4 we put= A1 − 7976.6 + 12 * A2 + 2.268e-8 * A2^4. In cell A5, we put = A1 − 600 * (A2 − A3). In cell A6, we put = A1 - 80 * A3 + 23808 - 4.8195e-8 * A3^4.

We put the criterion in cell A7. That is, cell A7 has = A4^2 + A5^2 + A6^2.

To get a solution, we call up *Solver* and tell it to make cell A7 zero by varying cells A1, A2, and A3. The Excel solution is

$$\left(\frac{q}{A}\right) = 3580 \text{ W/m}^2 \quad T_1 = 340.75 \text{ K} \quad T_2 = 334.78 \text{ K}$$

We used the Gauss–Seidel method to solve the program using Matlab. Eqs. (J.11)–(J.13) were rearranged so that each equation had one of the unknown parameters on its left side. The revised equations were

$$\left(\frac{q}{A}\right) = 7976.6 - 12T_1 - 2.268 \times 10^{-8} T_1^4 \tag{J.17}$$

$$T_1 = T_2 + \frac{(q/A)}{600} \tag{J.18}$$

$$T_2 = \frac{(q/A) + 23808 - 4.8195 \times 10^{-8} T_2^4}{80} \tag{J.19}$$

In the Matlab program, there is looping through the three equations. For each loop, the values of the three unknown parameters are updated. The values converged to a solution which was the same as the Excel solution. The Matlab m-file was the following:

```
% FHT Appendix J problem
% Gauss Seidel Solution
% Assign initial guess values
q_over_A=30000;
t1=400;
t2=350;
% Begin looping
for loop = 1:100
q_over_A=7976.6-12*t1-2.268e-8*t1^4;
t1=t2+q_over_A/600;
t2=(q_over_A+23808-4.8195e-8*t2^4)/80;
end
q_over_A, t1, t2
```

Solution of simultaneous nonlinear equations can also be obtained by using Matlab's *fsolve* function. Unfortunately, this function is in MathWorks Optimization Toolkit, which is not in the basic Matlab package.

# Appendix K
## Computer programs

This appendix contains Matlab m-files for the examples of, Chapter 5.

### Example 5.1 Matrix Inversion solution

```
% Example 5.1 Pin Fin (Matrix Inversion)
   clear
   clc
   a=zeros(10:10);
   % Enter non-zero elements of coefficient matrix "a"
   a(1,1)=-8.840; a(1,2)=4.418; a(10,9)=4.418; a(10,10)=-4.4217;
   for j=2:9
     a(j,j-1)=4.418; a(j,j)=-8.840; a(j,j+1)=4.418;
   end
   % Enter constants vector "b"
   b(1)=-883.66; b(2:9)=-0.06362; b(10)=-.05567;
   % Make "b" a column vector
   b=b';
   % Solve by matrix inversion for temperatures T 2 to T 11
   t= inv(a)*b
```

### Example 5.1 Gauss—Seidel solution

```
% Gauss-Seidel Solution for Example 5.1
   clear, clc
   t(1) = 200;
   % Start by giving initial guesses for the unknown temperatures
   t(2:11)=200;
   % A for-end structure will cause the looping. The loop index is "pass"
   for pass=1:500
     t(2)= 0.4998*t(3) + 99.962;
     for i=3:10
       t(i) = 0.4998 * (t(i-1) + t(i+1)) + 0.007197;
     end
     t(11) = 0.9992 * t(10) + 0.01259;
   end
   t
```

### Example 5.2 4-node model

```
% Example 5.2 by Gauss-Seidel
   clear, clc
   t1=200; t2=200; t3=200; t4=200;
   for loop=1:1000
     t1=(.5*t2+2*t3+350)/5;
     t2=(.5*t1+2*t4+400)/5;
```

```
    t3=(.5*t4+2*t1+1150)/5;
    t4=(.5*t3+2*t2+1200)/5;
  end
  t1,t2,t3,t4
```

### Example 5.2 10-node model

```
% Example 5.2 by Gauss-Seidel
  clear, clc
  t1=200;t2=200;t3=200;t4=200;t5=200;t6=200;t7=200;t8=200;t9=200;t10=200;
  for loop=1:1000
    t1=(300+100+t2+t6)/4;
    t2=(t1+100+t3+t7)/4;
    t3=(t2+100+t4+t8)/4;
    t4=(t3+100+t5+t9)/4;
    t5=(t4+500+t10)/4;
    t6=(300+t1+t7+500)/4;
    t7=(t6+t2+t8+500)/4;
    t8=(t7+t3+t9+500)/4;
    t9=(t8+t4+t10+500)/4;
    t10=(t9+t5+900)/4;
  end
  t1,t2,t3,t4,t5,t6,t7,t8,t9,t10
```

### Example 5.3

```
%Matrix Inversion Solution for Example 5.3
  % Flat Angled Plate
  clear, clc
  k=43;
  h=100;
  tinf=30;
  dx=0.08;
  qgen=0;
  % Define 12 x 12 matrix of zeroes
  a=zeros(12,12);
  % Define coefficient matrix
  a(1,1)=1;
  a(2,:)=[-.5 2 -.5 0 0 -1 0 0 0 0 0 0];
  a(3,:)=[0 -.5 2 -.5 0 0 -1 0 0 0 0 0];
  a(4,:)=[0 0 -.5 1 0 0 0 -.5 0 0 0 0];
  a(5,:)=[0 0 0 0 1 0 0 0 0 0 0 0];
  a(6,:)=[0 -1 0 0 -.5 2+h*dx/k -.5 0 0 0 0 0];
  a(7,:)=[0 0 -1 0 0 -.5 3+h*dx/k -1 -.5 0 0 0];
  a(8,:)=[0 0 0 -.5 0 0 -1 2 0 -.5 0 0];
  a(9,:)=[0 0 0 0 0 0 -.5 0 2+h*dx/k -1 -.5 0];
  a(10,:)=[0 0 0 0 0 0 0 -.5 -1 2 0 -.5];
  a(11,11)=1;
```

```
a(12,12)=1;
% Define constant vector
b(1)=500;
b(2)=qgen*dx^2/2/k;
b(3)=qgen*dx^2/2/k;
b(4)=qgen*dx^2/4/k;
b(5)=500;
b(6)=qgen*dx^2/2/k+h*dx/k*tinf;
b(7)=3/4*qgen*dx^2/k+h*dx/k*tinf;
b(8)=qgen*dx^2/2/k;
b(9)=qgen*dx^2/2/k+h*dx/k*tinf;
b(10)=qgen*dx^2/2/k;
b(11)=100;
b(12)=100;
% Make the constant vector a column vector
b=b';
% Solve for temperatures
temp=inv(a)*b
a, b
```

## Example 5.4

```
% Unsteady Example 5.4 for Example 5.2; How long for Node 1 to reach 150 C?
   % Stainless steel
   % Plate is initially at 30 C
   clear, clc
   k=15; rho=7900; c=477;
   dx=0.2; dy=0.1;
   alpha=k/rho/c;
   % Set time increment (seconds)
   dt=1;
   % Input initial temperatures
   t1=30; t2=30; t3=30; t4=30;
   % Start looping
   for loop=1:100000
     t1=t1*(1-2*alpha*dt/dx^2-2*alpha*dt/dy^2)+alpha*dt/dx^2*(300+t2)+alpha*dt/
   dy^2*(100+t3);
     t2=t2*(1-2*alpha*dt/dx^2-2*alpha*dt/dy^2)+alpha*dt/dx^2*(t1+400)+alpha*dt/
   dy^2*(t4+100);
     t3=t3*(1-2*alpha*dt/dx^2-2*alpha*dt/dy^2)+alpha*dt/dx^2*(300+t4)+alpha*dt/
   dy^2*(500+t1);
     t4=t4*(1-2*alpha*dt/dx^2-2*alpha*dt/dy^2)+alpha*dt/dx^2*(t3+400)+alpha*dt/
   dy^2*(t2+500);
     if t1>=150
       dt,loop,t1,t2,t3,t4
       break
     end
   end
```

## Example 5.5

```
% Example 5.5 Disk with Convection and Radiation Unsteady-State
clear, clc, format compact
k=15; rho=7900; c=477;
alpha=k/rho/c;
h=100;
eps=0.5;
% Kelvin temperatures since we have radiation
tsurr=293.15;
tinf=293.15;
dr=0.01;
b=0.01;
qin=300;
sigma=5.669e-8;
% Surface areas for the 5 nodes
a1=pi*(dr/2)^2; a2=pi*((3*dr/2)^2-(dr/2)^2); a3=pi*((5*dr/2)^2-(3*dr/2)^2);
a4=pi*((7*dr/2)^2-(5*dr/2)^2);
a5=pi*((4*dr)^2-(7*dr/2)^2);
% Flow areas for the 5 nodes
a12=b*2*pi*dr/2; a23=b*2*pi*3*dr/2; a34=b*2*pi*5*dr/2; a45=b*2*pi*7*dr/2;
% Volumes for the 5 nodes
v1=b*a1; v2=b*a2; v3=b*a3; v4=b*a4; v5=b*a5;
% Make the time step equal to one second
dt=1;
% Set initial temperatures at 20 C = 293.15 K
t1old=293.15; t2old=293.15; t3old=293.15; t4old=293.15; t5old=293.15;
% Begin looping; Each pass moves time ahead by dt
% We loop 240 times since we want the temperatures after 4 minutes = 240 s
for loop=1:240
t1new=t1old*(1-alpha*dt*a12/dr/v1-h*a1*dt/rho/c/v1-eps*sigma*a1*dt/rho/c/v1*t1ol-
d^3)+alpha*dt*a12/dr/v1*...
   t2old+h*a1*dt*tinf/rho/c/v1+eps*sigma*a1*dt/rho/c/v1*tsurr^4;
t2new=t2old*(1-alpha*dt*a12/dr/v2-alpha*dt*a23/dr/v2-h*a2*dt/rho/c/v2-
eps*sigma*a2*dt/rho/c/v2*t2old^3)+... alpha*dt*a12/dr/v2*t1old+alpha*dt*a23/dr/
v2*t3old+h*a2*dt*tinf/rho/c/v2+eps*sigma*a2*dt/rho/c/v2*tsurr^4;
t3new=t3old*(1-alpha*dt*a23/dr/v3-alpha*dt*a34/dr/v3-h*a3*dt/rho/c/v3-
eps*sigma*a3*dt/rho/c/v3*t3old^3)+... alpha*dt*a23/dr/v3*t2old+alpha*dt*a34/dr/
v3*t4old+h*a3*dt*tinf/rho/c/v3+eps*sigma*a3*dt/rho/c/v3*tsurr^4;
t4new=t4old*(1-alpha*dt*a34/dr/v4-alpha*dt*a45/dr/v4-h*a4*dt/rho/c/v4-
eps*sigma*a4*dt/rho/c/v4*t4old^3)+... alpha*dt*a34/dr/v4*t3old+alpha*dt*a45/dr/
v4*t5old+h*a4*dt*tinf/rho/c/v4+eps*sigma*a4*dt/rho/c/v4*tsurr^4;
t5new=t5old*(1-alpha*dt*a45/dr/v5-h*a5*dt/rho/c/v5-eps*sigma*a5*dt/rho/c/
v5*t5old^3)+... alpha*dt*a45/dr/v5*t4old+h*a5*dt*tinf/rho/c/v5+eps*sigma*a5*dt/rho/
c/v5*tsurr^4+qin*dt/rho/c/v5;
% Make the new values the old values before looping again
t1old=t1new; t2old=t2new; t3old=t3new; t4old=t4new; t5old=t5new;
end
```

```
t1new,t2new,t3new,t4new,t5new,loop,dt
t1c=t1new-273.15, t2c=t2new-273.15, t3c=t3new-273.15, t4c=t4new-273.15, t5c=t5new-
273.15
```

### *Example 5.6*

```
% Example 5.6
  % This is Example 4.8 Semi-Inf Slab water line depth done numerically (forward-
  difference)
  % For boundary condition, assumed ground is insulated at a depth of 20 ft.
  clear
  clc
  cond=2; rho=1800; c=2100; h=30; tinf=-18; ti=10;
  % x increment is 1 inch = 0.0254 m; time increment is 5 minutes = 300 s
  dx=0.0254; dt=300;
  alpha=cond/rho/c;
  adt=alpha*dt;
  % Let the number of nodes be 121. This makes the depth of the insulated node 10 feet
  n=121;
  % Set initial temperatures
  for i=1:n
    told(i)=ti;
  end
  % Start looping. Continue for 8064 time steps, which is 4 weeks since dt is 300 s
  for k=1:8064
    tnew(1)=told(1)*(1-2*h*dt/rho/c/dx-2*adt/dx^2)+2*h*dt/rho/c/dx*tinf+2*adt*told(2)/
  dx^2;
    tnew(n)=told(n)*(1-2*adt/dx^2)+2*adt/dx^2*told(n-1);
    for i=2:n-1
      tnew(i)=told(i)*(1-2*adt/dx^2)+adt/dx^2*(told(i+1)+told(i-1));
    end
    for j=1:n
      told(j)=tnew(j);
    end
  end
  tnew
  % Find x where freezing occurs (0 C)
  for i=1:n
    % Find depth where temp is greater than 0 C
    if tnew(i)>=0
      i
      disp('Burial depth is (inches)')
      disp(i*dx/.0254)
      break
    end
  end
```

### Example 5.7

```
% Example 5.7
  % Includes heat to fluid
  clc
  clear
  format short e
  cond=15;rho=7900;c=480;tinf=150;hside=100;htop=250;ti=1200;
  dr=0.005;dx=0.005;dt=1;
  alpha=cond/rho/c;
  adt=alpha*dt;
  % Set up matrices
  con=ones(11,6);told=ones(11,6);tnew=ones(11,6);v=ones(11,6);
  % Calculate volumes for nodes
  v(1,1)=pi*(dr/2)^2*(dx/2);
  v(11,1)=pi*((10*dr)^2-(19/2*dr)^2)*(dx/2);
  for i=2:10
    v(i,1)=pi*(((2*i-1)/2*dr)^2-((2*i-3)/2*dr)^2)*(dx/2);
  end
  for j=2:6
    for i=1:11
      v(i,j)=2*v(i,1);
    end
  end
  % Set initial temperature
  for i=1:11
    for j=1:6
      told(i,j)=ti;
    end
  end
  % Zero the heat to the fluid
  Q=0;
  % Start nodal equations forward difference
  for k=1:1000
    con(1,1)=1-htop*pi*(dr/2)^2*dt/rho/c/v(1,1)-2*pi*(dr/2)*(dx/2)*adt/dr/v(1,1)-
  pi*(dr/2)^2*adt/dx/v(1,1); tnew(1,1)=told(1,1)*con(1,1)+htop*pi*(dr/2)^2*dt*tinf/
  rho/c/v(1,1)+pi*(dr/2)^2*adt*told(1,2)/dx/v(1,1)+...
      2*pi*(dr/2)*(dx/2)*adt*told(2,1)/dr/v(1,1);
    con(11,1)=1-htop*pi*((10*dr)^2-(19/2*dr)^2)*dt/rho/c/v(11,1)-hside*2*pi*10*dr*dx/
  2*dt/rho/c/v(11,1)-...
      2*pi*19/2*dr*dx/2*adt/dr/v(11,1)-pi*((10*dr)^2-(19/2*dr)^2)*adt/dx/v(11,1);
    tnew(11,1)=told(11,1)*con(11,1)+htop*pi*((10*dr)^2-(19/2*dr)^2)*dt*tinf/rho/c/
  v(11,1)+... hside*2*pi*10*dr*dx/2*dt*tinf/rho/c/v(11,1)+2*pi*19/2*dr*dx/
  2*adt*told(10,1)/dr/v(11,1)+... pi*((10*dr)^2-(19/2*dr)^2)*adt*told(11,2)/dx/
  v(11,1);
    for i=2:5
      con(1,i)=1-2*pi*(dr/2)^2*adt/dx/v(1,i)-2*pi*dr/2*dx*adt/dr/v(1,i);
```

```
    tnew(1,i)=told(1,i)*con(1,i)+pi*(dr/2)^2*adt*(told(1,i-1)+told(1,i+1))/dx/
v(1,i)+2*pi*dr/2*dx*adt*...
    told(2,i)/dr/v(1,i);
  end
  con(1,6)=1-2*pi*(dr/2)^2*adt/dx/v(1,6)-2*pi*dr/2*dx*adt/dr/v(1,6); tnew(1,6)
=told(1,6)*con(1,6)+2*pi*(dr/2)^2*adt*told(1,5)/dx/v(1,6)+2*pi*dr/
2*dx*adt*told(2,6)/dr/v(1,6);
  for i=2:10
    con(i,1)=1-htop*pi*(((2*i-1)/2*dr)^2-((2*i-3)/2*dr)^2)*dt/rho/c/v(i,1)-
2*pi*(2*i-1)/2*dr*dx/2*adt/dr/v(i,1)-...
      2*pi*(2*i-3)/2*dr*dx/2*adt/dr/v(i,1)-pi*(((2*i-1)/2*dr)^2-((2*i-3)/2*dr)^2)
*adt/dx/v(i,1);
    tnew(i,1)=told(i,1)*con(i,1)+htop*pi*(((2*i-1)/2*dr)^2-((2*i-3)/2*dr)^2)
*dt*tinf/rho/c/v(i,1)+...
      2*pi*(2*i-1)/2*dr*dx/2*adt*told(i+1,1)/dr/v(i,1)+2*pi*(2*i-3)/2*dr*dx/
2*adt*told(i-1,1)/dr/v(i,1)+...
      pi*(((2*i-1)/2*dr)^2-((2*i-3)/2*dr)^2)*adt*told(i,2)/dx/v(i,1);
  end
  for i=2:5
    con(11,i)=1-hside*2*pi*10*dr*dx*dt/rho/c/v(11,i)-2*pi*19/2*dr*dx*adt/dr/v(11,i)-
2*pi*((10*dr)^2-...
      (19/2*dr)^2)*adt/dx/v(11,i); tnew(11,i)=told(11,i)
*con(11,i)+hside*2*pi*10*dr*dx*dt*tinf/rho/c/v(11,i)+2*pi*19/2*dr*dx*adt*...
      told(10,i)/dr/v(11,i)+pi*((10*dr)^2-(19/2*dr)^2)*adt*(told(11,i-
1)+told(11,i+1))/dx/v(11,i);
  end
  con(11,6)=1-hside*2*pi*10*dr*dx*dt/rho/c/v(11,6)-2*pi*19/2*dr*dx*adt/dr/v(11,6)-
2*pi*((10*dr)^2-...
    (19/2*dr)^2)*adt/dx/v(11,6); tnew(11,6)=told(11,6)
*con(11,6)+hside*2*pi*10*dr*dx*dt*tinf/rho/c/v(11,6)+2*pi*19/2*dr*dx*adt*...
    told(10,6)/dr/v(11,6)+2*pi*((10*dr)^2-(19/2*dr)^2)*adt*told(11,5)/dx/v(11,6);
  for i=2:10
    for j=2:5
      con(i,j)=1-2*pi*(2*i-3)/2*dr*dx*adt/dr/v(i,j)-2*pi*(2*i-1)/2*dr*dx*adt/dr/
v(i,j)-2*pi*(((2*i-1)/2*dr)^2-...
        ((2*i-3)/2*dr)^2)*adt/dx/v(i,j);
      tnew(i,j)=told(i,j)*con(i,j)+2*pi*(2*i-3)/2*dr*dx*adt*told(i-1,j)/dr/
v(i,j)+2*pi*(2*i-1)/2*dr*dx*adt*...
        told(i+1,j)/dr/v(i,j)+pi*(((2*i-1)/2*dr)^2-((2*i-3)/2*dr)^2)
*adt*(told(i,j-1)+told(i,j+1))/dx/v(i,j);
    end
  end
  for i=2:10
    con(i,6)=1-2*pi*(2*i-3)/2*dr*dx*adt/dr/v(i,6)-2*pi*(2*i-1)/2*dr*dx*adt/dr/
v(i,6)-2*pi*(((2*i-1)/2*dr)^2-...
      ((2*i-3)/2*dr)^2)*adt/dr/v(i,6);
    tnew(i,6)=told(i,6)*con(i,6)+2*pi*(2*i-3)/2*dr*dx*adt*told(i-1,6)/dr/
v(i,6)+2*pi*(2*i-1)/2*dr*dx*adt*...
```

```
        told(i+1,6)/dr/v(i,6)+2*pi*(((2*i-1)/2*dr)^2-((2*i-3)/2*dr)^2)*adt*told(i,5)/
dx/v(i,6);
  end
  % Calculating heat flow to fluid
  sum1=0; sum2=0;
  for i=2:5
    sum1=sum1+2*(tnew(11,i)-tinf);
  end
  for i=2:10
    sum2=sum2+(((2*i-1)/2*dr)^2-((2*i-3)/2*dr)^2)*(tnew(i,1)-tinf);
  end
  Qside=2*hside*dt*2*pi*10*dr*dx/2*(tnew(11,1)+tnew(11,6)-2*tinf+sum1);
  Qtop=htop*pi*dt*((dr/2)^2*(tnew(1,1)-tinf)+((10*dr)^2-(19/2*dr)^2)*(tnew(11,1)-
tinf)+sum2);
  Qbot=Qtop;
  Q=Q+Qside+Qtop+Qbot;
  if (tnew(1,6)<=500)
    tnew, k, tnew(7,4),Q
    disp('Time for center to reach 500 C is (sec)')
    k*dt
    break
  end
  for m=1:11
    for n=1:6
      told(m,n)=tnew(m,n);
    end
  end
end
```

# Index

Printed in the United States
By Bookmasters